AF624670

ADVANCED TOPICS IN BIOTECHNOLOGY AND PLANT BIOLOGY

ADVANCED TOPICS IN BIOTECHNOLOGY AND PLANT BIOLOGY

Edited By
C P MALIK
BHAVNEET KAUR
CHITRA WADHWANI
School of Life Sciences, Jaipur National University
Jaipur

MD Publications Pvt Ltd
New Delhi

Published by
MD Publications Pvt Ltd
"MD House" 11, Darya Ganj
New Delhi-110002
Tel. : +91-11-41563325, 41562846
E-mail : contact@mdppl.com
Website : www.mdppl.com

© 2008, All rights are strictly reserved.

ISBN : 978-81-7533-144-0

Published and Printed by Mrs. Rajni Gupta on behalf of **MD Publications Pvt Ltd** at Times Press.

CONTENTS

PREFACE

The volume entitled **Advanced Topics in Biotechnology and Plant Biology** provides a comprehensive survey of the major topics in the field of Biotechnology and biology. The book comprises 12 articles, written by experts in their respective disciplines and provides state-of-the-art information. The articles are put in two sections: Biotechnology, and response of plants to different stresses. The selected articles are most sought after by both researchers and students at the graduate as well as post-graduate level. Each article provides succinct information on central developments in the field and end with summaries and perspectives for the future.

The volume starts with an article by Malik, Kaur and Wadhwani. Biotechnology has made tremendous progress in recent years and has earned enormous level of public awareness and interest. The aim of this review is to provide enough and latest information and examples to provide the reader about recent advances in biotechnology. It is not intended to furnish an encyclopedic coverage of the subject, but give brief summaries of advances in the last two years. The current discussion entails notable discoveries made in the last one year. Year 2007 can easily be considered as the turning point for the life sciences industry. Article 2 discusses pattern of gene expression, regulation and properties for gene engineering with special reference to cold tolerance in maize *(Parvez A. Sofi, A.G. Rather and Shafiq A. Wani)*. The discussion is relevant since improvement in cold tolerance in crop plants has not been substantially studied through classical breeding approaches. A slight increase in tolerance to low temperature would have dramatic impact on maize yield. Enzymes have attracted the attention of scientists world over due to their wide range of physiological, analytical and industrial applications. Chandan discusses microbial production of industrial enzymes. Although enzymes have been isolated, purified and studied from microbial, animal and plant sources, microorganisms represent the most common source of enzymes due to their broad biochemical diversity, feasibility of mass culturing and ease of genetic manipulation. With the advent of molecular biology techniques, a number of genes of industrially important enzymes have been cloned and expressed in order to improve the production of enzymes, substrate utilization and other commercially useful properties. Special attention is being focused on enzymes isolated from thermophiles due to their inherent stability and industrial applications. In addition, a variety of methods have been employed to modify enzymes for their industrial usage including strain improvement, chemical modifications, expressed and various methods to control important diseases have been suggested. The changing environment and spread of diseases have caused serious economic and ecological losses. Among the disease agents of forest plants, fungal pathogens cause the maximum damage. Damping-off, root rot, cankers, vascular wilts and foliar diseases are serious in agroforestry plantations. Damping-off in nurseries can be effectively managed by treating the seeds, propagating materials and soil with fungicides and solarization. Root rots in plantations can be managed by proper sanitation, isolation of diseased trees, adequate spacing and by the use of biological control agents. Canker diseases can be controlled by the use of fungicides, proper silvicultural practices, avoiding injury and keeping the trees in more vigorous conditions. Mortality of important tree such as *shisham, kikar, khejri,* etc. cannot be attributed to single causing factor but a combination of factors, which has been discussed. Quarantine regulations, choice of species, selection of planting site and use of resistant plant material are some of the other measures employed for the management of tree diseases. Mercury is a very toxic metal whose concentration in the environment has increased in the recent past, leading to environmental pollution. As it is a very toxic metal, it causes various disorders in plants and animals. It is observed that it is more toxic in its ionic form and also in form of organic compounds. Various plants both terrestrial and aquatic are known which can accumulate mercury in their leaves. The physiological and biochemical mechanisms of plants may provide alternative strategies to adopt measures for purification of soil contaminated with heavy metals. Mer operon of bacterial genes is also known to confer resistance and help in hyperaccumulation of mercury. Hence, it is advisable to develop modified or transgenic plants that have bacterial genes in them to combat mercury toxicity. Shilpa Goel, Jahid Ali Malik and Harsh Nayyar in article 8 have discussed the physiological implications of this heavy metal. Article 9 describes Mechanism of arsenic tolerance and hyperaccumulation and is by Madhumita Dash and C. P. Malik. Arsenic is a very common element that ranks 20th in abundance in the earth's crust, 14th in the seawater, and 12th in the human body. Arsenic compounds are naturally present in the environment, and have been also used for many years in medicine, agriculture and forestry. Weathering of rocks converts arsenic sulfides to arsenic trioxide, which enters the arsenic cycle as dust or by dissolution in rain, rivers, or groundwater. Humans are exposed to this toxic arsenic primarily from air, food, and water. Thousands and thousands of people are suffering from the toxic effects of arsenicals in many countries all over the world due to natural groundwater contamination as well as industrial effluent and drainage problems. Hence the review is highly important. Sanjeev K Thind has discussed conceptual and metabolic engineering to analyze compartmentation and metabolite fluxes in C_3-C_4 and C_4 plants. The author has discussed carbon dioxide concentration mechanisms which mitigate the oxygenase activity of Rubisco, and potentially improves the efficiency of carboxylation of water use and nitrogen use in photosynthesis with several implications for plant survival. Interestingly carboxylation and decarboxylation processes are similar in C_4 and CAM plants but they operate on different anatomical, physiological and biochemical principles. Nitrogen fixation, Growth and Productivity of Legumes in response to Abiotic Stresses has been summarized by Pushp Sharma and Virender Sardana (article 10). Under stress conditions, reductions in biomass production to the extent of 30 to 60 % depending on intensity and duration of stress have been recorded. Traits like root and seed growth rate, membrane injury index and relative water content could be used in identifying suitable parents for breeding high yielding genotypes under stress environments. Regulation of processes limiting sink development, triggering the productivity by making critical use of nutrients, growth promoters and retardants which influence plant architecture and growth habits need manipulation for breaking the yield plateau. Areas where further understanding is particularly needed include research to define quantitatively and predicatively the environmental control of reproductive ontogeny, so that cultivars can be better matched to their environments; research to critically evaluate environmental and physiological constraints to partitioning of biomass and to seed, so that HI and NHI can be improved, research to explore the opportunities for exploiting physiological traits which assist survival during stress and/or basis of tolerance mechanisms against various abiotic stresses. Symbiotic, N_2 fixation by legume crops depends on interaction between plants and bacteria, which is affected by both above and below ground environment. Therefore, study is required to ascertain the reason for the failure of nodulation and/or N_2 fixation under certain environmental conditions, in order to minimize the detrimental effects on these processes. Further screening of germ plasm under various unfavorable abiotic conditions will be useful for better nitrogen fixation and yield of pulse crops. The field of molecular biology is developing very fast and, therefore, molecular markers closely linked with important sink development processes need to be identified. To achieve the goal of higher and stable productivity, the development of ideotypes and management technology should be location and season specific with major emphasis on the improvement of WUE and soil health. Article 11 nitrogen fixation : legume/rhizobium physiology phytohormones mediated response to nodulation and nitrogen fixation by Pushp Sharma and Poonam Sharma discuss role of various phytohormones on nitrogen fixation in this legume. There is a need for new and improved strains of both N_2 fixing bacteria which can be competive with native strains. Shorter shelf-life, contamination in the carrier and nonavailability of carrier material are the main reasons responsible for poor quality of *Rhizobium*. These improvements in strain quality can come through natural selection or by genetic engineering methods. Much current research is targeted at selecting or genetically manipulating *Rhizobium* strains to produce tailored strains that will out compete indigenous rhizobia. The lack of suitable methodology is one of the key factors limiting the proper assessment and success of individual *Rhizobium* strain in nodule occupancy with different cultivar and soil and agroclimatic conditions. The development and use of marker gene technology will enable soil ecologists to rapidly screen the competitive ability of a number of potential inoculum strain. In addition, application technique is being modified in order to deliver high viable inoculum at the site of seed germination.

The complete Bibliography is given at the end of the each chapter. The editors are grateful for the support of all the contributors. We are thankful to Dr Ruth Irwin, Editor ISB News Report for his generosity and full cooperation whenever requested for, We extend our thanks to all the learned contributors for their cooperation in compiling very useful information on various facets of Biotechnology and development biology. Special thanks go to various authors, each of who has endeavored to discuss up-to-date information of their area and expertise and has kindly spared valuable time and knowledge. We are sure the information compiled in this volume will prove helpful to those interested in biotechnology.

The book will serve as an excellent reference material for teachers, research workers, industrialists, biotechnologists and research students in the field of Applied Biology, Botany, Plant Physiology, Plant and Agriculture Science, Plant Breeding, Microbiology, Microbial Technology, Molecular and Cellular Biology, etc.

We apologize unreservedly for any mistakes, omissions or failure to acknowledge fully.

Finally, we would like to thank our families (CPM his spouse, Surakhsha; BK and CW their parents) for their continued support and encouragement throughout the completion of this book.

We wish to record our deep sense of gratitude and respect to our worthy Chancellor, Mr Sandeep Bakshi who always inspired and encouraged us, to Profs K L Sharma, Vice-Chancellor and H N Verma, Pro-Vice Chancellor, Jaipur National University for helpful suggestions and support. This reflects their incredible zest and zeal to promote academic growth in this portal of learning. Thanks are due to Mr Pranav Gupta and Ms Pragati of MD Publishers, New Delhi, who supported us fully for completing this project on time, with patience, care and support.

C P Malik

Bhavneet Kaur

Chitra Wadhwani

LIST OF AUTHORS

AGRAWAL, CHANDAN
SCHOOL OF LIFE SCIENCES, JAIPUR NATIONAL UNIVERSITY, JAIPUR

BANERJEE MEENAKSHI
LABORATORY OF ALGAL BIOTECHNOLOGY, DEPARTMENT OF BIOSCIENCES, BARKATULLAH UNIVERSITY BHOPAL-M.P. 462 026

DASH, MADHUMITA
SCHOOL OF LIFE SCIENCES, JAIPUR NATIONAL UNIVERSITY, JAIPUR

GAUTAM, ANURADHA
SCHOOL OF LIFE SCIENCES, JAIPUR NATIONAL UNIVERSITY, JAIPUR

GOEL, SHILPA
DEPARTMENT OF BOTANY, PUNJAB UNIVERSITY, CHANDIGARH

GOSAL, SS
DEPARTMENT OF PLANT BREEDING, GENETICS AND BIOTECHNOLOGY, PUNJAB AGRICULTURAL UNIVERSITY, LUDHIANA-141 004

KAUR, BHAVNEET
SCHOOL OF LIFE SCIENCES, JAIPUR NATIONAL UNIVERSITY, JAIPUR

KAUSHIK, J
DEPARTMENT OF FORESTRY, HARYANA AGRICULTURAL UNIVERSITY, HISAR

MALIK, CP
SCHOOL OF LIFE SCIENCES, JAIPUR NATIONAL UNIVERSITY, JAIPUR

MALIK, JAHID ALI
DEPARTMENT OF BOTANY, PUNJAB UNIVERSITY, CHANDIGARH

NAYYAR, HARSH
DEPARTMENT OF BOTANY, PUNJAB UNIVERSITY, CHANDIGARH

RATHER, A G
DIVISION OF PLANT BREEDING AND GENETICS, SKUAST, SHALIMAR, SRINAGAR

SHARMA, POONAM

DEPARTMENT OF PLANT BREEDING, GENETICS AND BIOTECHNOLOGY, PUNJAB AGRICULTURAL UNIVERSITY, LUDHIANA 141 004

SHARMA, PUSHP

DEPARTMENT OF PLANT BREEDING, GENETICS AND BIOTECHNOLOGY, PUNJAB AGRICULTURAL UNIVERSITY, LUDHIANA 141 004

SANDHU, J

DEPARTMENT OF PLANT BREEDING, GENETICS AND BIOTECHNOLOGY, PUNJAB AGRICULTURAL UNIVERSITY, LUDHIANA-141 004

SARDANA, VIRENDER

DEPARTMENT OF PLANT BREEDING, GENETICS AND BIOTECHNOLOGY, PUNJAB AGRICULTURAL UNIVERSITY, LUDHIANA 141 004

SOFI, PARVEZ A

DIVISION OF PLANT BREEDING AND GENETICS, SKUAST, SHALIMAR, SRINAGAR

THIND, SANJEEV

DEPARTMENT OF BOTANY, COLLEGE OF BASIC SCIENCES AND HUMANITIES, PUNJAB AGRICULTURAL UNIVERSITY, LUDHIANA 141004

TOKY, O, P

DEPARTMENT OF FORESTRY, HARYANA AGRICULTURAL UNIVERSITY, HISAR

WADHWANI, CHITRA

SCHOOL OF LIFE SCIENCES, JAIPUR NATIONAL UNIVERSITY, JAIPUR

WANI, A

DIVISION OF PLANT BREEDING AND GENETICS, SKUAST, SHALIMAR, SRINAGAR

WANI, SHABIR HUSSAIN

DEPARTMENT OF PLANT BREEDING, GENETICS AND BIOTECHNOLOGY, PUNJAB AGRICULTURAL UNIVERSITY, LUDHIANA-141 004

1

ADVANCES IN BIOTECHNOLOGY

C. P. MALIK, BHAVNEET KAUR AND CHITRA WADHWANI

Biotechnology has made tremendous progress in recent years and has earned enormous level of public awareness and interest. The aim of this review is to provide enough and latest information and examples to provide the reader a sound knowledge of biotechnology in all its various facets. It is not intended to furnish an encyclopedic coverage of the subject, but give brief summaries of advances in the last two years. The first review covered advances made during 2006-2007 (Kaur *et al.*, 2007). The current discussion entails notable discoveries made in the last one year. Year 2007 can easily be considered as the turning point for the life sciences industry. Worldwide there were a number of breakthroughs ranging across segments of pharma, biotech and medical technology.

A good measure of work happened in several countries. First, second phase of HapMap Project-created to identify and catalog genetic similarities and differences among populations around the world has been completed. An attempt has been made to search for genes associated with common diseases. This has led to the identification of 50 such disease-associated genes. Science unveiled realization that DNA differs from person to person much more than suspected before.

However, the biggest achievement is the "designer genome" brought into existence by famed scientist Dr. Craig Venter. Using lab-made chemicals, Dr. Venter's team stitched together a chromosome that is 381 genes long and contains 580,000 base pairs of genetic code. This has made artificial life a not-so-distant possibility.

Government of India approved the much awaited National Biotechnology Development Strategy which is an outcome of a two-year-long nationwide consultation process with multiple stakeholders including ministries concerned, universities, research institutes, private sector, civil society, consumer groups, non-government and voluntary organizations and international bodies.

It is suggested to form a National Biotechnology Regulatory Authority to provide a single window mechanism for biosafety clearance of genetically modified products and a high-powered Inter-ministerial Committee to effectively coordinate the development of the sector by addressing cross cutting issues. The policy also focuses on public-private partnership programs, Biotechnology Industry Partnership Programme (BIPP) for Advanced Technology. Approval is also accorded for the expansion of SBIRI during the 11th Plan.

There is added emphasis on nano sciences.

Dr. Mario R Capecchi, Dr. Martin J Evans and Dr. Oliver Smithies were honored with 2007 year's Nobel Prize for medicine for their discoveries of "principles for introducing specific gene modifications in mice by the use of embryonic stem cells". They have made a series of ground-breaking discoveries concerning embryonic stem cells and DNA recombination in mammals.

It is incredible that Indian-Australian stem cell scientist, Dr. Kuldip Sidhu has won the 2007 Top Invention Prize for his pioneering work in stem cells.

In the ensuing pages we recapture some of the breakthroughs, technologies, products and people who made news in 2007.

PLANT BIOTECHNOLOGY

1. **Researchers at the John Innes Centre in Norwich** under the leadership of **Dr. Michael Lenhard have discovered that cells at the margins of leaves and petals play a particularly important role in setting their size.** The uniformity of leaf and flower size delineates different species, such as daisies and marguerites. The cells at the margins seem to secrete a mobile growth signal that keeps the cells throughout the leaf dividing. The more of this signal is produced, the larger the leaves and flowers get. This signal seems to be distinct from the classical and well-studied plant hormones that are known to influence growth and development. The researchers suggested that the signal gets diluted as the leaf or petal grows, since it originates from the margins. Once the concentration falls below a certain threshold the cells in the leaf or petal stop dividing. The animals also seem to use the same principle of dilution for measuring size, for example of the wings in a fly, although the molecules used are very different. Efforts are under way to use this discovery to increase leaf growth in biofuel crops for the generation of sustainable energy and to boost the yield of fruits and seeds.

2. A study employing *Arabidopsis* as the model system has revealed the likely mechanism of flowering in plants. Earlier research had shown that plants' leaves perceived seasonal changes in day length, which triggered a long-distance signal to travel through the plant's vascular system from the leaf to the shoot apex, where flowering is induced. However, the identity of the long-distance signal remained unclear. The

new research, carried out by **scientists at Imperial College London and the Max Planck Institute for Plant Breeding Research in Cologne** led to the proposal that a protein known as **Flowering Locus T Protein** (FT protein), which is produced in leaves by the Flowering Locus T gene (FT gene) tells plants when to flower. It travels through the plant's vascular system to the shoot apex, where it activates other genes, causing the plant to flower. The research team was able to track the progress of the protein through the plant by tagging it with a green fluorescent protein originally isolated from jellyfish, allowing it to be detected in living tissues using highly sensitive microscope systems. The team then grafted two plants together, only one of which contained the gene for the fluorescent version of FT. This allowed them to show conclusively that FT protein moved from where it was produced in the leaves of one plant, across into the other plant. The FT protein is produced when the FT gene is switched on by another gene known as CONSTANS. This is a key gene expressed in leaves which reacts to changes in day length.

3. Scientists working at **Monsanto and Devgen have identified a new technology using RNAi to protect crops from insects.** The technology is expected to provide farmers with a new in-the-seed option to protect crop yields and present another tool to support agriculture's mission of meeting the world's growing food, feed and fuel demands. RNAi is a biological mechanism found in nature as a way to regulate gene expression. The scientists at Monsanto identified novel applications of RNAi, enabling plants to be better protected against insect pests that feed on crops and impact yield.

4. A group of **Malaysian scientists** have discovered a breakthrough **antiviral drug** from an **herbal extract** for the **treatment of dengue**. This could be the world's first dengue antiviral drug based on herbal extracts. Dengue is an acute infectious, eruptive, febrile disease caused by four antigenically related but distinct serotypes of the Dengue virus. It is transmitted by the bite of infected *Aedes* mosquitoes, especially *A. aegypti*. Scientists at the Medical Department of Universiti Malaya and the Universiti Malaya Medical Centre have developed the antiviral drug, known as **Dengimm2**, from extracts of **Artemisia** and eight other local herbs. Dengimm2 was proved safe when tested on infected cells and in mice. The plant extracts were supplied to Universiti Malaya Medical Centre (UMMC) by Autoimmune Sdn Bhd following a memorandum of understanding signed between the hospital and the company in 2004. Professor Sazaly Abu Bakar, Head of the Department of Microbiology at the university, stated that the discovery is just the first step and it would take another three or four more years before the product is ready for clinical use as it has to undergo toxicological, chemical and side-effects tests to attain approval. Currently the product is registered at the Malaysia Drug Control Authority.

5. Development of plants with enhanced **tolerance to multiple environmental stresses**: At **KRIBB's Environmental Biotechnology Research Center**, transgenic crops (e.g. sweet potato, potato) for sustainable agriculture were developed through the

expression of both CuZnSOD and APX in their chloroplasts under the control of an oxidative stress inducible SWPA2 promoter. These plants have strong resistance to multiple environmental stresses, including drought, extreme temperatures and air pollution.

6. **Cornell researchers** have finally proven a long-standing theory of **how plants transport their sugars, using genetic engineering techniques.** The findings will allow researchers to genetically engineer plants with increased photosynthetic rates, yields and carbon dioxide intake. The theory of transporting sugar, the polymer trap model, was first proposed in 1991 by **Robert Turgeon**, Cornell professor of plant biology. He is also the senior author of the latest research published in the Dec. 4 issues of the Proceedings of the National Academy of Sciences. **Ashlee McCaskill**, who worked in Turgeon's lab, is the paper's lead author. Turgeon's theory suggested that as sucrose, a form of sugar, accumulates in leaves as a product of photosynthesis, it diffuses into the plant's tube like transport tissue, called phloem, along with other nutrients to move to other areas of the plant. Once in the phloem, small molecules of sucrose polymerize, or combine, to form larger, more complex sugars, which become too large to flow back into the leaf. The polymerized sugars are then forced to move away from the leaf to parts of the plant where they may be used or stored.

To prove the theory, Turgeon and McCaskill genetically engineered a plant closely related to a member of the figwort family, purple mullein (*Verbascum phoeneceum* L.), so that two genes involved with polymerizing sucrose into larger molecules were silenced. When they did so, sugars backed up in the leaves. In normal plants, when sugars (made from water and carbon dioxide during photosynthesis) accumulate in the leaves, photosynthesis slows down, and the plant does not take in as much carbon dioxide from the air. As the sugars move out of the leaves, the rate of photosynthesis and carbon intake increases.

7. **Evolutionary biologists** have provided strong evidence that changes in global carbon dioxide levels probably had an important influence on the emergence of a specific group of plants, termed C4 grasses, which includes major cereal crops, plants used for **biofuels**, and species that represent important components of grasslands across the world. C4 plants are specially equipped to combat an energetically costly process, known as photorespiration that can occur under conditions of high temperature, drought, high salinity, and low carbon dioxide levels. Although a combination of any of these factors might have provided the impetus behind the evolution of the various C4 lineages, it had been widely speculated that a drop in global carbon dioxide levels, occurring approximately 30 million years ago during the Oligocene period, may have been the major driving force. Establishing the link between the two, however, has proven difficult partly because there are no known fossils of C4 plants from this period. **Enter Pascal-Antoine Christin and colleagues from the University of Lausanne, Switzerland,** decided to take an alternative approach to date a large group of grasses. By using a

"molecular clock" technique, the authors were able to determine that the Chloridoideae subfamily of grasses emerged approximately 30 million years ago, right around the time global carbon dioxide levels were dropping. Furthermore, a model of the evolution of these grasses suggests that this correlation is not a trivial coincidence and instead reflects a causal relationship.

As the authors noted in their study, many of the C4 grasses evolved after the drop in global carbon dioxide levels 30 million years ago. The authors speculate that while an atmosphere low in carbon dioxide established the basic conditions necessary for C4 evolution, other ecological factors might be at work. In light of this, the authors hope to investigate the role of other variables, such as drought, salinity, and flooding, in the evolution of C4 plants. In addition to improving our understanding of how climate changes influenced ecosystems in the past, such studies may allow predictions of how human activities could affect the planet in the future. Indeed, with regard to global carbon dioxide levels, Christin and colleagues write, "Besides its influence on climatic variables, increased CO_2 concentration could trigger important ecological changes in major terrestrial ecosystems by affecting the distribution of C4-dominated biomes and the affiliated flora and fauna." This implies that a reversal of the conditions that favored C4 plants could potentially lead to their demise—a startling prospect if one considers the human race's reliance on C4 crops like corn, sugarcane, sorghum, and millets.

8. **Biologists at the University of California, San Diego have discovered a gene in plants that disrupts fertilization only when mutations in the gene are present in both the female and male reproductive cells.** Their discovery has been named the **"abstinence by mutual consent"** mutation because of its unusual properties. "Mutations that do not allow fertilization are known in plants, but usually these mutations are caused either by a mutation in the female reproductive cells or by a mutation in the male reproductive cells," said **Julian Schroeder, a professor of biological sciences at UCSD** who **headed the study**. In this gene, when only the female carries the mutation, completely normal fertilization occurs, and when only the male carries the mutation, fertilization also occurs. But fertilization is completely disrupted when both male and female reproductive cells carry the mutation simultaneously.

The scientists say the discovery of new genes that control the ability of plants to undergo fertilization could have important applications to plant breeders and conservationists. Mutations that cause infertility in crops can provide a powerful tool for breeders who would like to avoid crossing of their plants to related species. Conversely breeders would at times like to breed crops by crossing them into distantly related species that however do not allow crossing due to infertility. For example, adding beneficial stress resistance genes from another species may not be possible if the male and female reproductive cells can't communicate properly. Understanding the mechanisms that mediate male-female communication during fertilization could help in circumventing the barrier of such interspecies crosses for breeding new varieties.

In animals and plants, fertilization relies on complex and specialized mechanisms that allow the precise delivery of the male reproductive sperm cell to the female egg cell. Fertilization requires fusion of the sperm and egg cells. In flowering plants, the male pollen tube carries sperm cells through the maternal tissues to deliver the sperm to the female reproductive cells. Once the pollen tube gets close to the egg cells, fertilization requires the bursting of the pollen tube. This pollen tube bursting expels sperm cells from inside the pollen tube, so sperm can then fertilize the female reproductive cell. In investigating why the mutation discovered caused disruption of fertilization in the mustard plant *Arabidopsis*, the scientists found that the pollen tubes did not burst when they came close to the female egg cells. So in the mutant plants, the sperm cells were not expelled from the pollen tubes toward the female cells. Instead the pollen tubes, in which the sperm reside, kept growing past the eggs cells.

"The abstinence by mutual consent gene is the first gene identified so far with a critical function at the same time in both plant male and female reproductive cells that is essential for the delivery of sperm cells to the egg cell," said **Boisson-Dernier**, postdoctoral scholar and first author of the study. "The abstinence by mutual consent mutant pollen just acts as if there were no egg cell around and keeps on growing," Schroeder added. "It's interesting because this lack of pollen bursting only happens if the mutation is carried by both the male and female, suggesting the abstinence by mutual consent gene somehow allows the male pollen and the female reproductive cells to talk to one another." The scientists found that the gene is responsible for producing a protein called peroxin that targets small organelles inside the cell called peroxisomes. Within the cell, peroxisomes are small organelles whose main functions are metabolizing fatty acids, protecting the cell from toxic free radicals and also generating a large range of signaling molecules. Boisson-Dernier's discovery demonstrated that peroxisomes play an unexpected key role in the dialogue between the male sperm carrier and the female egg cells. It also implies, the UCSD researchers say, that a diffusible signal generated in the peroxisomes of either the female or male reproductive cells is enough to allow the male-female communication to proceed. This signal coming from the peroxisomes of the male or female is sufficient to set off the eruption of the pollen tube allowing the sperm cells to be released.

9. A team of plant geneticists and crop scientists from **Cornell University, the University of Illinois, Boyce Thompson Institute, DuPont Crop Genetics Research, the University of North Carolina, the City University of New York, the International Maize and Wheat Improvement Center and the U.S. Department of Agriculture** has pioneered an economical approach to the **selective breeding of maize** that can **boost levels of provitamin A,** the precursors that are converted to vitamin A upon consumption. This innovation could help to enhance the nutritional status of millions of people in the developing world. The innovation involves a new approach for selecting

the parent stock for breeding maize, and significantly reduces the ambiguity and expense of finding varieties that yield the highest provitamin A content available. As part of this investigation, the researchers have identified a naturally mutated enzyme that enhances the provitamin A content of maize.

Vitamin A deficiency is a leading cause of eye disease and other health disorders in the developing world. Some 40 million children are afflicted with eye disease, and another 250 million suffer with health problems resulting from a lack of dietary vitamin A. "Maize is the dominant subsistence crop in much of Sub-Saharan Africa and the Americas," according to the researchers, where between 17 and 30 per cent of children under the age of 5 are vitamin A deficient.

The diversity in maize is tantalizing to those hoping to make use of desirable traits, but it also provides a formidable challenge in trying to understand the genetic basis of those attributes. One hurdle to increasing the provitamin A content of maize has been the expense of screening the parent stock and progeny of breeding experiments. High performance liquid chromatography (HPLC) can assess the provitamin A content of individual plant lines. However, screening a single sample costs $ 50 to $ 75.

The new approach uses much more affordable methods and gives a more detailed picture of the genetic endowment of individual lines. One technique the researchers employed, called quantitative trait loci (QTL) mapping, allowed them to identify regions of the maize chromosomes that influence production of the precursors of vitamin A. They also used association mapping, which involves studying variation in selected genes and tracking inheritance patterns to see which form of a gene coincides with the highest provitamin A content. Polymerase chain reaction (PCR) allowed them to amplify and sequence the different versions (alleles) of the genes of interest, to find the alleles that boosted levels of vitamin A precursors in the plant. This approach led to an important discovery. The team found a mutant form of an enzyme vital to the cascade of chemical reactions that produce the precursors of vitamin A in the plant. This mutant is transcribed in lower quantities than the normal allele and steers the biochemistry toward producing higher levels of vitamin A precursors. The study analyzed 300 genetic lines selected to represent the global diversity of maize, and identified some varieties that came close to the target amount of 15 micrograms of beta-carotene per gram. Current maize varieties consumed in Africa can have provitamin A content as low as 0.1 micrograms per gram.

10. Genetically modified Bt crops have no negative effect on the survival of honeybees, according to a study published by the online journal **PlosOne**. The meta-analysis, conducted by a group of **American scientists**, made use of 25 independent laboratory studies that focused on the effects of the transgenic Cry protein on honeybee larvae and adults. Although numerous studies have been performed on the effect of Bt proteins in honeybees, the short observation time and the low number of replicates limit these researches. By synthesizing the quantitative data obtained in these studies, researchers can greatly increase the statistical power of the experiment. The scientists further noted

that BT toxicity to honeybee is very unlikely, especially since studies typically expose honeybees to doses of Cry proteins that are ten or more times than those encountered in the field.

11. Researchers at the **Boyce Thompson Institute (BTI) for Plant Research on Cornell's campus** have identified a crucial player in the defense network that allows plants to respond to pathogens. These findings have opened a door for plant breeders to develop plants with better resistance to pathogens. Under normal circumstances, plants have a large number of resistance proteins, also known as R proteins, to respond to an attack by a pathogen. These proteins initiate such defense responses as releasing defense-signaling hormones and killing off cells surrounding an infection, which then keeps a pathogen from spreading. However, the researchers identified mutant *Arabidopsis* plants that did not trigger a normal cell death response to attack by pathogens. A close look at one of these plants revealed a mutation in a gene for an ATPase, an enzyme that is known to harness energy produced in cells to drive other chemical reactions. The researchers then discovered that this ATPase, called CRT1, interacts with and may mediate defense responses initiated by three different R proteins. "Because very few proteins that physically interact with resistance proteins have been identified to date, the identification and characterization of CRT1 should provide new insights in plant disease resistance and opportunities to enhance this resistance," said Hong-Gu Kang, a research associate at BTI.

12. Researchers in Germany (Bayer CropScience) have developed a way of engineering plants so they can flourish in difficult conditions without raising an ethical storm. The German firm has developed a way of engineering plants to withstand tougher conditions that could get round the ethical concerns posed by genetically modified plants. By increasing crop yields, the technique could help to feed more people more cheaply. It could also help to slow the pace of climate change by improving the yields of biofuels, and thereby cutting fossil fuel use. The breakthrough was needed because living threats such as pests and weeds do surprisingly little damage to crops. Most of the harm is done by "abiotic" factors such as heat, drought and flooding, which knock off about three quarters of the yield. This is because plants that are facing difficult environmental conditions spend too much energy on protecting their cells and not enough on growth. This severely limits the yields of many crops. Now, Bayer's scientists have developed a "genetic shock absorber" to right this balance and boost the growth of crops during periods of excessive heat, drought and cold. In stressful circumstances, plants protect themselves by switching on a gene called PARP. This produces a protein of the same name that repairs the plant's DNA and shields its cells from damage. Unfortunately, the protein needs a lot of energy to do its job, and because plants can't predict how long tough conditions will last, they end up over-producing it. If the hard times persist, the plant eventually runs out of energy. But **Michael Metzlaff**, head of Bayer's crop productivity research group, has developed a way to fine tune the repressive levels of PARP so the plant can continue to grow while still being protected from the elements.

He uses a technique called RNA interference, which employs a special molecule to block the message sent from the DNA that tells the plant to produce more of the PARP protein. The gene isn't completely shut off, but the plant's hyperactive stress response is toned down. Metzlaff's method can be very specifically targeted to a plant of choice, using parts of the PARP genetic sequence that do not exist in other species. The technique is not unnatural. Plants themselves use RNA interference to fine tune the activity of their genes and to protect themselves from viruses.

Metzlaff stresses that RNA interference only uses genes and proteins that are a natural part of the plant: it plays with the recipe, but never adds new ingredients to the mix. It's not classical GM. At first, Metzlaff tried the technique in mustard cress, a plant favoured by geneticists because it is easy to grow and manipulate. When subjected to heat, drought and cold, the modified plants showed an increase in yield of between 10 and 20 per cent compared with normal cress. In extreme cases, Metzlaff saw a 44 per cent improvement in growth. Since then, Bayer has successfully tested the RNA interference method in field trials with oilseed rape and canola, a crop that could be a significant source of biofuels. It is now conducting trials on cotton and rice. If successful, the hardier plants could be invaluable in coping with the difficult conditions that global warming will bring about, and so safeguarding tomorrow's harvests.

13. **Scientists in the United States and India** are reporting development of a **high-protein variety of rice**, dietary staple for half the world's population. Researchers have been trying to bolster the protein in rice for five decades. Rice already is a main source of calories as well as protein intake for billions of people, and its enrichment of protein would have a positive impact on millions of poor and malnourished people in developing countries. In the study, **Hari B. Krishnan and colleagues** created a hybrid by crossing a commonly cultivated rice species called *Oryza sativa* with a wild species, *Oryza nivara*. The product showed a protein content of 12.4 per cent, which is 18 per cent and 28 per cent higher than those of the parents. The results demonstrate the potential for wild rice's relatives for boosting the protein content in rice. The researchers conclude that the hybrid could serve as initial breeding material for new rice genotypes that could combine types with superior cooking quality with those of high protein content.

14. Genetically modified **Bt crops** have **no negative effect on the survival of honeybees**. The meta-analysis, conducted by a group of American scientists, made use of 25 independent laboratory studies that focused on the effects of the transgenic Cry protein on honeybee larvae and adults. Although numerous studies have been performed on the effect of Bt proteins in honeybees, the short observation time and the low number of replicates limit these researches. By synthesizing the quantitative data obtained in these studies, researchers can greatly increase the statistical power of the experiment. The scientists further noted that **BT toxicity to honeybee is very unlikely, especially since studies typically expose honeybees to doses of Cry proteins that are ten or more times than those encountered in the field.**

15. In order to meet the protein nutrition needs of the world population, greater reliance on plant protein sources will become necessary. The amino acid balance of most plant protein sources does not match the nutritional requirements of monogastric animals, limiting their nutritional value. In cereals, the essential amino acid lysine is deficient. Maize is a major component of human and animal diets worldwide and especially where sources of plant protein are in critical need such as sub-Saharan Africa.

To improve the amino acid balance of maize researchers have developed transgenic maize lines that produce the milk protein lactalbumin in the endosperm. Lines in which the transgene was inherited as a single dominant genetic locus were identified. Sibling kernels with or without the transgene were compared to determine the effect of the transgene on kernel traits in lines selected for their high content of lactalbumin. Total protein content in endosperm from transgene positive kernels was not significantly different from total protein content in endosperm from transgene negative kernels in three out of four comparisons, whereas the lysine content of the lines examined was 29-47% greater in endosperm from transgene positive kernels. The content of some other amino acids was changed to a lesser extent. Taken together, these changes resulted in the transgenic endosperms having an improved amino acid balance relative to non-transgenic endosperms produced on the same ear. Kernel appearance, weight, density and zein content did not exhibit substantial differences in kernels expressing the transgene when compared to non-expressing siblings. Assessment of the antigenicity and impacts on animal health will be required in order to determine the overall value of this technology.

16. The Defence Research Development Organisation (DRDO) has developed a transgenic tomato for growing in the cold desert regions of Ladakh. "This transgenic tomato is in the fourth generation and after its fifth generation, it would be subjected to controlled trials under the supervision of the Review Committee on Genetic Manipulation (RCGM)," said the chief controller of R&D in life sciences in DRDO, W Selvamurthy at a plenary session in the 95th Indian Science Congress. He said that care has been taken to make the transgenic tomato resistant to cold temperatures below 20°C and water stress conditions of the region. According to Selvamurthy, the Pusa Ruby tomato has been introgressed with Osmotin gene through *Agrobacterium* mediated genetic transformation to enhance inbuilt cold stress tolerance. *In vitro* selection against selectable marker Kanamycin and PCR analysis of regenerated tomato plantlets for promoter (CaMV), Kanamycin (npt-II) and Osmotin gene has confirmed the gene integration. Further confirmation of integration of Osmotic gene has been done by Southern Blot and Western Blot techniques. The T1 plants along with control have also been screened for tolerance to cold stress at a temperature of 3°C for three and half days and then exposed to ambient temperature. The morphological, physiological and biochemical studies are underway on these transformed plants. The confirmed lines are being

advanced and are under process of evaluation for different agronomically important parameters, he said.

17. A simple, cheap lab test developed at **Southern Illinois University Carbondale** can unerringly detect **Sudden Death Syndrome** (SDS), a costly fungal disease, in **soybean seedlings**. Once commercialized, its use will help breeders produce SDS-resistant soybean varieties much faster than they can now. The reliable procedure has been developed by **David A. Lightfoot**, a biotechnologist in SIUC's College of Agricultural Sciences. According to him the seedlings develop the leaf symptoms and the root rot or they don't. The procedure works every time, and the labor cost is very low – about $ 1 per assay. Lightfoot received a patent for the assay Oct. 31, which will allow the University to negotiate with private companies to license it. Lightfoot grows his fungus on a mix of cornmeal, sand, mineral salts and agar. Graduate students, armed with spades and pails of the mix, then dig that mix into piles of soil, turning it until the fungus-laden growing medium is distributed evenly throughout. Vulnerable seedlings planted in that infected soil contract the disease; seedlings with resistance potential do not.

18. Scientists are working towards developing drought resistant rice through genetic modification. Drought and a dwindling water supply are increasingly problematic for farmers worldwide. In October 2007, about one-third of the southeastern United States experienced exceptional drought, which caused widespread crop and pasture losses and a severe water shortage. It is the highest drought rating to date, according to the Department of Agriculture's drought monitor. In Eastern India, where rice is a main food staple, drought resulted in a loss of 5.4 million tons of rice between 1997 and 2003. For farmers, that can translate into an income loss of up to 50 per cent.

Some progress has been achieved in reducing the effects of drought according to **Andy Pereira**, a researcher from **Virginia Polytechnic Institute and State University**. Pereira and his team have developed a genetically modified (GM) strain of rice that is able to withstand drought conditions and grow with less water during normal conditions, which he believes will help countries like India cope with the harmful effects of drought. "Populations compete for water, for food, drinking, washing, industry, and rice uses most of the fresh water that we have," Pereira says. "And I, like many other people, think water is going to be scarce than it is right now." However, GM crops continue to raise a slew of concerns for people's health, loss of biodiversity and accessibility. So further testing on its safety and how it interacts with other plants must be done before it becomes commercially available. To create the GM rice, Pereira inserted a gene into the rice called the HARDY gene. The study, detailed in September 2007, Proceedings of the National Academy of Sciences, found that rice with the HARDY gene - when grown in drought conditions - had about 50 per cent more biomass than rice without the gene. Also, for drought and non-drought conditions, the modified rice used water about 65 per cent more efficiently.

19. The scientists of **MS Swaminathan Research Foundation (MSSRF)**, a Chennai-based foundation and other research organisations have come up with a solution using **crop biofortification** and producing **iron-rich rice**. Deficiency of micro-nutrients like iron, zinc and vitamin A is a common problem among various sections of the Indian populace. According to the scientists, nearly 30 per cent of the Indian population suffers from malnutrition especially due to lack of iron, zinc and Vitamin A. "Brown (normal) rice contains 8 mg iron per gram, but when it is polished to white rice, only 3 mg iron per gram remains. As our body needs at least 12 mg iron per gram of rice and polished rice cannot cater to the requirement, people do not get their daily dose of iron and hence suffer from iron deficiency," explains programme director of Biotechnology Department at MSSRF Ajay Parida. Various scientific research organizations took up research activity under the aegis of Department of Bio-Technology and developed two methods to tackle this problem after screening different varieties of rice available in the country. The biotechnology approach-where iron genes are injected into rice seed to further produce biofortified crop-yielded better results. Work began on this project in 2005 and success has been attained in developing third generation (level) of plant. The iron content level is improving with each generation and things are being tested. The biofortified variety of rice will be released commercially in a couple of years according to scientists. Efforts are on to follow similar methods to produce other bioforitified crops rich in nutrient content.

20. Scientists have created a **"tear-free" onion** using **Australian-developed biotechnology to switch off the gene behind the enzyme that makes us cry.** The **New Zealand Research Institute Crop and Food used gene-silencing technology** to make the breakthrough, which it hopes could lead to a prototype onion hitting the market in a decade's time. Colin Eady, the institute's senior scientist, said the project started in 2002 after **Japanese scientists located the gene responsible for producing the agent behind the tears.** The technology creates a sequence that switches off the tear-inducing gene in the onion so it doesn't produce the enzyme. Mr. Eady said that by stopping sulphur compounds from being converted to the tearing agent and redirecting them into compounds responsible for flavour and health, the process could even improve the taste of the onion. It is anticipated that the health and flavour profiles will actually be enhanced by what has been achieved. Mr. Eady, who has several model onion plants at the institute, said despite the excitement about the prospect of "no tears" onions in every home, it would be 10 to 15 years before this happened.

MEDICAL BIOTECHNOLOGY

1. The roundworm *C. elegans*, a staple of laboratory research, may be key in unlocking one of the central biological mysteries: why we sleep. **Researchers at the University of Pennsylvania School of Medicine** report that the round worm has a sleep-like state, joining most of the animal kingdom in displaying this physiology. This research has

implications for explaining the evolution and purpose of sleep and sleep-like states in animals. In addition, genetic work associated with the study provides new prospects for the use of *C. elegans* to identify sleep-regulatory genes and drug targets for sleep disorders. **David M. Raizen**, MD, PhD, Assistant Professor of Neurology, in collaboration with other researchers at the Penn Center for Sleep, showed that there is a period of behavioral quiescence during the worm's development called lethargus that has sleep-like properties. "Just as humans are less responsive during sleep, so is the worm during lethargus," explains Raizen. "And, just as humans fall asleep faster and sleep deeper following sleep deprivation, so does the worm." By demonstrating that worms sleep, Raizen and colleagues have not only demonstrated the ubiquity of sleep in nature, but also propose a compelling hypothesis for the purpose for sleep.

Because the time of lethargus coincides with a time in the round worms' life cycle when synaptic changes occur in the nervous system, they propose that sleep is a state required for nervous system plasticity. In other words, in order for the nervous system to grow and change, there must be down time of active behavior. Other researchers at Penn have shown that, in mammals, synaptic changes occur during sleep and that deprivation of sleep results in a disruption of these synaptic changes. In addition, the research team used *C. elegans* as a model system to identify a gene that regulates sleep. This gene, which encodes a protein kinase and is regulated by a small molecule called cyclic GMP, has been previously studied but not suspected to play a role in sleep regulation. The findings suggest a potential role for this gene in regulating human sleep and may provide an avenue for developing new drugs for sleep disorders. "It opens up an entire new line of inquiry into the functions of sleep," notes Penn Center for Sleep Director and co-author Allan I. Pack.

2. A team of **Vanderbilt researchers** have demonstrated that RNA interference can heal a genetic disorder in a live animal. The study shows that RNA interference can "rescue" a strain of mouse that has been genetically engineered to express a defective human hormone that interferes with normal growth. When the gene that produces defective human growth hormone is inserted into the mouse's genome, it also stunts the mouse's growth. But when a small snippet of RNA that interferes with the hormone's production is also added, the mouse is restored to normal. The blueprint for a protein like growth hormone is genetically encoded in exons which are copied into messenger RNA that is eventually converted to a sequence of amino acids called a protein. Normal growth hormone is produced by a series of five exons. The defective hormone is the result of a splicing error: It is made by combining the segments coded by the first two exons and the last two exons, mistakenly skipping the third exon. A transgenic mouse was created earlier at National Institute for Medical Research in London with the human growth hormone gene that duplicated growth hormone deficiency. Although the altered mice still contained the mouse growth hormone genes, it was found that high levels of the defective human growth hormone not only stunted their growth but actually killed the cells in the pituitary that produces growth hormone.

In the last 15 years, scientists have realized that short pieces of double-stranded RNA, called silencing-RNA, use a pathway that is normally used by cells to regulate genes. This has created an opportunity for developing highly targeted therapies for a number of genetic diseases including macular degeneration in the eye and to block viruses such as herpes and RSV respiratory viruses. The researchers realized that the messenger-RNA that produced the defective hormone had a unique signature created by skipping the third exon. A specific silencing-RNA was designed to bind uniquely with the defective messenger-RNA. Later this RNA was delivered to the pituitary gland which, in the case of the mouse, is the size of a grain of uncooked rice and is located at the base of the brain. As a proof of concept, the researchers decided to create a second strain of mouse which carried the special silencing-RNA and mate them with the growth deficiency strain. Their offspring should have both the genetic defect that produces the defective growth hormone and the silencing-RNA that should inhibit its production, allowing the mouse growth hormone to act. The experiment was successful. The offspring grew normally and showed no defects in their pituitaries. Now the researchers are investigating ways to deliver their silencing-RNA to the pituitary gland that would be suitable for treating humans. The cells that produce growth hormone have special receptors that signal the cells to release their stocks of growth hormone. If they can figure out a way to attach the silencing-RNAs to a compound that binds to this receptor, they should be able to deliver them to the cells where they can interfere with the activity of the defective growth hormone.

3. **Drug-resistant TB genome has been sequenced in South Africa.** This will help fight the deadly disease of tuberculosis. The genome of one strain of extensively drug-resistant tuberculosis (XDR-TB) has been sequenced using technology developed in the United States. This breakthrough will provide for more analysis of MDR (multi drug-resistant TB) and XDR-TB (extensively drug-resistant TB), and to understand the differences between them, according to Dr. Carl Montague of LifeLab working in a biotechnology project funded by the South African government. Elsewhere in the world, researchers are on way to develop two highly potent shorter-term tuberculosis drugs. **Two anti-TB drugs (Moxifloxacin and PA-824) have reached the advanced stages of clinical trial** and could be available worldwide by 2012-2014, raising hopes for shorter-duration treatment and fighting the bacterium that has become immune to the present drugs. Developed by TB Alliance, Moxifloxacin has reached the phase III stage of clinical trial and 900 patients were given the drug in the three phase II studies conducted on it. While the second drug, PA-824, has entered the phase II trials and will be now given to TB patients to study its short-term potency.

4. **Regenerative medicine** went a step further with **scientists discovering a key protein that helps newts regrow severed limbs.** Ability of newts and salamanders to renew damaged body parts has been the trigger for this research that holds the promise of limb regeneration in mammals. A British research team found that a protein nAG,

secreted by nerve and skin cells, plays a central role in producing a clump of immature cells, known as a blastema, which regrows the missing part. The importance of nAG was demonstrated by the fact that even when a nerve was severed below the stump tip, which would normally prevent regrowth, the scientists were able to coax regeneration by artificially making cells produce the protein. The focus now is on applying the finding in a mammalian context.

5. A privately held medical device company, **Tepha**, has developed **TephaFLEX Absorbable Suture** which is said to be the first medical device derived from a new class of biopolymers that is the product of patented recombinant DNA technology developed by Tepha and **licensed from the Massachusetts Institute of Technology (MIT)**. Compared to synthetic polymers such as polylactic acid (PLA) and polyglycolic acid (PGA), TephaFLEX material is tougher and more flexible, with an absorption rate and degradation profile that are compatible with human tissue repair and replacement applications. TephaFLEX polymer is said to be thermoplastic and can be fabricated into virtually any shape or form—including fibers, films, tubes, foams, textiles, microspheres, and molded constructs—using a wide range of conventional melt and solvent processing techniques. The TephaFLEX Absorbable Suture is engineered to be one of the strongest absorbable fibers known, offering up to 50 per cent greater tensile strength than currently marketed monofilament absorbable sutures. After the repair process, the biopolymers degrade in the body to natural metabolites in a biocompatible, cell-friendly manner.

6. To contain the possibilities of damage to the healthy tissues while **treating cancer cells**, the clinicians at the **BC Cancer Agency in Vancouver, Canada** used novel technique, called **volumetric arc therapy (VMAT.)** to deliver a single treatment in less than two minutes. The clinicians have treated a prostate cancer patient with a faster and a precise form of radiotherapy believed to be the first of its type in the world. The VMAT treatment is an advanced form of intensity modulated radiation therapy (IMRT), which has the potential to measurably improve treatment precision by sparing more healthy tissue in addition to making treatments much faster and more comfortable for patients. It was made possible by new software developed by clinicians at BC Cancer Agency together with an advanced medical linear accelerator and beam-shaping accessory from Varian Medical Systems. The software algorithm is designed to generate highly conformal dose distributions while taking advantage of the specific capabilities of the Varian linear accelerator. VMAT creates a precisely sculpted 3D dose distribution by tailoring the shape and intensity of the beam from a full 360 degrees around the patient. It provides more flexibility in maximizing the dose of radiation targeting the tumor while minimizing damage to surrounding healthy tissue.

7. A new **biochip technology** has been developed by scientists at **Taiwan's Academia Sinica** which has important implications in **cancer diagnostics**. Using a so-called **'glycan tagging system'**, the method allows for the monitoring of cell surfaces and intracellular trafficking of glycoproteins. It is known that more than 80 per cent of

human proteins have sugars attached, called glycoproteins. The process of protein glycosylation is known to play a central role in mediating protein function in living organisms. It is also known that altered glycosylation is often associated with inflammation and cancer metastasis. Therefore, the detailed correlation between glycosylation and biological or pathological status is of great interest and may provide information for disease diagnosis and treatment. According to the researchers, their new biochip technology can monitor this correlation and so will be useful for isolating and identifying the glycoproteins that are expressed differentially under various physiological or pathological conditions.

8. **Eli Lily** & Company simplified the **daily management of diabetes** by launching the **first insulin pen with memory, HumaPen MEMOIR**. MEMOIR is designed to meet the needs of people with diabetes who take several shots of mealtime insulin each day. It presents sophisticated technology and features in a consumer-friendly 'push-to-know' **digital display that allows patients to record and review their last 16 insulin doses, including the priming doses.**

9. **GlaxoSmithKline** launched its **first oral treatment for advanced breast cancer in Singapore**. The first oral ErbB1 (EGFR) and ErbB2 (HER2) dual inhibitor, **'lapatinib'**, received Health Sciences Authority (HAS), Singapore approval for use in combination with capecitabine for the treatment of patients with locally advanced or metastatic breast cancer whose tumors over expresses ErbB2 (HER2) and have received prior therapy, including trastuzumab. Lapatinib is a dual tyrosine kinase inhibitor targeting both ErbB1 (EGFR) and ErbB2 (HER2) receptors, and preventing the downstream signaling events essential for cancer cell proliferation and survival. This innovative mechanism of action is a new way to treat breast cancer and is different from current targeted therapy for ErbB2 positive disease.

10. Barely larger than a Blackberry and weighing just a little more than 700 grams, a new **portable ultrasound device Acuson P10 from Siemens Medical Systems**, supports medical personnel in emergency efforts. This handheld device is intended for complementary initial diagnostic care and triage, particularly in cardiology, emergency care and obstetrics. It is poised to change the physical exam by providing immediate information to health care providers at the earliest possible patient intervention points, such as intensive care units, ambulances and medical helicopters. In emergency situations, for example, this translates into earlier, faster and more accurate screening and triage decisions. **Its emergency applications include FAST (Focused Abdominal Sonography for Trauma) for detecting fluid, determining cardiac activity and pericardial effusion as well as for detecting abdominal aortic aneurysms and performing pelvic exams.**

11. Year 2007 saw India-based **Bharat Biotech International** launch **BioHib**, the first indigenously developed and manufactured *Haemophilus Influenza* type b (Hib) vaccine in India. It also launched its **tetravalent combination vaccine, Comvac4-HB** containing

Diphtheria, Pertussis, Tetanus and Hepatitis B in one vaccine. These two new vaccines BioHib and Comvac4-HB, enable the deployment of a combination pentavalent vaccine in a single injection. India's No. 1 biotechnology company and vaccine manufacturer, **Serum Institute of India**, earlier this year, launched its indigenously manufactured low-cost *Haemophilus influenza* type b conjugate vaccine under the brand name of HibPRO in India. **HibPRO** has been priced at $9 (Rs 375) for a monodose vial. HibPRO is a vaccine with clinically proven efficacy and safety at an affordable price. India's pharma major **Dr. Reddy's Laboratories** has launched **Reditux**, the Dr. Reddy's brand of rituximab, a monoclonal antibody (MAb) used in the treatment of Non-Hodgkin's Lymphoma. Talking about monoclonal antibody-based drugs, India's biotech major **Biocon's BIOMAb EGFR** for treating solid tumors of epithelial origin has extended applications. BIOMAb-EGFR that was launched in September 2006 has in 2007 seen an increase in the total number of registered patients to almost 1,000 in the country. The drug indicated for head and neck cancer is also being studied in global clinical trials for colorectal, lung cancer, glioma (brain cancer), and pancreatic cancers.

12. **Dr. Mario R Capecchi, Dr. Martin J Evans and Dr. Oliver Smithies were honoured with Nobel Prize for medicine (2007) for their discoveries of "principles for introducing specific gene modifications in mice by the use of embryonic stem cells"**. These Nobel Laureates have made a series of ground-breaking discoveries concerning embryonic stem cells and DNA recombination in mammals. Their discoveries led to the creation of an immensely powerful technology referred to as gene targeting in mice. It is now being applied to virtually all areas of biomedicine—from basic research to the development of new therapies. Gene targeting is often used to inactivate single genes. Such gene "knockout" experiments have elucidated the roles of numerous genes in embryonic development, adult physiology, aging and disease. To date, more than 10,000 mouse genes (approximately half of the genes in the mammalian genome) have been knocked out. Ongoing international efforts will make "knockout mice" for all genes available within the near future. With gene targeting it is now possible to produce almost any type of DNA modification in the mouse genome, allowing scientists to establish the roles of individual genes in health and disease. Gene targeting has already produced more than five hundred different mouse models of human disorders, including cardiovascular and neuro-degenerative diseases, diabetes and cancer.

13. Indian-Australian stem cell scientist, **Dr. Kuldip Sidhu has won the 2007 Top Invention Prize for his pioneering work in stem cells.** Dr. Sidhu, an Associate Professor at University of New South Wales, was acknowledged for the work on the derivation of a **new human embryonic stem cell line, Endeavour-1**. The prize is awarded by BioMed North Limited, a not-for-profit agency for the management and commercialization of intellectual property generated within the state of New South Wales (NSW). **Dr. Sidhu has produced a human embryonic stem cell line without the use of any animal product.** This breakthrough eliminates the risk of animal to-human contamination in potential stem cell therapy treatments. Dr. Sidhu's work could eventually lead to safer treatments

for conditions such as diabetes, Parkinson's disease, spinal cord injury and even breast cancer.

14. NZ's first Laureate Awardee **Professor Margaret Brimble**, Head of medicinal chemistry at Neuren Pharmaceuticals and researcher at the University of Auckland, is the first New Zealander to receive a **Laureate Award for Scientific Excellence** from the L'Oréal–United Nations Educational, Scientific and Cultural Organisation (UNESCO) for Women in Science Awards. The Award recognizes her research on the **synthesis of biologically active natural products that provide new lead compounds in the development of drugs.**

15. **Morphotek**, the US subsidiary of Japanese pharma giant Eisai, has received a US patent related to certain methods of **enhancing the affinity of monoclonal antibodies (mAbs)**. This indicates that its **Morphodoma technology**, a whole genome evolution platform can enhance cells producing mAbs to yield daughter cells producing antibodies with enhanced pharmacologic activity. Morphodoma technology enhances evolution within cells through the reversible regulation of the mismatch repair pathway, a highly conserved process employed by all living organisms to repair mutations that naturally occur during DNA replication.

16. Researchers at the **China Medical University Hospital (CMUH) based in Taichung** have developed an **autoimmune treatment for brain tumors** with a superior survival rate of 85.8 per cent. CMUH Neurosurgery Department Director Cho Der-yang said the new therapy employs dendritic cells extracted from a patient's blood to fight malignant gliomas cells in the brain – both to minimize side effects and increase the chances of survival. The dendritic cells extracted are first cultivated *in vitro* along with malignant gliomas cells, to better understand their enemy and trigger the immune mechanism against this type of tumor cell.

17. A neurology focused R&D company **NeuroDiscovery** has tied up with TetraQ of Australia to market the **safety pharmacology services**. TetraQ, based in the University of Queensland is a contract research organization in Australia providing integrated preclinical services to the global pharmaceutical and biotechnology industries. For this reason TetraQ has been chosen to market NeuroSolutions' safety pharmacology services in *Australasia*. This includes the Qpatch, an automated electrophysiology device to screen drugs for potential adverse cardiac affects (commonly known as a hERG screening).

18. **Mr. Vichai Jirathitikal**, a pharmacology graduate of **Mahidol University in Thailand**, has developed an **oral AIDS vaccine christened V-1 or V1**. V1 comprises pooled, inactivated HIV antigens derived from E and B clades of HIV-1 strain predominant in Thai population and is currently manufactured at a pharmaceutical plant owned by his family. The factory received GMP status from Thai FDA in July 2001. In contrast to other AIDS vaccines V1 is administered orally and not by injection. V1 is a specially formulated pill that contains HIV-1 viral antigens capable of withstanding the degradation in the stomach. As a result, V1 can pass into the gastrointestinal tract

and will be eventually presented by intestinal enterocytes to underlying mucosal immune cells. The V1 is presented as 850 mg pink colored oval-shaped pill. VI can be stable at ambient temperatures for more than a year. In the last few years, Mr. Jirathitikal has treated about 500 HIV positive patients at his clinic in Bangpakong near Bangkok City. Thailand has approximately one million people with HIV.

19. A team of **South Korean scientists** headed by Professor Oh Byung-ha at Pohang University of Science and Technology, Korea, has discovered how **a protein, TRIM5, blocks the progression of human immunodeficiency virus (HIV)**, which can develop into acquired immune deficiency syndrome (AIDS). This discovery could be a major breakthrough in the development of an AIDS cure. Prof Oh said that earlier in 2004 scientists had discovered that TRIM5 helps fight HIV, but they failed to discover how the protein works against the virus. Prof Oh has now identified a domain in TRIM5, called B30.2/SPRY, which blocks the progression of HIV and also brought out the three-dimensional structure of the domain. AIDS is the most severe manifestation of infection with HIV. It refers to a collection of symptoms and infections resulting from the depletion of the immune system caused by HIV. So far there are drugs that slow the progress of AIDS but no cures.

20. **GE Healthcare has launched Lullaby—a new blue compact fluorescent lamp (CFL) based Phototherapy System for the treatment of indirect hyperbilirubinemia, or jaundice in newborns, in India.** The Lullaby Phototherapy System was designed and developed at GE's research center in India and is part of their "In India, for India" program to make high-end technology more affordable, reliable and more accessible to larger sections of society. According to various neonatal experts, jaundice is one of the most common conditions requiring medical attention in newborns. There are 133 million newborn babies in the world diagnosed with jaundice annually. About 8-10 per cent of term, infants and 80 per cent of pre-term infants annually are required to undergo phototherapy for three-five days either at hospital or home.

21. Cuban scientists claim to have developed **a new therapeutic vaccine against AIDS**. Mr. Eduardo Martinez, the technological development director at **Havana-based Genetic Engineering and Biotechnology Center** said that the vaccine has been safely and successfully tested on animals and the next step is clinical trials. Mr. Martinez said the vaccine would be of immense help in improving the post treatment of AIDS patients whether or not the drug is used in conjunction with or without retroviral drugs. He also added that the outcome of the research would be discussed in the forthcoming Havana 2007 Biotechnology Congress that will see participation from 25 nations. Scientists and researchers across the world have been working on developing a vaccine to treat AIDS but have failed. There have been vaccines that prolong the lives of AIDS patients but those are again very expensive and out of bound for AIDS patients in countries such as India and South Africa, the two countries where the instances of AIDS are high. What

adds to problems of AIDS patients is the stigma attached to the disease that attacks and weakens the immune system that ultimately leads to death.

22. Scientist have created **Supermouse** in the lab whose genes were changed to enable it to **run at a speed of 20 meters for more than five hours non-stop.** Researchers at the **Case Western Reserve University at Cleveland, Ohio** in the US have genetically modified mouse to get top-notch physical endurance and athletic prowess, which they hope can be duplicated in humans to make them physically stronger—just by enhancing human physical ability. During their research, scientists also found out that the mouse lived longer and bred three times longer than a normal mouse and yet it was more aggressive for which the researchers don't have any answers. The mouse also ate more food but did not put on any weight. The research results were published in the *Journal of Biological Chemistry*. It elaborated on how the researchers managed to create the 'Supermouse' by altering a glucose metabolism gene, which in turn stimulated more efficient use of body fat to produce energy. The research also said the mouse dispersed lactic acid from its muscles, thus avoiding cramps during its endurance. Richard Hanson, the professor of biochemistry at Case western University, who created the "supermouse" said that gene that was modified was similar to the gene found in humans. Hanson said the strength of the mice is comparable only with very active and fit atheletes. This research will be useful for pharmaceutical companies who can develop new drugs that enhance muscle performance.

23. Monoclonal antibodies (mAbs) have been in the industry for three decades but now there is a major shift in progress. Beginning with murine monoclonal antibodies then progressing to humanized mAbs, the industry is now concentrating on **fully human mAbs**. According to a new report by *Insight Pharma* titled "*Monoclonal Antibodies: Pipeline Analysis and Competitive Assessment*" this major shift has ramifications for all industry professionals involved in the research and development of mAb therapeutics, as well as portfolio managers and marketers of approved products. Monoclonal antibodies have proven their value in the pharmaceutical industry with two drugs: **Remicade and Rituxan**. Combined these generated sales of about $ 4 billion each in 2006, while global sales for the entire mAb portfolio approached $ 20 billion. In the survey, 81 per cent of respondents indicated that their organization's pipeline of mAb products has increased over the past five years. Capturing that increased activity; *Monoclonal Antibodies: Pipeline Analysis and Competitive Assessment* identifies approximately 175 therapeutic products now in clinical trials that are largely concentrated in the areas of cancer, immunological and inflammatory diseases, and infectious diseases. Many more candidates are at the preclinical stage of development.

24. Biodegradable flaps of semi-transparent rubber are fused together to form structures snaked through with a network of interconnecting channels that could become the building blocks for the first 3D artificial organs according to **Jeff Borenstein, a micro-machinist at the Draper Laboratory in Cambridge, Massachusetts**. When

pumped with a solution of cells, nutrients and oxygen, these channels spring into life, forming a system of blood vessels and tissue that might one day be the basis of an artificial heart, kidney or liver. This technology could be used to build sophisticated artificial organs complete with their own network of blood vessels, or vasculature. So far simple organs including skin and bladders have been built by seeding a tissue scaffold with cells.

25. In a groundbreaking experiment **scientists from International Stem Cell (ISC) Corp** derived four unique embryonic stem cell lines that open the door for the creation of therapeutic cells that will not provoke an immune reaction in large segments of the population. The stem cell lines are "HLA-homozygous," meaning that they have a simple genetic profile in the critical areas of the DNA that code for immune rejection. The lines could serve to create a stem cell bank as a renewable source of transplantable cells for use in cell therapy to replace damaged tissues or to treat genetic and degenerative diseases. This research may lead to providing treatment of large numbers of patients by any one cell line. Jeffrey Janus, President of International Stem Cell and colleagues at the company and from the Russian Academy of Medical Sciences, described the successful parthenogenetic activation of human oocytes and the subsequent derivation of cell lines having the morphology and markers characteristic of human embryonic stem cells. The authors emphasize two key factors that would make this technology so valuable for future efforts to generate replacement tissues and organs and to use donor-derived cell repositories to develop cell-based therapies. First, the four human parthenogenetic stem cell lines, designated as HpSC-Hhom, are HLA (human leukocyte antigen) homozygous. This makes it possible to match the HLA types of a donor and recipient, reducing the chances of provoking an immune reaction against the transplanted donor cells. Second, the stem cells are derived from unfertilized donor eggs, not from fertilized embryos, so the technique does not carry the same ethical burden.

26. **Doris Taylor of the University of Minnesota in Minneapolis**, US, and colleagues have "decellularised" rat hearts by adding chemicals that break up cells, leaving behind just the connective tissue, which is made of proteins. The resulting scaffolds still have the heart's complex three-dimensional shape, including space for all the blood vessels. These rat hearts were repopulated with fresh cells, and were made to beat again in a pioneering experiment that could lead to novel treatments for human patients. The scientists seeded the scaffolds with blood vessel and heart muscle cells from newborn rats and flowed a soup of nutrients through the scaffold. The seed cells migrated throughout the structure and grew into muscle and blood vessels. The team also applied small electrical jolts to trigger beating. Within four days, the tissue had started to contract, and within just 8 days the new heart was pumping with 2% of the efficiency of an adult heart. Seeding could be done with a recipient's own cells, avoiding the immune response that complicates organ transplants. It can be applied to any organ taken from cadavers – that opens up new possibilities for saving lives.

27. **Stemagen, an embryonic stem cell research company, announced that it has become the first in the world to create a cloned human embryo using somatic cell nuclear transfer (SCNT).** This will be a major breakthrough in creating patient-specific and disease-specific stem cells for therapeutic use potentially including developing treatments for Parkinson's, Alzheimer's and other degenerative diseases. Five blastocysts were developed from 25 donated mature oocytes. Three were confirmed to be clones based on DNA fingerprinting demonstrating the presence of the skin cell donor DNA in the blastocyst, while one was further confirmed to be a clone by an additional mitochondrial DNA (mtDNA) analysis which revealed the presence of oocyte donor mtDNA without any oocyte donor nuclear DNA. For technical reasons, the genetic material in the remaining two blastocysts did not amplify to the extent required for analysis, and so while it is likely they were clones, the evidence required to claim that with certainty was not present. Thus, in this study, cloned blastocysts were successfully created from approximately 10% of all mature donated oocytes, an unexpectedly high rate given past research in this field.

The oocytes used in this study were donated, without compensation, by egg donors and intended parents undergoing egg donation cycles for reproductive purposes at the Reproductive Sciences Center in La Jolla, a leading fertility center specializing in egg donation and other advanced assisted reproductive technologies. All research procedures, including the culturing of the skin cells (fibroblasts) were performed under clinical laboratory conditions in close cooperation with the Assisted Reproductive Technologies (ART) Laboratory of the Reproductive Sciences Center. DNA fingerprinting is the scientifically accepted method for determining if an embryo is a true clone. All samples were subjected to this type of analysis to determine their true genetic makeup. For that, Stemagen turned to Genesis Genetics, a recognized worldwide leader in the field of reproductive embryonic analysis.

28. **Australian scientists are researching a possible way of making aggressive prostate cancer cells less invasive** after their discovery of a protein essential for the normal functioning of cells. Prof Robert Parton led a team of scientists from the **Institute for Molecular Bioscience at University of Queensland**, who discovered that the **protein PTRF-cavin is required for caveolae formation.** Caveolae are pits on the surface of cells, which are involved in many processes essential for the healthy functioning of the body, including tumour suppression. Scientists had already identified one protein involved in caveolae formation, called caveolin, however they were unaware that it functions in conjunction with PTRF-cavin. Caveolae formation is a fundamental process that affects every cell in the body, and importantly, has implications for prostate cancer diagnosis and therapy. In most cancers, the cells stop expressing caveolin, and the caveolae don't form. But some aggressive prostate cancer cells actually show much higher levels of caveolin than normal. The scientists found that although caveolin levels were high, caveolae still weren't forming, and further research found this is because

PRTF-cavin is missing. The team made this discovery by adding PTRF-cavin to a prostate cancer cell line, which prompted the formation of caveolae. They are currently studying if this change could alter the invasive properties of the cancer cells. This process could be developed for use in diagnosing, and eventually treating, prostate cancer acoording to Dr. Michelle Hill, a senior investigator on the team. Dr. Hill said, "The fact that a protein, caveolin, was associated with promoting cancer in one case and preventing it in another highlighted the need for different avenues of treatment for different cancers. The role of caveolin differs between cancers because it is in a different form at the cell surface, which highlights the importance of not only the protein, but also its interaction with other proteins and its location on the cell surface." Rather than current cancer treatments, which kill any cells that are growing quickly, this would be a more targeted, tailored treatment that should result in less side effects and a more effective therapy overall.

29. A new strategy to hunt for human viruses has revealed a previously unknown virus strongly associated with another rare but deadly skin cancer called Merkel cell carcinoma. **University of Pittsburgh Cancer Institute (UPCI)** researchers, Huichen Feng, Ph.D., Masahiro Shuda, Yuan Chang, and Patrick Moore, explain a nearly decade-long effort to harness the sequencing technology to identify the virus, which they call Merkel cell polyomavirus (MCV). The work does not prove MCV to be the cause of Merkel cell carcinoma however, it may lead to new cancer treatment and prevention options. "This is the first polyomavirus to be strongly associated with a particular type of human tumor," said Dr. Moore, professor of microbiology and molecular genetics at the University of Pittsburgh School of Medicine and leader of the molecular virology program at UPCI. "Although polyomaviruses have been studied in relation to cancer development for years, the weight of scientific evidence had been leaning toward the view that these viruses do not cause human cancers."

Polyomaviruses are a group of viruses that have been shown to cause cancers in animals for more than 50 years. But Dr. Moore noted that additional research is needed to determine what role, if any, MCV plays in human cancer development. A rare but extremely aggressive cancer that spreads rapidly into other tissues and organs, Merkel cell carcinoma (MCC) develops from specialized nerve cells that respond to touch or pressure. The incidence of MCC has tripled over the past 20 years to about 1,500 cases a year, especially among people whose immune systems are compromised by AIDS or transplant-related immunosuppressant drugs. About half of patients with advanced MCC live nine months or less, and some two-thirds of MCC patients die within five years.

Vaccines are now available against human papillomavirus (HPV) to prevent cervical cancer, noted Dr. Chang, professor of pathology. "MCV is another model that may increase our understanding of how cancers arise, with possibly important implications for non-viral cancers like prostate or breast cancer." MCV has additional similarities to

HPV since both viruses integrate into the tumor cell genome but not the genome of healthy cells. This integration destroys the virus's ability to replicate normally and may be the first critical step in MCC development.

The Pittsburgh team analyzed nearly 400,000 messenger RNA genetic sequences from four samples of MCC tumor tissue using a technique refined in their lab called **digital transcriptome subtraction (DTS)**. Comparing the sequences expressed by the tumor genome to gene sequences mapped by the Human Genome Project, the researchers systematically subtracted known human sequences, leaving a group of genetic transcripts that might be from a foreign organism. One sequence was similar to but distinct from all known viruses. The team went on to show that this sequence belonged to a new polyomavirus present in eight of 10 (80 per cent) Merkel cell tumors they tested but only five of 59 (8 per cent) control tissues from various body sites and four of 25 (16 per cent) control skin tissues. Although MCV is most commonly found in Merkel cell tumors, it also can be found in healthy people. The most important distinguishing feature is that MCV integrates into tumor cells in what is known as a monoclonal pattern, indicating that it infects the cell before the cell becomes cancerous. Tests on six of the eight MCV-positive tumors confirmed that viral DNA was integrated within the tumor genome in this monoclonal pattern, suggesting that infection with MCV could be a trigger for tumor formation. The Pittsburgh team subsequently confirmed these results with additional tumor specimens.

30. January 2008 Researchers at the **University of Southern California (USC)** have identified a novel **cyclic signaling in the dermis** that **coordinates stem cell activity and regulates regeneration in large populations of hairs** in animal models. The signaling switch involves **bone morphogenetic protein (Bmp) pathway**. Conceptually, the findings have important implications for stem cell research and in understanding how stem cell activity is regulated during regeneration according to Cheng-Ming Chuong, principal investigator and professor of pathology at the Keck School of Medicine of USC. The research presents a new dimension for the regulation of hair re-growth and ultimately organ regeneration. The hair is an important model for organ regeneration in mammals because it is one of the few organs that regenerate regularly. Recent work in the field has established hair cycling as one of the mainstream models for organ regeneration. However, most of these works focus on the cyclic regeneration of one single hair follicle. "Each of us has thousands of hair follicles. In our study, we were motivated to analyze the coordinative behaviour of cyclic regeneration in a population of organs," Chuong says.

The research team found that hairs, even in normal mice, regenerate in waves, rather than individually. The findings suggest that hair stem cells are regulated not only by the micro-environment within one hair follicle—as has previously been thought—but also by adjacent hair follicles, other skin compartments and systemic hormones, in a hierarchical order. At the molecular level, the findings showed that periodic expression of Bmp in the skin macro-environment appears to be at the center of the mechanism for coordinated hair stem cell activation. When many hairs regenerate, they must communicate activation signals among themselves. At different time points the macro-environment can be either permissive or suppressive for stem cell activation. This research shows that the formation of new tissues or organs from stem cells—such as the formation of new hairs—can be more robust if it occurs in a permissive macro-environment according to Maksim Plikus, Ph.D., a post-doctoral fellow and the first author of the study. The work also has critical implications for research using the mouse skin as a model for tumor growth or drug delivery.

31. Researchers at **University of Texas M. D. Anderson Cancer Center** have jumped a significant hurdle in the use of RNA interference (RNAi), believed by many to be the ultimate tool to both decode the function of individual genes in the human genome and to treat disease. Reporting in the journal Genes and Development, investigators have developed a simple way to use the RNAi approach to silence a selected gene in a specific tissue in a mouse to determine the function of that targeted gene. Previously reported approaches to achieve this were either technically cumbersome, not generally applicable, or only achieved transient knockdown of the target gene. Having a tool that will allow to knockdown the expression of any given gene in any specific tissue or cell type represents a significant advance in the field according to the study's lead investigator, **Miles Wilkinson, a professor in the Department of Immunology**. For example, this method could potentially be used in humans "to knockdown the expression of mutant or overexpressed genes that cause human diseases, including cancer," Wilkinson says. "Scientists and clinicians can use it to reduce the expression of the target gene in a single or limited numbers of cell types or tissues, thereby reducing side effects."

Equally, the technique can help researchers determine the function of a single gene in a single tissue, which "is potentially a powerful investigative tool," Wilkinson says. "By silencing a particular gene in a specific tissue, you can learn what the function of that gene is in that particular tissue without blocking its essential functions in other tissues." In their study, Wilkinson and his research team demonstrated how well their tool worked by silencing the WT1 tumor suppressor gene in the testes of mice. They found this gene is important in the production of healthy sperm by encoding a regulatory protein called a transcription factor that controls the formation of adherens junctions, or the cell-to-cell contacts between nurse cells and the germ cells that ultimately become sperm. Using RNAi to silence WT1, therefore, led to the discovery that WT1 is the first transcription factor shown to regulate the formations of these junctions, Wilkinson says. "It is a transcription factor that dictates both the formation of the testes in the embryo, and the function of the testes after birth."

Researchers worldwide have been trying to harness the power of RNAi since it was discovered in 1998. RNAi rocked the world of science because of the vast implications this natural cellular control system potentially offered medicine. RNAi is used by cells in many life forms to identify viral RNA, which often enters a cell as a double strand. The organism's own RNA, however, leaves the nucleus of cells (where it was produced by DNA) as a single strand, to be decoded by ribosomes into the proteins that perform all the work of the cell. RNAi, therefore, recognizes the double strand of viral RNA as different, and sets in motion cell machinery that destroys the invading RNA. This involves cutting up the double-stranded RNA, separating it into single strands, and destroying other single-stranded RNA molecules that are complementary to those small RNA segments. Scientists quickly realized that if they could produce a double-stranded RNA that mimics the RNA produced by a gene they wish to silence, RNAi would do the job for them. Producing the decoy "small RNAs" that trigger RNAi has become a fairly simple process, Wilkinson says, but the difficulty has been to use these matches only in specific tissues, and to figure out a way to make this "treatment" last. The work by the M. D. Anderson team has come up with a solution that solves both problems.

The technique they developed involves use of two different "modules" that can be swapped in and out of the backbone of a vector. One is a small stem loop designed to complement the RNA produced by the gene they wish to silence, and the other is a "promoter" that provides expression specific to the tissue they want to target. "There are whole batteries of different promoters, ones specifically for skin, or different parts of the brain, or whatever organ or tissue you are likely to want," Wilkinson says. "By swapping out either of these modules, you have the potential to silence any gene in any tissue you might want," he said. In their study, the researchers made a vector with a promoter specific for nurse cells in the testis, so that the small RNA that is associated with WT1 is only made in this cell type, thereby reducing WT1 levels that are only in the nurse cells. Other organs, like kidney, that need WT1 to function are not affected, Wilkinson says.

32. One way cancer arises is when tumor suppressor genes that normally keep cell growth in check are mysteriously turned off. Now, researchers at **Johns Hopkins** have discovered that at least one **tumor suppressor gene is in fact turned off by a "noncoding" single stranded RNA** nucleic acid similar to its double-stranded DNA cousin. The so-called antisense RNA is made by a gene on a neighboring strand of DNA. Most genes in the human genome have associated with them nearby antisense RNAs, which, as their name implies, are complementary to the amino acid sequences in a "sense" RNA to which they may bind and switch off. This is the first time an antisense RNA has been observed to silence a tumor suppressor through the means of epigenetic changes according to **Hengmi Cui**, Ph.D., assistant professor of molecular medicine at Hopkins. Epigenetic changes refer to heritable changes in genetic material that are not changes in the sequence of the DNA; these could include the addition of chemical tags onto DNA or otherwise altering how compressed the DNA is in a cell. The Johns Hopkins team notes that a similar phenomenon occurs in plants but until now has not been seen in any type of animal, including humans. Andrew Feinberg, M.D., M.P.H., professor of medicine, oncology and molecular biology and genetics and director of the Epigenetics Center at

Hopkins, says the results of the team's experiments "bring us closer to solving two outstanding mysteries in biology, namely what all those noncoding RNAs do in cells and how tumor suppressor genes get turned off. He also mentioned that noncoding RNAs may be important in silencing tumor suppressor genes. Following clues that suggested such a role for antisense RNA, the researchers first surveyed computer databases for tumor suppressor genes with known neighboring antisense RNAs. They found antisense counterparts to 21 well-known tumor suppressor genes and decided to further study one of them, p15. That gene is deleted or silenced in several types of human cancer, including melanomas, gliomas, lung, and bladder carcinomas and up to 60 per cent of leukemias.

The research team first analyzed leukemia cells for the presence of antisense p15. Of 16 patient samples, 11 showed an increase in antisense p15 and decreased p15. The researchers confirmed in other experiments that the more antisense p15 a cell contained the less sense p15 it was likely to have, strong evidence that the antisense was somehow turning down the normal, sense version. Chemically turning on the antisense gene, the team found, turned off the sense p15 gene. When they looked at the DNA around the p15 gene in cells, they found that the DNA was more compact and tightly packaged, which generally shuts off genes. "Somehow, the presence of the antisense RNA leads to the formation of this tightening of the chromosome to make heterochromatin around the p15 gene, turning it off," says Feinberg. Further characterization of the antisense RNAs, according to Feinberg, could lead to their use as markers for certain types of cancer as well as targets for cancer-specific drugs and therapies.

33. **UCLA researchers** have found that a **key protein in the body's dendritic cells can stop the virus that causes AIDS from "budding"**—part of the virus' life cycle that is crucial to its ability to replicate and infect other cells. "If we can block virus generation, then we can control the disease," said lead author **Shen Pang**, associate professor in the division of oral biology and medicine at the UCLA School of Dentistry and a member of the UCLA AIDS Institute. Dendritic cells are specialized white blood cells in the skin, mucosa and lymph nodes that kick-start a primary immune response to foreign invaders by activating lymphocytes, including the T cells that HIV targets. Though dendritic cells can be infected with HIV—and indeed play a crucial role in transmitting the virus to T cells—studies have shown that viral generation from these cells is nearly a hundred times lower than from infected T cells, indicating that the cells may possess some inhibiting property. Pang hypothesized that **DC-SIGN, a protein expressed in dendritic cells**, may be responsible for such inhibition. He and his colleagues found that DC-SIGN and a related protein, DC-SIGNR, both demonstrated 95 per cent to 99.5 per cent inhibition of viral production from host cells.

Very few cells are infected when HIV first enters the human body, but the virus rapidly creates new copies of itself, which in turn infect more cells. To achieve this, the virus, after infecting a cell, sends envelopes of protein to the cell's membrane. The viral genomes then combine with viral structural proteins and move into these envelopes. The envelopes bubble or bud outward, releasing viral particles that will infect more cells and start new viral life cycles. According to the researchers, DC-SIGN appears to block HIV generation by efficiently neutralizing an HIV glycoprotein on the surface of the HIV envelope known as gp120, a key to viral infection. In such cases, while some viral particles may still be released from the infected dendritic cells, the lack of gp120 in their envelopes means they are not infectious to CD4-positive T-lymphocytes and macrophages. In other words, these viral particles have been rendered uninfectious.

Current methods to interrupt the life cycle of the virus are limited because they generally target HIV at the stages of viral entry, reverse transcription and post-translational protein cleavages. Once the virus passes through these stages, treatment fails. The UCLA researchers, therefore, focused on halting the virus' generation at different stages in its life cycle. "The strong inhibition of viral production by DC-SIGN suggests the possibility of using this protein for treatment of HIV-infected patients," the researchers write. "Expression of this protein in various CD4-positive cells should inhibit viral production from infected cells. Because it can also enhance the immune response, DC-SIGN is expected to be useful for *in vivo* studies for developing an HIV vaccine."

34. Metabolic syndrome, a collection of related abnormalities like hypertension, obesity, insulin resistance, and excess cholesterol, poses a major risk for developing heart disease and diabetes. Individuals with a genetic predisposition to high cholesterol (familial hyperlipidemia or FH) can be especially vulnerable to metabolic syndrome, but researchers have now found that **blocking the enzyme stearoyl-CoA desaturase-1 (SCD1)**, which helps synthesize unsaturated fatty acids, greatly improves the profile of FH-mice affected by metabolic syndrome. Previous mouse studies on SCD1 found that blocking this enzyme could reduce obesity in normal mice; Michael Hayden and colleagues wondered whether such a protective effect would extend to mice with excess cholesterol levels. They mimicked FH in mice by knocking out the LDL receptor, causing a cholesterol buildup. When fed a high-fat Western diet, these mice develop obesity and diabetes in adulthood. However, when the researchers also knocked out SCD1, the mice improved dramatically, accumulating far less fat in their liver, which in turn led to fewer triglycerides in the blood, increased insulin sensitivity, and less weight gain. These results highlight, that SCD1 might be a potential drug target for FH individuals who have developed other components of metabolic syndrome.

35. Bacteria that eat sugar and release cavity-causing acid onto teeth may soon be made dramatically more vulnerable to their own acid. Researchers have identified key genes and proteins that, if interfered with, can take away the ability of a key bacterial species to thrive as its acidic waste builds up in the mouth. The ability of *Streptococcus mutans* (*S. mutans*) to survive in acid is one reason that the species is the main driver of tooth decay worldwide. Past research has shown that this ability has several components including a bacterial enzyme called **fatty acid biosynthase M (FabM)**, which when

shuts down, makes *S. mutans* almost precisely 10,000 times more vulnerable to acid damage.

In addition, early work suggests that FabM or one of its relatives may also help all *Streptococci* (strep) and *Staphylococci* (staph) infections to resist the human body's defenses, which include immune cells that subject bacteria to acid. Between them, "strep" and "staph" bacteria are responsible for meningitis, pneumonia, sepsis, methicillin-resistant staph aureus, the "flesh-eating" infection (fasciitis), as well as infections on heart valves and around stents.

While FabM represents a major target for the design of new drugs, the focus of the next round of work is to identify and rank every one of the 2,000 known *S. mutans* genes that contributes to its "fitness" (ability to survive, out-compete other strains and cause disease). A research team at the **University of Rochester Medical Center** announced that it has received a $ 3.6 million fitness profiling grant from the National Institute of Dental and Craniofacial Research (NIDCR), part of the National Institutes of Health (NIH). Grant-funded projects will seek to create a catalogue of proteins that, along with FabM, can serve as targets for a multi-pronged attack on bacteria that tend to evolve around single-thrust treatments.

S. mutans produces lactic acid as a waste product of fermentation, the process by which some ancient lifeforms convert sugar into energy for life without using oxygen. After a great many generations of exposure to its own acid waste, the membranes for this species have become "acid durable." The research team has shown that FabM contributes to this durability by making carbon chains, the main functional feature of *S. mutans* membrane fatty acids, grow longer. In fact as many as 60 per cent of the fatty acids in a bacterial outer membrane undergo this change as acidity increases. Researchers have already shown that such structural changes protect membranes, presumably by making it more difficult for acids to donate hydrogen ions to them, but they do not yet know why. Forcing hydrogen ions on other compounds gives acid its bad reputation. Remaining questions that the team will be seeking to solve over the next five years include how do longer fatty acids in membranes protect against acid specifically, and how do bacteria sense changes in acidity.

The team has genetically engineered the first and only mutant form of *S. mutans* with the FabM gene removed. This FabM "knockdown" mutant is a living model that shows the exact impact of the enzyme in live bacteria. Without FabM, the mutant fills its outer membrane with other, smaller fatty acids that are much less acid resistant than those normally created via FabM, but that still provide some protection from acid. Thus, a goal is to design a treatment that would prevent *S. mutans* from forming both straight chain and "smaller chain" fatty acids. As next-generation antibacterial drugs are designed, scientists are looking not just for a single way to stop the action of a single disease-causing enzyme, but how to shut down its three or four back-up systems. The process of cutting off genetic escape routes for bacteria applies to every trait central to the ability of the bacteria to survive and cause disease. Beyond acid durability, the team will also look at the genes and proteins that enable *S. mutans* to stick to teeth enamel like no other, which it does by producing a sugary polymer (plaque). Tooth decay is the result of plaque combined with acid.

The members of the team, Grayhack and Phizicky will create a library of mutant strains for the 2,000 known *S. mutans* genes, with each strain having just one of the 2,000 genes shut off. They will then subject the library to acid, for example, and see which strains thrive. Knowing which gene is missing from each strain, researchers will then be able to draw conclusions about each single gene's contribution to not only to acid durability, but also to many aspects of the strep bacteria's ability to survive and cause disease.

36. Scientists have genetically engineered a mosquito to release a sea-cucumber protein into its gut which impairs the development of malaria parasites, according to new research. Researchers say this development is a step towards developing future methods of preventing the transmission of malaria. The new study has focused on disrupting the growth and development of the malarial parasite with a lethal protein, CEL-III, found in sea cucumbers, to prevent the mosquito from passing on the parasite. The international team fused part of the sea cucumber lectin gene with part of a mosquito gene so that the mosquito would release lectin into its gut during feeding. The released lectin is toxic to the ookinete (invasive offspring) and therefore kills the parasite in the mosquito's stomach. In laboratory tests the research team showed that introducing lectin to the mosquito's gut in this way significantly impaired the development of malaria parasites inside the mosquito, potentially preventing transmission to other people. Early indications suggest that this sea cucumber protein could be effective on more than one of the four different parasites that can cause malaria in humans. Professor Bob Sinden from Imperial College London's Department of Life Sciences, one of the researchers said "The results are very promising and show that genetically engineering mosquitoes in this way have a clear impact on the parasites' ability to multiply inside the mosquito host."

37. Researchers at **Texas A&M University** have discovered that lower cholesterol levels can actually reduce muscle gain with exercising. Lead investigator **Steven Riechman,** assistant professor of health and kinesiology, and **Simon Sheather,** head of the Department of Statistics, along with colleagues from The **Johns Hopkins Weight Management Center and the Northern Ontario School of Medicine,** have recently had their findings published in the Journal of Gerontology. Further research in this area is needed before one can look differently at cholesterol, especially as it relates to a vigorous workout.

The team studied 55 men and women, ages 60-69, who were healthy non-smokers and were able to perform exercise testing and training. Three days a week for 12 weeks, participants performed several exercises, including stretching, stationary bike riding and vigorous weight lifting. Those who had to miss one or more sessions all conducted make-up sessions so that by the study's end, the entire group had engaged in uniform activities. Also, all participants consumed similar meals. At the conclusion of the study, the researchers found that there was a significant association of dietary cholesterol and change in strength. In general, those with higher cholesterol intake also had the highest muscle strength gain. Cholesterol circulating in the blood also appeared to have contributed to greater muscle gain in the participants. One possible explanation is through cholesterol's important role in the inflammation process according to the researchers. Riechman said that subjects who were taking cholesterol-lowering drugs while participating in the study showed lower muscle gain totals than those who were not. Combined with exercise, cholesterol appears to play a role in contributing to muscle gain. These findings show that the restricting of cholesterol - while in the process of exercising - appears to affect building muscle mass in a negative manner. Cholesterol may play a key role in muscle repair. An important point here is that cholesterol is negatively associated with cardiovascular health and hence, further study is required in this area.

38. Researchers are increasingly studying proteins as potential therapies for a variety of dread diseases because they can influence cell behavior by fueling or dampening certain molecular signals. **University of Florida** researchers have **coaxed liver and pancreatic cells within diabetic mice** into churning out insulin by injecting the animals with a naturally occurring protein called **Pdx1** opening up a new research avenue that someday could lead to safer treatments for type 1 diabetes. **Pdx1 activates the genes controlling the development of the pancreas cells that make and release insulin to maintain safe levels of glucose in the body**.

"Pdx1 is so special because it possesses a unique amino acid sequence that acts as a sort of molecular passport, allowing it to pass freely into cells, enter the nucleus and activate insulin production and release," said lead scientist **Dr. Li-Jun Yang**, an associate professor of pathology, immunology and laboratory medicine at UF's College of Medicine. Earlier research has shown that inserting the Pdx1 gene into liver or pancreas cells can induce insulin production, but most gene therapy methods use viruses to introduce a piece of genetically engineered DNA into cells. The disadvantage of such approaches is that researchers can never be certain the viruses are entirely harmless, Yang said. The idea with protein therapy is that eventually a person's own cells could be reprogrammed to naturally produce the hormone, restoring the body's ability to properly regulate blood sugar levels without having to use a potentially hazardous virus to slip corrective genes into the body or having to transplant pancreatic cells from someone else. That could eliminate the adverse effects sometimes associated with gene therapy and eliminate the need for lifelong suppression of the immune system so transplanted cells are not rejected, Yang said. The experiment was repeated six times, and reproducible results were obtained. The protein also promoted regeneration of insulin-producing cells in the pancreas, allowing the diabetic mice to become normal.

Yang said there is now reason to believe normal blood sugar levels can be maintained for long periods, suggesting that an infrequent Pdx1 injection might someday replace daily insulin injections. Even more importantly, the reprogrammed and regenerated cells should make and release insulin, automatically maintaining safe blood sugar levels, she said. "What's so innovative about UF's approach is the ability to normalize blood glucose levels in diabetic mice simply by delivering Pdx1 protein in the target cells, thus effectively eliminating the side effects associated with gene therapy," Yang said.

39. A new study finds that **gastric inhibitory polypeptide (GIP)** found in the body is **capable of promoting weight loss, improving insulin resistance and reversing diabetes in an animal model**. The study was jointly conducted at **School of Biomedical Sciences, University of Ulster, Coleraine, Northern Ireland, UK and Department of Medical Physiology, Panum Institute, University of Copenhagen, Copenhagen, Denmark**. GIP is a peptide hormone that is secreted in response to food. It inhibits the secretion of acids stimulates the released insulin as part of the digestive process in response to food. It is found in a variety of tissues, including the intestine, heart, stomach, brain and in adipose (fat). While the significance of its action is largely unknown, its potent and prolonged stimulation after a high-fat diet has led researchers to speculate it may play a key role in metabolizing fat. Research has shown that high fat feeding results in elevated circulating GIP concentrations, traits often found in patients who are obese with diabetes. GIP also effects the growth of fat cells. Other studies have shown that mice injected with the GIP receptor antagonist-(Pro3) GIP-can reverse or prevent many of the metabolic abnormalities associated with obesity.

The study examined whether prolonged GIP receptor antagonism using daily injections of (Pro3) GIP was able to reverse well established diet-induced obesity and related metabolic abnormalities. The researchers used a model for diet-induced obesity that has been used extensively alongside genetic models and has close parallels with obesity, increasingly found in humans who consume a high-fat, energy-rich diet. In this model, young (8-week old) male, age matched mice were age-divided into groups and housed individually in an air-conditioned room with a 12 hour light: 12 hour dark cycle. Experimental animals had free access to drinking water and a high fat diet (45 per cent fat, 20 per cent protein and 35 per cent carbohydrate; per cent of total energy of 26.15kj/g). Age-matched control mice from the same colony had free access to a standard rodent maintenance diet (10 per cent fat; 30 per cent protein; 60 per cent carbohydrate; per cent of total energy of 12.99kj/g.). The two were used for comparison purposes. Prior to the study, mice were maintained on a high fat diet for 160 days. In addition, a separate set of mice were maintained on a high fat diet for 112 days prior to measuring circulating GIP and GLP-1 levels. On both occasions, obesity and diabetes were clearly evident. The mice which had previously been fed a high fat diet for 160 days received only daily injections of either saline or (Pro3)GIP over a 50-day period. Food intake and body weight were recorded daily while plasma glucose and insulin concentrations were monitored at 5-7 day intervals. Blood was taken on day 50 to measure cholesterol, triglycerides, glucagon (the hormone involved in metabolizing carbohydrate), corticosterone (involved with carbohydrates in the liver) and circulating adipokines (which play a key role in obesity-related diseases). Glucose tolerance and insulin sensitivity tests were performed at the end of the study period. The metabolic response of both groups of mice was also analyzed.

The experiment generated some interesting findings:

(i) Compared with the standard rodent diet (control), the mice that were fed the high-fat diet for the previous 160 days exhibited increased body weight, energy intake, and circulating glucose concentrations. The levels remained elevated throughout the study. The cholesterol and triglycerides levels increased at day 50.

(ii) Consumption of the high fat diet resulted in progressive weight gain and elevations of plasma glucose and gyrated hemoglobin, leading to impaired insulin sensitivity and glucose intolerance by 10 days. Fat (adipose) tissue deposits were increased as were circulating cholesterol and triglyceride concentration levels.

(iii) (Pro3)GIP was able to counter many of the detrimental effects of high fat diet on body weight and indices of glucose and lipid metabolism.

The researchers concluded that blocking GIP activity using (Pro3)GIP in mice with established, high fat diet-induced obesity and diabetes results in significant weight loss, improvement of insulin resistance and amelioration of diabetes. These findings represent an interesting new approach to the treatment of obesity and metabolic disturbances.

40. Researchers have identified a **new gene linked to cerebral venous thrombosis, a condition that causes blood clots in the veins of the brain** that can lead to stroke. The condition is more common in young and middle-aged women. The study compared 78 people with cerebral venous thrombosis in Germany to 201 healthy people. Researchers found that a variant of the gene called factor **XII C46T** is more common in people with cerebral venous thrombosis than in healthy people. A total of 16.7 per cent of those with cerebral venous thrombosis had the gene variant, compared to 5.5 per cent of those without the condition. The results were the same after adjusting for factors that could affect blood clotting, such as age, gender, smoking, and use of oral contraceptives. "These results need to be confirmed, but it appears that people with cerebral venous thrombosis should be tested for this gene and should be considered for use of blood thinning medication to prevent future blood clots," said study author **Christoph Lichy, MD, of the University of Heidelberg in Germany**. Other genetic variants have also been linked to cerebral venous thrombosis. Cerebral venous thrombosis is a rare condition that is the cause of less than one per cent of strokes and other cerebrovascular disorders, but it results in death approximately 10 per cent of the time. Symptoms include headaches, seizures, visual problems, and motor and sensory problems. In addition to genetic factors, other factors that can cause cerebral venous thrombosis include head injury, infection, and certain drugs.

41. January 2008 **Stowers Institute's Baumann Lab** has identified the long-sought **telomerase RNA gene** in a single-cell research model. Chromosomes shorten with every cell division. In stem cells and in cancer cells, this shortening is compensated by telomerase, an enzyme that adds short repeat sequences to the ends of chromosomes to replenish lost DNA. As telomerase is required for the continued growth of most cancer cells, the enzyme is considered a promising target for new anti-cancer drugs. A correlation between telomere length and a variety of diseases has further intensified interest in understanding telomerase and its regulation. The RNA subunit of telomerase is of particular interest as it represents one of the two core components of telomerase and provides the template for the short repeats that are added to the ends of chromosomes. The Baumann Lab is working to understand how telomerase is assembled, how it is recruited to chromosome ends, and how its activity is regulated. These efforts may shed light on the sometimes surprising correlations between telomere shortening and stress, smoking, obesity, and a variety of diseases including cancer and coronary heart disease.

Telomerase RNA has been studied in a variety of simple model organisms, but telomere maintenance turned out to be quite different in these species compared to human cells. Recently, the Baumann Lab used a biochemical approach to identify and clone the RNA subunit of telomerase in *Schizosaccharomyces pombe*, or fission yeast."The identification of the fission yeast equivalent of the telomerase RNA gene provides us with a critical tool to study telomerase in a genetically tractable, single-cell organism with telomere maintenance machinery that shares many features with human cells," explained Peter Baumann, Ph.D., Assistant Investigator. "We and others had been studying telomerase activity, recruitment, and regulation for several years but the fact that the RNA subunit was unknown in our fission yeast model system severely limited our ability to make progress." The researchers are now turning their attention to how and where telomerase is assembled from its components in the cell and what processing it must undergo to become active.

42. Researchers from **Leiden University, the Netherlands** have worked with **tropical African butterflies** to find an interesting new insight into the evolution of lifespan. **Jeroen Pijpe** and his colleagues used **artificial selection** to create genetically long-lived butterflies of the species ***Bicyclus anynana***. In the field, the temperature experienced by the caterpillar sets up the butterfly to become the form that matches the season. This is called **phenotypic plasticity**, and in this species it has evolved as a response to the alternating seasons. The dry season form is long-lived and more starvation resistant. In the wet season, reproduction takes place. In this case the authors selected on starvation resistance under wet-seasonal conditions. They discovered that female butterflies had shifted their reproduction from quantity to quality of offspring. Next, they found that males and females had responded differently to the artificial selection: males seemed to have lowered their energy consumption, whilst females had increased the amount of energy to spend. The most important finding was that the artificial selection in a wet seasonal environment had produced butterflies that resemble the long-lived dry season

form. Pijpe said "In other words, we targeted genes that are needed to live longer, and the result is very much like how temperature induces butterflies to live longer in the dry season." It suggests that, in this species, the evolution of lifespan is closely associated with phenotypic plasticity. It is likely that the regulation of lifespan involves mechanisms of phenotypic plasticity during development, also in humans, according to the researchers.

43. There is new hope for patients with chronic wounds: **euroderm GmbH and the Fraunhofer Institute for Cell Therapy and Immunology IZI in Leipzig** have been granted approval to produce artificial skin from patients' own cells. These scientists have been given permission to grow dermal tissue for grafting onto chronic wounds such as open leg ulcers on diabetics patients according to IZI team leader Dr. Gerno Schmiedeknecht. Chronic wounds are treated by grafting on the patients' own skin, which is normally taken from the thigh. This leaves scars on both the thigh and the treated wound. If this skin is produced using the recently approved EpiDex® technique instead, the same can be achieved without hurting the patient. Moreover, the artificial skin grows onto the wound without scarring according to Dr. Andreas Emmendörffer, managing director of euroderm GmbH.

Another advantage is that the transplantation can be performed on an outpatient basis. A few days later, it is already possible to see whether the new skin has adhered to the wound. 72 days later, the grafted skin can no longer be distinguished from healthy skin. But how exactly is the new skin grown? A few hairs are pluck off the back of the patient's head and adult stem cells are extracted from their roots, which are then allowed to proliferate in a cell culture for about two weeks. Then the nutrient solution is reduced until it no longer covers the upper sides of the cells, exposing them to the surrounding air. The increased pressure exerted by the oxygen on the surfaces of the cells causes them to differentiate into skin cells. In this way, the researchers have grown numerous small pieces of skin, produced individually for each patient, which add up to a surface area of 10 to 100 square centimeters when pieced together. To ensure that they comply with the safety regulations at all times, the researchers are using new clean rooms at the IZI, a state-of-the-art facility for producing different kinds of cell therapeutics.

44. University of Pennsylvania researchers have used **gene therapy** to reduce the time it takes to breed large animals capable of producing **therapeutic proteins in their milk**, such as insulin or those that fight cancer. This represents a significant milestone in drug development, as current methods involve cloning, which takes more time and generally costs more. "Having an easier way to harness nature's power to produce large quantities of specific proteins in milk could increase the availability of drugs for people who could otherwise not afford these treatments," said Ina Dobrinski, one of the researchers on the study.

The study also is significant because it may also be a new way to eliminate diseases in future generations of animals, such as those used for livestock. Here's why: To get the goats to produce specific proteins, the researchers used radiation to kill a portion of a male goat's germ cells (the cells that produce sperm). Then they used a **modified adeno-associated virus** (a well studied and tolerated gene therapy vector) to insert a gene in the remaining cells. Once the new gene took hold in the germ cells, a predictable number of female offspring produced the desired protein in their milk. The advance is immediately valuable for pharmaceutical development and biology research, but a similar approach could be used to bolster the food supply by eliminating genetic disorders in animals over several generations. It is also possible that once perfected, this technique could eliminate disease genes in humans over several generations, assuming ethical concerns can be resolved adequately.

45. A research team at **Weill Cornell Medical College in New York City identified two genes that may be crucial to the production of an immune system cytokine called interleukin-10 (IL-10)**. The discovery fills in an important "missing link" in a biochemical pathway that's long been tied to disorders ranging from lupus and Type 1 diabetes, to cancer and AIDS. IL-10 production should be balanced for keeping good health. Too much IL-10 can leave the body more vulnerable to killers such as viruses and cancer, and to certain antibody-driven autoimmune diseases such as lupus, while too little can lead to run-away inflammatory pathology, according to Dr. Xiaojing Ma, Professor of Immunology and Microbiology in the Departments of Microbiology and Immunology and Pediatrics at Weill Cornell. Every second, millions of the body's cells undergo naturally programmed cell death—a process called apoptosis. In healthy individuals, these dying or dead cells are spotted and then quickly ingested and removed by immune system "scavenger" cells such as macrophages.

However, to prevent this type of clean-up from triggering a wider immune response, macrophages express the IL-10 cytokine in the presence of apoptotic cells. IL-10 suppresses the activity of immune system T-cells that might otherwise run amuck. Prior studies have shown that CD36—a protein receptor lying on the surface of the macrophage—is important for the recognition of apoptotic cells by macrophages. In this work, the researchers observed that CD36 also helped to trigger IL-10 production whenever apoptotic cells were around.

The team also studied various signals that lead to IL-10 production from CD36 present at the cell surface. The Weill Cornell group first exposed macrophages to apoptotic (dying) cells. They then used sensitive assays to look for key biochemical changes occurring downstream of CD36 activation. It was found that proteins in the cell nucleus that were binding to a site already known was critical for the production of IL-10 as macrophages encountered apoptotic cells. In subsequent biochemical experiments, the team identified the two genes responsible for the transcription (gene-directed production) of these proteins. These genes—pre-B transcription factor 1(Pbx-1) and Pbx-regulating protein 1 (Prep-1)—are best known to scientists as partners for their role in embryonic development and several forms of leukemia, with Pbx playing a major part in hematopoeisis, the production of new and myriad blood cell types. The findings could also reveal exciting new information as to how aberrant IL-10 expression contributes to disease.

46. **CAMBRIDGE, Mass.-MIT scientists have found a way to induce cells to form parallel tube-like structures that could one day serve as tiny engineered blood vessels.** The researchers led by Robert Langer, MIT Institute Professor found that they can control the cells' development by growing them on a surface with nano-scale patterning. Engineered blood vessels could one day be transplanted into tissues such as the kidneys, liver, heart or any other organs that require large amounts of vascular tissue, which moves nutrients, gases and waste to and from cells. This will provide a novel way to someday engineer tissues in the human body. The work focuses on vascular tissue, which includes capillaries, the tiniest blood vessels, and is an important part of the circulatory system. The team has created a surface that can serve as a template to grow capillary tubes aligned in a specific direction.

The researchers built their template using microfabrication machinery at Draper Laboratory in Cambridge. Normally such technology is used to build micro-scale devices, but the researchers adapted it to create nano-scale patterns on a silicone elastomer substrate. The surface is patterned with ridges and grooves that guide the cells' growth. The cells, known as endothelial progenitor cells (EPCs), not only elongate in the direction of the grooves, but also align themselves along the grooves. That results in a multicellular structure with defined edges, also called a band structure. Once the band structures form, the researchers apply a commonly used gel that induces cells to form three-dimensional tubes. Unlike cells grown on a flat surface, which form a network of capillary tubes extending in random directions, cells grown on the nano-patterned surface form capillaries aligned in the direction chosen by the researchers.

The researchers believe the technique works best with EPCs because they are relatively immature cells. Earlier attempts with other types of cells, including mature epithelial cells, did not produce band structures. Growing tissue on a patterned surface allows researchers a much greater degree of control over the results than the classic tissue engineering technique of mixing cell types with different growth factors and hoping that a useful type of tissue is produced.

GENOMICS

1. Researchers at **Stanford University** report Genome Biology, **a new approach to computationally predicting the locations and structures of protein-coding genes in a genome.** Gene finding remains an important problem in biology as scientists are still far from fully mapping the set of human genes. Furthermore, gene maps for other vertebrates, including important model organisms such as mouse, are much more incomplete than the human annotation. The new technique, known as **CONTRAST (CONditionally TRAined Search for Transcripts), works by comparing a genome of interest to the genomes of several related species.** CONTRAST exploits the fact that the functional

role protein-coding genes play a specific part within a cell and are therefore subjected to characteristic evolutionary pressures. For example, mutations that alter an important part of a protein's structure are likely to be deleterious and thus selected against. On the other hand, mutations that preserve a protein's amino acid sequence are normally well tolerated. Thus, protein-coding genes can be identified by searching a genome for regions that show evidence such patterns of selection. However, learning to recognize such patterns when more than two species are compared has proved difficult. Previous systems for gene prediction were able to effectively make use of one additional 'informant' genome. For example, when searching for human genes, taking into account information from the mouse genome led to a substantial increase in accuracy. But, no system was able to leverage additional informant genomes to improve upon state-of-the-art performance using mouse alone, although it was expected that adding informants would make patterns of selection clearer.

CONTRAST solves this problem by learning to recognize the signature of protein-coding gene selection in a fundamentally different way from previous approaches. Instead of constructing a model of sequence evolution, CONTRAST directly 'learns' which features of a genomic alignment are most useful for recognizing genes. This approach leads to overall higher levels of accuracy and is able to extract useful information from several informant sequences.

In a test on the human genome, CONTRAST exactly predicted the full structure of 59% of the genes in the test set, compared with the previous best result of 36%. Its exact exon sensitivity of 93%, compared with a previous best of 84%, translates into many thousands of exons correctly predicted by CONTRAST but missed by previous methods. Importantly, CONTRAST's accuracy using a combination of eleven informant genomes was significantly higher than its accuracy using any single informant. The substantial advance in predictive accuracy represented by CONTRAST will further efforts to complete protein-coding gene maps for human and other organisms.

2. Using computer simulations, **researchers at the University of Illinois** have demonstrated a strategy for **sequencing DNA by driving the molecule back and forth through a nanopore capacitor in a semiconductor chip**. The technique could lead to a device that would read human genomes quickly and affordably. Being able to sequence a human genome for $1,000 or less (which is the price most insurance companies are willing to pay) could open a **new era in personal medicine**, making it possible to precisely diagnose the cause of many diseases and tailor drugs and treatment procedures to the genetic make-up of an individual. Aksimentiev and collaborators describe the method in a paper accepted for publication in the journal Nano Letters, and posted on the journal's Web site. They have demonstrated that back-and-forth motion of a DNA molecule in a nanopore capacitor 1 nanometer in diameter produces an electrostatic fingerprint that can be used to read the genetic sequence through molecular dynamics simulations.

In the researchers' simulations, performed at the university's National Center for Supercomputing Applications, the nanopore capacitor consists of two conducting layers of doped silicon, separated by an insulating layer of silicon dioxide. As DNA passes through the nanopore, the molecule's electric field induces sequence-specific electrostatic potentials that can be detected at the top and bottom layers of the capacitor membrane. A semiconductor device capable of reading the electrostatic potentials and decoding the genetic sequence is within the grasp of current technology, Aksimentiev said. Nanometer pores in electronic membranes have been manufactured, and the voltage signals resulting from DNA movement through such pores have been recorded. The next big challenge, Aksimentiev said, is to minimize noise in the system, and reduce the speed of DNA molecules moving through the pore.

3. Researchers have solved the structure of a DNA-protein complex that is crucial in the spread of antibiotic resistance among bacteria. Knowing this structure also provides fundamental insight into how cells successfully divide into two new cells with intact DNA. Using X-ray crystallography, the team led by **structural biologists at The University of Texas M. D. Anderson, Maria Schumacher with colleagues at the University of Sydney, Australia** produced clear 3-D images of the structure that results when two proteins connect with a DNA site to "segregate" DNA during cell division. The researchers, in this study answer a basic science question and flag a possible target for clinical attack on antibiotic-resistant *Staphylococcus aureus*, a tenacious and often lethal staph infection.

"The plasmid segregation system we are working on, called pSK41, is found in *S. aureus* and confers resistance to multiple antibiotics, including the drug of last resort, vancomycin," Schumacher says. "Because the segregation systems are essential for the retention of these multidrug resistant plasmids, they represent wonderful drug targets." Plasmids are a great model for understanding cell division and segregation, Schumacher says, because plasmid segregation is relatively simple: two proteins connect to one DNA site to launch the process. Cells divide to multiply and it's crucial for this split to go smoothly so each daughter cell ends up with the DNA it needs to function. If these plasmids don't divide and go to the next generation of cells, those bacteria cells lose their drug resistance. These scientists capture the first structure ever solved of a segrosome complex that partitions and divides DNA.

A protein called ParR connects with a centromere DNA site, a round string of DNA repeats in the plasmid, to form the segrosome complex, which then completes itself by attracting filaments of another protein called ParM. The ParM filaments grow, literally pushing the two replicated plasmid segrosomes apart, resulting in two copies of the plasmid with intact DNA. The researchers used X-ray crystallography to discern the structure of both proteins connecting to the plasmid DNA. This technique first requires purification and crystallization of the protein or, in this case, the more challenging DNA-protein complex structure. Next, the crystal sample is mounted on a tiny loop in an X-ray diffractor and X-rays are beamed through it. Bending around the lattice of atoms in the crystal, the X-rays produce a diffraction pattern, which can be seen as a dot pattern on a monitor. A state-of-the-art X-ray diffractor shoots a different angle of the crystallized protein every few minutes. Taking between 180-300 images will produce 100,000 to 300,000 data points.

Mathematical analysis of those points reveals a three-dimensional structure. The first image is an electron density map, which illustrates a protein's hills and valleys based on the position of the electrons of the atoms that compose the complex. This leads to the more exotic ribbon diagram, a spiraling, looping almost skeletal representation of the complex's crucial components. The researchers used the Advanced Light Source (ALS) synchrotron at the U.S. Department of Energy's Lawrence Berkeley National Laboratory a fast, powerful diffractor that shoots a new diffraction pattern every half second, to solve this structure.

4. Viticulture, the growing of grapes (*Vitis vinifera*) chiefly to make wine, is an ancient form of agriculture, evidence of which has been found from the Neolithic and Early Bronze Ages. A detailed understanding of how nurture affects the qualities of a grape harvest leading to the concept of terroir (the range of local influences that carry over into a wine) is now possible by the publication of a high quality draft genome sequence of a Pinot Noir grape by an Italian-based multinational consortium.

The genome of the grape is spread over 19 pairs of chromosomes and is around 504.6 megabases in length. The team of researchers, led by **Dr. Riccardo Velasco of the Istituto Agrario di San Michele all'Adige, used a shotgun sequencing approach**, which has resulted in 10.7X coverage, 4.2X using pyrosequencing and 6.5X by Sanger sequencing. At the same time, the genome of the grape chloroplast was also sequenced and, remarkably, this was found to be identical to an independently determined sequence from a different strain of Pinot Noir that was published in 2006. The grape, therefore, has a relatively small genome for a crop plant, similar to that of rice or poplar trees and much smaller than that of wheat or maize. Nevertheless, sequencing the genome was complicated by the degree of heterozygosity between pairs of chromosomes, some 11.2% of the sequence differing between homologous regions. The team also discovered more than two million single nucleotide polymorphisms (individual letter changes in the grape's genetic blueprint) in 87% of the 29,585 identified genes. While this made sequencing the genome difficult, it now provides a massive library of inherent variation with which to investigate which genes influence which characteristics of the growing plant and in what ways.

The genome can also provide clues to the evolution of grapes. Many plant genomes, especially those of crop plants, have been produced by at least one duplication of a smaller ancestral genome. Whether this was true for grapes had been controversial but this study clearly shows that ten of the 19 chromosomes resulted from a duplication that occurred shortly after the lineage of grapes diverged from that of the model plants

Arabidopsis and poplar. The breeding of grape vines is difficult because they take several years to grow to maturity and domesticated grapes tend to have very low fertility. For this reason, grapes are usually propagated by cuttings or graftings so that vineyards are filled with hundreds of thousands of genetically identical clones. This leaves grapes highly susceptible to the emergence of aggressive microrganisms, such as phyloxera, which devastated European grape production in the 19th and early 20th century, and powdery mildew, which continues to threaten American harvests to this day.

The Pinot Noir genome will provide an invaluable tool for creating grape varieties resistant to such diseases without altering the quality of the resulting wine. Velasco and his colleagues have identified a large number of genes related to disease—resistance, 289 of which contain one or more SNPs. In spite of this, Pinot Noir remains susceptible to several fungi, bacteria and viruses possibly due to a defective system for recognition pathogen. Many of these disease-resistance genes are present in clusters whose associations with resistances or tolerances of different grape varieties to specific diseases can now be investigated. Also Pinot Noir can be crossed with many wild grape species providing a large reservoir of disease-resistance genes, which can be exploited with the aid of this genome road map. The genome sequence simultaneously identified hundreds of genes, which correspond to enzymes that produce flavor and aroma compounds. This will allow breeding for disease resistance to proceed without disturbing the biochemistry of taste and grape quality. This grape genome may also have implications beyond viticulture. Grapes can be both genetically transformed and micropropagated to produce hundreds of identical clones. With the sequencing of its relatively small genome, it is well placed to become a model organism for fruit trees in general. It is, however, in the safeguarding and improvement of grape stocks that the effects of this genome will be felt most strongly.

5. Estimates of the **number of genes in the human genome** have ranged wildly over the past two decades, from **20,000 all the way up to 150,000**. By the time the working draft of the human genome was published in 2001, the best approximation stood at 35,000, yet even that number has fallen. A new analysis, one that harnesses the power of comparing genome sequences of various organisms, now reveals that the true number of human genes is about **20,500**, thousands fewer than what is currently listed in human gene catalogs.

The work, led by researchers at the **Broad Institute of MIT and Harvard**, has implications beyond merely settling the debate over how many genes are in the human genome. An accurate gene count can help identify the locations of genes and their functions, an important step in translating genomic information into biomedical advances. Ironically, the way genes are recognized has triggered much of the confusion over the human gene count. Scientists on the hunt for typical genes – that is, the ones that encode proteins – have traditionally set their sights on so-called open reading frames, which are long stretches of 300 or more nucleotides, or "letters" of DNA, bookended by genetic start and stop signals. This method produced the most recent gene count of roughly 25,000, but the number came under scrutiny after the 2002 publication of the mouse genome revealed that many human genes lacked mouse counterparts and vice versa.

Such a discrepancy seemed suspicious in part because evolution tends to preserve gene sequences – genes, by virtue of the proteins they encode, usually serve crucial biological roles. But the 25,000 DNA sequences were already listed in the catalogs of human protein-coding genes, and skeptics had no systematic way to remove them. Far from blatant mistakes, non-gene sequences can masquerade as true genes if they are long enough and happen by chance to fall between start and stop signals. Despite having gene-like characteristics, these open reading frames may not encode proteins. Instead, they might have other functions or possibly none at all. To distinguish such misidentified genes from true ones, the research team, led by Clamp and Broad Institute director Eric Lander, developed a method that takes advantage of another hallmark of protein-coding genes: conservation by evolution. The researchers considered genes to be valid if and only if similar sequences could be found in other mammals - namely, mouse and dog. Applying this technique to nearly 22,000 genes in the **Ensembl gene catalog**, the analysis revealed 1,177 "orphan" DNA sequences. These orphans looked like proteins because of their open reading frames, but were not found in either the mouse or dog genomes. Although this was strong evidence that the sequences were not true protein-coding genes, it was not quite convincing enough to justify their removal from the human gene catalogs. Two other scenarios could, in fact, explain their absence from other mammalian genomes. For instance, the genes could be unique among primates, new inventions that appeared after the divergence of mouse and dog ancestors from primate ancestors. Alternatively, the genes could have been more ancient creations – present in a common mammalian ancestor – that were lost in mouse and dog lineages yet retained in humans.

If either of these possibilities were true, then the orphan genes should appear in other primate genomes, in addition to our own. To explore this, the researchers compared the orphan sequences to the DNA of two primate cousins, chimpanzees and macaques. After careful genomic comparisons, the orphan genes were found to be true to their name – they were absent from both primate genomes. This evidence strengthened the case for stripping these orphans of the title, "gene." After extending the analysis to two more gene catalogs and accounting for other misclassified genes, the team's work invalidated a total of nearly 5,000 DNA sequences that had been incorrectly added to the lists of protein-coding genes, reducing the current estimate to roughly 20,500. In addition to suggesting a major revision to the human gene count, this work provides a set of rules for evaluating any future proposed additions to the human gene catalog. It also underscores the benefit of genome sequencing projects.

More broadly, the research reveals that little invention of genes has occurred since mammalian ancestors diverged from the non-mammalian lineage. "There's no real creativity going on in the mammalian genome," explained Clamp. That means that the number, structure, and function of protein-coding genes are not expected to differ very much from mammal to mammal, so what makes humans different from mice and dogs likely lies outside this realm of the genome. Clamp and her Broad Institute colleagues are now peering into the genomes of many other mammals, in an attempt to explain what parts of our genome truly make us human.

6. **Norwegian and Swiss biologists** have made a startling discovery about the relationship between organisms that most people have never heard of. **The Tree of Life must be re-drawn, textbooks need to be changed, and the discovery may also have significant impact on the development of medicines.** The findings come from the largest ever genetic comparison of higher life forms on the planet. Of 5000 genes examined, researchers identified 123 common genes from all known groups of organisms; these common genes have been studied more closely. The results of the study were pretty astounding. All non-bacterial life on Earth—called eukaryotic life— can now be divided into four main groups instead of the five groups that have been known up to now according to Kamran Shalchian-Tabrizi, an associate professor from the University of Oslo's Department of Biology who has also worked with the Department of Zoology and Animal Biology and the Department of Genetic Medicine and Development, at the University of Geneva, Switzerland. The research group has found that brown algae and silica algae, and groups of single cell organisms like the malaria parasite, marine foraminifera, and the green sun animalcule (*Acanthocystis turfacea*) actually belong to the same group. Previously, these species were thought to be completely unrelated.

All life on Earth can be divided into two essentially different life forms—eukaryotes and prokaryotes. The eukaryotes gather their genetic material in a nucleus, while the prokaryotes (bacteria and archaea) have their genetic material floating freely in the cell. Eukaryotic organisms—such as humans—can, as a result of the new findings, be divided into the following four categories:

- Plants (green and red algae, and plants)
- Opisthokonts (amoebas, fungi, and all animals—including humans)
- Excavates (free-living organisms and parasites)
- SAR (the new main group, an abbreviation of Stramenophiles, Alveolates, and Rhizaria, the names of some of its members)

The SAR group has to some extent been identified earlier, but the scientists could not know if it was a correct observation because there was no statistical data available. To get that data the entire eukaryote tree was reconstructed with the help of these 123 genes. Chromalveolates and rhizaria were clearly separate groups. Three billion years ago, there was only bacteria and Archaea. Eukaryotic life, which comprises all multi-celled organisms, developed in the sea—probably between 1.2 and 1.6 billion years ago. It was not before about 500 million years ago that the first creatures crept onto land. By digging down into the historical layers with the help of phylogenetic reconstruction kinship between organisms at the genetic level can be found.

7. In a new report, scientists estimate that **more than 6,000 genes affect weight in mice**. Though those findings came from studies of mice, they may have meaning for people, notes researcher **Michael Tordoff, PhD, of Philadelphia's Monell Chemical Senses Center**. "Reports describing the discovery of a new 'obesity gene' have become common in the scientific literature and also the popular press," Tordoff says in a news release. "Our results suggest that each newly discovered gene is just one of the many thousands that influence body weight, so a quick fix to the obesity problem is unlikely." Tordoff's team reviewed previous studies of some 1,900 mouse genes. Most of those genes didn't affect the mice's weight, but 31% could boost weight and 3% could cut weight. In other words, weight-gain genes outnumbered weight-loss genes by 10 to one, "which might help explain why it is easier to gain weight than lose it," Tordoff says.

8. A study led by **McGill University** researchers has demonstrated that small differences between individuals at the DNA level can lead to dramatic differences in the way genes produce proteins. These, in turn, are responsible for the vast array of differences in physical characteristics between individuals. This study solves in part the mystery of how a relatively small number of differences within DNA protein coding sequences could be responsible for the enormous variety of phenotypic differences between individuals. It had previously been shown that individual differences reside in simple, relatively small variations in the DNA sequence called single nucleotide polymorphisms (SNPs), which exist primarily in the "junk code" of the DNA not previously known to have any profound genetic effect. "There are many SNPs," explained **Dr. Jacek Majewski** of McGill University. If all of them are added together then the two individuals would differ at more than a million of those positions. Majewski and his colleagues have demonstrated that the natural processing of messenger RNA (mRNA), via a process called splicing, is genetically controlled by these SNPs. The SNPs in certain individuals lead to changes in splicing and result in the production of drastically altered forms of the protein. These out-of-proportion consequences may lead to the development of genetic diseases such as cystic fibrosis and Type 1 diabetes.

9. **A preliminary assembly and annotation of the soybean genome, *Glycine max*, has been made available by the U.S. Department of Energy Joint Genome Institute (DOE JGI)**, to the greater scientific community to enable bioenergy research. The **large-scale shotgun DNA sequencing project** began in the middle of 2006 and will be completed in 2008. A total of about 13 million shotgun reads have been produced and deposited in the National Center for Biotechnology Information (NCBI) Trace Archive in accordance with the consortium's commitment to early access and consistent with the Fort Lauderdale genome data release policy. The current assembly (representing 7.23x coverage), gene, set, and browser are collectively referred to as "Glyma0". Glyma0

is a preliminary release, based on a partial dataset. This is expected to be replaced with an improved, chromosome-scale "Glyma1" version by the end of 2008. Early users of this data should track their favorite genes by saving local copies of the DNA sequences of these loci, and not by identifier or sequence coordinate, as these will change in future versions. DOE JGI's interest in sequencing the soybean stems from its role as a principal source of biodiesel, a renewable, alternative fuel with the highest energy content of any alternative fuel. Detailed knowledge of the soybean genetic code will enable crop improvements for more effective application of this plant for clean bioenergy generation. Knowing which genes control specific traits, researchers are able to change the type, quantity, and/or location of oil produced by the crop.

Through utilization of the sequence information generated by DOE JGI, it may be possible to develop a customized biomass production platform for combining oil seed production for biodiesel with enhanced vegetative growth for ethanol conversion, doubling the energy output of the crop. In 2004, over 3.1 billion bushels of soybeans were grown on nearly 75 million acres in the US, with an estimated annual value exceeding $ 17 billion, second only to corn, and about twice that of wheat.

The soybean genome project was initiated through the DOE JGI Community Sequencing Program (CSP) by a consortium led by DOE JGI's Dan Rokhsar, Stanford's Jeremy Schmutz, Gary Stacey of the University of Missouri-Columbia, Randy Shoemaker of Iowa State University, and Scott Jackson of Purdue University, with support from the U.S. Department of Agriculture and the National Science Foundation. Several other individuals, projects, grants, and agencies have made this monumental project possible. These included the four major projects: Public Expressed Sequence Tags (ESTs), SoyMap (which includes BAC libraries, modern physical mapping, and clone-based sequencing), and the Genetic Map with funding from USDA, NSF, United Soybean Board (USB), and the North Central Soybean Research Program (NCSRP). The original physical map development, funded by USB, was conducted by Jan Dvorak, from the University of California, Davis, along with the Washington University Genome Center in St. Louis, Missouri, and David Grant, USDA-ARS, Ames, Iowa.

10. Cells keep a close watch over the transcriptome—the totality of all parts of the genome that are expressed in any given cell at any given time. Researchers at the **Salk Institute for Biological Studies and the University of Missouri-Kansas City** teamed up to peel back another layer of **transcriptional regulation** and gain new insight into **how genomes work**. Converting the "genetic blueprint" into molecular building blocks requires two basic processes: transcription, which copies the information from DNA into RNA transcripts and takes place in the cell's nucleus, and translation, where the RNA serves as a template to manufacture proteins outside the nucleus. But before transcripts can guide protein synthesis or take on regulatory functions, they have to undergo a strict mRNA surveillance system that degrades defective, obsolete, and surplus transcripts. The scientists closely observed a specific subclass of transcripts that are under

the control of the exosome, a molecular machine in charge of controlled RNA degradation. They found evidence for widespread exosome-mediated RNA quality control in plants and a 'deeply hidden' layer of the transcriptome that is tightly regulated by exosome activity. Since the exosome is in the business of chewing things up, the scientists inactivated the multi-unit complex to bring its otherwise invisible substrates to the fore. Then they combed the transcriptional landscape for hitherto unseen peaks of transcripts that now were untouched by the degrading force of the exosome complex and came up with a genome-wide atlas of *Arabidopsis* exosome targets.

"Our careful design and rigorous validation of the system for conditionally and quickly inactivating the exosome turned out to be really crucial for homing in on its RNA targets," explains Dmitry A. Belostotsky of the University of Missouri-Kansas City. "On the other hand, genome-wide analyses of permanent genetic mutations often produce a complex mixture of direct and indirect effects, making it very hard to untangle. Thus, we think our strategy has a broadly-applicable value." Since the common notion is that the exosome plays a central role in bulk RNA turnover, the researchers say, they expected to find the levels of all transcripts increasing when they inactivated the exosome complex.

11. A group of scientists in **Princeton's Department of Ecology and Evolutionary Biology** has uncovered a new biological mechanism that could provide a clearer window into a cell's inner workings employing the singled-celled ciliate *Oxytricha trifallax* as their testbed. This mechanism could represent an "epigenetic" pathway—a route that bypasses an organism's normal DNA genetic program—for so-called Lamarckian evolution, enabling an organism to pass on to its offspring characteristics acquired during its lifetime to improve their chances for survival. The research also could have implications as a new method for controlling cellular processes, such as the splicing order of DNA segments, and increasing the understanding of natural cellular regulatory processes, such as which segments of DNA are retained versus lost during development.

Typical human cells each have one nucleus, serving as the control center for the cell, ciliate cells have two. One, the somatic nucleus, contains the DNA needed to carry out all the non-reproductive functions of the cell, such as metabolism. The second, the germline nucleus, like humans' sperm and egg, is home to the DNA needed for sexual reproduction. When two of these ciliate cells mate, the somatic nucleus gets destroyed, and must somehow be reconstituted in their offspring in order for them to survive. The germline nucleus contains abundant DNA, yet 95 per cent of it is thrown away during regeneration of a new somatic nucleus, in a process that compresses a pretty big genome (one-third the size of the human genome) into a tiny fraction of the space. This leaves only 5 per cent of the organism's DNA free for encoding functions. Yet this small hodgepodge of remaining DNA always gets correctly chosen and then descrambled by the cell to form a new, working genome in a process (described as "genome acrobatics") that is still not well understood, but extremely deliberate and precise.

Landweber and her colleagues have postulated that this programmed rearrangement of DNA fragments is guided by an existing "cache" of information in the form of a DNA or RNA template derived from the parent's nucleus. Through a series of experiments, the group tested out their hypothesis that DNA or RNA molecules were providing the missing instruction booklet needed during development, and also tried to determine if the putative template was made of RNA or DNA. DNA is the genetic material of most organisms, however RNA is now known to play a diversity of important roles as well. RNA is DNA's chemical cousin, and has a primary role in interpreting the genetic code during the construction of proteins.

First, the researchers attempted to determine if the RNA cache idea was valid by directing specific RNA-destroying chemicals, known as RNAi, to the cell before fertilization. This gave encouraging results, disrupting the process of development, and even halting DNA rearrangement in some cases. In a second experiment, Nowacki and Yi Zhou, both postdoctoral fellows, discovered that RNA templates did indeed exist early on in the cellular developmental process, and were just long-lived enough to lay out a pattern for reconstructing their main nucleus. In the third experiment, Nowacki, Zhou and Vijayan, a 2007 Princeton graduate in electrical engineering, constructed both artificial RNA and DNA templates that encoded a novel, pre-determined pattern; that is, that would take a DNA molecule of the ciliate's consisting of, for example, pieces 1-2-3-4-5 and transpose two of the segments, to produce the fragment 1-2-3-5-4. Injecting their synthetic templates into the developing cell produced the anticipated results, showing that a specified RNA template could provide a new set of rules for unscrambling the nuclear fragments in such a way as to reconstitute a working nucleus.

This wonderful discovery showed for the first time that RNA can provide sequence information that guides accurate recombination of DNA, leading to reconstruction of genes and a genome that are necessary for the organism. It reveals that genetic information can be passed on to following generations via RNA, in addition to DNA. The research team believes that if this mechanism extends to mammalian cells, then it could suggest novel ways for manipulating genes, besides those already known through the standard methods of genetic engineering. This could lead to possible applications for creating new gene combinations or restoring aberrant cells to their original, healthy state.

12. Researchers are poised to cross a dramatic barrier: **the creation of life forms driven by completely artificial DNA. Scientists in Maryland** have already built the world's first entirely handcrafted chromosome—a large looping strand of DNA made from scratch in a laboratory, containing all the instructions a microbe needs to live and reproduce. They hope to transplant it into a cell by 2009, where it is expected to "boot itself up," like software downloaded from the Internet, and cajole the waiting cell to do its bidding. And while the first synthetic chromosome is a plagiarized version of a natural one, others that code for life forms that have never existed before are already under

construction. The cobbling together of life from synthetic DNA, scientists and philosophers agree, will be a watershed event, blurring the line between biological and artificial—and forcing a rethinking of what it means for a thing to be alive.

At the core of synthetic biology's new ascendance are high-speed DNA synthesizers that can produce very long strands of genetic material from basic chemical building blocks: sugars, nitrogen-based compounds and phosphates. Experiments with "natural" DNA indicate that when a faux chromosome gets plopped into a cell, it will be able to direct the destruction of the cell's old DNA and become its new "brain"—telling the cell to start making a valuable chemical, for example, or a medicine or a toxin, or a bio-based gasoline substitute.

Unlike conventional biotechnology, in which scientists induce modest genetic changes in cells to make them serve industrial purposes, synthetic biology involves the large-scale rewriting of genetic codes to create metabolic machines with singular purposes. So far, synthetic biology is still semi-synthetic, involving single-cell organisms such as bacteria and yeast that have a blend of natural and synthetic DNA. The cells can reproduce a defining trait of life. But in many cases that urge has been genetically suppressed, along with other "distracting" biological functions, to maximize productivity.

13. The HIV virus contains just nine genes encoding 15 proteins, which wreak havoc on the human immune system. But this bare bones approach could have a fatal flaw. Lacking robust machinery, HIV hijacks human proteins to propagate, and these might represent powerful therapeutic targets. Drugs currently used to treat the viral infection interact directly with the virus itself, and it's quite simple for the rapidly mutating virus to avoid destruction by altering how it interacts with these chemicals. Patients use a cocktail of HIV inhibitors because the virus is less likely to evolve resistance to multiple drugs at the same time. But some HIV strains have still managed to evade particular drugs. These could eventually develop resistance to several drugs, especially among patients who don't adhere to their regimens.

A different approach has been followed by Harvard Medical School (HMS) Professor and senior author Stephen Elledge which centers on the human proteins exploited by the virus to get hold of it. The virus would not be able to mutate to overcome drugs that interact with these proteins.

Labs around the world have made impressive contributions to our understanding of the HIV life cycle. Over the last two decades, they've identified dozens of human proteins, or host factors, required for HIV propagation. The new study builds on this work, essentially quadrupling the list of host factors to include proteins involved with a surprising array of cellular functions ranging from protein trafficking to a type of programmed cell death called autophagy. The work began with a library of short interfering RNAs (siRNAs) targeting specific human genes. Each siRNA disrupts the gene's ability to produce a particular protein.

With the help of the staff at the Institute of Chemistry and Cell Biology at Longwood (ICCB-L), the siRNAs on thousands of human cells, with just one gene being targeted in each well of cells. Thus each well contained cells lacking a particular protein. Next, HIV were unleashed on the cells. If HIV replication was inhibited in a given well, it suggested the missing protein was involved. Of the 273 proteins, 36 proteins identified had been previously implicated in the HIV life cycle. Three of the other 237 proteins were picked up, and subjected them to a host of careful genetic experiments, proving they too truly play a role in HIV propagation. Immune cells—the very cells HIV attacks—contain high concentrations of many of the 273 host factors, validating the functions of these proteins. This is the first whole genome screen for human proteins required by HIV and is highly likely to play a role in HIV propagation.

CELL BIOLOGY

1. According to a new study by researchers at the **University of Massachusetts Medical School (UMMS)**, who gained new insights into **autophagy**—a cellular degradation process associated with a form of programmed cell death—by studying the salivary gland cells of the fruit fly. Since its initial discovery in the 1960s, programmed cell death has been a primary focus of studies for investigators across a wide array of scientific disciplines. An essential mechanism in development and homeostasis, programmed cell death allows for the clean intracellular destruction of unnecessary or damaged cells. While apoptosis is the most understood type of programmed cell death, recently scientists have begun to take a closer look at autophagy—a highly regulated, catabolic process that essentially allows a cell to eat itself. Paradoxically, autophagy is not only a major mechanism by which a starving cell reallocates nutrients to ensure survival, scientists are now demonstrating that autophagy also provides cells that cannot undergo apoptosis with an alternate form of self-destruction.

In "Growth arrest and autophagy are required for salivary gland cell degradation in Drosophila," published in the December 14 issue of Cell, **Eric Baehrecke, PhD, UMMS Associate Professor of Cancer Biology**, and colleagues examined fly salivary glands, which contain all of the machinery required to dismantle and recycle their own cellular components and thus provide a genetic model system for elucidating the complex functions of autophagy. The paper describes the cellular components required for autophagic cell death and defines multiple pathways that cooperate in the clearance of cells during fly development. Further, their findings demonstrate a critical relationship between growth and this form of cell death. Cell growth is arrested before cells die by autophagy. If growth is maintained by turning on certain genes, the death of these cells can be blocked. This ha. potential clinical implications. Further studies of autophagy could lead to new approaches to the treatment of human disease, such as cancer, Alzheimer's and Parkinson's.

2. Dr. David Silver and his colleagues have identified **two genes that are crucial for packaging fat into lipid droplets** at **Albert Einstein College of Medicine of Yeshiva University**. They called the genes FIT1 and FIT2 (for Fat-Inducing Transcripts 1 and 2). Both genes code for proteins that are more than 200 amino acids in length, and the two genes are 50 per cent similar to each other. The amino acid sequences of the FIT proteins do not resemble any other known proteins found in any species, indicating that the FIT genes comprise a novel gene family. The researchers conducted several different experiments to confirm the roles of FIT1 and FIT2 in fat storage. In one experiment, they overexpressed both FIT1 and FIT2 genes (i.e., inserted extra copies of them) in human cells. While the rate of fat synthesis stayed the same in both "overexpressed" and control cells, the number of lipid droplets in the "overexpressed" cells increased dramatically, between four-and six-fold.

Using a different tactic to evaluate FIT function, the researchers next "knocked down" FIT2 in mouse fat cells (FIT1 is not expressed in these cells). Their reasoning: If FIT2 is indeed essential for lipid droplet formation, then suppressing FIT2 expression should abolish lipid-droplet accumulation. Examination of these fat cells for lipid droplets revealed that cells with suppressed FIT2 expression had a drastic reduction in lipid droplets. Finally, the researchers carried out a similar FIT2 "knock down" experiment in a whole animal – the zebrafish. Zebrafish eggs were injected with a segment of DNA designed to interfere with FIT2 expression. Then, to induce lipid droplet formation in zebrafish larvae (where it is localized mainly in the liver and intestine), free-swimming six-day-old larvae were fed a high-fat diet for six hours. Although the larvae had exhibited normal feeding behavior, examination of their livers and intestines revealed a near-absence of lipid droplets.

PROTEOMICS

1. To find new **Phylomer drug candidates, Australian drug discovery company Phylogica** is searching the world's most diverse library of protein shapes which is derived from ancient bacteria, called **archaebacteria,** that have evolved over millions of years. Phylogica has patented this technology and is 'mining' this biological resource intensively as it turns out to be very rich in active peptide drug candidates. According to reports it is the first company to look at the importance of the shape within such proteins as a source of drug leads, independent of their original function. Phylogica has positive data from Phylomers developed for stroke and burns treatments and is developing other Phylomer candidates for inflammatory diseases such as diabetes and rheumatoid arthritis.

2. Even small mistakes made by cells during protein production can have profound disease effects, but the processes cells use to correct mistakes have been challenging to decipher. Recent work by scientists at **The Scripps Research Institute**, however, has uncovered two surprising new methods for such editing. The work, published in the January 3, 2008 issue of the Nature, and led by Professor Paul Schimmel of the Skaggs

Institute for Chemical Biology at Scripps Research, could help identify underlying causes of a range of diseases and, in time, even ways to correct the errors.

Producing proteins is an essential but complicated process involving a number of components. Messenger RNA within cells act as the instructions for protein synthesis. Ribosomes are cellular structures that read these instructions and follow their directions to bind with bits of transfer RNA carrying the amino acids needed for a given protein chain. In most cases, a single unique form of transfer RNA binds only to a single amino acid, and a specific enzyme called a synthetase is responsible for joining the two. In very rare cases—typically a fraction of a per cent—transfer RNA binds with the wrong amino acid. If this error, or mistranslation, is not corrected, that mistranslated amino acid is ultimately incorporated into a protein. Past research by the Schimmel team and others has shown that as little as one wrong amino acid can have profound consequences for health. "Even such a tiny defect can overwhelm a cell's ability to deal with misfolded proteins, ultimately causing specific neurological problems," says Scripps Research molecular biologist Kirk Beebe, first author of the new paper with Marissa Mock.

The prevailing thinking among researchers in the field has been that the portions of synthetases that recognize and bind transfer RNA and appropriate amino acids are so accurate that little editing is required. What little editing does occur was thought tied to the same checkpoints within the enzymes that perform recognition and binding. But the current study suggests a new perspective is needed, at least for one widely studied enzyme, a synthetase found in everything from bacteria to humans that binds the amino acid alanine. The Schimmel team's research revealed that a completely distinct segment of the enzyme acts as a second checkpoint responsible for identifying mistranslations and removing any amino acid besides alanine that might attach to the alanine transfer RNA. Remarkably, this second zone within the enzyme focuses its activity on the very same two nucleotides in the genetic code of the transfer RNA used by the first checkpoint, a guanine and uracil pair referred to as G3•U70. "The part that is astonishing is that the information that each of these checkpoints is looking for is embedded in the same transfer RNA molecule," says Schimmel, "There's no precedent for this that we're aware of."

The researchers were able to show that this editor checkpoint, even when separated from the rest of the enzyme, was able to efficiently cleave mistranslated amino acids from the alanine transfer RNA. Further experiments revealed that when the G3•U70 pair was transferred to a different type of transfer RNA, the editing unit still removed a non-alanine amino acid, showing clearly that the pair is the trigger for the activity.

The research also led to a surprising find regarding segments of genome-encoded fragments common in cells and sometimes referred to as freestanding domains, whose functions have been debated for many years. The team found that certain freestanding domains with a genetic sequence very similar to the second checkpoint in the alanine synthetase can independently remove mistranslated amino acids in test tube experiments.

This suggests that the fragments may act as yet another checkpoint to ensure that proteins are properly synthesized. The activity by the freestanding domain, called AlaXp, also targets the G3•U70 pair. Whether AlaXp actually edits within cells is not certain, and how they could perform such a function is also not yet clear. "That's still a question to be resolved," says Beebe, "But our work presents a great possibility for how things are likely to occur." The group has already begun new experiments to study AlaXp activity in mouse cells.

Beebe and Schimmel agree that the apparent triple redundancy the team discovered for preventing mistranslation highlights the importance of accurate protein synthesis. It is unlikely such a robustly redundant system would have evolved, they say, if this were not the case. The work's impact should go beyond adding to the limited understanding of how cells avoid errors during protein synthesis. "We do think there are probably powerful connections to disease," says Schimmel. Schimmel's and other groups have already shown clear connections between individual and cumulative errors in translation and a variety of problems tied to aging, such as neurodegenerative conditions. Now the Schimmel team has begun collaborative work with a number of other research groups to identify in synthetases telltale genetic mutations referred to as small nucleotide polymorphisms that can lead to mistranslation. Identifying such errors could reveal the underlying causes of diseases currently not understood and possibly identify potential paths for treatment.

ENVIRONMENTAL BIOTECHNOLOGY

1. Auckland-based **Ecodiesel** has commenced the **commercial production of liquid 'green fuel' from animal fat**. The plant will have the capacity to supply the government's biofuel obligation of a major oil company, with initial production of 20 million liters of biodiesel by the end of 2008, increasing to 40 million liters of biodiesel per annum by the end of 2009. **Biodiesel made from tallow has the lowest environmental footprint or Sustainable Process Index (SPI) of any biodiesel**, according to the Institute for Resource Efficient and Sustainable Systems, Graz, Austria. NZ firm produces commercial biodiesel TetraQ to sell pharmacology services.

2. Algae, like corn, soybeans, sugar cane and other crops, grow via photosynthesis (meaning it absorbs carbon dioxide) and can be processed into fuel oil. However, the slimy aquatic organisms yield 30 times more energy per acre than land crops such as soybeans, according to the **U.S. Department of Energy**. The reason: They have a simple cellular structure, a lipid-rich composition and a rapid reproduction rate. Many algae species also can grow in saltwater and other harsh conditions-whereas soy and corn require arable land and fresh water that will be in short supply as the world's population balloons. "If you replaced all the diesel in the U.S. with soy biodiesel, it would take half the land mass of the U.S. to grow those soybeans," says **Matt Caspari, chief executive of Aurora Biofuels, a Berkeley, Calif.-based private firm that specializes in algae oil technology**. On the other hand, the Energy Department estimates that if algae fuel

replaced all the petroleum fuel in the United States, it would require 15,000 square miles, which is a few thousand miles larger than Maryland. Algae can be grown just about anywhere in an enclosed space and this is being tested at several power plants across the nation as a carbon absorber. Smokestack emissions can be diverted directly into the ponds, feeding the algae while keeping greenhouse gases out of the atmosphere. Although processing technology for algae fuel-**"oilgae"** in some environmentalist circles-is improving, it's still years away from reaching your local gas pump. "It's feasible; it's just a question of cost, because no large-scale facilities have been built yet," Caspari says. Boeing and Air New Zealand recently announced a joint project with a New Zealand company to develop an algae-based jet fuel, while Virgin Atlantic is looking into the technology as part of a biofuels initiative.

3. Researchers at the **Biodesign Institute** are using the tiniest organisms on the planet 'bacteria' as a viable option to make electricity. **Andrew Kato Marcus and colleagues Cesar Torres and Bruce Rittmann** have gained critical insights that may lead to commercialization of a promising **microbial fuel cell (MFC) technology**. "We can use any kind of waste, such as sewage or pig manure, and the microbial fuel cell will generate electrical energy," said Marcus, a Civil and Environmental Engineering graduate student and a member of the institute's Center for Environmental Biotechnology. Unlike conventional fuel cells that rely on hydrogen gas as a fuel source, the microbial fuel cell can handle a variety of water-based organic fuels. "There is a lot of biomass out there that we look at simply as energy stored in the wrong place," said Bruce Rittmann, director of the center. "We can take this waste, keeping it in its normal liquid form, but allowing the bacteria to convert the energy value to our society's most useful form, electricity. They get food while we get electricity."

Bacteria have such a rich diversity that researchers can find a bacterium that can handle almost any waste compound in their daily diet. By linking bacterial metabolism directly with electricity production, the MFC eliminates the extra steps necessary in other fuel cell technologies. Bacteria provide a cheap source of electricitry. There are many types of MFC reactors and research teams throughout the world. However, all reactors share the same operating principles. All MFCs have a pair of battery-like terminals: an anode and cathode electrode. The electrodes are connected by an external circuit and an electrolyte solution to help conduct electricity. The difference in voltage between the anode and cathode, along with the electron flow in the circuit, generate electrical power. In the first step of the MFC, an anode respiring bacterium breaks down the organic waste to carbon dioxide and transfers the electrons released to the anode. Next, the electrons travel from the anode, through an external circuit to generate electrical energy. Finally, the electrons complete the circuit by traveling to the cathode, where they are taken up by oxygen and hydrogen ions to form water.

The MFC process is relatively stable, but one of the biggest questions is: How do the bacteria get the electrons to the anode? The bacteria depend on the anode for life. The bacteria at the anode breathe the anode, much like people breathe air, by transferring electrons to the anode. Because bacteria use the anode in their metabolism, they strategically position themselves on the anode surface to form a bacterial community called a biofilm. Bacteria in the biofilm produce a matrix of material so that they stick to the anode. The biofilm matrix is rich with material that can potentially transport electrons. The sticky biofilm matrix is made up of a complex of extracellular proteins, sugars, and bacterial cells. The matrix also has been shown to contain tiny conductive nanowires that may help facilitate electron conduction. The numerical model develops and supports the idea that the bacterial matrix is conductive. In electronics, conductors are most commonly made of materials like copper that make it easier for a current to flow through. "In a conductive matrix, the movement of electrons is driven by the change in the electrical potential." Like a waterfall, the resulting voltage drop in the electrical potential pushes the flow of electrons. The treatment of the biofilm matrix as a conductor allowed the team to describe the transport of electrons driven by the gradient in the electrical potential. The relationship between the biofilm matrix and the anode could now be described by a standard equation for an electrical circuit, Ohm's law. Within the MFC is a complex ecosystem where bacteria are living within a self-generated matrix that conducts the electrons. "The whole biofilm is acting like the anode itself, a living electrode," said Marcus.

The concept of the 'biofilm anode' allowed the team to describe the transport of electrons from bacteria to the electrode and the electrical potential gradient. The importance of electrical potential is well known in a traditional fuel cell, but its relevance to bacterial metabolism has been less clear. The next important concept the team had to develop was to understand the response of bacteria to the electrical potential within the biofilm matrix. Bacteria will grow as long as there is an abundant supply of nutrients. Jacques Monod, one of the founding fathers of molecular biology, developed an equation to describe this relationship. While the team recognized the importance of the Monod equation for bacteria bathed in a rich nutrient broth, the challenge was to apply the Monod equation to the anode, a solid. Previous studies have shown that the rate of bacterial metabolism at the anode increases when the electrical potential of the anode increases. The researchers thought of the electrical potential as fulfilling the same role as a bacterial nutrient broth. The team recognized that the electrical potential is equivalent to the concentration of electrons; and the electrons are precisely what the bacteria transfer to the anode.

Equipped with this key insight, the team developed a new model, the Nernst-Monod equation, to describe the rate of bacterial metabolism in response to the "concentration of electrons" or the electrical potential. In their model, the team identified three crucial variables to controlling an MFC: the amount of waste material (fuel), the accumulation of biomass on the anode, and the electrical potential in the biofilm anode. The third factor is a totally novel concept in MEC research. If the biofilm is too thick the electrons have to travel too far to get to the anode. On the other hand, if the biofilm is too thin, it has too few bacteria to extract the electrons rapidly from the fuel. To harvest the benefits of MFCs, the research team is using its innovative model to optimize performance and power output. The project, which has been funded by NASA and industrial partners OpenCEL and NZLegacy, lays out the framework for MFC research and development to pursue commercialization of the technology.

4. **Texas A&M University's** Scientist at the chemical engineering department envisions *E. coli* as a future source of energy, helping to power our cars, homes and more. By genetically modifying the bacteria, **Thomas Wood, a professor in the Artie McFerrin Department of Chemical Engineering**, has "tweaked" a strain of *E. coli* so that it produces substantial amounts of hydrogen. Specifically, **Wood's strain produces 140 times more hydrogen** than is created in a naturally occurring process, according to an article in "Microbial Biotechnology," detailing his research. Though Wood acknowledges that there is still much work to be done before his research translates into any kind of commercial application, his initial success could prove to be a significant stepping stone on the path to a hydrogen-based economy. Renewable, clean and efficient, hydrogen is the key ingredient in fuel-cell technology, which has the potential to power everything from portable electronics to automobiles and even entire power plants. Today, most of the hydrogen produced globally is created by a process known as "cracking water" through which hydrogen is separated from the oxygen. But the process is expensive and requires vast amounts of energy—one of the chief reasons why the technology has yet to catch on. Wood's work with *E. coli* could change that.

By selectively deleting six specific genes in *E. coli's* DNA, Wood has basically transformed the bacterium into a mini hydrogen-producing factory that's powered by sugar. Scientifically speaking, Wood has enhanced the bacteria's naturally occurring glucose-conversion process on a massive scale. These bacteria have 5,000 genes that enable them to survive environmental changes. The bacteria are made less competitive and less harmful because of what's been removed. With sugar as its main power source, this strain of *E. coli* can now take advantage of existing and ever-expanding scientific processes aimed at producing sugar from certain crops, such as corn according to Wood.

NANOTECHNOLOGY/ BIOMOLECULAR ENGINEERING

1. **Researchers in Japan** have developed a pair of **molecular-scale scissors that open and close in response to light**. The tiny scissors are the first example of a molecular machine capable of mechanically manipulating molecules by using light, the scientists say. The scissors measure just three nanometers in length, small enough to deliver drugs into cells or manipulate genes and other biological molecules, said principal investigator Dr. Takuzo Aida, professor of chemistry and biotechnology at the **University of Tokyo**. The scissors-like molecular machine uses a photo-responsive chemical group that extends or folds when light of different wavelengths falls upon it. Just like 'real' scissors, the molecular scissors consist of a pivot, blades and handles. The pivot part of the scissors is a double-decker structure made of chiral ferrocene, with a spherical iron (II) atom sandwiched between two carbon plates. The threepiece unit creates a shaft that allows the scissors to rotate and swivel. Scientists have long been looking for ways to develop

molecular-scale tools that operate in response to specific stimuli, such as sound or light. Biologists, in particular, are enthusiastic about development of such techniques because it would provide them with a simple way to manipulate genes and other molecules.

2. **Sydney-based biotechnology company, EnGeneIC**, has successfully created the world's first **nano-sized particle aimed at providing more effective and efficient drug delivery to cancer cells**. The new, patented delivery technology is derived from minicells of bacterial origin and is called EnGeneIC Delivery Vehicles (EDVs). EnGeneIC has achieved results in reducing tumors without toxic side effects in pre-clinical animal models of cancer. The results were published in the international journal of *Cancer Cell*. The drug delivery technology is applicable for a wide spectrum of cancer types. The nano cells containing chemotherapeutic drug are targeted at specific cancer cells via antibodies displayed on the surface of each particle. Once attached to the cancer cell, the particle is engulfed and the drug released directly into the cell. Two senior Australian scientists, Dr. Jennifer MacDiarmid and Dr. Himanshu Brahmbhatt, have led the EngeneIC research team. They founded the program in 2001 and are joint Managing Directors of the company. The researchers claim that for the first time it has been proved that this technology could lead to the **use of multi-drug combinations** and could eventually help in the development of custom-made therapies for cancer patients.

3. Scientists at **Arizona State University's Biodesign Institute** have developed the world's first **gene detection platform** made up entirely from **self-assembled DNA nanostructures** that could have broad implications for gene chip technology and may also revolutionize the way in which gene expression is analyzed in a single cell. Hao Yan is the researcher who led an interdisciplinary ASU team to develop a way to use structural DNA nanotechnology to target the chemical messengers of genes, called RNA. The team included: lead author and chemistry and biochemistry graduate student Yonggang Ke; assistant professor of chemistry and biochemistry Yan Liu; Center for Single Molecule Biophysics director and physics professor Stuart Lindsay; and associate professor in the School of Life Sciences, Yung Chang. The potential of structural DNA nanotechnology in biological applications has been underestimated and trillions of DNA nanostructures can form simultaneously in a solution of few microliters, and very importantly, they are biocompatible and water soluble. DNA chip and microarray technology have become a multi-billion dollar industry as scientists use it to examine thousands of genes at the same time for mutations or uncovering clues to disease. However, because DNA probes are pinned to the solid surface of the microarray chips, it is relatively slow process for the targets to search and find the probes. Also, it is hard to control the distances between the probes with nanometer accuracy.

In this work the scientists have developed a water soluble nanoarray that can take advantage of the DNA self-assembling process and also have benefits that the macroscopic DNA microchip arrays do not have. The arrays themselves are reagents, instead of solid surface chips. To make the DNA origami RNA probes, Yan has taken advantage of the basic DNA pairing rules in the DNA chemical alphabet ("A" can only form a zipper-like chemical bond with "T" and "G" only pair with "C"). By controlling the exact position and location of the chemical bases within a synthetic replica of DNA, Yan programmed a single stranded genomic DNA, M13, into nanotiles to contain the probes for specific gene expression targets. Yan refers to the self-assembled DNA nanoarrays as nucleic acid probe tiles, which look like a nanosized postage stamp. In a single step, the M13 scaffold system can churn out as many as 100 trillion of the tiles with close to 100 per cent yield. Yan's team designed three different DNA probe tiles to detect three different RNA genes along with a bar code index to tell the tiles apart from each other. The group uses a powerful instrument, atomic force microscopy (AFM), which allows the researchers to image the tiles at the single molecule level. On the surface of each DNA probe tile is a dangling single stranded piece of DNA that can bind to the RNA target of interest. Each probe actually contains two half probes, so when the target RNA comes in, it will hybridize to the half probes and turn the single stranded dangling probes into a stiff structure. When it is stiffened, it will be sensed by the atomic force microscope cantilever, and a bright line, can be seen that is a height increase. The result is a mechanical, label-free detection. The technology is able to detect minute quantities of RNA. Since the DNA-RNA hybridization has such a strong affinity, in principle, a single molecule would be able to hybridize to the probe tile.

4. **The goal of being able to program biochemical reactions as precisely and easily as computers crunch numbers and process words has moved a giant step closer** with the creation of a **DNA-based fabricator** at the **California Institute of Technology**, led by **biomolecular engineer Niles Pierce**. This is a system that allows the team to specify a piece of DNA with a desired shape and function, and then execute a molecular program to assemble it in a test tube. As an example, they used their system to construct a piece of DNA that walks along another strip of DNA. Just as computer languages let programmers create any number of applications, the researchers behind the approach predict that biochemical programming "languages" inspired by their work could let bioengineers create any number of desired molecular products and processes. **Other researchers have created a variety of DNA-based objects, including self-replicating nanostructures, movable arms, tweezers, nanoscale faces, maps and other structures, and even game-playing computers.** But Pierce and his group have gone even further with their biochemical programming. At the heart of the group's system are hairpin-shaped strands of DNA each about 10 nanometers long with three specific binding sites called "toeholds". These hairpins can snap together in specific ways. When a hairpin is closed, for example, two out of its three binding sites are unavailable. But, if a suitable strand of DNA docks with it, the hairpin springs open. A reaction between two DNA strands can also free up the exposed site on an attached hairpin, causing it to close once more. In computer terms, the hairpins act as interconnected logic gates. "This elementary unit has one input port and two output ports," says Pierce. "And they can interact - the input port of one can receive an input from the output port of another."

The group has also developed a graphical way to represent the state of these molecular building blocks and the step-by-step interactions between them. These "reaction graphs" allow them to map out the assembly and disassembly steps needed to produce a desired sequence of reactions. The necessary reactions are then translated into specific sequences of complementary DNA base pairs with the requisite binding characteristics. Finally, the program runs in a test tube that contains the specified mix of molecules. To demonstrate the power and flexibility of their programming approach, the group used it to produce several different reactions including, most dramatically, two-legged DNA molecules that walk along a ladder-like track. Pierce's group also demonstrated other reactions that could be useful for practical tasks, such as biosensing. The two-legged walker, says Pierce, was inspired by the vital intracellular transport protein, kinesin, which walks along microtubules at about 100 steps per second.

REFERENCES

www.abc.net.au/ra/innovations/stories/s1729809.htm
www.academic.marist.edu/VBSC/secondfloor/biosciencesjournals.htm
www.advancedbiotech.in
www.africancrops.net/news
www.agriscape.com/journals/biotechnology
www.ausbiotech.org/.../details.asp?memberId=151624&returnUrl=%2Fabndirectory
www.bio-link.org/resFree.htm
www.biomedical-market-news.com/
www.bionews.in
www.bioone.org/perlserv/?request=get-document&doi=10.1614%2FWS-04-194
www.biosource-technologies.com
www.bspp.org.uk/ppigb/journals.htm
www.eevl.ac.uk/scitechnews/13sep01.html
www.ejcc.org.uk/.
www.epicyte.com
www.ericse.org/digests/dse96-5.html
www.eurekaalert.org
www.haccpnow.co.uk
www.intelliwareint.com/Links.html
www.internet-work.com/a2z/links/director/discrete/science/93.htm
www.jbc.org/cgi/content/full/277/25/22985
Kaur, B Malik CP and Wadhwani Chitra 2008 Biotechnology:Perspectives and Prospects ed Malik *et al.*, Pp317, MD Publishers, New Delhi
www.lib.unc.edu/prices/1999/PRIC238.HTML
www.mathtools.net/Applications_and_Industries/
www.molecularfarming.com
www.montegen.com
www.organicconsumers.org/biod/biotechind102904.cfm_
www.ourfood.com
www.pk.uni-bonn.de/ppigb/society
www.reedsmith.com/library/search_library.cfm

2

COLD TOLERANCE IN MAIZE: PATTERNS OF GENE EXPRESSION, REGULATION AND PROSPECTS FOR GENETIC ENGINEERING

PARVEZ A SOFI, AG RATHER AND SHAFIQ A WANI

Maize, even though being of sub-tropical origin, was transported to diverse environmental zones of the world including the true highlands with cold climate. In the highlands of Mexico, South and Central America winter is severe with below 0° C temperatures frequently. Thus maize is exposed to cold temperature at early growing stages such as germination and early seedling growth and at the time of grain filling or ripening (Granados, 1991). Real adaptation to low temperature seems to be rather difficult in maize. Such an adaptation requires a host of component features such as resistance to frost, resistance to chilling, resistance to fungi during germination at low temperature and the ability to germinate, grow and mature at low temperatures (Miedema, 1982). Especially in areas with shorter growing season such as those found in higher altitudes, it is necessary to have genotypes with inherent cold tolerance with high emergence and rapid growth under low early spring temperatures.

Improvement in cold tolerance in crop plants has not been substantial through classical breeding approaches (Fowler and Gusta, 1979; Hoard and Crosbie, 1985). In fact, a modest increase in tolerance to low temperature (chilling and freezing) would have a dramatic impact on agricultural profitability. Cold tolerance, when achieved, would serve as an economical and reliable mechanism to reduce crop losses due to low temperatures and greatly reduce the use of costly practices aimed at manipulation of microclimatic (Steponkus *et al.*, 1998). However, the progress in enhancing suboptimal temperature in maize by conventional breeding approaches has been limited by use of "yield" as the main selection index (Greaves, 1996). To add to the problem is the fact that heritability of yield is reduced in stress conditions, and the selected genotypes perform differentially in different cycles of selection. In maize, earlier studies in traditional genetic analysis of cold tolerance mainly focussed on elucidation of inheritance pattern, gene action and identification of component traits for seeking improvement in cold tolerance through indirect selection. These studies implicated additive, dominance and epistatic generations in the inheritance additive components (Eagles, 1988; Revilla *et al.*, 2000; Rather *et al.*, 2004). The traits like early emergence and rapid growth under cold temperatures were identified as important traits but these traits do not themselves confer cold tolerance and do not ensure early vigour (Revilla *et al.*, 1998). Correlation studies also revealed that emergence index was not a reliable trait to improve cold tolerance (Gul Zaffar *et al.*, 2005), but seedling emergence per cent and seedling survival can be potential traits for the purpose (Yu *et al.*, 2004). Although these traits are measurable in practical breeding programmes, the efficiency of selection is seriously limited due by various factors such as diurnal variation in temperature, yearly variation in temperature, field variations, limited selection time, difficulty in selecting for component traits, subjective selection criteria and possibility of negative linkage between tolerance traits and other agronomic traits. Moreover, since yield is the overriding factor in selection, the opportunity for stacking component traits of cold tolerance can be missed in favour of selecting exclusively for yield (Greaves, 1996). The general observation that not much progress has been made in improving cold tolerance point towards the qualitative nature of the trait. The reports of continuous variations and transgressive segregants also support this idea (Guy, 1990).

Damages caused by cold temperatures

Low temperature (chilling and freezing) is a major stress limiting crop productivity in maize, especially in temperate regions where low temperatures (0-12° C) are quite common during growing season (Allen and Ort, 2001). Two primary problems arise due to exposure to low temperature. One is maintaining fluidity of membrane systems to enable them to resist Subzero temperatures (Nishida and Murata, 1996). The second is the thermodependency of photosynthetic election transport and CO_2 fixation, due to vulnerability of enzyme systems to drop in temperature (Huner et al., 1998). Miedema (1982) reviewed various physiological and morphological damages to maize crop by low temperature. The visual manifestations of chilling damage in vegetative tisssues include surface lesions and necrotic areas, waterlogged appearance, internal discolouration, wilting, enhanced senescence and a general reduction in growth (Saltviet and Morris, 1990). At the cellular level, low temperatures results in reduced metabolic rate and membrane fluidity and denaturation of thermolabile proteins (Levitt, 1980). The generalised effects of such cellular and molecular changes are observed in key life processes such as respiration, photosynthesis, sugar and phenolic metabolism and redox regulation (Markhart, 1986). Plasma membrane is the primary site of damage due to chilling injury. The freezing causes formation of extra-cellular ice crystals that induce dehydrative stress on cells (Hughes and Dunn, 1996). The resulting lesions in plasma membrane may cause loss of osmotic responsiveness during subsequent thawing (Steponkus, 1984).

The optimum temperature for growth process in maize including germination, shoot elongation, leaf appearance, leaf expansion and dry matter accumulation is around 30° C (Meidema *et al.*, 1987) where as optimum temperature for maximum grain yield is about 25° C (Keeling and Greaves, 1990). The thermal threshold is the temperature below which the plants ability to perform to its potential is reduced. In maize, the level of induced stress increases as the temperature drops below 20° C (Greaves, 1996). Even though there seems to be some degree of genetic variability for tolerance/susceptibility to sub-optimal temperature ranges, below 20° C, but it is probably limited (Gardner *et al.*, 1987). Khehra *et al.*(1988) reported that low temperature injury is expressed as chlorotic stripes and drying of leaves. Cold temperatures especially at sowing delays and reduces early growth (Khehra *et al.*, 1991). Chlorosis is usually found at emergence stage which results in reduced photosynthesis. Even though the chlorotic seedlings recover after rise in temperature but they still lag behind the non-chlorotic seedlings in dry matter accumulation (Miedema *et al.*, 1982). Verheul *et al.* (1995) compared maize inbred lines Z7 (cold tolerant) and Penjalinan (cold sensitive) usuing suboptimal (15/13° C) and near optimal (24/22° C) temperatures and found that transferring maize inbred lines to suboptimal temperatures caused reduction in absolute and relative growth rates. However, the reduction was less in case of Z7 (CT). There were marked differences for net photosynthesis rates in lines acclimated at sub-optional temperatures but oxygen evolution was unaffected indicating that the PSII system was not damaged. Persistance of low temperatures during day caused an additional decline in photosynthesis due to photo-inhibition in maize (Baker *et al.*, 1983). Studies of Lopez *et al.* (1990), Fryer *et al.* (1995) and Haldimann *et al.* (1996) in maize reported that chilling induces a decline in CO_2 assimilation of photosynthesis involving both dissipation of excitation energy in PSII and photo damage to PSII reaction centre. There is a decreased activity of Calvin cycle enzymes (Smith *et al.*, 1997) and poor development of photosynthetic apparatus (Nie *et al.* 1995). Such a reduced CO_2 assimilation has been found to increase the electron flux to O_2 resulting in formation of reactive oxygen species such as superoxide, hydrogen peroxide and hydroxyl radicles which are highly detrimental to lipids, proteins and pigments, if not rapidly scavenged by enzyme systems such as SOD, GTR, DHAR, MDHAR and APX (Asada, 1996).

Fryer *et al.* (1998) studied the CO_2 assimilation, electron transport and O_2 metabolism in stressed maize leaves and found a reduction in carbon assimilation, probably due to less efficient CO_2-comentrating mechanism and a consequent oxygenation of Ribulose-1, 5-Bisphosphate. Further, all antioxidant enzymes were found in elevated levels during chilling period. There were also higher levels of lipoxygenase activity that suggests that lipoxygenase mediated lipid peroxidation of membrane lipids causes oxidative damage in cold stressed leaves. This also supported the suggestion that higher O_2 is the possible candidate as electron acceptor due to increased electron flux relative to CO_2 assimilation rate.

Maize being a C_4 plant suffers from the decline in quantum yield of photosynthesis with decreasing temperatures, a notion that was put forth to explain the dominance of C_4 plants in warm climates (Sage *et al.*, 1999). Since the CO_2 assimilation process is a function of both light saturated and light limited rates of photosynthesis, it can be assumed that one or more of the carbon-concentrating mechanisms must be sensitive to low temperature. (Long, 1999). Infact enzymes such as PPDK (Pyruvate orthophosphate dikinase) and PEPc (Phosphoenolpyruvate carboxylase) have been found to dissociate below 8-12° C *in vitro* (Krall and Edwards, 1993). Later on Rubisco was proposed as a rate of limiting factor for photosynthesis at low temperatures. In maize, reduced activity of Rubisco at suboptimal temperatures was reported by Smith *et al.* (1997). Pittermann and Sage (2000) demonstrated in *Bouteloua gracilis* that there was high correlation between net photosynthesis rate and Rubisco activity below 17° C . Rubisco activity was also found to be a rate limiting factor for photosynthesis at low temperatures in another C_4 plant *Muhlenbergia glomerata* (Kubien and Sage, 2004). These results were further validated by Kubien *et al.* (2003) by studying the pattern of control exerted by Rubisco on C_4 photosynthesis over a range of temperatures (6-40° C). They transformed C_4 dicot *Flaveria bidentis* with antisense constructs of Rubisco to reduce Rubisco content and found that it becomes a major controlling factor over photosynthesis at lower temperature as the *in vivo* rate of photosynthesis at low temperature matched the *in vitro* Rubisco activity. The reduced activity of Rubisco results in leakage of CO_2 from bundle sheath cells. They concluded that Rubisco exercises major control over photosynthetic rates of suboptimal temperatures whereas; at near optimal temperatures the control is shared between Rubisco and other enzymes. A major hurdle in accepting such propositon is the relative availability of CO_2 and CO_2 specificity of Rubisco increases at lower temperatures but such a situation is more advantageous to C_3 plants which have 3-4 times more Rubisco than C_4 plants (Long, 1999). In fact, C_3 plants use increased Rubisco level as an acclimation mechanism at low temperature whereas, in C_4 plants, such responses are variable (Hurry *et al.*, 1995). *Miscantheus x giganticus*, a perrenial C_4 grass, is an exception to such down-regulation of photosynthesis at low temperature. It has efficient photosynthetic rates at low temperatures due to higher levels of enzymes PPDK, Rubisco and PEPC (Naida *et al.*, 2003). Even though it is taxonomically similar to maize, *Miscanthus x giganticus* has remarkably greater cold resistance due to comparatively higher levels of activity of these enzymes. In maize, the quantity of such enzymes decreases at sub-optimal temperatures, especially PPDK, which is a key enzyme of C_4 cycle at lower temperatures (Ward, 1987). Thus in maize PPDK is a key control factor in C_4 photosynthesis at low temperature probably due to its dissociation. In *Flaveria brown* a cold-tolerant C_4 plant, PPDK has been found to have specific amino acid residues that confer thermostability upon it (Ohta *et al.*, 1996). Naidu *et al.* (2003) observed 80 per cent reduction in photosynthetic activity at lower temperature but concluded that the reduction could not be wholly attributed to reduction in phososynthetic enzymes. They proposed that other factors such as production of reactive oxygen species, decrease in the thylakoid proteins and disruption of thylakoid and chloroplast structure may also have a significant role by further reducing the activity of key C_4 enzymes. They could not find any obvious differences in PPDK sequences of *M* x *giganteus* and maize that could explain higher protein stability at lower temperatures and attributed the differences in photosynthetic ability to the corresponding differences in levels of PPDK and Rubisco in leaves grown at cool temperatures. Tsonev *et al.* (2003) studied the down regulation of photosynthesis in French bean (*Phaseolus vulgaris*) and reported that the low temperature induced down regulation mainly occurs due to reduced capacity of plants to utilise incident radiation. They found that a 10-fold increase in incident radiation at low temperature caused just doubling of net assimilation compared to nine-fold increase at 22° C. They further outlined that other factors that causes photosynthetic downregulation include changes in respiratory metabolism, decrease activity of Xanthophyll cycle, greater membrane damage by ROS due to insufficient antioxidant activity and a reduced rate of enzymatic PSII repair.

Pattern of gene expression in cold stress and acclimation

Plant body temperature changes with ambient temperature, although a few plants can control their temperature by several degrees above and below optimal by alternate oxidase respiration and transcriptional cooling respectively (Vishwanathan and Zhu, 2002). The tropical and subtropical crops like rice, maize, tomato, soybean and cotton are severely affected even by non-freezing temperatures and are simultaneously incapable of cold acclimation. In contrast, herbaceous plants from temperate regions do have capacity to survive freezing temperatures (–5 to –30° C) and some perennial forest trees can tolerate below 30° C freezing (Thomashan, 1998). The level of tolerance is greatly enhanced by exposure that probably induces resistance mechanisms in plants by regulation of specific gene systems. This is much like a situation, wherein, a weakened virus is used as a vaccine against that virus as it is non-potential enough to cause the infection but is sufficient enough to induce production of body defence to combat the viral attack.

Considerable body of knowledge has been accumulated regarding patterns of gene expression in cold acclimation, signal perception and its transduction, that have significantly contributed to our understanding of crop response to low temperature. Such an elucidation of various genetic factors underlying cold tolerance is imperative to isolation and characterisation of candidate genes for engineering maize for cold tolerance. Weiser (1970) was first to propose that cold acclimation may require activation of genes, not expressed otherwise coupled with synthesis of certain new proteins. Guy (1990) reviewed the earlier work in identification of genetic and molecular factors underlying cold response. He outlined various factors such as low-molecular weight proteins that act as cryoprotectants (polyol; sorbitol; proline, polyamines, glycinebetaine), carbohydrates (raffinose, sucrose), antioxidants (ascorbate, glutathione), membrane lipids, non-enzymatic proteins (insect thermal hysteresis proteins, fish antifreeze proteins). Guy *et al.* (1985) reported increase in translatable mRNA content in spinach exposed to 5° C that resulted mediating the observed increase in polypeptides, which probably mediated increased freezing tolerance. In maize, exposure to 10° C, resulted in rapid increase (8 folds) in alcohol dehydrogenase-1 in RNA than seedlings grown at optimal temperatures (Christie *et al.*, 1991). They proposed that cold induces root specific transcription of ADH1 gene and that deviated levels are maintained during prolonged cold stress.

The plant stress hormone (ABA) has also generated a lot of interest for its possible role in cold tolerance. Two observations were significant in this regard. One the levels of ABA were higher in plants exposed to low temperatures and other, that exogenous application of ABA confers chilling tolerance to maize cell suspension cultures (Xin and Li, 1992). In fact Li (1991) found that induction of endogenous ABA by mefluidide increases chilling tolerance in maize, probably by altered gene expression (Xin and Li, 1993). The important gene involved in ABA mediated cold response is catalase. Anderson *et al.* (1994) isolated three cold acclimation responsive genes (Car 333, Car 30 and Car 757). The sequence analysis identified Car 333 as Cat-3 which is a maize mitochondrial catalase-3. The catalase acts by scavenging active oxygen species such as H_2O_2 that cause bulk of membrane damage during chilling. The elevated levels of ABA may not be direct response accompanying acclimation but probably chilling induced water stress (Pardossi *et al.*, 1992).

The cold responsive genes (COR genes) are a set of genes that have been implicated in enhanced cold tolerance in crops like *Arabidopsis*. A large number of COR genes have been identified and characterised in various crop species and their gene products usually share homology with cryoprotective proteins that protect the thylakoids from freeze induced disorganisation. Most of these COR genes have in their promoters the dehydration responsive elements (DRE) which is a 9 bp element with the common 5 bp core motif (5'-CCGAC-3'). The proteins encoded by these COR genes are highly hydrophyllic and have simple amino acid sequences, thus cannot be enzymes (Thomashow, 1998).

A number of COR genes such as COR 6.6, COR 15a, COR 47 and COR 78 are upregulated in cold acclimation (Steponkus *et al.*, 1998). COR 15a synthesises a 15 kd polypeptide that is transported to chloroplasts where the signal sequence is removed resulting in mature 9.4 kd polypeptide that increases the cryostability of plasma membrane (Lee *et al.*, 2005). COR 15a is extremely hydrophilic and remains soluble upon boiling. The mature 9.4 kd polypeptide is rich in Alanine, Lysine, Glutamate and Asparagine residues that constitute more than 60 per cent of the protein. It is devoid of proline, methconine, tryptophan, cystine, arginine, glutamine and histidine and has imperfect tetra-repeats of 13-amino acid sequences. The COR 15a genes are concert with other COR genes. Simultaneously expression of all COR genes enhances freezing tolerance more than COR genes enhances freezing tolerance more than COR 15a alone (Jaglo-Ottosen *et al.*, 1998). In barley, COR 14b also encodes a protoplast targeted protein in response to cold stress (Crosatti *et al.*, 1998). Most of these COR genes are upregulated by trascription factors DREB1 (Dehydration responsive element binding factor –1) or CBF (C-repeat binding factor). Some of these activators are induced by cold stress and upregulate the expression of genes with DRE/CRT motifs in promoter sequences (Medina *et al.*, 1999). These include CBF1/ (DREB 1B/A), CBF2 (DREB1C) and CBF (DREB1). Recently, Chinnusamy *et al.*, (2003) identified a transcription factor ICE1 (inducer of CBF expression 1) that binds to CBF3 promoter and activates CBF3 expression. These genes are potential candidates for engineering cold tolerance in crop plants, like maize which is not capable of acclimation.

In addition to COR genes, a number of genes are activated in cold stress response. These include late embryogenesis active genes (LEA3-L1, LEA3-L-2, LEA-3) of barley (Dong *et al.*, 2002), FAD 8 gene of *Arabidopsis* (Gibson *et al.*, 1994). Receptor protein kinases (Osakabe *et al.*, 2005), LOS4 (Gong *et al.*, 2002), RAB (Zereal and McBride (2001), Dhr-2 (Cicek and Esen, 1998), Sensitive to freezing (SFR2) of *Arabidopsis* (Thorlby *et al.*, 2004), COR 14b of barley (Crosatti *et al.*, 1999), COR14a of wheat (Tsventanov *et al.*, 2000), FRY of *Arabidopsis* (Xiong *et al.*, 2001). Besides these a number of transcription factors that regulate transcription of cold-responsive genes or have other diverse functions have been identified and characterised. These include RING finger (Saurin *et al.*, 1996), Zufinger (Cook *et al.*, 1999). MYB (Jin and Martin, 1999), NAC (Xic *et al.*, 2000), bzip (Jakoby *et al.*, 2002) Hos1 (Inshitani *et al.*, 1998), HOS9 (Zhu *et al.*, 2002), AP2/ERF (Fowler and Thomashow, 2002) and ARF7 (Harper *et al.*, 2000).

Recently, antifreeze proteins (AFP's) that act via thermal hysterisis have also been proposed to have possible role in cold acclimation (Griffith *et al.*, 1997). These AFP's decrease the temperature at which ice is formed without affecting the melting point of the solution (Thomashov, 1998). Almost six AFP's have been identified in winter rye and share homology with PR-proteins, a situation that is explained by the fact that winter cereals are more resistant to fungal diseases after cold acclimation. These AFP's are similar to chitinases, glucanases and thaumatin like proteins. More recently Thorlby *et al.*, (2004) identified and characterised a sensitive to freezing gene SFR-2 which is cold inducible and encodes a β-glucosidase. The glucosidases are ubiquitons and have been implicated in preparing cells for physical deformations associated cells for physical deformations associated with freezing, release of active molecules from mert precursors, phytohormones and antioxidants. Since, in Thorlby's study, SRF2 protein was detected in all green tissues and very low in roots, it was proposed that SFR2 protein plays a role in freezing tolerance by protecting photosynthetic tissues such as plastids.

Membrane lipid composition is an important component of cold tolerance that affects the membrane fluidity. A number of lipid transfer protein (LTP's) are expressed in aleurone cells during germination. The LTP gene products help transport of lipids between membranes especially mitochondria and liposomes. Various such genes have been identified in barley and brocolli (LTP, LTP 2, LTP 3 and LTP 4). Some of the LTPs such as blt 4.9 of barley have been found to be expressed in epidermal cells of leaf sheath. Recently on LTP isolated from brocolli leaves is also expressed in leaf cells (Pyee and Kolattukudy, 1995) during cold acclimation.

Some groups of small proteins are also implicated in cold response of plants. Danyluk *et al.* (1996) identified a group of action-depolymerising factors (ADF's) in wheat which are associated with freezing tolerance in wheat (Quellet *et al.*, 2001). These ADF's share high homology across plant kingdom and include proteins such as cofilin, destin, depactin and actophorins. These proteins modulate cell actin cytoskeleton dynamics. Overexpression of ADG1 in *Arabidopsis* enhances synthesis of actin filaments (Carlier *et al.*, 1997). Other protein groups that are highly expressed during cold acclimation include lipid transfer protein (LTP), cryoprotectants and rare cold induced genes. Cryoprotectants are boil-stable proteins isolated from cabbage leaves of acclimated plants only (Hincha *et al.*, 2001). Similarly, RCIA and RC12A are homologous cold and ABA induced genes identified in *Arabidopsis* (Capel *et al.*, 1997). RCBA and RC12B encode small hydrophobic proteins that are localised in plasma membranes.

Regulation of cold-induced genes

The genes induced by low temperature are collectively called as cold responsive (COR) genes or responsive to dehydration (RD) or early responsive to dehydration (ERD) or low temperature induced (LTI) or cold induced (KIN) genes. The regulation of these genes is mediated by two pathways (Nordin *et al.*, 1991). One is ABA dependent and other is ABA-independent. In the ABA-dependant pathway, the expression of cold responsive genes is accompanied by a transient increase in ABA levels. The high levels of ABA may not be a direct response to cold stress but chilling induced dehydration stress. Genome-wide analysis of *A. thaliana* has revealed that more than 1300 genes are responsive to ABA (Kreps *et al.*, 2002). The promoter regions of ABA-induced genes such as RAB 18, LT165, RD29A and RD29B have a ABA responsive element (ABRE) with common motif of (ACGTGGC). In fact the transcription of RAB18 and LT165 was severely diminished in aba mutants of *Arabidopsis* (Lang and Palva, 1992). Ishitani (1997) isolated mutants with over expressor or low expression of RD29A in response to cold and ABA. The expression of genes with ABRE is regulated by bZ1P transcription factors (Jakoby *et al.*, 2002). The bZIP proteins have been implicated in stress signalling and isolated in rice (Aguan *et al.*, 1993), maize (Kusano *et al.*, 1995) and *Arabidopsis* (Sakamote *et al.*, 2000). The N-terminis of these ABF's have one or two serine or threonine residues that are potential targets for certain protein kinases. A large number of such kinases belong to the calcium-dependent protein kinase family and have been implicated in regulation of ABA/stress inducible genes : In maize, one such protein kinase designated as CPK30 (CdPK1a) has been found to activate ABA/stress inducible promoter in leaf protoplasts (Sheen, 1996). Other CDPK's shown to mediate ABA/stress responsive gene expression are CDPK1 (*Arabidopsis*), CDPK7 (rice) and CDPK 2 (tobacco). Recently Choi *et al.* (2005) identified CPk32, a protein kinase, that interacts with ABF4 which is a transcriptional regulator of ABA responsive gene expression. Similarly, Zn finger proteins (Cook *et al.*,

1999) also regulate ABRTE containing genes. In soybean, a Zn finger protein SCOF1 was cloned by Kim *et al.* (2001) and found to be induced by exogenous ABA application. The ScoF1 does not directly bind to ABRE but enhances DNA binding capacity of SGBF1 (a bzip protein) which is a cold inducible transcription factor. Similarly HOS 1 and ESK1 are negative regulators of ABRE (Islutani *et al.*, 1998). In maize, Anderson *et al.* (1994) reported strong correlation between elevated levels of ABA and expression of mitochondrial catalase gene (Cat-3) involved in cold acclimation. Inhibition of ABA biosynthesis by fluridone abolished acclimated driven cold tolerance which was reversed by exogenous ABA application. *Arabidopsis* ITIL 40 gene is also induced by ABA. In fact, ABA induced COR gene expression was significantly low in abi (ABA insensitive) mutants (Gilmour and Thomashow, 1991). Recently Rabbani *et al.* (2003) analysed expression profile of cold inducible genes in rice and reported almost 43 ABA inducible genes involved in stress tolerance. They obtained promoter sequence of such genes and found that many of these promoter sequences did not contain ABRE motifs. Lee *et al.* (2005) identified two ABRE binding factors ABF1 and ABF4 that were cold inducible. Therefore the genes regulated by these transcription factors can also be induced by cold stress. However, they reported that none of the ABA biosynthesis genes are cold regulated thereby implying that ABA synthesis as such is not involved in cold response. This is consistent with the fact that endogenous levels of ABA do not increase to a great extent during cold stress (Lang *et al.*, 1994).

Another pathway of cold responsive genes is independent of ABA mediation. As has been demonstrated that a large number of genes activated in cold stress have in their promoter sequences, a dehydration responsive element with a C-repeat (CRT/DRE). These elements possess a common core motif of CCGAC. A number of transcription factors designated as DRE binding factors (DREB) or C-repeat binding factors (CBF) upregulate the expression of such cold responsive (COR) genes. The first such factor CBF1 was identified in *Arabidopsis* by Stockinger *et. al.* (1997). This was followed by identification and characterisation of other DREB's like DREB1 (DREB1A, DREB1B, DREB1C), DREB2 (DREB2A, DREBB), CBF2 and CBF3 (Liu *et. al.*, 1998; Gilmour *et al.*, 1998). These factors are induced by cold temperature and chilling induced dehydration. Cold responsive genes such as KIN1, COR 15a, COR 47, COR 78 and ERD 10 have been found to be regulated by CBF/DRE factors (Thomashow, 2001). Jaglo Ottosen *et al.* (1998) found that constitutive overexpression of CBF1 induced expression of entire battery of COR genes (COR15 a, COR 6.6, COR 47 and COR 78) in *Arabidopsis* resulting in increased freezing tolerance than any of the individual genes. The activation of DRE elements by CBF's or DREB's is accompanied by modifications in chromation structure of transcriptional complex (Stockinger *et al.*, 2001). In rice, Rabbani *et al.* (2003) analysed promoter sequence of 30 cold induced genes and found that 22 contained DRE within 1-4 kb region of their promoters. Recently, Lee *et al.* (2005) reported in *Arabidopsis* that out of 74 cold induced genes, 50 are CBF2 upregulated, 18 are target genes for both CBF 2 and CBF 3 and 6 are CBF 3 specific.

The regulatory network of DREB and CBF genes is an important area of study in understanding molecular basis of cold tolerance. This is especially important for engineering cold tolerant phenotypes in crop plants for which pattern of expression of transgene along with its regulatory mechanisms needs to be elucidated. As in case with ABA-dependent gene expression in cold stress where a Zn finger protein SCOF1 regulates the DNA binding ability of various factor to ABRE element, similar mechanisms have been reported in ABA independent gene expression. (Ishitani *et al.*, 1998) identified a cold induced locus HOS1 that regulates gene expression in cold. The HOS1-1 mutants experience a dramatically higher expression of RD29A, COR47, ADH, RAB18, RD29B, COR15a and KIN1. It acts as a negative regulator of RD29, COR47 and ADH but positively upregulates RD29B, COR15a and KIN 1. Similarly, Zhu *et al.* (2002) reported that HOS 1 and HOS 2 act as negative regulators of RD29A via their down regulation of CBF/DREB1 genes. The HOS1 encodes RING finger protein that have role in protein-protein interaction (Saurini *et al.*, 1996). Even though SFR 6 gene encoding a β-glucosidase negatively regulate cold and ABA induction of genes with CRT/DRE motifs it has no effect on CBF gene expression (Xin and Browse, 1998). Gong *et al.* (2002) reported a positive regulator of RD29A, LOS4 that encodes a RNA helicase possibly involved in RNA metabolism.

Another positive of CBF3/DREB1A was identified using map based cloning by (Chinnusamy *et al.*, 2003) and was designated as 1CE 1 (inducer of CBF expression 1). ICE1 encodes a MYC type basic helix-loop-helix (bHLH) transcription factor that specifically activates DREB1A/CBF3. The expression of ICE1 is cold inducible and overexpression of ICE1 improves freezing tolerance in *Arabidopsis*. Similarly, ERD1 gene is also regulated by ABA-independent pathway. The expression of ERD1 is positively regulated by transcription factors such as NAC including RD26 (Tran *et al.*, 2004) and Zn finger homeodomain proteins (Nakashima and Yamaguchi-Shinozaki, 2005) under dehydration stress.

Despite two distinct regulation pathways for expression of cold inducible genes, there are pressing evidences that there are interactions between DREB and ABRE regulations. Ishitani *et al.* (1997) first reported enhanced or diminished expression of RD29A in response to cold or ABA and concluded that both cold and ABA dependent mechanisms cross-talk at some nodes of transduction process. A number of cold inducible genes have in their promoter sequences both DRE and ABRE notifs. In case of RD29A, both these elements were found to function synergistically implying that there are interactions between these distinctly regulated pathways. The DREB1/CBF genes are mostly cold inducible by ABA-independent pathway, but a drought induced gene DREB1D/CBF4 was identified in *Arabidopsis*, and was regulated by ABA-dependent pathway. In maize, DBF1 and DBF2, DRE binding proteins activate RAB17 by ABA dependent pathway (Kizis and Pages, 2002), while as Zm DREB1A is cold induced (Quin *et al.*, 2004). This implies that genes with DRE motifs can also be regulated by ABA. The interaction seems to by way of transactivation or inactivation thereby up-or down-regulating the target gene. Warren *et al.* (1996) proved this fact in his classical study of sfr6 mutants in *Arabidopsis*. The sfr mutants witnessed inactivation of both cold inducible COR15a, KIN1 and RD29A and ABA-inducible Kin1 even in presence of stress. Thus, SFRC acts by way of transactivating DRE/CRT or ABRE elements.

More recently, small non-coding RNA species designated as S1RNA (small interfering RNAs) have been implicated in regulation of stress inducible genes. Sunkar and Zhu (2004) surveyed small RNAs under stress in *Arabidopsis* and reported 26 micro RNA and 102 novel small RNAs that are inducible by abiotic stress. Rhoades and Bartel (2004) used comparative genomics approach to predict function of micro RNAs and found that the potential targets for miRNAs are superoxide dismutase (SOD), laccases and ATP sulfurylases and proposed that miRNAs can be stress inducible. These miRNAs exert their regulatory influences by affecting translation process or alternative splicing of mRNA (Jen *et al.*, 2005). Therefore, the regulatory network of stress inducible genes seems to be more complex than is perceived in terms of ABA-dependent or independent mechanisms.

Future prospects

The success of any scientific development relies on how speedily it is able to meet the set objectives in a problem area. Abiotic stresses have been a major limiting factor in enhancing and sustaining crop productivity and has been therefore a compelling area of research in basic and applied aspects of stress biology, physiological malformations associated with a particular stress and patterns of gene expression and various regulatory factors underlying such differential expression. Low temperature stress has received justifiable focus from scientific community in terms of physiological damage, gene identification and characterisation and possible extension of results from model species such as *Arabidopsis* to non-acclimable crops like maize. Maize is inherently viz., sensitive to chilling and freezing and invariably experiences sub-optimal temperatures especially at early growth stages.

Since, many of the cold acclimation events are under transcriptional control, the transcription factors are more than ideal candidate genes for engineering enhanced cold tolerance in crop plants (Zhang *et al.*, 2004). From the initial attempts to validate the assumption that differential gene action is associated with enhanced chilling tolerance (Artus *et al.*, 1996), the focus is now shifted towards characterising genes with key roles in cold acclimation process in terms of mode of action and their feasibility to develop transgenic in crops like maize (Thomashow, 1998). The results of Jaglo-Ottosen *et al.* (1998) with overexpressing entire battery of DREB regulated COR genes to engineer freezing tolerant *Arabidopsis* plants gave impetus to the drive. Thus today we have a large array of stress inducible genes with varied functions that finally culminate in increased stress tolerance. Recently Lee *et al.* (2005) identified 655 cold upregulated genes in *Arabidopsis*. The stress inducible genes and their regulons are ideal candidate genes for use in genetic engineering. DREB/CBF genes have specially received greater focus in view of their conservation across plant kingdom and orthologous genes of DREB1/CBF have been found in number of crops like canola, broccoli, tomato, alfalfa, wheat, barley, maize and rice. An important finding in this regard was reported by Nakashima and Yamaguchi-Shinozaki (2005). They isolated rice orthologs of *Arabidopsis* DREB1 and DREB2 that are stress inducible. The overexpression of these rice orthologs in *Arabidopsis* has a function similar to *Arabidopsis* homologs indicating that similar transcription factors function in monocots and dicots. Similarly, Qin *et al.*, (2004) isolated a maize Ortholog of DREB1A and its overexpression in *Arabidopsis* gave similar results. However, significant negative impact on plant phenotype caused by constitutive over expression of DREB/CBF genes is dwarfing or stunted growth. This problem has been overcome by use of constructs with stress inducible promoters having low background expression under normal growing conditions. Tobacco and wheat plants have been transformed using DREB/CBF genes driven by RD29A promoters resulting in stress inducible over expression of target genes (Lee *et al.*, 2003; Pellegrineschi *et al.*, 2004). Despite being an attractive target for biotechnological intervention, not all TF's cause similar enhancement in cold tolerance levels. DREB, for example caused weak induction of downstream genes and therefore, low levels of increase in cold tolerance (Liu *et al.*, 1998). This implies that several post-transcriptional alterations are also involved in activity of these proteins in transgenic plants (Shinozaki and Yamaguchi-Shinozaki, 2000).

Apart from DREB/CBF, ROS scavengers are ideal candidate genes. This is because oxidative stress signalling is a key component of freezing tolerance. Foyer *et al.* (1995) first reported results on engineering cold tolerance in maize by manipulation of antioxidant defence system. The ROS scavenging enzymes like glutathione reductase, ascorbate peroxidase and superoxide dismutase were found to be differentially localised in mesophyll and bundle sheath cells that could affect the protection of enzymes from oxidative inactivation. Superoxide dismutase (SOD) is a key ROS scavenging enzyme, that dismutates superoxide into H_2O_2 and O_2. In maize ten isozymes of SOD have been identified (4 cytosolic cu/zu SODs; 4 mitochondrial Mn SODs; one chloroplastic Cu/Zn SOD, and one novel Fe SOD. Breusegen *et al.* (1999) used a *Nicotiana* Mn SOD gene driven by 35S promoter in maize and found enhanced chilling tolerance resulting from higher antioxidant capacity of leaves. The resulting H_2O_2 from dismutase activity induces synthesis of heat shock proteins (HSPs) that confer protection from low temperature stress (Shon *et al.*, 2004). They also proposed and validated that hypothesis that H_2O_2 increases activity of MAPKKK (nitrogen activated protein kinase, kinase, kinase). Maize plants transformed with tobacco ortholog NPK1 of *Arabidopsis* MAPKKK caused enhanced chilling tolerance by about 2° C as compared to non-transgenic plants (Shon *et al.*, 2004). This strategy is advantageous as it is not associated with any growth retardation problem. Wang (2005) also used NPK1 gene in maize and found that some transgenic lines could tolerate –5° C for 3 hours. They also reported a positive correlation between enhanced freezing tolerance and total soluble sugars that presumably act as cryprotectants. The overexpression of NPK 1 also upregulated two important oxidative signalling pallen genes namely GSt (glutathione S. transferase and HSP 17.8).

More recently, high end technologies like genomics, proteomics and metabolimics are finding their way and beginning to have an impact on our understanding of chilling tolerance and also our ability to dissect seemingly complex traits into finer details (Bohnert *et al.*, 2006). It has been a steady progress from gene identification to its functional characterisation and then its possible use in engineering stress tolerant plants. All through this, there has been a shift of focus on upstream elements of a signalling pathway that can trigger the whole of downstream cascades rather than targeting the end points (Vincour and Altiman, 2005). The advances is structural, functional and comparative genomics have provided insights into the adaptive diversity in stress tolerance. There has been focus on relatives of model crops such as *Arabidopsis thaliane* and *Thellungiela halophia* in order to exploit genomic resources available in those crops. Ideally breeder would like to identify those alleles that confer stress tolerance without actually compromising on growth and yield attributes. Substantial variability in such adaptive diversity has been studied in various model system such as *Lycopersicon pennelii* (Fridman *et al.*, 2004), *Hordeum spontaneum* (Baum *et al.*, 2003), *T. halophila* (Gong *et al.*, 2005.), *Zea mays* ssp. *parviglumis* (Bomblies and Doebly, 2005) and *A. Thalania* (Koornneef *et al.*, 2005).

Proteomics or proteome analysis that refers to monitoring changes in protein expression in response to changing environments. The information generated helps in knowing protein sequence and sequence homology, thereby predicting function of unknown protein under different conditions. This in turn will help in their molecular cloning. Proteomics has the potential to study the protein expression induced under stress condition such as chilling and freezing. The knowledge about tissue specificity, developmental stage and physiological conditions will help identify the role of these proteins expression induced under stress condition such as chilling and freezing. The knowledge about tissue specificity, developmental stage and physiological conditions will help identify the role of these proteins. The importance of proteome analysis is important given the fact that it is the central link between genome and cell. On one hand it is the culmination of gene expression and on other hand it heralds the beginning of biochemical activities constituting cellular response to stresses (Brown, 2002). However, Peck (2005) opined that large scale proteomics analysis studies are still limited especially under stress situations apparently due to problems in isolation or quantification or our ability to recognise protein modifications induced under stress.

REFERENCES

Agnan K Sugawara K Suzuki N and Kusano T 1993 Low temperature dependent expression of a rice gene encoding a protein with *Leucine zipper* motif *Mol Gen Genet* **24** : 1-8

Allen D and Ort D 2001 Impacts of chilling temperature on photosynthesis in warm-climate plants *Trends Plant Sci* **6** : 36-42

Anderson M Prasad T Martin B and Stewart C 1994 Differential gene expression in chilling acclimated maize seedlings and evidence for involvement of abscisic acid in chilling tolerance *Plant Physiol* **105** : 331-339

Artus N Uemura M Steponkus P Gilmour S Lin C and Thomashow M 1996 Constitutive expression of cold regulated *Arabidopsis thaliana* COR 15a gene affects both chloroplast and protoplast freezing tolerance *Proc Natl Acad Sci* **93** : 13404-13409

Asada K 1996 Radical production and scavenging in chloroplasts In: Photosynthesis and Environment ed Baker N K Kluver Academic Publishers The Netherlands 123-150

Baker N East T and Long S 1983 Chilling damage to photosynthesis in your *Zea mays* II Photochemical function of thylakoids *in vivo J Expt Bot* **34** : 189-197

Baum M Grando S Backes G Jahoor A and Sabbagh A 2003 Qtl's for agronomic traits in the mediterranian environment identified in recombinant inbred inbred lines of cross Arta x *H spontaneum* 41-1 *Theor Appl Genet* **107** : 1215-1225

Bohnert H Gong Q Li P and Ma S 2006 Unravelling abiotic stress tolerance mechanisms – getting genomics going *Curr Opin Plant Biol* **9** : 180-188

Bomblies K and Doebley J 2005 Plaeotropic effects of duplicate maize FLORICAULA/LEAFY genes Zfl 1 and Zfl2 on traits under selection during maize domestication *Genetics* **172** : 519-531

Breusegam F Slooten L Stassart J Botlerman J Moens T Mantagu M and Inze D 1999 Effects of overproduction of tobacco Mn Sod in maize chloroplats on foliar tolerance to cold and oxidative stress *J Expt Bot* **50** : 71-78

Brown T 2002 Genomes John Wiley and Sons Inc USA pp 572

Capel J Jarillo J Salinas J and Zapater J 1997 Two homologous low temperature inducible genes from *Arabidopsis* encode highly hydropholoic protein *Plant Physiol* **115** : 569-576

Chinnusamy V Ohta M Kanrar S lee B Hong X Agarwal M and Zhu J 2003 ICE 1 : A regulator of cold induced transcriptome and freezing tolerance in *Arabidopsis Genes Dev* **17** : 1043-1054

Choi H Park H Park J Kim S Im M Seo H Kim Y Hwang I and Kim S Y 2005 *Arabidopsis* calcium dependent protein kinase CPK32 interacts with ABF4 a transcriptional regulator of Abscisic acid responsive gene expression and modulates its activity *Plant Physiol*

Christie P Hahn M and Walbot V 1991 Low temperature accumulation of alcohol dehydrogenase-1 mRNA and protein activity in maize and rice seedlings *Plant Physiol* **95** : 699-706

Cicek M and Esen A 1998 Structure and expression of dhurrinase (glucosidase) from Sorghum *Plant Physiol* **116** : 1469-1478

Cook T Gebelien B Belal M Mesa K and Urrutia R 1999 Three conserved transcriptional repressor domains are a defining feature of the TIEG subfamily of sp1 like zinc finger proteins *J Biol Chem* **274** : 29500-29504

Crosatti C Laureto P Bassi R and Cattivelli L 1998 The interaction between cold and light controls the expression of cold regulated barley gene COR 14b and the accumulation of corresponding protein *Plant Physiol* **119** : 671-680

Dong C Danyluk J Wilson K Pocock T Norman P Huner A and Sarhan F 2002 Cold regulated cereal chloroplast late embryogenesis abundant – like proteins Molecular characterisation and functional analysis *Plant Physiol* **129** : 1368-1381

Eagles H 1988 Inheritance of emergence time at low temperature in segregating generations of maize *Theor Appl Genet* **76** : 459-464

Fowler S and Thomashov M 2002 *Arabidopsis transcriptome* profiling indicates that simple regulatory profiling indicates that simple regulatory pathways are activated during cold acclimation in addition to CBF cold response pathways *Plant Cell* **14** : 1675-1690

Foyer C Smith H and Harbinson J 1995 Engineering low temperature tolerance in maize by manipulation of antioxidant defences *J Expt Bot* **46** : 64

Fridman E Carrari F Liu Y Fernie A and Zamir D 2004 Zooming in on a quantitative trait for tomato yield using interspecific introgressions *Science* **305** : 1786-1789

Fryer M Oxborough K Martin B Ort D and Baker N 1995 Factors associated with depression of photosynthetic quantum efficiency in maize at low growth temperatures *Plant Physiol* **108** : 761-767

Fryer M Andrews J Oxborough K Blowers D and Baker N 1998 Relationship between CO_2 assimilation photomyuthetic electron transport and active O_2 metabolism in leaves of maize in field during periods of low temperature *Plant Physiol* **116** : 571-580

Gardner C Compton M Gocken T and Eichel berger K 1987 Selection for cold and freeze tolerance in corn : evaluations of original and selected populations *Proc 42nd Ann Corn and Sorghum Res Conf* 104-118

Gibson S Arondel V Iba K Somerville C 1994 Cloning of a temperature regulated gene encoding a chloroplast omega-3 desaturase from *Arabidopsis Plant Physiol* **106** : 1615-1621

Gilmour S and Thomashow M 1991 Cold acclimation and cold regulated gene expression in ABA mutants of *Arabidopsis thaliana Plant Mol Biol* **17** : 1233-1240

Gilmour S Zakura D Stockinger E Salazar M Houghton J and Thomashow M 1998 Low temperature regulation of *Arabidopsis* CBF family of AP2 transcriptional activators as an early step in cold induced COR gene expression *Plant J* **16** : 433-442

Gong Q Li P Ma S Rupassara S and Bohnert H 2005 Salinity stress adaptation competence in the Extremophile *Thellungiela halophila* in comparison with its relative *Arabidopsis thaliana Plant J* **44** : 826-839

Gong Z Lee H Xiong L Jagendrof a Stevenson B and Zhu J 2002 RNA helicase like protein as an early regulator of transcription factors for plant chilling and freezing tolerance *Proc Natl Acad Sci* **99** : 11507-11512

Granados G 1991 Genetics of abiotic stress tolerance in maize *Maize Genet Persp* 180-195

Greaves J 1996 Improving sub-optimal temperature tolerance in maize In search of variation *J Expt Bot* **296** : 307-323

Griffith M Antikainlu M Hon W Maunsback K Yu X Chum Y and Yang S 1997 Antfreeze proteins in winter rye *Plant Physiol* **100** : 327-332

Gul Zaffar Shikari A and Noor-ul-Saleem 2005 Correlation studies among cold tolerance traits in highland maize *SKUAST J Res* 7 : 295-301

Guy C 1990 Cold acclimation and freezing stress tolerance : Role of protein metabolism *Ann Rev Plant Physiol Plant Mol Biol* **41** : 187-223

Guy C Niemi K and Brambl R 1985 Altered gene expression during cold acclimation of spinach *Proc Natl Acad Sci* **82** : 3673-3677

Haldimann P Frachebond Y and Stamp P 1996 Photosynthetic performance and resistance to photo inhibition of *Zea mays* L leaves grown at sub-optimal temperatures *Plant Cell and Environ* **19** : 85-92

Harper R Evans E Leusse D Muto H Tatematsu K Watahiki M Yamamoto K and Liscum E 2000 The NPH_4 locus encodes the auxin receptor factor ARF7 a conditional regulator of differential growth in aerial *Arabidopsis Plant Cell* **12** : 757-770

Hoard K and Crosbie T 1985 S_1 line recurrent selection for cold tolerance in two maize populations *Crop Sci* **25** : 1041-1045

Hughes M and Dunn M 1996 The molecular biology of plant acclimation to low temperature *J Expt Bot* 291-305

Huner N Oquist G and Sarhan F 1998 Energy balance and cold acclimation to light and cold *Trends Plant Sci* **3** : 224-230

Hurry V Keerbergy O Parnik T Gardestron P and Oquist G 1995 Cold hardening results in increased activity of enzymes involved in carbon metabolism in leaves of winter rye (*Secale cereale*) *Planta* **195** : 554-562

Ishitani M Xiong L Stevenson B and Zhu J 1997 Genetic analysis of osmotic and cold stress signal transduction in *Arabidopsis* : Interactions and divergence of ABA-dependent and ABA-independent pathways *Plant Cell* **9** : 1935-1949

Ishitani M Xung L lee H Stevenson B and Zhu J K 1998 HOS1 a genetic locus involved in cold responsive gene expression in *Arabidopsis Plant Cell* **10** : 1151-1162

Jaglo-Ottosen K Gilmour S Zarka D Shabenberger O and Thomashaw M 1998 *Arabidopsis* CBFI overexpression induces COR genes and enhances freezing tolerance *Science* **280** : 104-106

Jakoby M Weisshar B Carbajosa J Tiedman J Kroj T and Parey F 2002 bZIP transcription factors in *Arabidopsis Trends Plant Sci* 7 : 106-111

Jen C Michalopoulos I Westhead D and Meyer P 2005 Natural antisense transcripts with coding capacity in *Arabidopsis* may have a regularity role that is not linked to double stranded RNA degradation *Genome Biol* **6** : 51

Jin H and martin C 1999 Multifunctionality and diversity within the plant MYB gene family *Plant Mol Biol* **41** : 577-584

Keeling P and Greaves J 1990 Effects of temperature stress on corn-opportunities for breeding and biotechnology *Proc Ann Corn and Sorghum Res Conf* 29-42

Khehra A Saxena V and Kapoor W 1991 Genetics of cold tolerance in maize *Maize Genet Persp* pp 196-208

Kim J 2001 A novel cold induced zinc finger protein from soybean SCOF-1 enhances cold tolerance in transgenic plants *Plant J* **25** : 247-259

Kizis D and Pages M 2002 Maize DRE-binding proteins DBF1 and DBF2 are involved in rab17 regulation through drought-responsive element in an ABA-dependent pathway *Plant J* **30** : 679-689

Kornneef A Blanco C and Vreudenhil D 2004 Naturally° C curring genetic variation in *Arabidopsis thaliana Ann Rev Plant Biol* **55** : 141-172

Krall J and Edwards G 1993 PEP carboxylases from two C_4 species of paniom with markedly different susceptibilities to cold inactivation *Plant Dell Physiol* **34** : 1-11

Kreps J Wu Y Chang H Zhu T Wang X and Harper J 2002 Transcriptome changes for *Arabidopsis* in response to salt osmotic and cold stress *Plant Physiol* **130** : 2129-2141

Kubien D and Sage R 2004 Low temperature photosynthetic performance of a C_4 grass and a co-occurring C_3 grass native to high altitudes *Plant Cell Environ* **27** : 907-916

Kubien D Caeninerer S Furbank R and Sage R 2003 Cu photosyntheis at low temperature A study using transgenic plant with reduced amounts of Rubico *Plant Physiol* **132** : 1577-1585

Kusano T Berberich T Harada M Suzuki N and Sugawara K 1995 A maize DNA-binding factor with a bZIP motif is induced by low temperature *Mol Gen Genet* **248** : 507-517

Lang V Mantyla E Welin B Sundberg B and Palva E 1994 Alterations in water status endogenous abscisic acid content and expression of rab 18 gene during development of freezing tolerance in *Arabidopsis thaliana Plant Physiol* **104** : 1341-1349

Lee B Handerson D and Zhu J 2005 The *Arabidopsis* cold responsive transcriptome and its regulation by ICE 1 *Plant Cell* **17** : 3155-3175

Lee J 2003 Expression of *Arabidopsis* CBF1 regulated by an ABA/stress inducible promoter in transgenic tomato confers stress tolerance without affecting yield *Plant Cell Environ* **26** : 1181-1190

Levitt J 1980 Responses of plants to environmental stress **In:** chilling freezing and high temperature stress Academic Press N Y

Liu Q Kasuga M Sakuma Y Abe H Miura S and Shinozaki K 1998 Two transcriptional factors DREB1 and DREB2 with an EREBP/AP2 DNA binding domains separate two cellular signal transduction pathways in drought and low temperature responsive gene expression respectively in *Arabidopsis Plant Cell* **10** : 139-1401

Long S 1999 Environmental responses **In:** C_4 Plant Biology Academic Press San Deigo 215-249

Long V and Palva E 1992 The expression of rab gene rab 18 is induced by absicis acid during cold acclimation process *Plant Mol Biol* **20** : 951-962

Lopez A Nie G Ort D and Baker N 1990 The involvement of photo inhibition of photo system II and impaired membrane energization in the reduced quantum yield of carbon assimilation in chilled maize *Planta* **181** : 78-84

Markhart A 1986 Chilling injury : a review of possible causes *Hort Sci* **21** : 1329-1333

Medina J Bargues M Terd J Alonso M and Salinas J 1999 The *Arabidopsis* CBF gene family is composed of three genes encoding AP2 domain containing proteins whose expression is regulated by low temperature but not by Abscisic acid or dehydration *Plant Physiol* **119** : 463-469

Meidema P Post and Groot P 1987 The effects of low temperature on seedling growth of maize genotypes Agri Res Rep No 926 Wageningen

Miedema P 1982 The effects of low temperatures on *Zea mays Adv Agron* **35** : 93-129

Naidu S Moose S Shoaibi A Raines C and Long S 2003 Cold tolerance of C_4 photosynthesis in *Miscanthus x giganticus* : Adaptation in amounts and sequence of C4 photosynthetic enzymes *Plant Phyisol* **132** : 1688-1697

Nakashima K and Yamaguchi-Shinozaki K 2005 Molecular studies on stress-responsive gene expression in *Arabidopsis* and improvement of stress tolerance in crop plants by regulon biotechnology JARQ **39** : 221-229

Nie G Robertson E Fryer M Leech R and Baker N 1995 Response of photosynthetic apparatus in maize leaves grown at low temperatures to transfer to normal growth temperature *Plant Cell Environ* **18** : 1-12

Nishida I and Murata N 1996 Chilling sensitivity in plants and cyanobacteria : the crucial contribution of membrane lipids *Ann Rev Plant Physiol Plant Mol Biol* **47** : 541-568

Nordin K Heino P and Palva E 1991 Separate signal pathways regulate the expression of low temperature induced gene in *Arabidopsis thaliana Plant Mol Biol* **16** : 1061-1071

Ohta S Usma S Ueki J Kumashiro T and Burnel J 1996 Identification of amino acid residues responsible for cold tolerance in *Flaveria brownii pyrunate* orthophosphate dikinase FEBS Lett 396 : 1520-156

Osakabe Y Maruyama K Seki M Satou M Shinozaki K and Yamaguchi-Shinozaqki K 2005 Leucine-rich repeat recepter like kinase 1 is a key membrane and bound regulator of abscisic acid early signalling in *Arabidopsis Plant Cell* **117** : 1105-1109

Ouellet F Carpenter E Cope M Monroy A and Sarhan F 2001 Regulation of a wheat actin depolymerising factor during cold acclimation *Plant Physiol* **125** : 360-368

Pardossi A Vervieri P and Tagnoni F 1992 Involvement of absicic acid in regulating water status in *Phaseolus vulgaris* during chilling *Plant Physiol* **100** : 1243-1250

Peck S 2005 Update on protemics in *Arabidopsis* : where do we go from here? *Plant Physiol* **138** : 591-599

Pellegrinsdoi A 2004 Stress induced expression in wheat of the *Arabidopsis thaliane* DREB1A gene delays water stress symptoms under green house conditions *Genome* **47** : 493-500

Pittermann J and Sage R 2000 Photosynthetic performance at low temperature of *Bouteloua gracilis* a high altitude C4 grass from rocky mountains *Plant Cell Environ* **23** : 81-823

Pyee J and Kolattukudy P 1995 The gene for the major culticular wax-associated protein and three homologous genes from brocolli (*Brassica oleracea*) leaves and their expression pattern *Plant J* 7 : 49-59

Qin F Sakuma Y Li J Liu Q Li Y Shinozaki K and Yamaguchi Shinozaki K 2004 Cloning and functional analysis a novel DREB1/CBF transcription factor involved in cold responsive gene expression in *Zea mays Plant Cell Physiol* **45** : 1045-1051

Rabbani M Maruyama K Abe H Kha M Katsura K Ito Y Yoshiwara K Seki M Shinozaki K and Yamaguchi-Shinozaki K 2003 Monitoring expression profiles of rice genes under drought salinity cold and high salinity stress and abscisic acid application using CDNA microassay and RNA gel-blot analysis *Plant Physiol* **13** : 1755-1767

Revilla P Malavar R Cartea M Burton A and Ordas A 2000 Inheritance of cold tolerance at emergence and early season growth in maize *Crop Sci* **40** : 1579-1585

Rhoades M and Bartel D 2004 Computational identification of plant micro RNAs and their targets including stress induced m1RNA *Mol Cell* **14** : 787-799

Rvilla P Malvar R Cartea M and Ordas A 1998 Identifying open pollinated populations of filed corn as a sources of cold tolerancek for improving sweet corn *Euphytica* **101** : 239-247

Sage R Wedin D and Li M 1999 The biogeography of C4 photosynthesis patterns and controlling factors In C4 Plant Biology ed Sage R and Monson R Academic Press 313-373

Sakamoto H Araki T Meshi T and Iwabuchi M 2000 Expression of a subset of *Arabidopsis* cys/His type zinc finger protein gene family under water stress *Gene* **248** : 23-32

Saurin A Borden K Buddy M and Freemount P 1996 Does this have a familiar RING *Trends Biochem Sci* **21** : 208-214

Sheen J 1996 Ca dependent protein kinase and stress signal transduction in plants *Science* **274** : 1900-1902

Shinozaki K and Yamaguchi-Shinozaki K 2000 Molecular responses to dehydration and low temperature differences and cross talk between two stress signalling pathways *Curr Opin Plant Biol* **3** : 217-223

Shou H Bordallo P Fan J Yeakley J Bibikova M Sheen J and Wang K 2004 Expressionn of an active tobacco mitogen-activated protein kinase kinase kinase enhances freezing tolerance in transgenic maize *Proc Natl Acad Sci* **101** : 3298-3303

Smith A Harbinson J and Fryer W 1997 Effect of chilling on carbon assimilation enzyme activation and photosynthetic electron transport in the absence of photoinhibition in maize leaves *Plant Physiol* **114** : 1039-1046

Steponkus P 1984 Role of plasma membrane in freezing injury and cold acclimation *Ann Rev Plant Physiol* **35** : 543-548

Steponkus P Uemmura M Joesph R Gilmour S and Thomashaw M 1998 Mode of action of the COR 15A gene on freezing tolerance of *Arabidopsis thaliana Proc Natl Acad Sci* **95** : 14570-14575

Stockinger E Gilmour S and Thomashov M 1997 *Arabidopsis* CBF1 encodes on AP2-domain containing transcriptional activator that binds to C-repeat/DRE cis-acting element that stimulates transcription in response to low temperature and water deficit *Proc Natl Acad Sci* **94** : 1035-1040

Stockinger E Mao Y Regier M Treizenbergy S and Thomashon M 2001 Transcriptional adaptor and histone acetyltransferase proteins in *Arabidopsis* and their interaction with CBF1 a transcriptional activator involved in cold regulated gene expression *Nucl Acids Res* **29** : 1524-1533

Sunkar R and Zhu J 2004 Novel and stress regulated micro RNAs and other small RNAs from *Arabidopsis Plant Cell* **16** : 2001-2019

Thomashow M 1998 Role of cold responsive genes in plant freezing tolerance *Plant Physiol* **118** : 1-8

Thomashow M 2001 So what is new in the field of plant cold acclimation? Lots *Plant Physiol* **125** : 89-93

Thorlby G Fourier N and Warren G 2004 The sensitive to freezing 2 gene required for freezing tolerance in *Arabidopsis thaliana* encodes a β-glucosidase *Plant Cell* **16** : 2929-2203

Tsonev T Velikov V Georgieva K Hyde P and Jones H 2003 Low temperature enhances photosynthetic down regulation in French bean (*Phaseolus vulgaris* L) plants *Annuals of Botany* **91** : 343-352

Tsvetanov S Ohno R Tsuda K Takumu S Mori N Atanasov A and Nakamura C 2000 A cold responsive wheat gene Wcor 14 identified in winter hardy cultivar Mironovska 888 *Genes Genet Syst* **75** : 49-57

Verheul M Hassel P and Stamp P 1995 Comparison of maize inbred lines differing in low temperature tolerance : Effect of acclimation at suboptimal temperatures on chloroplast functioning *Annals of Bot* **76** : 7-14

Vincour B and Altman A 2005 Recent advances in engineering plant tolerance to abiotic stress : achievements and limitations *Curr Opin Biotechnol* **16** : 123-132

Viswanathan C and Zhu J K 2002 Molecular genetic analysis of cold regulated gene transcription *Phil Trans R Soc Land* **37** : 877-886

Wang K 2005 Genetic acclimation for freezing tolerance www.Isb.com

Ward D 1987 The temperature accilimation of photosynthetic response to CO_2 in *Zea mays* and its relationship with activities of photosynthetic enzymes and CO_2-concentrating mechanisms of CO_2 photosynthesis *Plant Cell Environ* **10** : 407-411

Warren G Mckown R Martin A and Teutonico V 1996 Isolation of mutations affecting development of freezing tolerance in *Arabidopsis thaliana Plant Physiol* **111** : 1011-1019

Xie Q Furgis G Colgan D and Chua N 2000 *Arabidopsis* NAC1 transduces auxin signal down stream of T_1R_1 to promote lateral root development *Genes Dev* **14** : 3024-3036

Xin Z and Browse J 1998 Eskimo1 mutants of *Arabidopsis* are constitutively freezing tolerant *Proc Natl Acad Sci* **95** : 7799-7804

Xin Z and Li P 1992 Abscisic acid induced chilling tolerance in maize suspension culture cells *Plant Physiol* **99** : 707-711

Xin Z and Li P 1993 Alteration of gene expression associated with abscisic acid induced chilling tolerance in maize suspension culture cells *Plant Physiol* **101** : 277-284

Xiong L Lee B Ishitani M Lee H Zhang C and Zhu J K 2001 FIERY1 encoding an inositol polyphosphate 1-phosphatase is a negative regulator of absicis acid and stress signalling in *Arabidopsis Genes Dev* **15** : 1971-1984

Yu J Tuinstra M Classen M Gordon W and Wilt M 2004 Analysis of cold tolerance in sorghum under controlled environmental conditions *Field Crops Res* **85** : 21-30

Zerial M and McBride H 2001 Rab proteins as membrane organizers *Nat Rev Mol Cell Biol* **2** : 107-117

Zhang J Creelman R and Zhu J K 2004 From laboratory to field Using information from *Arabidopsis* to engineer salt cold and drought tolerance in crops *Plant Physiol* **135** : 615-621

Zhu J Shi H Lee B Damaz B Chang S Stirm V Zhu J K Hasegawa P and Bressan R 2002 An *Arabidopsis homeodomain* transcription factor gene HOS9 mediates cold tolerance through a CBF independent pathway *Proc Natl Acad Sci*

3

MICROBIAL PRODUCTION OF INDUSTRIAL ENZYMES-CURRENT STATUS

CHANDAN AGARWAL

Enzymes have attracted the attention of scientists world over due to their wide range of physiological, analytical and industrial applications. Although enzymes have been isolated, purified and studied from microbial, animal and plant sources, microorganisms represent the most common source of enzymes due to their broad biochemical diversity, feasibility of mass culturing and ease of genetic manipulation. With the advent of molecular biology techniques, a number of genes of industrially important enzymes have been cloned and expressed in order to improve the production of enzymes, substrate utilization and other commercially useful properties. Special attention is being focused on enzymes isolated from thermophiles due to their inherent stability and industrial applications. In addition, a variety of methods have been employed to modify enzymes for their industrial usage including strain improvement, chemical modifications, modification of the reaction environment, immobilization and protein engineering (Chand and Mishra, 2003).

Enzymes are proteins that speed up biochemical reactions without being consumed or changed by the reaction. They are found throughout nature. Enzymes enhance chemical and biochemical reactions and this process is called catalysis. The substance upon which enzyme works when performing catalysis are known as substrate, and each enzyme will only fit and act upon a specific set of substrates. In the absence of enzymes, chemical reactions occur only when molecules collide while in proper alignment with each other. Because molecules are bumping into each other randomly, chemical reactions are essentially due to chance events. This sometimes results in reactions that occur very slowly, or reactions that do not occur at all. Enzymes act like tiny molecular machines to ensure that molecules come into contact with each other and react.

HISTORY OF ENZYMES

The history of enzyme technology began in 1874 when the Danish Chemist Christian Hansen produced the first specimen of rennet by extracting dried calves' stomach with saline solution. Apparently this was the first enzyme preparation of relatively high purity used for industrial purposes.

This significant event had been preceded by a lengthy evolution. Enzymes have been used by man throughout the ages, either in the form of vegetables rich in enzymes, or in the form of microorganisms used for a variety of purposes, for instance in brewing processes, in baking and in the production of alcohol. It is known that enzymes were also used in the production of cheese since old times. Enzymatic processes, particularly fermentation, were the focus of numerous studies in the 19th century and many valuable discoveries in this field were made. A particularly important experiment was the isolation of the enzyme complex from malt by Payen and Persoz in 1833. This extract, like malt itself, converts gelatinized starch into sugars, primarily into maltose and was termed, 'diastase'.

Development progressed during the following decades, particularly in the field of fermentation where the achievements by Schwann, Liebig, Pasteur and Kuhne were of great importance. The dispute between Liebig and Pasteur concerning the fermentation process caused much heated debate. Liebig claimed that fermentation resulted from a chemical process and that yeast was a non-viable substance continuously in the process of breaking down. Pasteur, on the other hand, argued that fermentation did not occur unless viable organisms were present.

The dispute was finally settled in 1897, after the death of both adversaries, when the Buchner brothers demonstrated that cell-free yeast extract could convert glucose into ethanol and CO_2 just like viable yeast cells. In other words, the conversion was not ascribable to yeast cells as such, but to their non-viable enzymes. In 1876, William Kuhne proposed that the name 'enzyme' be used as the new term to denote phenomenon previously known as 'unorganized ferments', that is, ferments isolated from the viable organisms in which they were formed. The word itself means 'in yeast' and is derived from the Greek 'en' meaning 'in', and 'zyme' meaning 'yeast' or 'leaven'.

During the early part of this century, enzyme technology was also developing slowly but surely outside Europe. In the far East, an age old tradition prevailed where mold fungi, the so called koji, were (and still are) used in the production of certain food stuffs and flavour additives based on soya proteins (shoyu, miso, tempeh) and fermented beverages. Koji is prepared from steamed rice into which a mixture of mold fungi is inoculated the composition of the mixture being passed down from generation to generation. This formed the basis on which the Japanese scientist Takamine developed a fermentation process for the industrial production a fungal amylase; the process included the culture of *Aspergillus oryzae* on moist rice or wheat bran. The product was called 'Takadiastase' and it is still used as a digestive aid. The method of fermentation suggested by Takamine, the 'surface culture' or 'semi solid culture' is still actively used in the production of various enzymes.

Investigation carried out by the German chemist and industrial magnate Otto Röhm before World War I were of great importance for the further development of the industrial use of enzymes. Among other things, he studied the so called 'bating' process, a step in the preparation of hides and skins prior to tanning.

According to the traditional method, bating required the excrement of dogs and pigeons, a fact that did not improve the image of tanning which was considered a stinking and unpleasant activity. Röhm's theory was that these excrements exerted their effect because they contained residual amounts of the animals' digestive enzymes. If this was so, it might be possible to use extracts of the pancreas directly for bating. Such extracts were tried and produced the expected positive results. Naturally, Röhm accepted this as conformation of the correctness of his theory, but later experiments showed that it was not the animals' enzymes that were active, but rather enzymes of bacteria growing in the intestinal tract.

Parallel to his studies of the problems involved in tanning, Röhm investigated other processes where enzymes would prove even more valuable. Nevertheless, his efforts were not to score a success until 50 years later.

Röhm actually developed the first method for washing protein stained clothes in detergents containing enzymes and manufactured the first detergent containing enzymes in Germany in 1914 (Bhat, 2000). This trypsin enzyme isolated from animals degraded proteins and was used as a detergent. It proved to be so powerful compared to traditional washing powders that the original small package size made the German housewives suspicious so that the product had to be reformulated and sold in large packages. The real breakthrough of enzymes occurred with the introduction of microbial proteases into washing powders. The first commercial bacterial *Bacillus* protease was marketed in 1954 and became big business when Novozymes in Denmark started to manufacture it and major detergent manufacturers started to use it around 1965 (Godfrey and West, 1996).

In food industry—in addition to cheese manufacturing, enzymes were used already in 1930 in fruit juice manufacturing. These enzymes clarify the juice. They are called pectinases, which contain numerous different enzyme activities. The major usage of microbial enzymes in food industry started in the 1960s in starch industry. The traditional acid hydrolysis of starch was completely replaced by α-amylase and glucoamylases, which could convert starch with over 95% yield to glucose. Starch industry became the second largest user of enzymes after detergent industry.

ADVANTAGES OF MICROBIAL ENZYMES VERSUS ANIMAL ENZYMES

There are many advantages of using supplemental microbial enzymes as opposed to animal derived enzymes. Microbial enzymes are specially selected on the basis of each enzyme's unique characteristics. Fermented enzymes exhibit broad ranges of pH, temperature and substrate specificities.

Supplemental microbial enzymes are choosen on their ability to work within the gastrointestinal system of mammals. Especially selected for compatibility with the body's temperature, microbial enzymes also exhibit activity across a broad pH range. Unlike supplemental enzymes of animal origin, microbial enzymes work at the pH found in the upper stomach. Food sits in the upper portion of stomach for as long as an hour before gastric secretions being to act. Several studies have shown that the enzymes in the saliva continue their digestive activity in the upper stomach and can digest up to 30% of the ingested protein, 60% of the ingested starch and 10% of ingested fat during the first 30 to 60 minutes after consumption (Beazell, 1941).

Though salivary enzymes accomplish a significant amount of digestion, their activity is limited to a pH level above 5.0. Supplemental microbial enzymes are active in the pH range of 3.0 to 9.0 and can facilitate the utilization of a much larger amount of proteins, carbohydrates and fats before hydrochloride is secreted in sufficient amounts to neutralize their activity. In contrast, supplemental enzymes of animal origin are destroyed by the low pH within the stomach unless they are enterically coated. Yet, this coating can prevent the dissolution of the enzymes and prevent any digestive benefit. Studies have shown that non-enteric coated products can be more effective then coated products (Schneider, 1985). Further, animal-based enzymes function only at the narrow pH ranges found at specific anatomical sites.

Another advantage of microbial enzymes is the variety of enzymes available for supplementation. While pancreatin offers only protease, lipase and amylase activities, microbial enzymes offer protease, peptidase, lipase, amylase, glucoamylase, invertase, malt diastase, lactase, α-galactosidase, cellulase, hemicellulase, pectinase and phytase activities. Pancreatin is a pre-defined blend and the only alternatives are proportional increases in total activity that may not be necessary. Customization and flexibility to match the appropriate enzymes to the diet is another strength of supplemental microbial enzymes (Schwimmer, 1981).

Yet, another advantage of microbial enzymes is that they are animal friendly and cruelty free.

Table 1 : Application of A Range of Bulk Microbial Enzymes (Michael, 2001).

Proteases, mostly neutral proteases or zinc metalloproteases from species of *Aspergillus* and *Bacillus*, along with some alkaline serine proteases.

(a) Biological detergents, e.g. the alkaline protease subtilisin EC 3.4.4.16 from *Bacillus licheniformis* and *B. subtilis*.

(b) Baking-dough modification, gluten weakening and flavour improvement.

(c) Beer brewing, e.g. 'chill proofing' of beer to remove protein haze.

(d) Leather bating and tendering.

(e) Cheese manufacture-clotting of milk protein and promotion of ripening, e.g. an aspartic proteases (EC 3.4.23.6) from *Rhizomucor* species and recombinant *Kluveromyces* species that produce calf chymosin. Aminopeptidase from *Lactococcus lactis* are also used to enhance flavour development.

(f) Meat tenderization and removal of meat from bones.

(g) Flavour control and production in food products.

(h) Waste treatment, e.g. recovery of silver from the spent photographic films.

Lipases EC3.1.1.3, primarily from species of *Bacillus, Aspergillus, Rhizopus* and *Rhodotorula*.

(a) Biological detergents, also available from recombinant *Aspergillus oryzae* that produces *Humicola* lipase.

(b) Leather processing-fat removal.

(c) Production of flavour compounds and acceleration of ripening in dairy and meat products.

Carbohydrases

α-amylase EC3.2.1.1, mostly derived from species of *Aspergillus* and *Bacillus*.

(a) Starch processing-liquefaction of starch in the production of sugar syrups.

(b) Baking-partial starch degradation in flour modification and the generation of fermentable sugars, and for improvement of crust colour.

(c) Brewing- Starch hydrolysis during wort preparation.

(d) Biological detergents-starch removal from food stains.

(e) Textile manufacture-desizing.

β-amylase EC 3.2.12 from *Bacillus* species such as *B. polymyxa* and species of *Streptomyces* and *Rhizopus*.

(a) Maltose syrup production.

(b) Brewing-increasing wort fermentability.

α-myloglucosidase (Glucoamylase) EC 3.2.1.3 from *Aspergillus niger* and *Rhizopus niveus.*

(a) Glucose syrup production-complete starch saccharification.

(b) Baking-improved bread crust colour.

(c) Brewing-dextrin saccharification in the production of low carbohydrate beer.

(d) Wine and fruit juice-starch removal.

β-galactosidase (Lactase) EC3.2.1.23 from species of *Bacillus, Kluveromyces* and *Candida.*

(a) Whey syrup production-hydrolysis of lactase to glucose and galactose to give greater sweetness.

(b) Milk and dairy product processing-lactose reduction/removal for those who are lactose intolerant.

(c) Ice cream manufacture, prevention of 'sandy' texture caused by lactose crystals.

Glucose isomerase EC 5.3.1.5 from species of *Actinoplanes, Arthrobacter* and *Streptomyces.*

(a) Manufacture of high fructose syrup as 'high sweetners'-glucose conversion to fructose.

Glucose oxidase EC 1.1.3.4 from *Aspergillus niger* and *Pencillium notatum.*

(a) Antioxidant, often used along with catalase to remove oxygen from food products.

Inulinase EC 3.2.1.7 from *Aspergilus niger, Candida species* and *Kluveromyces* species.

(a) Processing of Jerusalem artichoke tubers-hydrolysis of polyfructans and levans.

Invertase EC 3.2.1.26 from species of *Saccharomyces* and *Kluveromyces.*

(a) Sweets and confectionery production -liquefaction of sucrose, e.g.-soft centred chocolates.

(b) Invert sugar production-sucrose conversion to glucose and fructose.

Pullulanase EC 3.2.1.41 from *Klebsiella pneumoniae* and *Bacillus acidopullulyticus.*

(a) Starch processing-used for debranching of starch in sugar syrup manufacture and brewing.

Pectinases-a mixture of enzymes such as polygalacturonase EC 3.2.1.15, produced by *Aspergillus niger* and *Penicillium* species often by solid substrate fermentation.

(a) Plant juice and oil extraction

(b) Fruit juice and wine clarification

(c) Coffee bean fermentation

Hemicellulases and **β-glucanases**/cellulases β-galucanases, e.g. cellulases EC 3.2.1.4 from *Trichoderma reesei, Penicillium funiculosum, Aureobasidium pullulans* and *Bacillus* species, and specific β-glucanases from *Penicillium emersonii.*

(a) Fruit juice and olive production and processing.

(b) Wine and beer production and processing-hydrolysis of β-glucan gums to improve filtration rates.

(c) Malting-speeds modification of grains.

(d) Textile processing-biopolishing of cellulose fibres.

(e) Wood pulp processing.

Hemicellulases from the yeasts *Cryptococcus* and *Trichosporon*, e.g. Xylanase EC 3.2.1.32

(a) Baking-pentonases are used to alter dough consistency.

(b) Brewing

(c) Animal feed stuffs

(d) Neutraceutics

(e) Woodpulp processing

Miscellaneous enzymes

Catalase EC 1.11.1.6 from *Aspergillus niger, Corynebacterium glutamicum* and *Microoccus lysodeikticus*

(a) Bleaching textiles

(b) Cheese manufacture

Phytase EC3.1.3.8

(a) Addition to feeds for monogastric animals, in order to release phosphate from phytic acid, e.g. from recombinant *Pichia angusta.*

Urease EC 3.5.1.5 from *Lactobacillus fermentum*

(a) Wine production

(b) Ceramics manufacture, to enhance the manufacture and quality of modern ceramics which are of importance for the computer and aeronautics industries or for orthopaedics.

Enzymes in biotransformation processes

(a) **Penicillin and Cephalosporin acylases** (amidases) EC 3.5.2.6, e.g. *Bacillus megaterium* and *Escherichia coli* produce penicillin acylase and *Pseudomonas* species produce cephalosporin acylase. Used for antibiotic conversion to remove the side groups of penicillins or cephalosporins to form 6-amino penicillanic acid or 7-aminocephalosporanic acid, respectively, the base molecules for the synthesis of semisynthetic penicillins and cephalosporins.

(b) **L-amino acid acylases** EC 3.5.1.14 from *Aspergillus* species, used for the resolution of L-amino acids from acylated racemic mixtures of D,L-amino acid mixture, e.g. lysine production.

(c) **Steroid Biotransformation**, e.g. 11-β-hydroxylase (monooxygenase) EC 1.14.15.4 from *Curvularia lunata* for the production of prednisolone from 11-deoxycortisone.

Global Market for Industrial Enzymes Estimated Worth $2.7 Billion by 2012

According to a new technical market research report 'Enzymes for Industrial Applications' (BIO030E) from BCC Research, the global market for industrial enzymes will be worth $2.3 billion in 2007. This is expected to increase to over $2.7 billion by 2012, a compound average annual growth rate (CAGR) of 4%.

The market, is broken down into application of technical, food and animal feed enzymes. Of these, technical enzymes have over 50% of the market. Valued at nearly $1.1 billion in 2007, this segment is expected to be worth $1.4 billion by 2012, a CAGR of 3.5%.

The second largest segment, food enzymes, was worth an estimated $830 million in 2007 and will reach $1.0 billion by 2012. This represents a CAGR of 4%. The animal feed enzyme segment is currently worth $280 million and will be worth $375 million in 2012, a CAGR of 4%. The higher growth in this sector will be helped in part by the increased use of phytase enzymes to fight phosphate pollution.

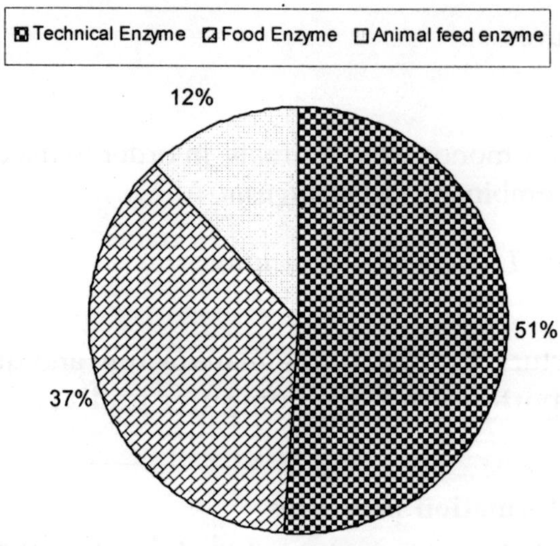

Enzymes for Industrial Applications (BIO 030E)

With environmental and cost issues in conventional chemical processes being subjected to considerable scrutiny, biotechnology is gaining rapid ground as it offers several advantages over conventional technologies. Industrial enzymes represent the heart of biotechnology.

The field of industrial enzymes is now experiencing a major research and development initiative, resulting in the development of a number of new products and an improvement in the process and performance of several existing products.

Table 2: Global Enzyme Markets Based on Application Sectors ($ millions)

Application Sector	2005	2006	2007	2012	CAGR% 2007-2012
Technical Enzymes	1075	1105	1140	1355	3.5
Food Enzymes	775	800	830	1010	4.0
Animal Feed Enzymes	240	260	280	375	6.0
Total	2090	2165	2250	2740	4.0

(BCC Research, 2007)

Use of Microorganisms for Enzyme Production

Presently more than 2000 different enzyme activities have been isolated and characterized. The sequence information of a growing number of organisms opens the possibility to characterize all the enzymes of an organisms on a genomic level. The smallest known organism *Mycoplasma genitalium*, contains 470 genes out of which 145 are related to gene replication and transcription. Baker's yeast has 7000 genes coding for about 3000 enzymes. Thousands of different variants of the natural enzymes are known. The number of reported three dimensional enzyme structure is rapidly increasing. In the year 2000, the structure of about 1300 different proteins were known. Of the hundred or so enzymes being used industrially, over half are from fungi and yeast. A major proportion of industrially useful enzymes is also obtained from bacteria with the remainder divided between animal (8%) and plant (4%) sources. Plant derived commercial enzymes include proteolytic enzymes papain, bromelain and ficin and some other speciality enzymes like lipoxygenease from soyabeans. Animal derived enzymes include proteinases like papain and renin. No source provides a larger proportion of this at the present time. Microbes are preferred to plants and animals as sources of enzymes because :

(1) They are generally cheaper to produce,

(2) Their enzyme contents are more predictable and controllable,

(3) Reliable supplies of raw material of constant composition are more easily arranged and

(4) Plant and animal tissues contain more potentially harmful materials than microbes, including phenolic compounds (from plants), endogenous enzyme inhibitors and proteases.

It can, therefore, be seen that suspended or immobilized microbial cells and spores may be employed as biocatalyst in some industrial bioconversions, particularly where coenzymes are necessary. However, in many cases this has limitations because:

(1) The cells 'waste' energy and resources in growth and/or maintenance activities,

(2) Side reactions may lead to a reduction in the potential yield of the target product,

(3) Conditions for microbial growth, where required, may be different from those necessary for optimum product formation and

(4) Difficulties may be encountered in the isolation and purification of the product from the cells or spent fermentation medium.

To overcome such problems, the use of partially purified 'bulk' microbial enzyme preparation is often preferred for numerous and varied industrial processes.

In some cases, whole conventional microbial fermentation processes may eventually be replaced by multienzyme systems that could provide more efficient substrate utilization, higher yields and greater product uniformity.

Most of the enzymes are produced by microorganisms in submerged cultures in large reactors called fermentors. The enzyme production process can be divided into the following phases :

(1) Selection of an enzyme

(2) Selection of production strain

(3) Construction of an overproducing strain by genetic engineering

(4) Optimization of culture medium and production conditions

(5) Optimization of recovery process (and purification if needed)

(6) Formulation of a stable enzyme product.

Criteria used in the selection of an industrial enzyme include specificity, reaction rate, pH and temperature optima and stability, effect of inhibitors and affinity to substrates. Enzymes used in paper industry should not contain cellulose degrading activity as a side activity because this activity would damage the cellulose fibres. Enzymes used in animal feed industry must be thermotolerant to survive in the hot extrusion process used in animal feed manufacturing. The same enzyme must have maximal activity at the body temperature of the animal.

Enzymes used in industrial applications must usually be tolerant against various heavy metals and have no need for cofactors. They should be maximally active already in the presence of low substrate concentration so that the desired reaction proceeds to compeletion in a realistic time frame.

In choosing the production strain several aspects have to be considered. Ideally the enzyme is secreted from the cell. This makes the recovery and purification process much simpler compared to production of intracellular enzymes which must be purified from thousands of different cell proteins and other components. Secondly, the production host should have a GRAS status, which means that it is Generally Regarded As Safe. This is especially important when the enzyme produced by the organism is used in food processes.

Table 3 : Examples of Microorganisms Classified as GRAS

Bacteria
Bacillus subtilis
Lactobacillus bulgaricus
Lactococcus lactis
Leuconostoc oenos
Yeasts
Candida utilis
Kluveromyces marxianus
Kluveromyces lactis
Saccharomyces cerevisiae
Filamentous fungi
Aspergillus niger
Aspergillus oryzae
Mucor javanicus
Penicillium roqueforti

Thirdly, the organism should be able to produce high amount of the desired enzyme in a reasonable time frame. The industrial strains typically produce over 50 g/l extracellular enzyme proteins. Most of the industrial enzymes are produced by a relatively few microbial hosts like *Aspergillus* and *Trichoderma,* fungi, *Streptomyces* fungi imperfect and Bacillus bacteria. Yeasts are not good producers of extracellular enzymes and are rarely used for this purpose. Most of the industrially used microorganisms have been genetically modified to over produce the desired activity and not to produce undesired side activities.

Table 4 : Examples of Microbial Enzymes Used for Analytical, Diagnostic and Therapeutic Purpose.

Enzyme	Source	Use
α-amylase EC3.2.1.1	*Aspergillus* and *Bacillus* species	Digestive aids
Alcohol dehydrogenases	*Saccharmyces cerevisiae*	Ethanol testing
Alcohol oxidase EC1.1.3.13	*Candida boidinni* and *Pichia pastoris*	
Asparaginase EC3.5.1.1	*Escheria coli, Serratia marcescens* and *Erwinia carotovora*	Maintains low asparagine levels in treating cancers (lymhomas and leukemias) whose cells can not synthesize the amino acids
Catalase EC 1.11.1.6	*Aspergillus niger, Corynebacterium glutamicum Micrococcus lysodeikticus*	Contact lens cleaning systems
Cholesterol esterase EC 3.1.1.13	*Pseudomonas fluorescens*	Monitoring serum cholesterol levels
Creatinnase EC 3.5.2.10	*Pseudomonase putida*, recombinant *Escherichia coli*	Determination of serum cretinine levels
Glucose oxidase EC 1.1.3.4	*Aspergillus niger*	Analysis of blood glucose levels
β-Lactamase EC 3.5.2.6	*Bacillus cereus* and *Escherichia coli*	Treatment of pencillin allergy
Proteases (various)	Several bacterial and fungal sources	Used as digestive aids for wound debridement and contact lens cleaning
Rhodanse EC 2.8.1.1	*Trichoderma* species	Treatment of cyanide poisoning
Streptokinase EC 3.4.22.10	Various haemolytic *streptococii*	To breakdown blood clots
Urease EC 3.5.1.5	*Lactobacillus fermentum*	Removal of urea from blood in renal failure
Uricase EC1.7.3.3	*Arthrobacter globiformis Candida utilis*	Used in treatment and diagnosis of gout
DNA directed Microbial Enzymes		
DNA ligases EC 6.5.1.1	*Escherichia coli*	Joining pieces of DNA
DNA polymerase EC2.7.7.7	*Thermus aquaticus*	DNA synthesis, e.g. in PCR for forensic analysis
Restriction endonucleases e.g. Bam H1 and Eco R1	*Bacillus amyloliquefaciens Escherichia coli* Ry13	'Cutting' DNA at specific base sequences
RNA polymerase EC2.7.7.6	*Salmonella typhimurium*	RNA synthesis

PRODUCTION OF SOME MICROBIAL ENZYMES

1. PROTEASES

Proteases represent the class of enzymes which occupy a pivotal position with respect to their physiological roles as well as their commercial applications. They perform both degradative and synthetic functions. Since they are physiologically necessary for living organisms, proteases occur ubiquitously in a wide diversity of sources such as plants, animals and microorganisms. Microbes are an attractive source of proteases owing to the limited space required for their cultivation and their ready susceptibility to genetic manipulation. There is a renaissance of interest in using proteolytic enzymes as targets for developing therapeutic agents. Protease genes from several bacteria, fungi and viruses have been cloned and sequenced with the prime aims of :

(1) Overproduction of the enzymes by gene amplification,

(2) Delineation of the role of the enzymes in pathogenicity, and

(3) Alteration in enzyme properties to suit its commercial application.

Protein engineering techniques have been exploited to obtain proteases which show unique specificity and/or enhanced stability at high temperature or pH or in the presence of detergents and to understand the structure-function relationships of the enzymes. Protein sequences of acidic, alkaline and neutral proteases from diverse origins have been analyzed with the aim of studying their evolutionary relationships. Despite the extensive research on several aspects of proteases, there is a paucity of knowledge about the roles that govern the diverse specificity of these enzymes.

Sources of Proteases

Advances in analytical techniques have demonstrated that proteases conduct highly specific and selective modifications of proteins such as activation of zymogenic forms of enzyme by limited proteolysis, blood clotting and lysis of fibrin clots, and processing and transport of secretory proteins across the membrane. The functions of proteases extend from cellular level to the organ and organism level, to produce cascade systems such as hemostasis and inflammation. They are responsible for the complex processes involved in the normal physiology of the cell as well as in abnormal pathophysiological conditions. Their involvement in the life cycle of disease causing organisms has led them to become a potential target for developing therapeutic agents against fatal diseases such as cancer and AIDS. Their application in the leather industry for dehairing and bating of hides to substitute currently used toxic chemicals is a relatively new development and has conferred added biotechnological importance (Mata *et al.*, 1998) The vast diversity of proteases in contrast to the specificity of their action, has attracted world wide attention in attempts to exploit their physiological and biotechnological applications (Poldermans, 1990).

(1) Plant Proteases

The use of plants as a source of proteases is governed by several factors such as the availability of land for cultivation and the suitability of climatic conditions for growth. Moreover, production of proteases from plants is a time consuming process. Some of the well known proteases of plant origin are as follows :

Papain-Papain is a traditional plant protease and has a long history of use (Schechler and Berger, 1967). It is extracted from the latex of *Carica papaya* fruits. The performance of the enzyme depends on the plants source, the climatic conditions for growth, and the methods used for its extraction and purification. The enzyme is active between pH 5 and 9 and is stable upto 80 or 90° C in the presence of substrates.

Bromelain-Bromelain is prepared from the stem and juice of pineapple. The major supplier of the enzymes is Great Food Biochem., Bangkok, Thailand. The enzyme is characterized as a cysteine protease and is active from pH 5 to 9. Its inactivation temperature is 70° C, which is lower than that of papain.

Keratinases-Some of the botanical groups of plants produce proteases which degrade hair. Digestion of hair and wool is important for the production of essential amino acids such as lysine and for the prevention of clogging of waste water systems.

Ficin-Obtained from figs.

(2) Animal Proteases

The most familiar proteases of animal origin are pancreatic trypsin, chymotrypsin, pepsin and rennis (Boyer, 1971; Hoffman, 1974). These are prepared in pure form in bulk quantities. However, their production depends on the availability of livestock for slaughter, which in turn is governed by political and agricultural policies.

Trypsin-Trypsin is the main intestinal digestive enzyme responsible for the hydrolysis of food proteins. It is a serine protease and hydrolyzes peptide bonds in which the carboxyl groups are contributed by the lysine and arginine residues. Based on the ability of protease inhibitors to inhibit the enzyme from the insect gut, this enzyme has received attention as a target for biocontrol of insect pests. Trypsin has limited applications in the food industry, since the protein hydrolysates generated by its action have a highly bitter taste. It is used in the preparation of bacterial media and in some specialized medical applications.

Chymotrypsin-Chymotrypsin is found in animal pancreatic extract. Pure chymotrypsin is an expensive enzyme and is used only for diagnostic and analytical applications. It is specific for the hydrolysis of peptide bonds in which the carboxyl groups are provided by one of the three aromatic amino acids, i.e., phenylalanine, tyrosine or tryptophan. It is used extensively in the deallergenizing of milk protein hydrolysates. It is stored in the pancreas in the form of a precursor, chymotrypsinogen and is activated by trypsin in a multistep process.

Pepsin-Pepsin is an acidic protease that is found in the stomachs of almost all vertebrates. The active enzyme is released from its zymogen, i.e., pepsinogen, by autocatalysis in the presence of hydrochloric acid. Pepsin is an aspartyl protease and resembles human immunodeficiency virus type 1(HIV-1) protease, responsible for the maturation of HIV-1. It exhibits optimal activity between pH 1 and 2. Pepsin inactivated above pH 6.0. The enzyme catalyzes the hydrolysis of peptide bonds between two hydrophobic amino acids.

Rennin-Rennet is a pepsin-like protease that is produced as an inactive precursor, prorennin, in the stomachs of all nursing mammals. It is converted to active rennin by the action of pepsin or by its autocatalysis. It is used extensively in the dairy industry to produce a stable curd with good flavor. The specialized nature of the enzyme is due to its specificity in cleaving a single peptide bond in k-casein to generate insoluble para-k-casein and C-terminal glycopeptide.

(3) Microbial Proteases

The inability of the plant and animal proteases to meet the current world demands has led to an increased interest in microbial proteases. Microorganisms represent an excellent source of enzymes owing to their broad biochemical diversity and their susceptibility to genetic manipulation. Microbial proteases account for approximately 40% of the total worldwide enzyme sales (Godfrey and West, 1996). Proteases from microbial sources are preferred to the enzymes from plant and animal sources since they possess almost all the characteristics desired for their biotechnological applications.

Bacteria-Most commercial proteases, mainly neutral and alkaline, are produced by organisms belonging to the genus *Bacillus*. Bacterial neutral proteases are active in a narrow pH range (pH 5 to 8) and have relatively low thermotolerance. Due to their intermediate rate of reaction neutral proteases generate less bitterness in hydrolyzed food proteins than do the animal proteinases and hence are valuable for use in the food industry. Neutrase, a neutral protease, is insensitive to the natural plant proteinase inhibitors and is therefore useful in the brewing industry. The bacterial neutral proteases are characterized by their high affinity for hydrophobic amino acid pairs. Their low thermotolerance is advantageous for controlling their reactivity during the production of food hydrolysates with a low degree of hydrolysis. Some of the neutral proteases belong to the metalloprotease type and require divalent metal ions for their activity, while others are serine proteinases, which are not affected by chelating agents.

Bacterial alkaline proteases are characterized by their high activity at alkaline pH, e.g., pH 10 and their broad substrate specificity. Their optimal temperature is around 60° C. These properties of bacterial alkaline proteases make them suitable for use in the detergent industry.

Fungi-Fungi elaborate a wider variety of enzymes than do bacteria. For example, *Aspergillus oryzae* produces acid, neutral and alkaline proteases. The fungal proteases are active over a wide pH range (pH 4 to 11) and exhibit broad substrate specificity. However, they have a lower reaction rate than the bacterial enzymes. Fungal enzymes can be conveniently produced in a solid state fermentation process. Fungal acid proteases have an optimal pH between 4 and 4.5 and are stable between pH 2.5 and 6.0. They are particularly useful in the cheese making industry due to their narrow pH and temperature specificities. Fungal neutral proteases are metalloproteases that are active at pH 7.0 and are inhibited by chelating agents. In view of the accompanying peptidase activity and their specific function in hydrolyzing hydrophobic amino acid bonds, fungal neutral proteases supplement the action of plant, animal and bacterial proteases in reducing the bitterness of food protein hydrolysates. Fungal alkaline proteases are also used in food protein modification.

Viruses-Viral proteases have gained importance due to their functional involvement in the processing of proteins of viruses that cause certain fatal diseases such as AIDS and cancer. Serine, aspartic and cysteine peptidases are found in various viruses (Rawlings and Barrett, 1993). All of the virus encoded petidases are endopeptidases, there are no metallopeptidases. Retroviral aspartyl proteases that are required for viral assembly and replication are homodimers and are expressed as a part of the polyprotein precursor. The mature protease is released by autolysis of the precursor. An extensive literature is available on the expression, purification, and enzymatic analysis of retroviral aspartic protease and its mutants (Kuo and Shafer, 1994). Extensive research has focused on the three dimensional structure of viral proteases and their interaction with synthetic inhibitors with a view to designing potent inhibitors that can combat the relentlessly spreading and devastating epidemic of AIDS.

Genetic Engineering of Microbial Proteases

Gene cloning is a rapidly progressing technology that has been instrumental in improving our understanding of the structure-function relationship of genetic systems. More than 50% of the industrially important enzymes are now produced from genetically engineered microorganisms (Hodgson, 1994). Several reports have been published in the past decade on the isolation and manipulation of microbial protease genes with the aim of :

(1) Enzyme overproduction by the gene dosage effect,

(2) Studying the primary structure of the protein and its role in the pathogenicity of the secreting microorganism, and

(3) Protein engineering to locate the active site residues and/or to alter the enzyme properties to suit its commercial applications.

Protease genes from bacteria, fungi and viruses have been cloned and sequenced.

1. Bacteria

The objective of cloning bacterial protease genes has been mainly the overproduction of enzymes for various commercial applications in the food, detergent, and pharmaceutical industries. The virulence of several bacteria is related to the secretion of several extracellular proteases. Gene cloning in these microbes was studied to understand the basis of their pathogenicity and to develop therapeutics against them. Proteases play an important role in cell physiology, and protease gene cloning, especially in *E. coli.* has been attempted to study the regulatory aspects of proteases.

Bacilli

(i) *B. subtilis* as a host for cloning of protease genes from *Bacillus* spp.-The ability of *B. subtilis* to secrete various proteins into the culture medium and its lack of pathogenicity make it a potential host for the production of foreign polypeptides by recombinant DNA technology. Several *Bacillus* spp. secrete two major types of protease, a subitilisin or alkaline protease and a mealoprotease or neutral protease which are of industrial importance.

(ii) *B. subtilis-B. subtilis* 168 secretes at least six extracellular proteases into the culture medium at the end of exponential phase. The structural genes encoding the alkaline protease (*apr*) or subtilisin, neutral protease A and B (*npr A* and *npr B*) (Henner *et al.*, 1985), minor extracellular protease (*epr*) (Bruckner *et al.*, 1990), bacillopeptidase F(*bpr*) (Rudolph *et al.*, 1990), and metalloprotease (*mpr*) have been cloned and characterized. To increase the expression of subtilisin and neutral proteases, Henner *et al.*, replaced the natural promoters of *apr* and *npr* genes with the amylase promoter from *B. amyloliquefaciens* and the neutral protease promoter from *B. subitilis* respectively (Henner *et al.*, 1985).

(iii) Alkalophilic *Bacillus* spp.-Bacillus proteases with an extermely alkaline pH optimum are generally used in detergent powders and are preferred over the subtilisins (optimal pH 8.5-10.0). Kaneko *et al.*, cloned and sequenced the *ale* gene, encoding alkaline elastase YaB, a new subtilisin from an alkalophilic *Bacillus* strain (Kaneko, 1989). The deduced amino acid sequence of the highly alkaline serine protease from *B. alcalophilus* PB92, showed considerable homology to YaB.

(iv) Other *bacilli*-A gene encoding the highly thermostable neutral proteinase (*Npr*) from *Bacillus* sp. strain EA1was shown to be closely related to an *npr* gene from *B. caldolyticus* YP-T, except for a single amino acid change in the gene product. The enzyme from *Bacillus* sp. strain EA1 was more thermostable than the enzyme from *B. caldolyticus* YP-T; this can be attributed to the single amino acid change.

Lactococci-(*Lactococcus lactis* subsp. *lactis* and *cremoris* previously *Streptococcus lactis* and *Streptococcus cremoris*). On the basis of differences in caseinolytic specificity, the lactococcal proteases have been classified into two main groups, the PI-type protease, which degrades predominantly β-casein, and the P-III-type protease which degrades αS1-, β- and K-casein. Most of the genetic studies have focussed on the PI-type protease genes. Lactococcal protease genes are located mostly on plasmids.

Curing experiments have suggested that plasmid pWV05 of *S. cremoris* Wg2 specifies proteolytic activity. The entire plasmid was subcloned in *E.coli* (Kok *et al.*, 1985). A 4.3

MDa *Hind* III fragment of the plasmid, specifying the proteolytic activity, was cloned in *B. subtilis* and in a protease deficient *S.lactis* strain.

L. lactis subsp. *cremoris* H2 carries plasmid pD121 containing the gene for the protease positive phenotype (Prt^+).

S. lactis NCDO763 harbors plasmid pLP763, containing the gene for Prt^+, which enables it to grow to a higher density in milk.

The P-III type protease is found only in *L. lactis* subsp. *cremoris* AM1 and SK11. They both contain the proteases encoded by the 78-kbp plasmid psk 111.

Streptomyces-*S. griseus*, an organism used for the commercial production of pronase, secretes two extracellular serine proteases : proteases A and B. The genes encoding protease A (*spr A*) and protease B (*spr B*)were isolated from the *S. griseus* genomic library, and their proteolytic activity was demonstrated in *S. lividans* (Henderson *et al.*, 1987).

Serratia-These gram negative bacteria belonging to the family *Enterobacteriaceae* are known to secrete large amounts of extracellular proteases into the surrounding medium. The gene encoding the protease from *Serratia* sp. strain E-15 was expressed both in *E. coli* and *S. marcescenes* (Nakahama *et al.*, 1986).

Pseudomonas-*P. aeruginosa* is an opportunistic pathogen and its virulence is related to the secretion of several extracellular proteins (Liu, 1974). *P. aeruginosa* secretes two proteases, an alkaline protease and elastase. The alkaline proteases genes (*apr*) from *P. aeruginosa* 1FO 3455 and PAO1 were cloned in *E. coli* (Atsumi *et al.*, 1989; Guzzo *et al.*, 1990). The Las A protease (elastin degrading) of *P. aeruginosa* is also an important contributor to the pathogenesis of this bacterium. The *Las A* gene from strain FRD1 was overexpressed in *E. coli* (Gustin, 1996).

Miscellaneous

Aqualysin I, an alkaline serine protease, is secreted into the culture medium by *Thermus aquaticus* YT-1. Aqualysin I shows high DNA sequence homology to the subtilisin type serine proteases, especially in the regions containing the active site residues (Asp 32, His 64 and Ser 221) of subtilisin BPN' (Kwon *et al.*, 1988). The nucleotide sequence also revealed that the enzyme is produced as a large precursor, containing the N-terminal portion, the protease and the C-terminal portion.

Immunoglobulin A1 (IgA1) proteases form a very heterogenous group of extracellular endopeptidases produced by a number of bacterial pathogens that colonize the human mucosal surfaces. The enzymes specifically cleave human IgA1, which participates in the immune system surveillance in the human mucosa. A number of reports (Fishman *et al.*, 1985) on the cloning of the *IgA* gene, encoding the *IgA1* protease from Neisseria gonorrhoeae are available. Nucleotide sequence analysis revealed that the enzyme is produced as a large precursor with three functional domains, i.e., the N terminal leader peptide, the protease, and the carboxyl terminal 'helper' domain-cloning of streptococcal

IgA1... genes from streptococcus genes from *Streptococcus sanguis* ATCC 10556 (Gilbert *et al.*, 1988) and *S. pneumoniae* (Poulsen *et al.*, 1996) has been reported.

2. Fungi

As in bacteria, cloning of the protease genes of fungi has been attempted from both the commercial and pathogenicity points of view.

(i) Acidic proteases

(a) *Mucor*-Two closely related species of Zygomycete fungus, *Mucor pusillus* and *Mucor miehei*, secrete aspartate proteases, also known as mucor rennins into the medium. Sequencing of the cloned gene encoding *M. pusillus* rennin (*MPR*) revealed an *ORF* without introns, encoding possible pre-pro sequences (66 aa) unpstream of the mature MPR sequence. The deduced amino acid sequence showed a high degree of homology to that of *M. miehei* rennin (*MMR*). The gene encoding *M. miehei* aspartyl protease (*MMAP*) has also been cloned and sequenced (Gray *et al.*, 1986).

(b) *Rhizopus*-*Rhizopus niveus*, belonging to the Zygomycete class, also secretes aspartyl protease abundantly. The gene encoding *R. niveus* aspartic protease (*RNAP*) has been cloned and sequenced (Hoiuchi *et al.*, 1988).

(c) *Aspergillus*-The *pep A* gene encoding the aspartic protease, aspergillopepsin A, from *A. awamori* (Berka, 1990), the *pep A* gene from *A. oryzae* (Gomi *et al.*, 1993) and the cDNA coding for an elastinolytic aspartic protease, aspergillopepsin F, from *A. fumigatus* (Lee, 1995) have been cloned and sequenced.

(ii) Alkaline Proteases

(a) *Aspergillus*-Alkaline protease (Alp) produced by *A. oryzae*, a filamentous ascomycete used in the manufacture of soy sauce, is considered to play an important role in producing the flavour of soy sauce. A cDNA library of *A. oryzae* ATCC 20386 was constructed in PUC 119 to isolate a cDNA (1100 bp) encoding the mature region of Alp. The nucleotide sequence of the cDNA lacked most of the DNA sequences corresponding to the pre pro region. The entire cDNA coding for pre pro-Alp was cloned and expressed in *S. cerevisiae*. The character of the Alp secreted from *S. cerevisiae* was shown to be identical to that of native Alp (Tatsumi *et al.*, 1989).

(b) *Acremonium*-*A. chrysogenum* ATCC 11550 produces a considerable amount of extracellular Alp. The cDNA and genomic DNA encoding Alp were isolated from the *A. chrysogenum* cDNA and genomic DNA libraries, respectively (Isogai *et al.*, 1991). The nucleotide sequence of the gene was determined the deduced amino acid sequence showed 57% homology to that of *A. oryzae* Alp. Cloning of the entire cDNA encoding *A. chrysogenum* Alp into *S. cerevisiae* directed the secretion of enzymatically-active Alp into the culture medium.

(c) *Fusarium*-The transfer of the *Fusarium* alkaline protease gene (Kitano *et al.*, 1992) into *A. chrysogenum* resulted in transformants producing large amount of Alp.

(iii) Serine Protease

(a) ***Tritirachuim*****-**Proteinase K is a serine endoproteinase excreted by the fungus *Tritirachium album* Limber. The enzyme is able to hydrolyze native proteins rapidly and is active in the presence of detergents (urea, sodium dodecyl sulphate, etc.) making proteinase K one of the most useful tools in molecular biology. The genomic DNA as well as the cDNA encoding proteinase K from *T. album* Limber have been cloned in *E. coli*, and the entire nucleotide sequence of the coding region, including the 5′ and 3′ flanking regions, has been determined (Gunkel and Gassen, 1989).

3. Viruses

Gene cloning of viral proteases has been undertaken for the isolation and overexpression of the gene and for subsequent screening of inhibitory compounds that may be used in the development of chemotherapeutic agents. Viral protease is responsible for processing of polyprotein precursors into the structural proteins of the mature virion. Among viruses, reports on cloning of protease genes are limited mainly to animal viruses such as Herpesviruses, Adenoviruses, Retroviruses, Picornaviruses etc. Studies have also been conducted on the plant virus-Potyvirus.

Current Problems and Potential Solutions

Proteases are a complex group of enzymes which differ in their properties such as substrate specificity, active site and catalytic mechanism. Despite the extensive research on several aspects of proteases from ancient times, there are several gaps in our knowledge of these enzymes and there is tremendous scope for improving their properties to suit projected applications. The future lines of development would include:

(i) Genetic approaches to generate microbial strains for hyperproduction of the enzymes,

(ii) Application of site directed mutagenesis to design proteases with unique specificity and increased resistance to heat and alkaline pH,

(iii) Synthesis of peptides (synzymes) to mimic proteases,

(iv) Production of abzymes (catalytic antibodies) with specific protease activity, and

(v) Understanding of the structure-function relationship of the enzyme,

Some of the important problems faced in the desired usage of proteases and the possible solutions to overcome these hurdles are discussed below :

i) Enhancement of Thermostability

The industrial use of proteases in detergents or for leather processing requires that the enzyme be stable at higher temperatures. One of the common strategies to enhance the thermostability of the enzyme is to introduce disulphide bonds into the protease by site directed mutagenesis. Introduction of a disulphide bond into subtilisin from *B. subtilis* resulted in an increase of 4.5° C in the T_m of the mutant enzyme without causing any change in its catalytic efficiency (Takagi *et al.*, 1990). Enhanced stability of subtilisin was observed as a result of mutations of Asn^{109} and Asn^{218} to Ser. The analog containing both the mutations showed on additive effect on the thermal stability. The optimal temperature of the alkaline protease, from *A. oryzae*, was enhanced from 51 to 56°C by the introduction of a disulphide (Cys^{169}-Cys^{200}) bond. Another strategy for improving the stability of the protease was by replacing the polar amino acid groups by hydrophobic groups. Replacement of Lys by Ser or Asp resulted in an increase in the thermostability of the neutral protease from *B. subtilis* in the range of 0.3 to 1.2° C.

ii) Prevention of Autoproteolytic Inactivation

Subtilisin, an extensively studied protease, is widely used in detergent formulations due to its stability at alkaline pH. However, its autocatalytic digestion presents a major problem for its use in industry. It was deduced that there is a correlation between the autolytic and conformational stabilities. Computer modelling followed by introduction of a Cys^{24} or Cys^{87} mutation resulted in destabilization of subtilisin from *B. amyloliquefaciens*. Introduction of a disulphide bond increased the stability of the mutant to a level less than or equal to that of the wild type enzyme. It appears logical that mutations in the amino acids involved at the site of autoproteolysis may prevent the protease inactivation caused by self digestion.

iii) Alteration of pH Optimum

Different applications of proteases require specific optimal pHs for the best performance of the enzyme e.g., the use of the proteases in the leather industry requires an enzyme with an alkaline pH optimum, whereas the use in the cheese making industry requires an acidic protease. Protein engineering enables us to tailor the pH dependance of the enzyme catalysis to optimize the industrial processes. A change of Asp^{99} to Ser in subtilisin from *B. amyloliquefaciens* has demonstrated the potential of altering the optimal pH of the enzyme by systematic multiple mutations on the surface of protein (Thomas, *et al.*, 1985).

iv) Changing of Substrate Specificity

The properties needed for industrial applications of proteases differ from their physiological properties. The natural substrates of the enzyme are usually different from those desired for their industrial applications.

Despite extensive research on proteases, relatively little in known about the factors that control their specificities towards non physiological substrates. Strategies involving site directed mutagenesis are being explored to tailor these specificities at will.

v) Improvement of Yield

The cost of enzyme production is a major obstacle in the successful application of proteases in industry. Protease yields have been improved by screening for hyperproducing strains and/or by optimization of the fermentation medium. Strain improvement by either conventional mutagenesis or recombinant DNA technology have been useful in improving the production of proteases.

2. Amylase

Amylase is the name given to glycoside hydrolase enzymes that break down starch into maltose molecules. Although the amylases are designated by different Greek letters, they all act on α-1, 4-glycosidic bonds. Under the original name of diastase, amylase was the first enzyme to be found and isolated. Although the amylases can be derived from several sources such as plants, animals and microbes; the enzymes from microbial sources generally meet the industrial demands. With the advent of new frontiers in biotechnology, the spectrum of amylase application has widened in many other fields, such as clinical, medicinal, analytical, starch analytical chemistries as well as in starch saccharification, textiles, food, brewing and distilling industries.

The major classes of amylase that have identified are as follows :

1. α-amylase : EC 3.2.1.1. Also called as 1, 4-α-d-glucan glucanohydrolase; glycogenase.
2. β-amylase : EC 3.2.1.2. Also called as 1, 4-α-d-glucan maltohydrolase; glycogenase; saccharogen amylase.
3. γ-amylase : EC 3.2.1.3. Also called as glucan, 1, 4-α-glucosidase; glucoamylase; amyloglucosidase; Exo-1,4-α-glucodase; lysosomal α-glucosidase; 1, 4-α-d glucose glucohydrolase.

Production of Amylases

Screening of microorganisms to isolate amylase producing strains is a challenging task as a single strain can sometimes produce more than one enzyme, i.e. α-amylase as well as glucoamylase. The environmental factors employed during production of amylases affect the growth of microorganisms. These factors include nutrient supplementation, pH of medium, osmotic relationship, degree of aeration and temperature. Maintaining the purity of the medium also becomes important, especially if the fermentation is carried out under aerobic conditions.

There are generally two methods that are used for amylase production : solid substrate fermentation and submerged fermentation. Of the two methods, the solid substrate fermentation is the most widely used. It involves the growth of microorganisms on solid, normally organic materials in the absence or near absence of free water. The substrates used are often cereal grains, bran, legumes and lignocellulosic material (straw, wood chippings) etc. The solid substrate fermentations lack the sophisticated control mechanisms that are usually associated with the submerged fermentations. Also there can be problems with heat build up as the substrate moisture level can be difficult to control.

Since thermostability is a feature of most of the enzymes sold in bulk for industrial applications, thermophillic microorganisms are of special interest for the production of amylases. Facultative thermophiles with their growth optima at 45°C are also capable of growing well at higher temperature, thus covering both the mesophillic and thermophillic ranges. However, not much is known about the process for enzyme production using such microorganisms. Comparative analysis of the structure and physico-chemical properties of these two groups may provide some information about the molecular basis of the high temperature tolerance ability of thermostable amylases (Dawlathabad *et al.*, 2007).

Production of α-Amylase :

α-amylase may be derived from several bacteria, yeasts and fungi. Strains of *Aspergillus* sp. (Crueger and Crueger, 1989) and *Bacillus* (Kikuchi *et al.*, 1993; Outtrup, 1989; Mamo and Gessesse, 1997) mainly *Bacillus amyloliquefaciens* and *B. licheniformis* are employed for commercial applications. Generally bacteria are preferred over fungi for amylase production. Carbon sources as lactose, dextran and soluble starch are suitable for production of α-amylase but the highest enzyme yield was obtained when glucose was used (Salva and Moraes, 1995). High aeration rates were found to be essential for good yields of the enzyme (Babu and Satyanaryan, 1995). Control of dissolved oxygen tension, however, was not found to be advantageous. Since filamentous fungi are generally considered to be the most prolific producers of extracellular enzymes, attempts have been made to study α-amylase production by using them. The thermophillic fungus Thermomyceslanuginosa was reported to be an excellent producer of α-amylase (Insen and Olsen, 1992; Arnesen *et al.*, 1998). Increased production of the enzyme could be obtained by manipulating the growth conditions and medium composition.

The cell immobilization technique has also been employed for α-amylase production. Lvanova *et al.*, compared various immobilization techniques including entrapment gels such as calcium alginate, α-carrageenan, agar and their combinations with polyethylene oxide and fixation on formaldehyde activated acrylonitrile/acrylamide membranes for α-amylase production (Lvanova *et al.*, 1995).

Production of β-Amylase :

β-amylase is usually of plant origin. Some microorganisms such as *Bacillus polymyxa*, *B. cereus*, *B. megaterium*, *Streptomyces* sp., *Pseudomonas* sp. and *Rhizopus japonicus* have been studied for their ability to produce this enzyme. Ray *et al.*, compared the production of β-amylase from starch waste by a hyper amylolytic strain of *B. megaterium* B_6 mutant UN12 in submerged and solid substrate fermentation. The starchy wastes used as substrates were from arrowroot, arum, pulses, rice husk, rice, tamarind, maize, cassava, water chestnut and wheat bran. The highest yields were obtained when wheat bran and arum were used as substrate (Ray *et al.*, 1997).

Production of γ-Amylase

Extensive work has been carried out on the production of glucoamylase in solid cultures using *A. niger*. A large number of agro-industrial residues such as wheat bran, rice bran, rice husk, gram flour, corn flour, coconut waste etc. have been screened as substrates for the production of gluco-amylase (Selva Kumar *et al.*, 1998). Design and configuration of fermentor, its mode of operation and fermentation influence the enzyme production. These included flasks, trays, rotary reactors, horizontal and vertical columns. Enzyme production in trays occurred in maximum quantities in 36 hrs compared with 96 hrs typically required in flasks.

Table 5 : Properties of Amylases

Class	Molecular weight (KDa)	pH optimum	Temperature optimum (°C)	Cofactor requirement
α-Amylase	50-60	5.5-8.0	45	Ca^{+2}
β-Amylase	35-80	6.8	55	--
γ-Amylase	27-112	4.5-5.0	40-60	--

Current Problems and Potential Solutions

Genetic engineering has been used extensively for cloning of amylase producing strains, mainly α-amylase and glucoamylase, in order to achieve desirable characteristics in the cloned host. Gene cloning can be used for the higher enzyme productivity, expression of thermostable enzymes and co-expression of the two enzymes by the same organism. A great deal of work has been done on the cloning of α-amylase genes in different microbes, mostly in *Escherichia coli* and *Saccharomyces cerevisiae*. Apparently, not much attention has been paid to the cloning of genes for β-amylase of microbial origin, although a few reports are available on its expression in *E. coli* mostly from *Bacillus* species.

3. Lipases

With the realization of the biocatalytic potential of microbial lipases in both aqueous and non aqueous media in the last one and a half decades, industrial fronts have shifted towards utilizing this enzyme for a variety of reactions of immense importance. Lipases have gained importance to a certain extent over proteases and amylases, specially in the area of organic synthesis. Lipases have found a number of applications, for example in the resolution of chiral drugs, biofuels, fat modification, as flavour enhancers, synthesis of cocoa butter substitutents and for synthesis of personal care products owing to their enantioselective and regioselective nature (Priest, 1992; Gerhartz, 1990). Lipases (triacylglycerol acylhydrolases) do not require any co-factor as they belong to the class of serine hydrolases. Triacylglycerols having very low solubility in water are the natural substrates of lipases. Lipases catalyse the hydrolysis of the ester bonds at the interface of an insoluble substrate phase and the aqueous phase in which the enzyme is dissolved. However, in the absence of water, they catalyse the reverse reaction which leads to the formation of glycerides from fatty acids and glycerol by esterification. The usual industrial lipases are special classes of esterase enzymes that act on fats and oils, and hydrolyse them initially into substituted glycerides and fatty acids, and finally into glycerol and fatty acids upon total hydrolysis (Ghosh *et al.*, 1996).

In nature, the lipases available from various sources have considerable variation in their reaction specificities (enzyme specificity). It has been observed that from the fatty acid side, some lipases have affinity for short chain fatty acids (acetic, butyric, capric, caproic, caprylic etc.), some have preference for unsaturated fatty acids (oleic, linoleic, linolenic etc.), while many others are non specific and randomly split the fatty acids from the triglycerides. From the glycerol side of the triglycerides, the lipases often show positional specificity and attack the fatty acids at 1 or 3 carbon position of glycerol or at both the positions but not the fatty acid at the 2 position of the glycerol molecule. However, through random acyl migration, the 2-fatty acid monoglyceride undergoes rearranagement pushing the fatty acid to the 1 or 3 position of the glycerol molecule. Since acyl migration is a slow process, the rate of hydrolysis also decreases as the lipase does not act on 2-mono glycerides and awaits the completion of the acyl migration which allows the lipase to attack the glyceride at the 1 and/or the 3 position.

It is also interesting to note that the lipases function at the oil-water interface and the activity of the enzyme is determined by the amount of the oil available. This interface area can be increased substantially to its saturation limit by the use of emulsifiers or by agitation. As lipases act on ester bonds, they can have diverse application as in the splitting of fat, transesterification, development of flavours in cheese, improving the pallatability of beef fat, synthesis of value added esters from organic acids and alcohols in water deficient organic solvents, in detergents etc.

Lipases can be obtained from plants, animals as well as microorganisms. Among the microorganisms, attention has been focused on the bacterial and fungal lipases.

Bacterial Lipases : These are glycoproteins by nature but some extracellular bacterial lipases are lipoproteins. It has been reported that lipase production in most bacteria is affected by certain polysaccharides. Most of the bacterial lipases reported so far are constitutive and are non specific in their substrate specificity. A few bacterial lipases are thermostable also.

Bacteria such as *Achromobacter* sp., *Alcaligenes* sp., *Arthrobacter* sp., *Pseudomonas* sp., *Staphylococcus* sp. and *Chromobacterium* sp., have been exploited for the production of lipases (Godfredson, 1990). Staphylococcal lipases are lipoproteins in nature. Lipases purified from *S. aureus* and *S. hyicus* show molecular weights ranging between 34-46 kDa. They are stimulated by divalent calcium ions and inhibited by EDTA. The optimum pH varies between 7.5 and 9.0. The lipase genes from *S. hyicus* and *S. aureus* have been cloned, sequenced and compared with other lipases. This revealed two conserved domains separated by 100 amino acids which are likely to form the active sites (Brune and Gotz, 1992). Putative active site residues around His 269 and Ser 369 of the *S. hyicus* lipase are highly conserved the two *S. aureus* lipases and in several eukaryotic lipases. Lipases from different species of *Pseudomonas* have also been purified. The purified lipases of *P. fragi*, *P. flourescens* and *P. aeurginosa* were found to be monomeric with molecular weight of 33 KDa, 45 KDa and 29 KDa respectively. The enzyme was inhibited by Zn^{++}, Fe^{++} and Al^{+++} and activated by Ca^{++}. The lipase gene of *P. fragi* has been cloned and sequenced.

Fungal Lipases : Fungal lipases are being exploited due to their low cost of extraction, thermal and pH stability, substrate specificity and activity in organic solvents. Among Mucorales, the lipolytic enzymes of the molds *Mucor hiemalis, M. miehei, M. lipolyticus, M. pusillus, Rhizopus japonicus, R. arrhizus, R. delemar, R. nigricans, R. nodosus, R. microsporus* and *R. chinesis* have been studied in great detail (Lazar and Schroder, 1992). *M. pusillus* produces a thermostable extracellular lipase.

Due to the 1,3-(regio) specificity of *Rhizopus*, lipases that are especially suited for the conversion of triglycerides to their corresponding monoglycerides and interesterification reactions of fats and oils, have food and pharmaceutical application. The lipases of various *Rhizopus* species show maximum activity towards medium chain fatty acids (C_8-C_{10}). *R. japonicus* lipase has been used to produce hard butter suitable for chocolate manufacture by interesterification of palm oil with methyl stearate (Matsuo *et al.*, 1981).

Lipase producers within the order Entomophthorales include *Entomephthora apiculata, E. coronata, E. thaxteriana, E. virulenta, Basidiobolus* sp. and *Conidiobolus* sp. The genera *Pichia, Hansenula* and *Saccharomyces* are also reported to produce lipase. Lipases are reported from *Candida curvata, C. tropicalis* and *C. pellioculosa* are non-specific towards the different ester bonds in triglycerides with the exception of *C. deformans* (Lazar and Schroder, 1992). The imperfect fungus *Geotrichum candidum* is responsible for acid formation in dairy products by lipolyzing fat.

The intracellular and extracellular lipases of *Aspergillus niger* are regiospecific. *A. oryzae* was reported to be an efficient host for the heterologous expression of the lipase from *R. miehei* and *Humicola lanuginosa* (Lazar and Schroder, 1992). The lipase of *Penicillium roqueforti* is used for flavour production of blue cheese. Lipolytic activity has also been detected in *P. camemberti*, the white surface mould of Brie and Camembert cheese. The lipase of *H. lanuginosa* DSM 3819 is suitable as a detergent additive because of its thermostability, high activity at alkaline pH and stability towards anionic surfactants (Huge and Gormsen, 1987).

Lipase Production

Various environmental factors such as temperature, pH, nitrogen, carbon, lipid sources, agitation and dissolved oxygen influence the production of lipases. Lipase production is usually coordinated with and dependent on the availability of triglycerides. Free fatty acids, hydrolysable esters, bile salts and glycerol also stimulate lipase production. The exact growth requirements however, vary from organism to organism. High production of lipase in case of *P. fragi* occurs in peptone supplemented medium, although different peptones vary in their effectiveness. Though *Pseudomonas* species grow in a basal medium with ammonium sulphate, glucose, citrate or pyruvate, it requires an organic nitrogen source for lipase production. A mixture of arginine, lysine and glutamic acid in medium was observed to be effective for lipase production. Peptone stimulates lipase production by *A. wentii, M. hiemalis, R. nigricans, M. racemosus* and *Pseudomonas* species. Soyabean meal extract if added in the medium stimulates lipase production by *Rhizopus oligosporus*. While glucose is essential for production of lipases by some microorganisms, other carbon sources such as lactose, mannose, xylose, fructose, rhamnose etc. can also be used. Polysaccharides such as glycogen, hyaluronate, laminarin, gum arabic and pectin have been reported to stimulate lipase production in *Serratia marcescens* and *Saccharomycopsis lipolytica*.

The initial pH of the growth medium also affects lipase production. Lipase production at alkaline pH (7-9) was observed in *P. fragi* and *P. aeuroginosa*. Maximum growth at acidic pH (4.0-6.5) was observed in *S. lipolytica, M. caseolyticus, R. nigricans* and *A. wentii*.

The temperature for the growth of microbes and lipase production also exhibits a wide range, for example: 8° C temperature is suitable for the growth of *Pseudomonas* sp. and 45° C for *T. emrsonii*. Lipases of thermotolerant *H. lanuginosa* and *P. nitroreducens* are stable at 60ºC and 70ºC respectively.

Aeration also has variable effect on lipase production by different organisms. The degree of aeration appears to be critical in some cases since shallow layer cultures (moderate aeration) produced much more lipase than shake cultures (high aeration).

Lipases have been purified from animal, plant, fungal and bacterial, sources by different methods involving ammonium sulphate precipitation, gel filteration and ion exchange chromatography. In recent years, affinity chromatographic techniques have come into use as these techniques decrease the number of steps necessary for lipase purification as well as increase specificity. Currently, reversed micellar and two phase systems, membrane processes and immunopurification are being used for purification of lipases.

Current Problems and Potential Solutions

Lipases are amongst the most important bicatalysts that carry out novel reactions in both aqueous and non-aqueous media. Lipases have the remarkable ability to carryout a wide variety of chemo, regio and enatioselective transformation. Thus lipases are the tools of choice for organic chemists. Their general ease of handling, broad substrate tolerance, high stability towards temperature and solvents, high enantioselectivity and convenient commercial availability have all added to their wide spread popularity among organic chemists. Today, lipases find immense applications in various areas of industrial microbiology and biotechnology. Lipases show immense versatility regarding their catalytic behaviour. Therefore, there is a lot of scope to search for newer lipases with desired selectivity and substrate tolerance.

To widen the usage level of lipases, there is an urgent need to understand the mechanisms behind the lipase catalysed reactions. The unique interfacial activation of lipases has always fascinated enzymologists and recently biophysicists and crystallogaphers have made progress in understanding the structure-function relationships of these enzymes. However, complete understanding of the lipase molecule requires greater input of research effort.

4. Phytase

Since its discovery in 1907, a complex of technological developments has created a potential $ 500 million market for phytase as an animal feed additive (Mullaney *et al.*, 2000). Of all the sources of phytase that have been studied (plant, animal and microorganisms), the highest yields are produced by a wild type strain *A. niger* NRRL 3135 (12.7 mg P/hr/ml = 6.8 microns P/ml/min = 113.9 nKat/ml) in a mineral salt medium in which total phosphate (4 mg%) is limiting for growth and corn starch and glucose are the carbon sources. Synthesis of the enzyme is repressed by phosphate in the wild type strains (Wodzinski and Ullah, 1996). During the last 30 years, research has led to an increased use of soya bean meal and other plant material as protein sources in animal field. One problem that had to be overcome was the presence of antinutritional factors, including phytate, in plant meal. Phytate phosphorus is not digested by monogastric animals (hogs and poultry), and in order to supply enough of this nutrient, additional phosphate was required in the feed ration. Rock phosphate soon proved to be a cost effective means of supplying this additional phosphorus, and the excess phytin phosphorus could be disposed off easily with the animals' manure. However, this additional phosphorus creates a massive environmental problem when the land's ability to bind it is exceeded. Over the last decade, numerous feed studies have established the efficacy of a fungal phytase, *A. niger* NRRL 3135, to hydrolyze phytin phosphorus in an animal's digestive tract, which benefits the animal while reducing the total phosphorus levels in manure.

Aspergillus niger NRRL 3135 produces two phytases : One with pH optima at 2.5 and 5.5 (*phy A*) and the other with an optimum at pH 2.0 (*phy B*). It also produces a pH 6.2 optimum phosphatase that has no phytase activity. These three glycoproteins have been purified to homogeneity, characterized, sequenced and cloned. The sequences have been compared to each other, other phytases and to known phosphatases. Their homology has been determined. The active sites of phytases show remarkable homology to the active site residues of the members of a particular class of acid phosphatases (histidine phosphatases). The most conserved sequence is RHGXRXP. Phytase has been covalently immobilized on Fractogel TSK HW-75F and glutaraldehyde-activated silicate. It has been immobilized on agarose. Losses of activity have been noted on immobilization but these may be minimized by future research.

Phy A from *A. niger* NRRL 3135 has been cloned into an *A. niger* glucoamylase producing strain CBS 513.88 using a construct that has a glucoamylae promoter and an *A. niger* NRRL 3135 leader sequence, and that is devoid of phosphate repression. The yield of the secreted enzyme was increased 52-fold above that of wild type *A. niger* NRRL 3135. The bioengineered organism produces 270 microns P/ml/min which is approximately 7.9 g/liter in the medium. Commercial preparations of the cloned enzyme are available. Phytase A (*phy A*) has been cloned into tobacco and canola. The enzyme is localized in the seed and expressed at high levels. Feeding of the seed to animals has made the phytin–P in the commercial diets available to the animals. The efficacy of feeding phytase to monogastric animals (poultry and swine) has been established. The amount of enzyme that is necessary to be added to commercial diets has been titred for broilers, layers, turkeys, ducks and swine. The units of enzyme required are related to the amount of phytin–P in the diet. The use of the enzyme as a feed additive has been cleared in 22 countries. If phytase were used in the diets of all the monogastric animals reared in the US, it would release phosphorus that has a value of $ 1.68 x 10^8 per year. The FDA has approved the enzyme preparation as GRAS.

The effect of feeding phytase to animals on pollution has been quantitatively determined. If phytases were used in the diets of all of the monogastric animals reared in the United States, it would preclude 8.23 10^7 kg P from entering the environment.

Current Problems and Potential Solutions

Perhaps the pivotal events in the last century that created the need for phytase was the development of modern technology and a tremendous increase in food grain production. These large quantities of cereals and meal in turn led to the transition of one segment of agriculture into 'animal agriculture'. The huge volumes of manure spawned by these production units in time exceeded both the capacity of their crops and crop lands to utilize or bind the increased amount of phosphorus. Nutrient run off from such land has now been linked to a number of blooms of toxin producing microbes. Fish kills associated with these blooms have attracted the public and governmental concern, as well as greater interest in phytase as a means to reduce this phosphorus pollution.

Phytase research efforts now are focused on the engineering of an improved enzyme. Improved heat tolerance to allow the enzyme to survive the brief period of elevated temperature during the pelletization process is seen as an essential step to lower its cost in animal feed. Information from the X-ray crystal structure of phytase is also relevant to improving the pH optimum, substrate specificity and enzyme stability. Several studies on new strategies that involve synergistic interactions between phytase and other hydrolytic enzymes have shown positive results. Several studies have already investigated the use of various yeast expression systems as an alternative to the current production method for phytase using overexpression in filamentous fungi. Expression in plants is underway as a means to commercially produce phytase, as in biofarming in which plants such as alfaalfa are used as 'bioreactors' and also by developing plant cultivars that would produce enough transgenic phytase so that additional supplementation of their grains or meals is not necessary. Ultimately, transgenic hogs and poultry may produce their own digestive phytase.

As attempts are made to widen the use of phytase, it is also important, that extended exposure and breathing its dust be avoided as prudent safety measures to avoid possible allergic responses.

5. Cellulase

Successful utilization of cellulosic materials as renewable carbon sources is dependent on the development of economically feasible process technologies for cellulose production and for the enzymatic hydrolysis of cellulosic materials to low molecular weight products such as hexoses or pentoses.

Cellulose is a linear polysaccharide of glucose residues connected by β-1,4 linkages. Like chitin, it is not cross linked. Native crystalline cellulose is insoluble and occurs as fibers of densely packed, hydrogen bonded, anhydroglucose chains of 15 to 10000 glucose units. Its density and complexity make it very resistant to hydrolysis without preliminary chemical or mechanical degradation or swelling. In nature, cellulose is usually associated with other polysaccharides such as xylan or lignin. It is the skeletal basis of plant cell walls. According to some researchers cellulose is the most abundant organic source of food, fuels and chemicals. However, its usefulness is dependent on its hydrolysis to glucose. Acid and high temperature degradation are unsatisfactory in that the resulting sugars are decomposed; also waste cellulose contains impurities that generate unwanted byproducts under these harsh conditions.

A cellulasic enzyme system consists of three major components :

1. Endo-β-glucanase (EC 3.2.1.4); 1,4-β-D-glucan glucano-hydrolase; CMCase; Cx : "random" scission of cellulose chains.
2. Exo-β-glucanase (EC 3.2.1.91); 1,4-β-D glucan cellobiohydrolase; Avicelase; C1: exo-attack on the non reducing end of celulase with cellobiose as the primary structure.
3. β-glucosidase (EC 3.2.1.21); cellobiase : hydrolysis of cellobiose to glucose.

Although cellulases are distributed throughout the biosphere, they are most common in fungi and microbial sources. The enzymatic mechanism whereby certain micro-organisms can quite rapidly and completely degrade cellulose is not yet understood. It is believed that two steps are involved in the breakdown of cellulose : first, a prehydrolytic step wherein anhydroglucose chains are swollen or hydrated and secondly, hydrolytic cleavage of the now susceptible polymers either randomly or end wise. The first step would involve an enzyme designated as C_1 and the second, hydrolytic enzymes termed as Cc. The C_1 component attacks the highly ordered (crystalline) cellulose i.e. cotton fibers or avicel, but has little effect on soluble derivatives such as carboxymethyl cellulose (CMC). The C_1 decrystallizes or hydrates cellulose chains whereas Cc consists of exo and endo β-1, 4 glucanases that attack soluble derivatives or cellulose that has been acid or alkali swollen. The cellulose complex of *Trichoderma reesei* (formerly named *T. viride*) has been most thoroughly studied.

Cellulase Producing Microorganisms

Although a large number of microorganisms are capable of degrading cellulose, only a few of these microorganisms produce significant quantities of cell free enzymes capable of completely hydrolysing crystalline cellulose *in vitro*. Fungi are the main cellulose producing microorganisms, though a few bacteria and actinomycetes have also been reported to possess cellulase activity.

Table 6 : Microorganisms Capable of Cellulase Production

Fungi	
Aspergillus acculeatus	*Sporotrichum cellulophilum*
Aspergillus fumigatus	*Talaromyces emersonii*
Aspergillus niger	*Trichoderma koningii*
Fusarium solani	*Trichoderma reesii*
Irpex lacteus	*Ruminococcus albus*
Penicillium fummiculoseum	*Schizophyllum commune*
Sclerotium rolfsii	*Thielavia terrestris*
Bacteria	
Clostridium cellulovorans	*Streptomyces aureofaciens*
Clostridium cellulosi	*Streptomyces flavogriseus*
Clostridium cellulolyticum	*Streptomyces lividans*
Clostridium thermocellum	*Thermobifida alba*
Bacteroides cellulosolvens	*Microbispora bispora*
Acetivibrio cellulosolvens	*Fibrobacter succinogenes*
Acidothermus cellulolyticus	*Cytophaga* sp.
Cellulomonas cellulans	*Flavobacterium johnsoniae*
Cellulomonas fimi	*Archromobacter piechaudii*
Curtobacterium falcumfaciens	*Xanthomonas* sp.
Micromonospora melonosporea	*Cellvibrio gilvus*
Streptomyces reticuli	*Cellvibrio mixtus*
Streptomyces alboguseolus	*Pseudomonas mendocina*

Current Problems and Potential Solution

The commercial use of cellulases is dependent on the following :

1. High titer and good enzymatic activity,
2. Low production cost, and
3. Feasible mass production.

Therefore, strain improvement for cellulose production should lay emphasis on the following features :

1. Development of a process for high titer cellulose production,
2. Cellulase production at low cost
3. Potential for mass production of cellulase

6. Glucose Isomerase

D-glucose/xylose isomerased (D-xylose ketol isomerase, EC 5.3.1.5), commonly referred to as glucose isomerase (GI) is one of the three highest tonnage value enzymes, amylase and protease being the other two. It catalyzes the reversible isomerization of D-glucose and D-xylose to D-fructose and D-xylulose respectively. Interconversion of xylose to xylulose serves a nutritional requirement in saprophytic bacteria that thrive on decaying plant material and also helps in the bioconversion of hemicellulose to ethanol. Isomerization of glucose to fructose is of commercial importance in the production of high fructose corn syrup (HFCS). Sucrose derived from sugar beet (40%) and sugar cane (60%) was the main sweetener in the world until 1976. The production of HFCS was developed first in Japan and later in the United States. GI gained commercial importance in the United States because of the lack of supply of sucrose after the Cuban revolution of 1958, and it continues to be one of the most important industrial enzymes to this day (Snehlata *et al.*, 1996).

The origin of today's successful development of fructose syrup products lies in the discovery of glucose isomerizing enzymes. Historically, four different types of enzymes have been termed glucose isomerases. The discovery by Marshall and Kooi in 1957 of the enzyme from *Pseudomonas hydrophila* was the starting point of the exploitation of this enzyme for the manufacture of HFCS as a substitute for cane sugar (Marshall and Kooi, 1957). Production of the enzyme required xylose in the growth medium and was enhanced in the presence of arsenate. Later, a xylose isomerase activity, which was independent of xylose was found in *Escherichia intermedia* (Natake and Yoshimura, 1964). This enzyme was a phosphoglucose isomerase (EC 5.3.1.9) which could isomerise the unphosphorylated sugar only in the presence of arsenate. Takasaki and Tanabe isolated from *Bacillus megaterium* AI a glucose isomerase (EC 5.3.1.18) which was NAD linked and was specific for glucose (Takasaki and Tanabe, 1962). A similar glucose isomerase activity, which catalyzes the isomerization of both glucose and mannose to fructose, was isolated from *Paracolobacterium aerogenoides* (Takasaki and Tanabe, 1964). Of these glucose isomerizing activities, xylose isomerase (EC 5.3.1.5) is the most suitable for commercial applications as it is heat stable and does not require expensive cofactors as NAD^+ and ATP for its activity.

Enzymatic glucose isomerization was first accomplished on an industrial scale in 1967 by Clinton Corn Processing Co. in the United States. Immobilized GI was commercially available by 1974. The demand for HFCS in the food industry increased, and by 1980 practically all major starch processing companies were resorting to the GI technology. The most common raw material used for the production of HFCS in the United States is the cornstarch manufactured by the wet milling process. The production of HFCS from starch comprises three major processes :

1. Liquefaction of starch by α-amylase,
2. Saccharification of starch by the combined action of amyloglucosidase and debranching enzyme, and
3. Isomerization of glucose by GI.

The final product is a corn syrup containing a mixture of glucose and fructose and hence with a greater sweetening capacity than that of sucrose. Other sources of starch such as wheat, tapioca, rice etc. can also be used. The byproducts of the corn milling industry are important in deciding the economics of HFCS production. At present, HFCS has almost completely replaced sucrose in the United States and a moderate growth rate of 4% is expected in its production on a global basis. The price of HFCS is also 10 to 20% lower than that of sucrose. HFCS is preferred by the food industry since it does not pose the problem of crystallization as is the case with sucrose. Moreover, D-fructose plays an important role as a diabetic sweetener because it is only slowly reabsorbed by the stomach and does not influence the glucose level in the blood. A large number of noncalorlfic and noncarbohydrate artificial sweeteners such as saccharine, cyclamate, aspartame and thaumatin have been discovered and dismissed on the basis of health concerns or other reasons.

Microbial Production of Glucose Isomerase

GI is widely distributed in prakoryotes. After its discovery in *Pseudomonas hydrophila*, a large number of bacteria and actinomycetes were found to produce GI that is active in the absence of arsenate.

Table 7 : Glucose Isomerase Producing Organisms

Actinomyces olivocinereus, A. phaeochromogens
Actinoplanes missouriensis
Aerobacter aerogens, A. cloacae
Arthrobacter sp.
Bacillus stearothermophilus
Corynebacterium sp.
Escherichia freundii, E. coli, E. intermedia
Lactobacillus brevis,L. fermenti, L. buchneri, L. pentosus
Leuconostoc mesenteroides
Microbiospora rosea
Mycobacterium sp.
Nocardia asteroides, N. corallia
Pseudomonas hydrophila
Sarcina sp.
Staphylococcus bibila, S.echinatus
Streptococcus achromogenes, S. venezaelie, S. olivaceus
Streptomyces olivochromogenes, S. rubiginosus, S. albus, S. griseus
Thermopolyspora sp.
Thermus sp.
Xanthomonas sp.
Zymononas mobilis

Table 8 : Commercial Producers of Glucose Isomerase

Organism	Trade Name	Manufacturer
Actinoplanes missousriensis	Maxazyme	Gist Brocades and Anheuseur-Busch Inc.
Bacillus coagulans	Sweetzyme	Novo-Nordisk
Streptomyces rubiginosus	Optisweet	Miles Kali-Chemie
	Spezyme	Finnsugar
Streptomyces phaeochromogenes	Swetase	Nagase
Arthrobacter sp.		Reynolds Tobacco
Streptomyces olivaceus		Miles Laboratories Inc.

Production of Glucose Isomerase

The cost of production of the enzyme is an important factor in evaluation of its suitability for industrial application. Intensive efforts have been made to optimize the fermentation parameters for the production of GI with a view to developing economically feasible technology. Research is focused on three majors aspects : (i) improvement of the yields of GI, (ii) optimization of the fermentation medium with special reference to the replacement of xylose by a cheaper substitute and elimination of requirement for Co^{2+} ions, and (iii) immobilization of the enzyme.

GI is generally produced by submerged aerated fermentation. Optimization of the fermentation medium has been extensively studied with a view to development of an economically viable fermentation technology for the production of GI. Efforts have been directed mainly towards :

1. Replacement of xylose by another inexpensive inducer : Most GI producing organisms have an obligate requirement for D-xylose to induce production of the enzyme. However, xylose is expensive and impractical for use on a commercial scale. Starch, glucose, sorbitol or glycerol can be used in place of 75% of xylose (Drazii *et al.*, 1980).
2. Evaluation of the effect of cheaper nitrogen sources on the yield of enzyme : The nitrogen source is a critical factor which needs to be optimized for each source of enzyme. Although complex nitrogen sources are usually used for GI production, the requirement for a specific nitrogen supplement differs from organism to organism. Peptone, yeast extract and ammonium salts can be used in the medium. Corn steep liquor was found to be a cheap and suitable source of nitrogen by some workers but its use is limited by its seasonal and interbatch variability. Addition of certain amino acids also imporoves the yield of the enzyme.
3. Optimization of pH and temperature for maximum enzyme production : The nature of the nitrogen source affects the pH and consequently the yield of the enzyme. Most GI producing fermentations are carried out between pH 7.0 and 8.0. The period of fermentation varies from 6 to 48 hrs depending on the types of culture used for GI production.
4. Substitution of Co^{2+} by other divalent metal ions in the fermentation medium: Divalents cations are required in the fermentation medium for optimum production of GI. However, the requirement for specific metal ions depends on the source of the enzyme. Co^{2+} was essential for GI production by *Streptomyces* strain YT-5 (Takasaki and Tanabe, 1966), whereas *Bacillus coagulans* required Mn^{2+} or Mg^{2+} for production of the enzyme (Yoshimura *et al.*, 1966). Generally cobalt salts are used in the medium of mesophillic *Streptomyces* species but not

thermophillic species. It is important to reduce the addition of Co^{2+} to the medium because of the health hazards related to the human consumption of HFCS containing Co^{2+} and the environmental pollution problem related to the disposal of the spent medium.

Genetic Engineering of Glucose Isomerase

Recombinant DNA technology provides an excellent means for isolating and manipulating the gene of a desired protein. One of the ways to increase the production of GI is to identify the GI gene and clone it on a multicopy vector containing a strong promoter such as *lac*, *tac* or P_L. The GI gene has been cloned from several microorganisms with the primary aims of (i) over production of the enzyme by gene dosage effect, (ii) direct conversion of xylose to ethanol by yeasts, and (iii) engineering of the protein to alter its properties to suit its biotechnological applications. Molecular cloning and expression of GI have been carried out in both homologous and heterologous hosts and in yeasts.

Engineering proteins by manipulation of their genes is at present a viable approach which complements the structure—function studies performed by pre-existing methods and allows production of tailor-made proteins with desirable properties to give a complete insight into the mechanism of the enzyme. These studies lead to a hypothesis which can be verified by protein engineering. Site directed mutagenesis of GI has been carried out with several objectives, such as (i) increasing the thermal stability, (ii) lowering of the pH optimum, (iii) changing of the substrate preference, (iv) deducing the functional role of the important amino acid residues, and (v) studying the subunit interactions. These studies have contributed substantially to our knowledge about the molecular mechanisms of GI and have created new possibilities of producing an enzyme with properties that are better suited for biotechnological applications.

Current Problems and Potential Solutions

The ideal GI should possess a lower pH optimum, a higher temperature optimum, a resistance to inhibition by Ca^{2+}, and a higher affinity for glucose than do presently used enzymes. Introduction of all these properties into a single protein is a difficult task, which has been an obstacle in the development of an economically feasible commercial process for enzymatic isomerization of glucose to fructose. Advances in recombinant DNA technology and protein engineering have opened new and encouraging possibilities of combining the desirable properties in a single organism to produce a tailor made protein. Reduction of enzyme cost by amplification of GI gene may cause an increase in fermentation productivity. Isolation of a mutant for constitutive production of GI and elimination of the requirement of metal ions will contribute significantly to the improvement of the existing processes for HFCS production. Combination of saccharification of starch with isomerization will result in shortening of reaction time and lead to a major saving in terms of equipment cost. However, the major drawback in the development of the uni-pH process is that the wide difference in optimum reaction conditions for the two enzymes tends to lower the efficiency of a simultaneous system. Efforts to produce thermostable and acid stable GI with higher affinity for glucose by site directed mutagenesis of the GI gene are already under way, with a view to evolving a GI preparation suitable for biotechnological applications.

7. Pectinase

Pectinase is a general term for enzymes that breakdown pectin, a polysaccharide substrate that is found in the cell walls of plants. One of the most studied and widely used commercial pectinases is polygalacturonase. It is useful because pectin is the jelly-like matrix which helps cement plant cells together and in which other cell wall components, such as cellulose fibrils are embedded. Therefore, pectinase enzymes are commonly used in processes involving the degradation of plant materials, such as speeding up the extraction of fruit juice from fruits. Pectinases have also been used in wine production since the 1960s.

Like all enzymes, pectinases have an optimum temperature and pH at which they are most active. For example, a commercial pectinase might typically be activated at 45 to 55° C and work well at a pH of 4.8 to 5.0. If it is hotter, it will make the reaction go more quickly. However, the higher temperature will gradually denature the enzyme, eventually stopping it from working. Pectinases also break down pectin best at a pH of 4.8 to 5.0. Too high or too low a pH would also result in enzyme denaturation. The pectinase enzymes act in different ways on the pectin which are found in the primary cell walls and in the middle lamella. Pectins are large polysaccharide molecules, mainly made up of chains of several hundred galacturonic acid residues. Enzymes in this pectinase group include polygalacturonases, pectin methyl esterase and pectin lyases.

Pectinases are produced during the natural ripening process of some fruits, where together with cellulases, they help to soften their cell walls. These enzymes are also secreted by some plant pathogens such as the fungus *Monilinia fructigena* and the soft rot bacterium *Erwinia carotovora*, as part of their strategy for penetrating the plant cell walls. The products of such enzyme assaults (oligasaccharins) act as a signal which induces uninfected cells to defend themselves. The major commercial source of pectinase is *Aspergillus*. These fungi produce a complex of enzymes, including endo and ectopectinase, pectiesterase and pectin lyases. These enzymes are used industrially to weaken cell walls and enhance extraction of fruit juice. They help to decrease viscosity and prevent gelation in the extracts.

Table 9 : Pectinase Producing Microorganisms

Source	Protein
Actinidia chinensis	Polygalacturonase A
	Polygalacturonase C
	Polygalacturonase B
	Pectinesterase inhibitor
Actinidia deliciosa	Pectin methylesterase inhibitor
	Polygalacturonase
Actinobacillus succinogenes 130Z	Pectate-disaccharide lyase
	Oligogalacturonide lyase
Agrobacterium tumefaciens	Polygalacturonase
Agrobacterium vitis	Endo-polygalacturonase precursor
Alternaria alternata, *Alternaria arborescens,* *A. citri, A. mali, A. tangelonis,* *A. toxicogenica*	Endo-polygalacturonase
Arabidopsis thaliana	Probable pectate lyase 3, 18, 15, 14, 12, 11, 10, 21, 13, 22 and 9 precursor, polygalacturonase A, C, B, exopolygalacturbonase, pectin methylesterase, putative pectin acetylesterase, oectinesterase
Aspergillus clavatus NRRL1	Polygalacturonase (EC3.2.1.15)
A. flavus, A. awamori,	Endo-polygalacturonase
A. fumigatus Af293, *A kawachii, A. niger,* *A. nidulans, A. oryzae,* *A. parasiticus, A. tubingensis*	Pectin methylesterase, pectin lyase
Azospirillum irakense	Pectatelyase

Current Problems and Potential Solutions

Creation of thermostable and pH tolerant pectinases is a major challenge to increase the commercial application of this enzyme. These objectives can be fulfilled by site directed mutagenesis and techniques of protein engineering.

INDUSTRIAL USES OF ENZYMES

The enzymes can be used in a number of industries. Large scale application of the enzymes is possible in the following areas :

Detergent industry

Proteases are one of the standard ingredients of all kinds of detergents ranging from those used for household laundering to reagents used for cleaning contact lenses and dentures. The use of proteases in laundry detergents accounts for approximately 25% of the total world wide sales of enzymes. The preparation of the first enzymatic detergent, "Burnus", dates back to 1913; it consisted of sodium carbonate and a crude pancreatic extract. The first detergent containing the bacterial enzyme was introduced in 1956 under the trade name BIO-40. In 1960, Novo Industry introduced alcalase, produced by *Bacillus licheniformis*; its commercial name was BIOTEX. This was followed by Maxatase, a detergent made by Gist-Brocades. The biggest market for detergents is in the laundry industry, amounting to a world wide production of 13 billion tons per year. The ideal detergent protease should possess broad substrate specificity to facilitate the removal of a large variety of stains due to food, blood and other body secretions. Activity and stability at high pH and temperature and compatibility with other chelating and oxidizing agents added to the detergents are among the major pre-requisites for the use of proteases in detergent. The key parameter for the best performance of a protease in a detergent is its pI. It is known that a protease is most suitable for this application if its pI coincides with the pH of the detergent solution. Esperase and Savinase T (Novo Industry), produced by alkalophilic *bacillus* sp. are two commercial preparations with very high isoelectric points (pI=11); hence, they can withstand higher pH range. Due to the present energy crisis and the awareness for energy conservation, it is desirable to use proteases that are active at low temperatures.

A combination of lipase, amylase and cellulase is also added as it enhances the performance of protease in laundry detergents. Lipases decompose fats into more water soluble compounds by hydrolyzing the ester bonds between the glycerol backbone and fatty acid. The most important lipase in the market was originally obtained from *Humicola ranuginose*. It is produced at commercial levels by *Aspergillus oryzae* host after cloning the *Humicola* gene into this organism.

Amylases are used in detergents to remove starch based stains. Amylases hydrolyze gelatinized starch, which tends to stick to the textile fibers and bind other stain components. Cellulases have also been a part of detergents since the early 90s. Cellulase is actually an enzyme complex capable of degrading crystalline cellulose to glucose. In textile washing, cellulases, remove cellulose microfibrils, which are formed during washing and the use of cotton based cloths. This can be seen as colour brightening and softening of the material. Alkaline cellulases are produced by *Bacillus* strains and neutral and acidic cellulases by *Trichoderma* and *Humicola* fungi.

All detergent proteases currently used in the market are serine proteases produced by *Bacillus* strains. Fungal alkaline proteases are advantageous due to the ease of down stream processing to prepare a microbe-free enzyme. An alkaline protease from *Conidiobolus coronatus* was found to be compatible with commercial detergents used in India (Phadatare *et al.*, 1993) and retained 43% of its activity at 50ºC for 50 min in the presence of Ca^{2+} (25 mM) and glycine (IM).

Leather Industry

Leather processing involves several steps such as soaking, dehairing, bating and tanning. The major building blocks of skin and hair are proteinaceous. The conventional methods of leather processing involve hazardous chemicals such as sodium sulphide, which create problems of pollution and effluent disposal. The use of enzymes as alternatives to chemicals has proved successful in improving leather quality and in reducing environmental pollution. Proteases are used for selective hydrolysis of non collagenous constituents of the skin and for removal of non fibrillar proteins such as albumins and globulins. The purpose of soaking is to swell the hide. Traditionally, this step was performed with alkali. Currently, microbial alkaline proteases are used to ensure faster absorption of water and to reduce the time required for soaking. The use of nonionic and, to some extent, anionic surfactants is compatible with the use of enzymes. The conventional method of dehairing and dewooling consists of development of an extremely alkaline condition followed by treatment with sulphide to solubilize the proteins of the hair root. At present, alkaline proteases with hydrated lime and sodium chloride are used for dehairing, resulting in a significant reduction in the amount of waste water generated (Mala *et al.*, 1998).

Earlier methods of bating were based on the use of animal faeces as the source of proteases. These methods were unpleasant and unreliable and were replaced by methods involving pancreatic trypsin. Currently, trypsin is used in combination with other *Bacillus* and *Aspergillus* proteases for bating. The selection of the enzyme depends on its specificity for matrix proteins such as elastin and keratin, and the amount of enzyme needed depends on the type of leather (soft or hard) to be produced. Increased usage of enzymes for dehairing and bating not only prevents pollution problems but is also effective in saving energy. Novo Nordisk manufactures three different proteases – Aquaderm, NUE and Pyrase for use in soaking, dehairing and bating respectively.

Dairy Industry

The use of enzymes in the food industry dates back to antiquity. The major application of proteases in the dairy industry is in the manufacture of cheese. The milk coagulating enzymes fall into three main categories : (i) animal rennet, (ii) microbial milk coagulants, and (iii) genetically engineered chymosin. Both animal and microbial milk coagulating proteases belong to a class of acid aspartate proteases and have molecular weights between 30000 to 40000. Rennet extracted from the fourth stomach of unweaned calves contains the highest ratio of chymosin (EC 3.4.23.4) to pepsin activity. A world shortage of calf rennet due to increased demand for cheese production has intensified the search for alternative microbial milk cogulants. The microbial enzymes exhibited two major drawbacks, i.e. (i) the presence of high levels of non specific and heat-stable proteases which led to the development of bitterness in cheese after storage, and (ii) a poor yield. Extensive research in this area has led to the production of enzymes that are completely in activated at normal pasteurization temperatures and contains very low levels of non specific proteases. In cheese making, the primary function of proteases is to hydrolyze the specific peptide bond (Phe^{105} - Met^{106} bond) to generate para-K-casein and macropeptides. Chymosin is preferred due to its specificity for casein, which is responsible for its excellent performance in cheese making. In 1988, chymosin produced through recombinant DNA technology was first introduced to cheese makers for evaluation. Genencor International increased the production of chymosin in *Aspergillus niger* var. *awamori* to commercial levels.

Whey is a by-product of cheese manufacture. It contains lactose, proteins, minerals and lactic acid. The insoluble heat-denatured whey protein is solubilized by treatment with immobilized trypsin.

Lipases are also used extensively in the dairy industry for the hydrolysis of milk fat. Current applications include flavour enhancement of cheese, acceleration of cheese ripening, lipolysis of butter fat and cream etc. While addition of lipases primarily release short chain (C4 and C6) fatty acids that leads to the development of sharp, tangy flavour; the release of medium chain (C12 and C14) fatty acids tends to impart a soapy taste to the product. In addition, the free fatty acids take part in simple chemical reactions where they initiate the synthesis of other flavour ingredients such as aceto-acetate, β-keto acids, methyl ketones, flavour esters and lactones. More recently, a whole range of microbial lipase preparations have been developed for cheese manufacturing industry, such as those of *Mucor miehei, Aspergillus niger* and *Aspergillus oryzae*. Lipases are widely used for imitation cheese made from ewe's or goat's milk. Addition of lipases to cow's milk, generates a flavour rather similar to that of ewe's or goat's milk. This is used for producing cheese - the so called enzyme modified cheese (EMC). EMC is a cheese that has been incubated in the presence of enzymes at elevated temperatures in order to produce a concentrated flavour for use as an ingredient in other products such as dips, sauces, soups and snacks.

Baking Industry

Wheat flour is a major component of baking processes. It contains an insoluble protein called gluten, which determines the properties of the bakery doughs. Endo-and exproteinases from *Aspergillus oryzae* have been used to modify the wheat gluten by limiting proteolysis. Enzymatic treatment of the dough facilitates its handling and machining and permits the production of a wider range of products. The addition of enzymes reduces the mixing time and results in increased loaf volumes. Bacterial proteases are used to improve the extensibility and strength of the dough. α-amylase is also used to improve bread quality and its shelf life. Both bacterial and fungal amylases are used. The use of xylanase decreases the water absorption and thus reduces the amount of added water needed in baking. This leads to a more stable dough.

Manufacture of Soy Products

Soyabeans serve as a rich source of food, due to their high content of good quality protein. Proteases have been used from ancient times to prepare soy sauce and other soy products. The alkaline and neutral proteases of fungal origin play an important role in the processing of soy sauce. Proteolytic modification of soy proteins helps to improve their functional properties. Treatment of soy proteins with alcalase at pH 8 results in soluble hydrolysates with high solubility, good protein yield, and low bitterness. The hydrolysate is used in protein fortified soft drinks and in the formulation of dietic feeds.

Debittering of Protein Hydrolysates

Protein hydrolysates have several applications, example, as constituents of dietetic and health products, in infant formulae and clinical nutritional supplements, and as flavouring agents. The bitter taste of protein hydrolysates is a major barrier to their use in food and health care products. The intensity of the bitterness is proportional to the number of hydrophobic amino acids in the hydrolysate. The presence of a proline residue in the center of the peptide also contributes to the bitterness. The peptidases that can cleave hydrophobic amino acids and proline are valuable in debittering protein hydrolysates. Amino peptidases from lactic acid bacteria are available under the trade name Debitrase. Carboxypeptidase A has a high specificity for hydrophobic amino acids and hence has a great potential for debittering. A careful combination of an endoprotease for the primary hydrolysis and an aminopeptidase for the secondary hydrolysis is required for the production of a functional hydrolysate with reduced bitterness.

Plant Juice Production

Several microbial enzymes are employed in fruit juice processing, but probably the most important are pectinases. These processing aids are often produced by solid-substrate fermentation of *A. niger* or *Pencillium* species. Fruits and berries contain varying amounts of pectin, which acts as a binding layer between plant cells to hold adjacent cell walls together. Pectin is a heteropolymer of galacturonic acid, methyl esters of galacturonic acid and other sugar residues. In plant juice production, some of this pectin is extracted during pressing. It causes an increase in juice viscosity, leading to difficulties in obtaining optimal juice yields, and in juice clarification and filteration. These problems can be overcome by adding pectinase preparations to the fruit pulp before pressing. Similar enzyme treatment is used to increase the yields of oils from olive pulp, palm fruit and coconut flesh.

The commercial pectinase is not one enzyme, but is usually a complex cocktail of enzymes (pectin methyl esterases, polygalacturonases and pectin lyases) capable of attacking a variety of bonds in correspondingly diverse pectin molecules. Commercial pectinase proprations vary considerably in the proportions of these different enzymes. Fungal pectinases are also available that remain active in very acidic juices from citrus fruits, where the pH can be as low as 2.2-2.8. They may be used in the enzymic peeling of citrus fruits for canning.

The polysaccharide araban, a polymer of the pentose arabinose, is also an important component of fruit cell walls. Like pectin, it is often extracted during pressing of some fruits, especially pears. This may lead to haze formation, but it can now be eliminated by using commercial arabanases. Some extracts, such as apple juice, contain starch, which must be degraded to produce clear juices and concentrates. This is achieved by adding amylases, along with pectinases, during depectinization of the juice.

In wine making, commercial enzymes are used for several purposes, as well as for juice extraction. For red wines, colour extraction from the grape skins during pressing can be promoted by the addition of commercial cellulases, e.g. from the fungus *Trichoderma reesei*. In addition, glycosidases can be employed to hydrolyse terpenyl glycosides, releasing the terpenes that are important consistituents of the wine bouquet. Wine prepared from grapes allowed to undergo attack by the fungus *Botrytis dnerea* before harvesting are often difficult to clarify and filter, due to the presence of fungal β-glucans. These polymers can be degraded by adding a specific microbial β-glucanase to the wine.

Citrus fruits, such as grape fruit and bitter oranges, contain the bitter tasting fruit flavonoid called naringin. The bitterness of products derived from these fruits can be adjusted using naringinase (α-rhamnosidase+β-glucosidase) from *A. niger*. It first converts naringin to the less biter compound prunin, then on to non bitter naringenin. The level of naringin hydrolysis may be controlled by regulating the flow rate of fruit juice through a column of immobilized naringinase.

Glucose oxidase from *A. niger* or *Penicillium* species, often coupled with catalse, can be employed to remove molecular oxygen from wine, beer, fruit juices and soft drinks. This prevents potentially damaging oxidation that otherwise affects product quality.

Animal feed

Intensive study to use enzymes in animal feed started in early 80s. The first commercial success was addition of β-glucanase into barley based feed diets. Barley contains β-glucan, which causes high viscosity in the chicken gut. The net effect of enzyme usage in feed has been increased animal weight gain with the same amount of barley resulting in increased feed conversion ratio. Finnfeeds International was the pioneer in animal feed enzymes.

Enzymes were later tested in wheat based diets also. Xylanase enzymes were found to be effective. Usually, a feed enzyme preparation is a multienzyme cocktail containing glucanases, xylanases, proteinases and amylases. Enzyme addition reduces viscosity which increase absorption of nutrients, liberates nutrients either by hydrolysis of non-degradable fibers or by liberating nutrients blocked by these fibres, and reduces the amount of faeces.

Another example of important feed enzyme is phytase marketed by DSM in Netherlands. Phytase is a phosphoesterase which liberates phosphate from phytic acid which is a common compound in plant based feed materials. The net effect is reduced phosphorus in faeces resulting in reduced environmental pollution. The use of phytase reduces the need to add phosphorus to the feed diet.

Enzymes have become an important aspect of animal feed industry. In addition to poultry, enzymes are used in pig feeds and turkey feeds. They are added as enzyme premixes (enzyme flour mixture) during the feed manufacturing process, which involves extrusion of wet feed mass in high temperature (80-90° C). Therefore, the feed enzymes need to be thermotolerant during the feed manufacturing and operative at the animal body temperature.

Wood Pulp and Paper Industry

Paper manufacture is a major industry. In the United States alone, over 70 million tonnes of paper and paperboard are manufactured each year, which have a value in excess of US $50 billion. Paper manufacture involves the processing of mostly virgin fibres, but there is increasing utilization of some recycled materials or secondary fibres. Recycling saves both trees and energy, decreases waste effluents and reduces landfill loads.

Traditionally, wood pulp processing has involved the extensive use of chemicals, which can lead to problems with effluent treatment and environmental pollution. Thus, the development of enzyme based technology for pulp processing and paper manufacture has major advantages. Microbial enzymes can be used in several stages of pulp and paper processing to :

1. Enhance pulp digestion,
2. Improve drainage rates in water removal during paper formation,
3. Increase fibre flexibility,
4. Selectively remove xylan without affecting other components,
5. Remove resins,
6. Enhance bleaching,
7. Remove contaminants, such as in the de-inking of high quality waste paper; and
8. Fibrillate or increase interfibre bonding in chemical pulps and herbaceous fibres.

The major application is the use of xylanases in pulp bleaching. Xlylanases liberate lignin fragments by hydrolyzing residual xylan. This reduces considerably the need for chlorine based bleaching chemicals. Enzymes are also used to remove fine particles from pulp.

In the use of secondary (recycled) cellulose fibre the removal of ink is important. The fibre is diluted to 1% concentration with water, flocculating surfactants and ink solvents added and the mixture is aerated. The ink particles float to the surface. This process is facilitated by addition of cellulase enzymes.

In paper making, enzymes are used especially in modification of starch, which is used as an important additive. Starch improves the strength, stiffness and erasability of paper. The starch suspension must have a certain viscosity, which is achieved by adding amylases in a controlled manner.

Lipases can also be used to remove a sticky substance called pitch in softwoods. It is composed of lipids and is a special problem when pulp of red wood is used as a raw material. Pitch causes problems in paper machines and can be removed by lipases.

Textile industry

The use of enzymes in textile industry is one of the most rapidly growing fields in industrial enzymology. Starch has for a long time been used as a protective glue of fibres in weaving of fabrics. This is called 'sizing'. Enzymes are used to remove the starch in a process called 'desizing'. Amylases are used in this process since they do not harm the textiles fibre.

Enzymes have replaced the use of volcanic stones which were used in the preparation of Denim. This process was called 'stone washing' and was carried out to achieve a high degree of dye fading. However the stones caused considerable damage to fibres and machines. The same effect can be obtained by 'bio stone washing' which makes use of the cellulase enzyme. The faded look is a result of alternating cycles of desizing and bleaching enzymes and chemicals in washing machines.

Recently, hydrogen peroxides have been tested as bleaching agents to replace chlorine based chemicals. Catalase enzyme, which destroys hydrogen peroxide, may than be used to degrade excess peroxide. Another recent approach is to use oxidative enzymes directly to bleach textile. Laccase, a polyphenol oxidase from fungi, is a new candidate in the field. Laccases are produced by white rot fungi, which use them to degrade lignin. It is a copper containing enzyme which can degrade many different types of molecules like dye pigments when it is oxidized by oxygen.

Speciality Enzymes

In addition to large volume enzyme applications, there are a large number of speciality applications for enzymes. These include use of enzymes in analytical applications, flavour production, protein modification, personal care products and DNA technology. These are as follows :

Enzymes in Analytics

Enzymes are widely used in the clinical analytical methodology. Contrary to bulk industrial enzymes these enzymes need to be free from side activities. This means that elaborate purification processes are needed. Normally automatic analysers carry out these measurements. The reactions normally involve either changes in NAD(P)/NAD(P)H proportions, which can be detected spectrophotometrically or production of H_2O_2 which can be detected in peroxidase catalysed reactions leading to coloured products, which can be easily quantified spectrophotometrically.

Immunoassays are based on detection of target molecules by specific antibodies. The detection of the antibody-antigen complex is usually based on enzymes linked to the antibodies. This enzyme is either an alkaline phosphatase, which can be detected in colour forming reaction by p-nitrophenyl phosphate or peroxidase, which is detected in the presence of H_2O_2 with a colour forming substrate.

An important development in analytical chemistry is biosensors. They are based on H_2O_2 producing oxidative enzymes. Two different types of electrodes, one based on peroxide detection and the other based on oxygen consumption, can be used to quantify the analyte in question. The most widely used application is a glucose biosensor involving glucose oxidase catalysed reaction :

$$\text{Glucose} + O_2 + H_2O \rightarrow \text{gluconic acid} + H_2O_2$$

Several commercial instruments are available which apply this principle for measurement of molecules like glucose, lactate, lactose, sucrose, ethanol, methanol, cholesterol and some amino acids.

Enzymes in Personal Care Products

Personal care products are a relatively new area for enzymes and the amounts used are small but worth mentioning. One application is contact lens cleaning. Proteinase and lipase containing enzyme solutions are used for this purpose. Hydrogen peroxide is used in disinfection of contact lenses. The residual hydrogen peroxide after disinfection can be removed by a haeme containing catalase enzyme, which degrades hydrogen peroxide.

Some toothpastes contain glucoamylase and glucose oxidase. The reasoning behind this practice is that glucoamylase liberates glucose from starch based oligomers produced by α-amylase and glucose oxidase converts glucose to gluconic acid and hydrogen peroxide which function as disinfectants.

Dentures can be cleaned with protein degrading enzyme solutions. Enzymes are also studied for their applications in skin and hair care products.

Enzymes in DNA Technology

DNA technology has revolutionized both traditional biotechnology and opened totally new fields for scientific study. Most traditional enzymes are produced by organisms, which have been genetically modified to overproduce the desired enzyme. Recombinant DNA technology allows one to produce new enzymes in traditional over producing and safe organisms. The DNA modifying enzymes can be divided into two classes :

1. Restriction enzymes recognize specific DNA sequences and cut the chain at these recognition sites.
2. DNA modifying enzymes synthesize nucleic acids, degrade them, join pieces together and remove parts of the DNA.

Restriction enzymes recognize a specific code sequence in the DNA. This is usually 4-8 nucleotides long sequence. Their role in nature is to cut foreign DNA material. These enzymes do not cut the cells's own DNA because its recognition sites are protected. More than 150 different restriction enzymes have been isolated from several bacterial species and they are used in cutting the DNA in question at specific points. DNA polymerases synthesize new DNA chains. Many of them need a model template, which they copy. Nucleases hydrolyze the phosphodiester bonds between DNA sugars. Kinases add phosphate groups and phosphatases remove them from the end of DNA chain. Ligases join adjacent nucleiotides together by forming phosphodiester bonds between them.

In the cell these enzymes are involved in DNA replication, degradation of foreign DNA, repairing of mutated DNA and in recombining different DNA molecules. The enzymes used in gene technology are produced like any other enzyme but their purification needs extra attention. Many restriction enzymes from different sources are produced in *Escherichia coli* by recombinant DNA technology. They are often labile and therefore are preserved at -20^{0}C in buffered glycerol solution.

Enzymes in Fine Chemical Production

Biocatalysis has been used in fine chemical production for a long time. Usually the catalyst has been a living organism. Ethanol, acetic acid, antibiotics, vitamins, pigments, solvents are but a few examples of biotechnological products. One of the reasons to use whole cell catalysts lies in the need to combine chemical energy source (in the form of ATP) or reducing/oxidizing power (in the form of NAD (pH) to the production process. This is elegantly done in a living cell. *Candida* yeast can reduce the five carbon sugar xylose to xylitol by a xylose reductase enzyme :

Xylose + NADH $\rightarrow$ xylitol + NAD

The enzyme can be isolated and the reaction proceeds easily in a test tube. However, the reducing power of NADH has to be regenerated for the reaction to proceed. This is done in a living cell by other reactions, which reduce NAD back to NADH. One can isolate another enzyme, which does the same and couples two reactions together. One suitable enzyme is formate dehydrogenase.

Xylose + NADH $\rightarrow$ Xylitol + NAD

Formate + NAD $\rightarrow$ CO_2 + NADH

Coupled enzymatic reactions have been extensively studied but only few commercial examples are known. Leucine dehydrogenase is used commercially to produce L-tert-leucine with a concomitant cofactor recycling using the formate reduction for cofactor regeneration. Commercial production of chemicals by living cells using pathway engineering is still in many cases the best alternative to apply biocatalysis. Isolated enzymes have, however, been successfully used in fine chemical synthesis. Some of these applications are as follows :

Chirally Pure Amino Acids and Aspartame

Natural as well as synthetic amino acids are widely used in the food, feed, agrochemical and pharmaceutical industries. Many proteinogenic amino acids are used in infusion solutions and essential amino acids as animal feed additives. Aspartic acid and phenyl alanine methyl ester are combined to form the low calorie sweetener aspartame. In addition to natural amino acids, synthetic ones are also intermediates in the production of pharmaceuticals and agrochemicals. For example, several thousand tons of D-phenylglycine and D-p-hydroxyphenylglycine are produced annually for the synthesis of the broad spectrum antibiotics-amplicillin, amoxicillin, cefalexin and others. Natural amino acids are usually produced by microbial fermentation. Novel enzymatic resolution methods have been developed for the production of L-as well as for D-amino acids. The concept is based on the specificity of enzymes to detect only one of the two chiral molecules of amino acid derivatives. Aspartame, the intensive non-calorie sweetener, is synthesized in non-aqueous conditions by thermolysin, a proteolytic enzyme, from N-protected aspartic acid and phenylalanine methyl ester. The enzyme catalyzes not only a typical condensation reaction in the absence of water but also shows remarkable selectivity in forming the correct bond to form aspartame. After the condensation reaction the protective group is removed.

Rare Sugars

Non natural monosaccharides are needed as starting materials for new chemicals and pharmaceuticals, for example, L-ribose, D-psicose, L-xylose, D-tagatose and others. Some of the sugars are presently produced by chemical isomerization or epimerization. Recently enzymatic methods have been developed to manufacture practically all D- and L-forms of simple sugars. Glucose isomerase is one of the important industrial enzymes used in fructose manufacturing. Recently it has been shown that it can catalyse previously unknown conversions. For example L-arabinose is isomerized to L-ribulose and slowly also to L-ribose. D-xylose is isomerized to D-xylulose and slowly to D-lyxose. Also four carbon sugars are also good substrates for the production of rare sugars.

Semi-Synthetic Penicillins

Penicillin is produced by genetically modified strains of *Penicillium.* Most of the penicillin is converted by immobilizd acylase enzyme to 6-amino penicillanic acid which serves as a backbone for many semisynthetic penicillins. These can be synthesized by chemical or enzymatic methods.

Lipase Based Reactions

In addition to detergent applications, lipases can be used in versatile chemical reactions since they are active in organic solvents. Thus, water can be replaced by other nucleophiles like alcohols. The transferase activity of lipases is used to convert low value fats into more valuable ones in transesterification reactions. This occurs when low value fats are incubated in the presence of lipases and fatty acids. Lipases have also been used to form aromatic and aliphatic polymers. The enzyme can be used for enantiomeric separation of alcohols. In place of alcohols, amines can be used as the nucleophile. This makes it possible to separate racemic amine mixtures. Chirally pure amines can be used as building blocks for bioactive molecules. Several other intensively studied synthetic reactions are possible in lipase catalyzed reactions.

Enzymatic Oligosaccharide Synthesis

The chemical synthesis of oligosaccharides is a complicated multistep effort. The saccharide building blocks must be selectively protected, then coupled and finally deprotected to obtain desired stereochemistry and regio-chemistry. Biocatalytic synthesis with isolated enzymes like glycosyltransferases and glycosidases or engineered whole cells are powerful alternatives to chemical methods.

Glycosyltransferases catalyze the transfer of monosaccharides from a donor to saccharide acceptors. Typically the donor is a nucleotide. The type of donor that the enzyme utilizes and the position and stereochemistry of the transfer to the acceptor classify these enzymes.

These enzymes can also be extracellular. *Leuconostoc* lactic acid bacteria produce an enzyme called dextran sucrase. It converts sucrose into fructose and a glucose polymer called dextran. Dextran is used in biomedical applications and as a matrix in separation processes. The enzyme can use other molecules also as acceptor and thus novel oligomers with (example) : antibacterial properties can be produced. Glycosidases are hydrolytic enzymes, which can also be used for synthetic reactions. Oligosaccharides have found applications in cosmetics, medicines and as functional foods.

Development of Natural Flavours in Foods

The interest in developing natural food flavour has paved way for biotechnological processes. Two broad strategies are adopted for flavour development in foods. Microorganisms or enzymes are used to generate flavouring complexes or multi-component flavour systems as practiced in the production of dairy products. Alternatively, biosynthesis may be facilitated to develop single flavour compounds.

Multi-component flavour systems are developed by the use of specific strains of lactic acid bacteria and fungi to arrive at specific cheese flavours. For example, the specific flavour of blue cheese is developed by *Penicillium* sp., mainly *P. roqueforti* which converts the free fatty acids formed by lipolytic reactions in cheese to methylketones, the key flavour components of blue cheese. A variety of methylketones are formed in a two step process of β-oxidation of the fatty acids to β-keto acids which undergo decarboxylation to give methylketones of which the main products include 2-pentanone, 2-heptanone and 2-nonanone. Another example of the development of multi-component flavour systems involves the use of short chain fatty acid specific lipases and proteases for the production of enzyme modified cheeses.

Biosynthesis and isolation of individual flavour compounds is the second strategy involving biotransformation through hydration/dehydration or oxidation/reduction reactions. For example, the buttery flavour of fermented dairy products such as sour cream and butter milk is primarily due to diacetyl, biosynthesized from citric acid naturally present in milk. The pleasant coconut and peach like fruity flavours natural to fruits, vegetables, nuts and milk products are due to lactones such as γ-decalactone formed by yeast species (*Candida*) and fungi (*Sporobolonyces odorus, Trichoderma viride*) using ricinoleic acid (12-hydroxyoctadecyl-9-enoic acid) as the substrate.

Biosynthesis of esters such as ethyl acetate, ethyl butyrate, ethyl isovalerate and ethyl hexenoate responsible for the fruity aroma and flavour is facilitated by the use of species of *Pseudomonas, Lactococcus, Lactobacillus and Candida utilis*. Pyrazines and their derivatives are associated with roasted and nutty flavours (found naturally in green beans and potatoes) and are biosynthesized by species of *Lactococcus lactis, Pseudomonas perotens, Bacillus subtilis* and *Corynebacterium glutamicum*. The pungent taste of mustard, horse radish and cress is due to the formation of isothiocyanates from glucosinolates by the action of the enzyme myrosinase. The flavour and aroma of cucumber, apple and tomato are identified with green flavour components such as nonenal, nonadienal and hexanal. These compounds are formed by degradation of unsaturated fatty acids via a lipoxygenase catalyzed formation of hydroperoxides followed by cleavage by a hydroperoxide lyase.

Flavour enhancers such as IMP and GMP can be produced by degradation of RNA with 5′-phosphodiesterase. RNA is a by-product of plant processing and single cell protein production (Sivasankar, 2003).

Some microorganisms that are used for flavour production in different foods by fermentation are as follows :

Table 10 : Fermentation by Microorganisms for Flavour Production

Food Item	Microorganism	Flavour Compounds Produced
Buttermilk	*Streptococcus lactis, S. cremoris L. bulgaricus*	Lactic acid, diacetyl, acetaldehyde
Yoghurt	*Streptococcus thermophilus L. bulgaricus*	Acetaldehyde, acetoin
Sauerkraut	Mixed culture of *L. brevis, L. plantarum, Leuconostoc mesenteroides*	Acetate and short chain fatty acids
Soyabean milk	*Lactobacillus* sp.*S. thermophilus*	Aldehydes including pentanal
Soya sauce	*A. oryzae, Saccharomyces rouxii Lactobacillus* spp.	Organic acids, phenols, pyrazines
Tempeh (wheat based product made in Indonesia)	*Rhizopus* sp.	Fatty acids
Cocoa	*Saccharomyces* sp. *Acetobacter* sp. *Lactobacillus* sp.	Fatty acids and aromatic acids
Bread	*S.cerevisiae*	Ethanol
Swiss cheese	*Propionibacteriumshermanii*	Propionic acid

Future Trends in Industrial Enzymology

The industrial enzyme market is growing steadily. The reason for this lies in improved production efficiency resulting in cheaper enzymes, in new application fields and in new enzymes from screening programmes or in engineered properties of traditional enzymes. New applications are to be expected in the field of textile, new animal diets like ruminant and fish feed. It can be expected that breakthroughs in pulp and paper industry will materialize. The use of cellulases to convert waste cellulose into sugars and further to ethanol by fermentative organisms has been a major study topic for years. Increasing environmental pressures and energy prices will make this application a real possibility in the near future.

Tailoring enzymes for specific applications will be a future trend with continuously improving tools and understanding of structure-function relationships and increased search for enzymes from exotic environments. New technical tools to use enzymes as crystalline catalysts, ability to recycle cofactors and engineering enzymes to function in various solvents with multiple activities are important technological developments, which will steadily create new applications (Godfrey and West, 1996).

Two major developments that made it possible to change the structure of macromolecules in a predetermined way leading to their desired functional role are :

1. First, is the possibility of site directed mutagenesis, and

2. Second is the availability of computational and graphical tools which allow, on the screen, the display and exploration of the three dimensional structure of proteins.

These developments help us to decide the three dimensional structure of a molecule before using recombinant DNA technology. This allows production of novel proteins through genetic engineering.

For its industrial application, an enzyme should have some features additional to those produced within the cell. These include :

1. The enzyme should have a long life,
2. The enzyme should be able to use the substrate supplied in the industry, and
3. The enzyme should be able to work under the given concentrations of pH, temperature and substrate concentration.

Various properties of the enzyme that can be altered by genetic engineering in a predictable manner are : the kinetic properties of enzyme turn over and Michaelis constant, thermostability, stability and activity of enzymes in non aqueous solvents, substrate and reaction specificity, co-factor requirements, allosteric regulation, protease resistance, subunit structure, molecular weight etc. It is not always possible to get an enzyme with a combination of all the desired properties. Therefore, a compromise in properties may have to be made if we have to select an enzyme from the available variability or even if we use mutagenesis to do so. However, if the structure – function relationship of an enzyme is known, the structural features for the desirable function may be combined and protein engineering may then be used to create a novel enzyme exhibiting a combination of all desirable functional properties. Mutagenesis and selection can also be effectively used for improving a specific property of an enzyme as has been done in the case of enzymes anthranilate synthetase (the altered form is insensitive to tryptophan inhibition), lactate dehydrogenase (a single amino acid substitution of gly. → arg. changes the nature of the enzyme) etc. If an improvement requires changes in several amino acids, such a mutant will be rare or non existent and a modification of this type would be possible through gene modification techniques. *In vitro* mutagenesis using synthetic oligonucleotides as well as synthesis of the complete modified gene can also be done de novo.

Enzymes should however, not be considered alone but rather as a part of a biocatalyst technology. Whole cell catalysts, increased ability to engineer metabolic pathways and a combination of specific biocatalytic reactions with organic chemistry from a basis to develop new technologies for chemical production.

REFERENCES

Arnesen S Eriksen SH Olsen J and Jensen B 1998 Amylase production from thermophilic fungi *Eng Microb Technol* **23** : 249-252

Atsumi Y Yamamoto S Morihara K Fukushima J Takeuchi H Mizuki N Kawamoto S and Okuda K 1989 Cloning and expression of the alkaline protease gene from *Pseudomonas aeruginosa* 1F03455 *J Bacteriol* **171** : 5173-5175

Babu KR and Satyanarayan T 1995 Use of solid substrate fermentation for α-amylase production by *Bacillus* species *Process Biochem* **30** : 305-309

BCC Research Press Release 2007 Press Room Email : Editor@bccresearch com

Beazell JM 1941 A reexamination of the role of the stomach in the digestion of carbohydrate and protein *Am J Physiol* **132** : 42-50

Berka RM Ward M Wilson LJ Hayenga KJ Kodama KH Carlomango LP and Thomson SA 1990) Molecular cloning and deletion of the gene encoding aspergillopepsin A from *Aspergillus Awamori Gene* **86** : 153-162

Bhat M 2000) Cellulase and related enzymes in biotechnology *Biotechnology Advances* **18** : 355-383

Boyer PD 1971 In-the Enzymes 3rd edition Academic Press Inc New York NY

Brueckner R Shoseyor O and Doi RH 1990) Multiple active forms of a novel serine protease from *Bacillus subtilisin Mol Gen Gent* **221** : 486-490

Brune AK and Gotz F 1992 In – Microbial Degradation of Natural products (ed Winklemann G) VWeinheinm : 243-263

Chand S Mishra P 2003 Research and application of microbial enzymes – India's contribution *Adv Biochem Eng Biotechnol* **85** : 95124

Cruger W and Cruger A 1989 In-Industrial Microbiology Sinauer Association Sunderland : 189-218

Dowlathabad Muralidhara Rao Swami AVN Siva RamKrishna G 2007 Bioprocess technology strategies production and purification of amylase-An overview *The Int J of Genomics and Proteomics* **2** : 1540

Fishman Y Bricker J Gilbert JV Plant AG and Wright A 1985 Cloning of the type 1 immunology A1 protease from *Neisseria gonorrhoeae* and secretion of the enzyme from *E coli American Society for Microbiology* – Proceeding of the 4th In Symposium : 164-168

Gerhartz W 1990 In-Enzyme In Industry : Production and Application VCH Weinheim

Ghosh PK Saxena RK Gupta R Yadav RP and Davidson S 1996 Mechanism of action of lipases *Sci Prog* **79** : 119-121

Gilbert JV Pant AG Fishman Y and Wright A 1988 Cloning of the gene encoding streptococcal immunoglobulin A1 preotease and its expression in *E. coli Infect Immun* **56** : 1961-1966

Godfredson SE 1990) In-Microbial Enzymes and Biotechnology (eds Fogarty WM and Keely E. Elsevier Applied Science the Netherlands : 255-273

Godfrey T and West S 1996 In-Industrial Enzymology (authored by industrial experts; www novozymes com is a web page of world's largest enzyme company second largest is Genencor Int Inc www.genencor.com) Macmillan Publishers New York

Gomi K Arikawa K Kamiya N Kitamoto K and Kumagat C 1993 Cloning and nucleotide sequence of the acid protease encoding gene (*pep A*) from *A. oryzae Biosci Biotechnol Biochem* **57** : 1095-1100

Gray GL Hayenga K Cullen D Wilson LJ Norton S 1986 Primary structure of *Mucor miehei* aspartyl protease : evidence for a zymogen inter mediate *Gene* **48** : 41-53

Gunkel PA and Gassen HG 1989 Proteinase K from *Critirachium album* Limber : Characterization of the chromosomal gene and expression of the cDNA in *E. coli Eur J Biochem* **179** : 185-194

Gustin JK Kessler E Ohman DE 1996 A substitution at His-120 in the Las A proteases of *P aeuroginosa* blocks enzyme activity without affecting propetide processing or extracellular secretion *J Bacteriol* **78** : 6608-6617

Guzzo I Murgier M Filloux A and Lazdunski A 1990) Cloning of the *P aeruginosa* alkaline protease gene and secretion of the protease into the medium by *E. coli J Bacteriol* **172** : 942-948

Henderson G Krygsman P Liu CJ Davey CC and Malek LT 1987 Characterization and structure of genes for protease A and B from *S. griseus J Bacteriol* **169** : 3778-3784

Henner DJ Yang MB and Wells JA 1985 Expression of cloned protease genes in *B subtilis* In Proceedings of 9th *Int Spore Conference* : 95-103

Hodgson J 1994 The changing bulk catalysis market : recombinant DNA techniques have changed bulk enzyme production dramatically *Biotechnology* **12** : 789-790

Hoffman T 1974 Food Related Enzyme *Adv. Chem. Ser.* **136** : 146-185

Horuchi H Yanai K Okazaki T Takagi M and Yano K 1988 Isolation and sequencing of genomic clone encoding the aspartic proteinase of *Rhizopus niveus J Bacteriol* **170** : 272-278

Huge JB and Gormsen E 1987 EP Patent 258068

Isogai T Fukagawa M Kojo H Kohsaka M Aoki H and Imanaka H 1991 Cloning and nucleotide sequences of the complementary and genomic DNAs for the alkaline protease from *Acremonium chrysogenum Agri Biol Chem* **55** : 471-477

Jenson B and Olsen J 1992 Thermophillic fungus *Thermomyceslanuginosa* as a producer of α-amylase *Enz Microb Technol* **14** : 112-116

Kaneko R Koyama N Tsai YC Juang RY Yoda K and Yanasaki M 1989 Molecular cloning of the structural gene for alkaline elastase *YaB* a new subtilisin produced by alkalophilic *Bacillus* strain *J Bacteriol* **171** : 5232-5236

Kikuchi S Shibayama J Yasui M Takeuchi T Yamamoto I and Sato M 1993 *Bacillus* species for α-amylase production *Biosci Biotechnol Biochem* **57** : 845-846

Kitano K Morita S Kuriyama M and Maejima K 1992 Cloning and expression of a *S cremoris* proteinase in *B subtilis* and *S lactis Appl Environ Microbiol* **50** : 94-101

Kuo LC and Shafer JA 1994 Retroviral proteases *Methods Enzymol* **241** : 3-178

Kwon ST Terada I Matsuzawa H Ohta T 1988 Nucleotide sequence of the gene for aqualysin I of *Thermus aquaticus* YT-1 and characterization of the deduced 1° structure of the enzyme *Eur J Biochem* **173** : 491-497

Lazar G and Schroder FR 1992 In-Microbial Degradation of Natural Prodcuts (ed Winkelmann VCH Weinheim : 267-291

Lee JD and Kolattukudy PE 1995 Molecular cloning of the cDNA and gene for an elastinolytic aspartic proteinase from *A fumigatus* and evidence of its secretion by the fungus during in vasion of the host lung *Infect Immun* **63** : 3796-3803

Liu PV 1974 Extracellular toxins of *P. aeruginosa J Infect Dis* **130** : 94-99

Luanova V Stefanova M Tonkva A Doberva E and Spassova D 1995 Comparison of immobilization techniques for α-amylase *Appl Biochem Biotechnol* **50** : 305-316

Mala B Rao Aparna MT Mohini SG and Vasanti VD 1998 Molecular and biotechnological aspects of microbial proteases *Microbiol and Mol Bio Reviews* **62**(3) : 597-635

Mamo G and Gessesse A 1997 Production of a-amylase by *Bacillus Bitotechnol Tech* **11** : 447-450

Marshall RO and Kooi ER 1957 Enzymatic conversion of D-glucose to D-fructose *Science* **125** : 648-649

Matsuo T Sawamura N Hashimoto Y and Hashida W 1981 European Patent Application EP 0883

Michael JW Neil IM John SR Gray H 2001 In-Industrial Microbiology Panima Educational Book Agency New Delhi

Mullaney EJ Daly CB and Ullah AH 2000) Advances in phytase research *Adv Appl Microbiol* **47** : 157-99

Nakahama K Yeshimuza K Maramoto R Kikuchi M Lee I Hase T and Matsubara H 1986 Cloning and sequencing of *Serratia* protease gene *Nucleic Acids Res* **14** : 5843-5855

Natake M and Yoshimura S 1964 Studies on glucose isomerase of bacteria mode of action of glucose isomerase from *Escherichia* intermedia strain HN-500 *Agric Biol Chem* **28** : 510-516

Ottrup H and Norman BE 1989 α-amylase by *B. amyroliquefaciens* and *B. licheniformis Starch/Sta 4 Ke* **36**: 247-251

Phadatare SO Srinivasan MC and Deshpande VV 1993 High activity alkaline protease from *Conidiobolus coronatus* (NCCL 86 8 20) : enzyme production and compatibility with commercial detergents *Enz Microb Technol* **15** : 72-76

Poldermans BC1990) Proteolytic Enzymes In-Proteolytic enzyme in industry : production and application VCH Publishers Winheim Germany

Poulsen K Reinholdt J Kilian M 1996 Characterization of *S. pneumoniae* IgA1 protease gene (*iga*) and its translation product *Infect Immun* **64** : 3957-3966

Priest FG 1992 In-Encyclopedia of Microbiology Academic Press New York Vol 2

Roa MB and Deshpande VV In-Microbes for health wealth and sustainable environment Malhotra Publishing House New Delhi

Rawlings ND and Barrett AJ 1993 Evolutionary families of peptidases *Biochem J* **290** : 205-218

Ray RR Jana SC and Nanada G 1997 Production of β-amylase from starch waste by a hyper amylolytic strain of *B. megaterium* B6 mutant UN12 *Indian J Exp Biol* **35** : 285-288

Rudolph CF Rufo GA Sullvian BJ Theriault KA Ally D 1990) Gene encoding a novel extracellular metalloprotease in *B. subtilis J Bacteriol* **172** : 1025-1029

Sivasankar B 2003 In-Food Processing and Preservation Prentice Hall of India New Delhi

Salva TJG and Moraes IO 1995 Effects of different carbon sources on α-amylase production *Rev Microbiol* **26** : 46-51

Schechler I and Berger A 1967 The size of active site in protease I papain *Biocehm Biophys Res Commun* **27** : 157-162

Schneider MU 1985 Pancereatic enzyme replacement therapy : comparative effects of conventional and enteric coated microspheric pancreatin and acid stable fungal enzyme preparation in chronic pancreatitis *Hepato Gastroexterol* **32** : 97-102

Schwimmer S 1981 In-source book of food enzymology Westport CT : AVI Publishing Company

Selvakumar P Ashakumary L and Pandey A 1998 Screening of agro industrial residues for gluco amylase production by *A niger* in solid cultures *Biores Technol* **65** : 83-85

Snehlata H Mala BR and Vasanti VD 1996 Molecular and industrial aspects of glucose isomerase *Microbial Review* **60** : 280-300

Takagi H Takahashi T Momose H Inouye M Maeda Y Matsuzawa H and Ohta T 1990 Enhancement of thermostability of subtilisin E by introduction of a disulphide bond engineered on the basis of structural comparison with a thermophilic serine protease *J Biol Chem* **265** : 6874-6878

Takasaki Y and Tanabe O 1962 Forenation of fructose from glucose by bacteria; properties of glucose isomerase *Agric Biol Chem* **20** : 449-455

Takasaki Y and Tomabe O 1964 NAD linked D-glucose isomerizing and D mannose isomerizing enzyme from *Paracolobacterium aerogenoides Agric Biol Chem* **28** : 740-741

Takasaki Y and Tambe O 1966 Studies on sugar isomerization by enzyme production and utilization of glucose isomerase from *Streptomyces* species *Agric Biol Chem* **30** : 1247-1253

Tatsumi H Ogawa Y Murakami S Ishida Y Murakami K Masaki A Kawabe H Arimura H Nakano E and Motai H 1980) A full length cDNA clone for the alkaline protease from *A. oryzae* : structural analysis and expression in *S. cerevisiae Mol Gen Genet* **219** : 33-38

Thomas PG Russel AJ and Fersht AR 1985 Tailoring the pH dependence of enzyme catalysis using protein engineering *Nature* **28** : 375-376

Yoshimura S Danno G and Natake M 1966 Studies on D-glucose isomerizing activity of D-xylose grown cells from *B. coagulans* strain HN-68 *Agric Biol Chem* **30** : 1015-1023

4

BIOTECHNOLOGICAL APPROACH TO THE STUDY OF BIOFILMS

MEENAKSHI BANERJEE

INTRODUCTION

In most natural environments, association with a surface in a structure known as a biofilm is the prevailing microbial lifestyle. Surface association is an efficient means of lingering in a favorable microenvironment rather than being swept away by the current. Taken to the extreme, we may view the planktonic or free-swimming microbial phase primarily as a mechanism for translocation from one surface to another. Genetic studies of single-species biofilms have shown that they form in multiple steps (Watnick and Kolter, 1999), require intercellular signaling (Davies *et al.*, 1998), and demonstrate a profile of gene transcription that is distinct from that of planktonic cells (Prigent-Combaret *et al.*, 1999). In natural environments, however, the biofilm is almost invariably a multispecies microbial community harboring cyanobacteria and bacteria that stay and leave with purpose, share their genetic material at high rates, and fill distinct niches within the biofilm. Thus, the natural biofilm is less like a highly developed organism and more like a complex, highly differentiated, multicultural community much like our own city.

For quite some time we have known that bacteria can adhere to solid surfaces and form a slimy, slippery coat. These bacterial biofilms are prevalent on most wet surfaces in nature and can cause environmental problems. Perhaps because many biofilms are sufficiently thick to be visible to the naked eye, these microbial communities were among the first to be studied by the late-developing science of microbiology. However, it was not until the 1970s that we began to appreciate that bacteria in the biofilm mode of existence, sessile bacteria, constitute a major component of the bacterial biomass in many environments (Costerton *et al.*, 1978), and it was not until the 1980s and 1990s that we began to appreciate that attached bacteria were organized in elaborate ways (Lawrence *et al.*, 1991)

Definition of Biofilm

Because bacterial biofilms can cause environmental problems and studies of biofilms have required the development of new analytical tools, many recent advances have resulted from collaborations between microbial ecologists, environmental engineers, and mathematicians. These efforts have led to our current definition of a bacterial biofilm as a structured community of bacterial cells enclosed in a self-produced polymeric matrix and adherent to an inert or living surface. Biofilms constitute a protected mode of growth that allows survival in a hostile environment. The structures that form in biofilms contain channels in which nutrients can circulate (DeBeer *et al.*, 1994), and cells in different regions of a biofilm exhibit different patterns of gene expression (Davies *et al.*, 1993). The complexity of biofilm structure and metabolism has led to the analogy of biofilms to tissues of higher organisms (Costerton *et al.*, 1995). These sessile biofilm communities can give rise to nonsessile individuals, planktonic bacteria that can rapidly multiply and disperse. The common view is that planktonic bacteria must expose themselves to deleterious agents in their environment, be they phage or amoeba in nature, biocides in industrial settings, or potent antimicrobial agents in a clinical setting. In this light, it is not surprising that an impressive number of chronic bacterial infections involve bacterial biofilms, which are not easily eradicated by conventional antibiotic therapy.

The Basis of Biofilm Development

First, the bacterium approaches the surface so closely that motility is slowed. The bacterium may then form a transient association with the surface and/ or other microbes previously attached to the surface. This transient association allows it to search for a place to settle down. When the bacterium forms a stable association as a member of a microcolony, it has chosen the neighbourhood in which to live. Finally, the buildings go up as a three-dimensional biofilm is erected. Occasionally, the biofilm-associated bacteria detach from the biofilm matrix. The genetic basis of the steps in biofilm formation has been investigated for a number of bacterial species, including *Escherichia coli* (Pratt and Kolter, 1998), *Pseudomonas aeruginosa* (O' Toole and Kolter, 1998a) and *Vibrio cholerae* (Watnick and Kolter, 1999). For these studies, a simple genetic screen was utilized in which random transposon mutants are grown in 96-well plates (Cowan and Fletcher, 1987; Genevaux *et al.*, 1996; O'Toole and Kolter, 1998b). After removal of the planktonic cells, the remaining biofilm-associated cells are stained with crystal violet. Those wells with no crystal violet staining correspond to mutants that are defective in biofilm formation. These genetic screens for biofilm-defective mutants have shown that the initial interaction with the surface is accelerated by force-generating organelles such as type IV pili and flagella. Once temporary contact with the surface is made, bacteria use either flagella or type IV pili to move along the surface in two dimensions until other bacteria are encountered and microcolonies are formed or enlarged (O'Toole and Kolter,

1998a; Pratt and Kolter, 1998; Watnick and Kolter, 1999). Finally, exopolysaccharide production is necessary to stabilize the pillars of the biofilm (Watnick and Kolter, 1999). Competition studies between wild-type *V. cholerae* and pilus or flagellar mutants show that these structures provide a great advantage in surface colonization. Thus, speed of attachment may be an important factor in garnering an apartment in the microbial city.

Genes involved in biofilm formation

Evidence exists that different genes are transcribed in the planktonic and biofilm-associated phases of the bacterial life cycle. This is again reminiscent of a developmental process. Prigent-Combaret *et al.* (1999), performed a screen for genes in *E. coli* that are differentially expressed in biofilm-associated cells, using a library of random insertion mutants generated with a MudX transposon carrying a promoterless *lacZ* gene. One interesting finding from this study is that flagellin synthesis is decreased in biofilm-associated cells, while production of colanic acid, an exopolysaccharide made by *E. coli*, is increased. The situation appears to be similar in *P. aeruginosa*. Alginate is an exopolysaccharide that is found in *P. aeruginosa* biofilms (Gacesa, 1998). Transcription of *algC*, a gene involved in the production of alginate, is increased approximately fourfold in biofilm-associated cells as compared with planktonic cells (Garrett *et al.*, 1999). Furthermore, for many years, researchers have noted that pulmonary isolates of *P. aeruginosa* are mucoid due to production of copious amounts of alginate (Gacesa 1998). Also Garrett and coworkers noted that flagella are absent from these mucoid isolates (Garrett *et al.*, 1999). In addition, they showed by mutational analysis that while alginate synthesis is positively regulated by the alternative sigma factor σ^{22}, this sigma factor negatively regulates the synthesis of the flagellum. This suggests that when synthesis of the exopolysaccharide, alginate, is increased in biofilm-associated cells, flagellar synthesis decreases. Thus, to become a productive member of a biofilm community, the bacterium must differentiate into a biofilm-associated cell by repressing synthesis of the flagellum that might destabilize the biofilm and producing exopolysaccharide that will reinforce the biofilm structure.

Some genes may be expressed in response to a specific surface on which the bacterium has chosen to settle. For instance, chitin, a polymer of *N*-acetylglucosamine, is a component of crustacean and insect exoskeletons. Attachment to and degradation of chitin for use as a nutrient source is an important part of survival for many marine *Vibrio* species (Colwell and Spira, 1992; Keyhani and Roseman, 1996). The structural genes that are important for attachment to chitin differ from those required for attachment to abiotic, nonnutritive surfaces such as plastic and glass (Pruzzo *et al.*, 1996). Furthermore, although liquid medium that is rich in nutrients primes many bacteria for attachment to any local surface (O'Toole and Kolter 1998a; Pratt and Kolter, 1998), the bacteria will attach to chitin, but not plastic or glass, even when surrounded by a nutrient-poor medium (Watnick *et al.*, 1999). In some marine bacteria, it has been shown that chitinase and chitin-binding genes are expressed selectively in the presence of chitin (Montgomery and Kirchman, 1994; Stretton *et al.*, 1998). Thus, when the bathing medium is rich in nutrients, a bacterium will attach to any available surface, while in a nutrient-poor environment the bacterium will attach preferentially to a nutritive surface. This adaptation ensures that the bacterium will maximize access to nutrients in both nutrient-poor and nutrient-rich aqueous medium.

Occurence of Biofilms

Specific coaggregation of oral bacteria is thought to determine the distribution of bacteria within multispecies dental biofilms known as plaque. These interactions are thought to be essential for successful plaque formation (Klier *et al.*, 1998; Kolenbrander *et al.*, 1995; Whittaker *et al.*, 1996). Furthermore, the environment in a biofilm is not homogeneous. Microelectrode measurements have shown that the oxygen concentration and pH fall in a biofilm as the substratum is approached (Okabe *et al.*, 1999; Xu *et al.*, 1998). In single-species biofilms, the biofilm-associated bacteria alter gene expression to maximize survival in their particular microenvironment (Huang *et al.*, 1998, Xu *et al.*, 1998). In mixed biofilms, which are more representative of biofilms occurring in nature, bacteria distribute themselves according to who can survive best in the particular microenvironment and also based on symbiotic relationships between the groups of bacteria (Moller *et al.*, 1996; Moller *et al.*, 1998; Okabe *et al.*, 1999). Thus, the bacteria in a multispecies biofilm are not randomly distributed but rather organized to best meet the needs of each.

Communication of Biofilms

Bacteria also communicate with each other by a method called quorum sensing. Intercellular communication between bacteria is generally carried out by bacterial products that are able to diffuse away from one cell and enter another cell. It is difficult to envision this as an effective means of communication between planktonic bacteria in natural, aquatic environments, since molecules are likely to be carried off in the aqueous phase with a very small probability of reaching neighbouring bacteria. Rather, this method of intercellular signaling seems ideally suited for bacteria in a diffusion-limited environment such as the biofilm. Production of the quorum-sensing molecules known as acyl-homoserine lactones (acyl-HSLs) has been demonstrated in both natural and cultured biofilms (Allison *et al.*, 1998; Davies *et al.*, 1998; McLean *et al.*, 1997; Stickler, 1998). The importance of acyl-HSLs in single-species biofilms has been clearly demonstrated. In *P. aeruginosa*, acyl-HSLs are responsible for defining the separations between bacterial pillars in the three-dimensional structure of the biofilm (Davies *et al.*, 1998). *P. aeruginosa* mutants that do not produce acyl-HSL form biofilms in which the cells are closely packed together and are easily disrupted by sodium dodecyl sulphate. Acyl-HSLs are also mediators of surface attachment in *Pseudomonas fluorescens* (Allison *et al.*, 1998). Extracellular signals, therefore, enforce the zoning laws in single-species biofilms. Although little is known of the role of intercellular signaling in multispecies biofilms, it is suspected that it may differ significantly from that observed in single-species biofilms. These signals are expected to be especially important in favorable environments where surfaces are heavily colonized and competition for attachment to the surface is fierce. These signals can be defined as any actively or passively transported bacterial products that alter the state of neighbouring microbes. These might include bacterial metabolites, acyl-HSLs, secreted proteins, genetic material such as DNA or RNA, or as yet undiscovered bacterial products. These signals might alter the distribution of specific bacterial species in the biofilm, alter protein expression in neighbouring cells, introduce new genetic traits into neighbouring cells, or lure and incorporate bacteria into the biofilm for subsequent consumption. The last function of intercellular communication in multi-species biofilms is both fascinating and as yet uncharted. There are, however, laboratory models of lethal interspecies bacterial communication (Riley, 1998; Shi and Zusman, 1993). *Myxococcus xanthus*, for instance, is known to prey on *E. coli*. On soft agar plates, *E. coli* moves towards *M. xanthus*. Its chemotaxis machinery is required for this directed movement. The hypothesis is that *M. xanthus* secretes a signal that lures *E. coli* to its death (Shi and Zusman, 1993; Dunny and Leonard, 1997). The bacteriocins are another example of cell-cell signals that result in lethal interspecies interactions. These are bacterially derived antibacterial proteins that act against closely related species (Riley, 1998; De Beer *et al.*, 1994). In fact, mathematical models predict that bacteriocin production would be most advantageous in a spatially structured environment such as a biofilm (Durrett and Levin, 1997), suggesting that these secreted proteins may have evolved specifically for the biofilm environment. The impact of intercellular communication on multispecies biofilms is potentially far reaching and intercellular signaling, whether beneficial or detrimental to the recipient, is a critical factor in the diversity and distribution of bacteria in a biofilm. The thick biofilm is like a densely settled area. The buildings are back to back, and they are filled with people. It is difficult to imagine how bacteria can divide in such an environment. Thus, zero population growth may be the norm because the spatial constraints are such that cell division is impeded by surrounding exopolysaccharide. Such a situation may be akin to that of the polymer-encased bacteria that are used for biocatalytic engineering applications (Lyngberg *et al.*, 1999). Although these bacteria do not divide, they are viable and culturable once freed from the plastic encasement. Thus, one of the dictates of planktonic bacterial life, that consumed nutrients are funneled into procreation, may not apply to biofilm-associated cells. One possibility is that cell division is infrequent in a mature biofilm, and instead excess energy is used to make exopolysaccharide, an edible scaffold, that the cell can digest and use in time of need. As an example of this, production of an exopolysaccharide lyase has been shown for *P. fluorescens* under starvation conditions (Allison *et al.*, 1998). This enzyme degrades the biofilm-associated exopolysaccharide for consumption and frees cells from the biofilm scaffold to seek more favorable environments. Both these functions seem adaptive during times of starvation.

One advantage of biofilm living is the ability to acquire transmissible, genetic elements at accelerated rates. There are many reports of accelerated rates of conjugation in bacterial biofilms (Angles *et al.*, 1993). This suggests that evolution by horizontal transfer of genetic material may occur rapidly in a biofilm, making it the perfect milieu for emergence of new pathogens by acquisition of antibiotic resistance, virulence factors, and environmental survival capabilities. There are other advantages to living in a city. People live together because this is advantageous in times of adversity. Similarly, biofilm-associated cells are more resistant to many toxic substances such as antibiotics, chlorine, and detergents (Costerton *et al.*, 1987). There is evidence that decreased diffusion into the biofilm (Hausner and Wuertz, 1999; Suci *et al.*, 1994), decreased bacterial growth rate in a biofilm (Evans *et al.*, 1990), biofilm-specific substances such as exopolysaccharide (Yildiz and Schoolnik, 1999), and the quorum-sensing specific effects (Davies *et al.*, 1998; Hassett *et al.*, 1999) may be reasons for this resistance. This property of biofilms, thus, is most likely multifactorial.

If the bacteria were unable to escape the biofilm, the biofilm would, like an old apartment building, become a death trap when the nutrient supply was exhausted, environmental conditions became unfavorable, or an unfriendly neighbour entered the community. Once the bacterium is encased in exopolysaccharide, however, abandoning the biofilm becomes a significant task. At such times, a polysaccharide lyase may provide the bacterium with an escape. The multispecies bacterial biofilm can be compared to a city where bacteria settle selectively, limit settlements of new bacteria, store energy in exopolysaccharide, and transfer genetic material horizontally all for the good of the many. A genetic and biochemical understanding of the interactions between species in a biofilm, complex though they may be, is critical to our understanding of how the biofilm city functions and survives. It is predicted that in multiple-species biofilms many different types of soluble biofilm-specific signals will be discovered whose influence on dissimilar bacterial neighbours will be sometimes helpful and sometimes detrimental or even fatal. When conditions in the biofilm change, such interactions may determine which cells survive, which perish, and which move on. An understanding of the relationships among species in the biofilm city is essential to our appreciation of the benefits of biofilm-associated living.

BIOTECHNOLOGICAL ADVANTAGES OF BIOFILMS

There has been some speculation about the advantages of forming a biofilm versus living as individual cells. Although it is difficult to test these speculations experimentally, in the section below some reasons about the adoption of biofilm strategy by many microbes have been outlined. Bacteria experience a certain degree of shelter and homeostasis when residing within a biofilm, and one of the key components of this microniche is the surrounding extrapolymeric substance matrix. This matrix is composed of a mixture of components, such as EPS, protein, nucleic acids, and other substances. The best studied of these components is EPS. Most bacteria are able to produce polysaccharides, either as wall polysaccharides (capsules) or as extra cellular excretions into the surrounding environment (EPS). Some bacterial species, such as *Klebsiella aerogenes*, appear to be limited in the types of polymers that they can synthesize, but members of some genera (e.g., *Streptococcus pneumoniae*) are able to produce EPS with a wide range of different components (Sutherland 1985). Hence, broad generalizations about the function of EPS may be misleading. It is most likely that EPS plays various roles in the structure and function of different biofilm communities. Moreover, it is quite possible that EPS plays a different role in similar microbial communities under different environmental conditions. The EPS matrix also has the potential to physically prevent access of certain antimicrobial agents into the biofilm by acting as an ion exchanger, thereby restricting diffusion of compounds from the surrounding milieu into the biofilm (Gilbert *et al.*, 1997). This characteristic largely depends on the nature of both the agent and the EPS matrix. The effect appears to be most pronounced with antibiotics that are hydrophilic and positively charged, such as the aminoglycosides (Nichols, 1988; Nichols *et al.*, 1989; Nickel *et al.*, 1985). EPS has also been reported to sequester metals, cations, and toxins (Decho, 1990). Therefore, not only are biofilms able to accumulate metals, they can also be a key link in the transfer of metals through an ecosystem. EPS has also been reported to provide protection from a variety of environmental stresses, such as UV radiation, pH shifts, osmotic shock, and desiccation. One study reported by Elasir and Miller (1999) utilized a whole-cell bioluminescent biosensor to investigate the response of *Pseudomonas aeruginosa* biofilms to UV radiation.

Nutrient Availability and Metabolic Cooperatively

The highly permeable water channels interspersed throughout the biofilm in the areas surrounding the microcolonies have been compared to a primitive circulatory system. They provide an effective means of exchanging nutrients and metabolites with the bulk aqueous phase, enhancing nutrient availability as well as removal of potentially toxic metabolites (Costerton, 1995) Although it is not always the case, in multimember films, microcolonies often consist of a mixture of species (Hirsch, 1984; MacLeod *et al.*, 1990). These multispecies microconsortia can result from an association between metabolically cooperative organisms, and their proximity facilitates interspecies substrate exchange and the removal or distribution of metabolic products. Biofilms provide an ideal environment for the establishment of syntrophic relationships. Syntrophism is a special case of symbiosis in which two metabolically distinct types of bacteria depend on each other to utilize certain substrates, typically for energy production. Syntrophic associations have been well studied with regard to methanogenic degradation (Schink, 1997).

Acquisition of New Genetic Traits

Horizontal gene transfer is important for the evolution and genetic diversity of natural microbial communities. The importance of studying gene transfer in natural environments has recently been emphasized by the emergence of multidrug-resistant bacteria, the extensive use of antibiotics to promote growth in domestic animals (Witte, 1998), and the use of genetically engineered microorganisms in industrial processes

(Jansson and Prosser, 1997; Sayler and Ripp, 2000). The prevalence of plasmids in bacteria from diverse habitats is well established, and gene transfer by conjugation is one of the best-understood mechanisms for dissemination of genetic information. Since most bacteria in natural settings reside within biofilms, it follows that conjugation is a likely mechanism by which bacteria in biofilms transfer genes within or between populations. Although there may be fewer incidences of mating events within a biofilm, the "fixed" close quarters are likely to favor conjugation, especially within a population. Researchers have investigated the role that conjugation may play in the spread of genetic information in biofilms. Another recent study examined gene transfer in a microcosm dental plaque. These experiments were performed by creating a *Streptococcus* biofilm in a constant-depth fermentor (a device which uses physical scraping of the biofilm to keep the community at a constant depth). A *Bacillus subtilis* strain harboring a conjugative transposon that confers resistance to tetracycline was introduced to the system, and the resistance profile of the biofilm bacteria was assessed (Roberts, 1999). It was found that transfer of the conjugative transposon occurred within the biofilm, resulting in a *Streptococcus* species that now harbored the transposon. This was the first demonstration of gene transfer in an oral microbe growing in a biofilm, and these findings indicate that nonoral bacteria have the potential to transfer genes to oral commensals.

Another research group has investigated the transfer of the TOL plasmid, carrying genes for the degradation of toluene and benzyl alcohol, into a biofilm community growing on benzyl alcohol as the sole carbon and energy source (Christensen *et al.*, 1998) In this study, the biofilm community consisted of three different organisms, *P. putida, Acinetobacter*, and an unidentified isolate. All three isolates are able to mineralize benzyl alcohol, but only *P. putida* is able to propagate the TOL plasmid. To monitor the occurrence and growth of *P. putida* transconjugants, a *gfp*-tagged TOL plasmid was created. In addition, the *lacI* gene was inserted into the chromosome of the donor strain, also a *P. putida* strain, resulting in repression of *gfp* expression from the plasmid in this strain. Consequently, expression of the *gfp* gene was induced only if the plasmid was transferred to the recipient *P. putida* strain that does not contain the *lacI* gene encoding the repressor (zygotic induction of fluorescence). Confocal Scanning Laser Microscopy was used to identify the specific starting strains and transconjugants in the community by using 16S rRNA hybridization probes and expression of green fluorescence. Using these tools, this group was able to monitor conjugation in the biofilms.

Virus-mediated transduction is another mode of gene transfer. In the late 1980s, it was discovered that there is a very high abundance of viruses (as high as 10^8 per ml) in both limnetic (fresh water) and marine systems and that the majority of these viruses are bacteriophages (Bergh et al., 1989; Proctor and Fuhrman, 1990). Various procedures have been used to evaluate the impact of viruses on microbial mortality and gene transfer. The data indicate that viral lysis is a major contributor to bacterial mortality (Wommack and Colwell, 2000).

ROLE OF SURFACE-ATTACHED BACTERIA IN DISEASE

Bacterial Biofilm Infections

One of the greatest accomplishments in modern medicine has been our progress against infectious disease. As a result of scientific ingenuity, most modern-day acute infections can be treated effectively with antibiotics. However, there are two important exceptions to this rule. First are bacteria that are innately antibiotic resistant, and the second pertains to bacteria that reside within a biofilm. Biofilm bacteria can be up to 1,000-fold more resistant to antibiotic treatment than the same organism grown planktonically (Gilbert *et al.*, 1997), but the mechanisms by which the biofilm-grown bacteria attain this resistance are still a matter of speculation. Mechanisms of resistance that are considered likely include (i) phenotypic changes in bacteria resulting in resistance occurring within the biofilm environment, (ii) inactivation of the antibiotics by extra cellular polymers or modifying enzymes, and (iii) nutrient limitation resulting in slowed growth rate (Gilbert *et al.*,1997).

Implant-Based Infections

The role of biofilms in the contamination of medical implants has been well established. Early electron microscopy studies of medical implants revealed signs of bacteria residing in biofilms on these abiotic surfaces (Busscher *et al.*, 1998; Gristina *et al.*, 1988; Nickel *et al.*, 1989). Millions of catheters (e.g., central line, intravenous, and urinary catheters) are inserted into patients every year, and these implants serve as a potential surface for biofilm formation. Biofilm formation can also occur on contact lenses, and these biofilms are thought to contribute to keratitis (Gorlin *et al.*, 1996; Elder *et al.*, 1995). Overall, it is thought that upwards of 60% of all nosocomial infections are due to biofilms.

Biofilms and Pathogenesis

Although the role of biofilms in implant infections has been demonstrated in numerous systems, the role of biofilms in non-implant disease is less well established. One example of a disease in which biofilms are thought to play a prominent role is the occurrence of lung infections by *P. aeruginosa* in patients with cystic fibrosis (CF). Two lines of evidence are consistent with a role for *P. aeruginosa* biofilms in the CF lung. Microcolonies of bacteria have been observed in sections of lung from CF patients, and these may represent biofilm-grown cells (Govan and Deretic, 1996). Second, many isolates of *P. aeruginosa* from the CF lung are mucoid due to the over expression of an EPS called alginate.

Another example of a likely biofilm-mediated infection is chronic ear infection (otitis media). These infections are often caused by biofilm bacteria (Dingman *et al.*, 1998). In addition, it has been a quandary for some time why patients with chronic otitis media do not appear to have infections when tested by routine culture methods. This is likely due to the fact that biofilm bacteria can be difficult to culture by routine methods

(Costerton *et al.*, 1995). Periodontitis is another example of a biofilm-mediated disease that results in chronic inflammation of the tissue supporting the gums and can eventually lead to tooth loss. The main microbe associated with this disease is *Porphyromonas gingivalis* (Lamont and Jenkinson, 1998).

CONCLUSION

Although researchers recognized and studied surface-attached bacteria almost 70 years ago, we are only just beginning to fully realize the significance of biofilm communities. Moreover, viewing bacteria from the perspective of multicellular behavior is, in essence, altering our view of microbiology. It is evident that bacterial cells have the ability to aggregate into particular three-dimensional assemblages, differentiate and hence divide labor within these assemblages, and then disperse as part of their life cycle. Moreover, whether single-species or mixed-species biofilms, intercellular interactions and communication, are undoubtedly required for biofilm development and persistence. Dissecting these interactions provides one of the future challenges in biofilm research. Particularly challenging is the attempt to understand the complexity of the interactions within the biofilm community. Communication between species may include extracellular compounds whose sole role is to influence gene expression, metabolic cooperativity and competition (possibly encompassing global changes in gene expression and metabolism).

Another class of organisms, which can contribute greatly to our understanding of biofilms, is the cyano bacteria that are present in almost all ecological niches. Cyan bacteria are highly motile, and their localization in a given space indicates a clear physiological preference for environmental conditions of that particular area or depth where they form biofilms. Different cyanobacteria have different preferential placement and an understanding of the environmental conditions that these various cyanobacteria are adapted to and selecting for will allow us far more certainty in back-interpreting environmental conditions. Few literatures can be found about biofilm formation, and quorum sensing related to unicellular cyanobacteria like *Synecococcous* but in general this area is untouched for cyanobacteria. This aspect could be highly developed in the endolithic cyanobacteria of Antarctic or other hot deserts. The dominant organism within these rocks, the unicellular *Chroococcidiopsis* is found as a naturally immobilized layer or biofilm and never as single cells. Their movement is through bayocytes that squeeze out of the colony or layer when the gelatinous sheath surrounding it swells due to water absorption and colonize elsewhere. When does the colony sense the need for this and how these cells survive in the harshest of environments on earth is an intriguing matter? Such a survival is only possible when all cells in that layer coordinate and this may be through quorum sensing. Studies can be also be conducted in cyanobacteria like *Aulosira, and Mastigocladus* which are found in diverse habitats like rice fields, or the thermal springs to see how quorum sensing controls the formation of biofilms in cyanobacteria or natural immobilization as mats. If it does then is it done in the same way as in bacteria but which genes are involved in cyanobacteria needs to be deciphered. We can find answers to these questions once the molecular work related to biofilm formation and quorum sensing is done in selected species of cyanobacteria.

REFERENCES

Allison DG Ruiz B SanJose C Jaspe A Gilbert P 1998 Extracellular products as mediators of the formation and detachment of *Pseudomonas fluorescens* biofilms *FEMS Microbiol Lett* **167** 179–184

Angles ML Marshall KC Goodman AE 1993 Plasmid transfer between marine bacteria in the aqueous phase and biofilms in reactor microcosms *Appl Environ Microbiol* **59** 843–850

Bergh O Borsheim KY Bratbak G Heldal M 1989 High abundance of viruses found in aquatic environments *Nature* **340** 467–468

Busscher HJ Bruinsma G Weissenbruch RV Leunisse C Mei HCVD Dijk F Albers FWJ 1998 The effect of buttermilk consumption on biofilm formation on silicone rubber voice prostheses in an artificial throat *Eur Arch Otorhinolaryngol* **255** 410–413

Christensen BB Sternberg C Andersen JB Eberl L Moller S Givskov M Molin S 1998 Establishment of new genetic traits in a microbial biofilm community *Appl Environ Microbiol* **64** 2247–2255

Colwell RR Spira WM 1992 The ecology of *Vibrio cholerae* In: Barua D Greenough W B I (eds) Cholera New York NY Plenum pp 107–127

Costerton JW Cheng KJ Geesey GG Ladd TI Nickel JC Dasgupta M Marrie TJ 1987 Bacterial biofilms in nature and disease *Annu Rev Microbiol* **41** 435–464

Costerton JW Geesey GG Cheng KJ 1978 How bacteria stick *Sci Am* **23** 86-95

Costerton JW Lewandowski Z Caldwell DE Korber DR Lappin-Scott HM 1995 Microbial biofilms *Annu Rev Microbiol* **49** 711–745

Cowan MM Fletcher M 1987 Rapid screening methods for detection of bacterial mutants with altered adhesion abilities *J Microbiol Methods* **7** 241–249

Davies DG Chakrabarty AM Geesey GG 1993 Exopolysaccharide production in biofilms: substratum activation of alginate gene expression by *Pseudomonas aeruginosa* Appl *Environ Microbiol* **59** 1181–1186

Davies DG Parsek MR Pearson JP Iglewski BH Costerton JW Greenberg EP 1998 The involvement of cell-to-cell signals in the development of a bacterial biofilm *Science* **280** 295–298

De Beer D Srinivasan R Stewart PS 1994 Direct measurement of chlorine penetration into biofilms during disinfection *Appl Environ Microbiol* **60** 4339–4344

Decho AW 1990 Microbial exopolymer secretions in ocean environments: their role(s) in food webs and marine processes *Oceanogr Mar Biol Annu Rev* **28** 73–153

Dingman JR Rayner MG Mishra S Zhang Y Ehrlich MD Post JC Ehrlich GD 1998 Correlation between presence of viable bacteria and presence of endotoxin in middle-ear effusions *J Clin Microbiol* **36** 3417–3419

Durrett R Levin S 1997 Allelopathy in spatially distributed populations *J Theor Biol* **185** 165–171

Elasir MO Miller RV 1999 Study of the response of a biofilm bacterial community to UV radiation *Appl Environ Microbiol* **65** 2025– 2031

Elder MJ Stapleton F Evans E Dart JKG 1995 Biofilm related infections in ophtalmology *Eye* **9** 102–109

Evans DJ Brown MRW Gilbert P 1990 Susceptibility of bacterial biofilms to tobramycin: role of specific growth rate and phase in the division cycle *J Antimicrob Chemother* **25** 585–591

Gacesa P 1998 Bacterial alginate biosynthesis: Recent progress and future prospects Microbiology **144** 1133–1143

Garrett ES Perlegas D Wozniak DJ 1999 Negative control of flagellum synthesis in *Pseudomonas aeruginosa* is modulated by the alternative sigma factor AlgT (AlgU) *J Bacteriol* **181** 7401–7404

Genevaux P Muller S Bauda P 1996 A rapid screening procedure to identify mini-Tn10 insertion mutants of *Escherichia coli* K-12 with altered adhesion properties FEMS *Microbiol Lett* **142** 27–30

Gilbert P Das J Foley I 1997 Biofilms susceptibility to antimicrobials *Adv Dent Res* **11** 160–167

Gorlin AI Gabriel MM Wilson LA Ahearn DG 1996 Effect of adhered bacteria on the binding of *Acanthamoeba* to hydrogel lenses *Arch Ophthalmol* **114** 576–580

Govan JRW Deretic V 1996 Microbial pathogenesis in cystic fibrosis: mucoid *Pseudomonas aeruginosa* and *Burkholderia cepacia Microbiol Rev* **60** 539–574

Gristina AG Dobbins JJ Giammara B Lewis JC DeVreies WC 1988 Biomaterial-centered sepsis and the total artificial heart *JAMA* **259** 870– 874

Hassett DJ Ma JF Elkins JG McDermott TR Ochsner UA West SE Huang CT Fredericks J Burnett S Stewart PS McFeters G Passador L Iglewski BH 1999 Quorum sensing in *Pseudomonas aeruginosa* controls expression of catalase and superoxide dismutase genes and mediates biofilm susceptibility to hydrogen peroxide *Mol Microbiol* **34** 1082–1093

Hausner M Wuertz S 1999 High rates of conjugation in bacterial biofilms as determined by quantitative in situ analysis *Appl Environ Microbiol* **65** 3710–3713

Hirsch P 1984 Microcolony formation and consortia *In* Marshall KC (ed) Microbial adhesion and aggregation Springer Berlin Germany 373–393

Huang CT Xu KD McFeters GA Stewart PS 1998 Spatial patterns of alkaline phosphatase expression within bacterial colonies and biofilms in response to phosphate starvation *Appl Environ Microbiol* **64** 1526–1531

Jansson JK JI Prosser 1997 Quantification of the presence and activity of specific microorganisms in nature *Mol Biotechnol* **7** 103–120

Keyhani NO Roseman S 1996 The chitin catabolic cascade in the marine bacterium *Vibrio furnissii J Biol Chem* **271** 33414–33424

Klier CM Roble AG Kolenbrander PE 1998 *Actinomyces* serovar WVA963 coaggregation-defective mutant strain PK2407 secretes lactose-sensitive adhesin that binds to coaggregation partner *Streptococcus oralis* 34 *Oral Microbiol Immunol* **13** 337–340

Kolenbrander PE Parrish KD Andersen RN Greenberg EP 1995 Intergeneric coaggregation of oral *Treponema* spp. with *Fusobacterium* spp. and intrageneric coaggregation among *Fusobacterium* spp. *Infect Immun* **63** 4584–4588

Lamont RJ Jenkinson HF 1998 Life below the gum line: pathogenic mechanisms of *Porphyromonas gingivalis Microbiol Mol Biol Rev* **62** 1244–1263

Lawrence JR Korber DR Hoyle BD Costerton JW Caldwell DE 1991 Optical sectioning of microbial biofilms *J Bacteriol* **173** 6558-6567

Lyngberg OK Thiagarajan V Stemke DJ Schottel JL Scriven LE Flickinger MC 1999 A patch coating method for preparing biocatalytic films of *Escherichia coli Biotechnol Bioengg* **62** 44–55

MacLeod FA Guiot SR Costerton JW 1990 Layered structure of bacterial aggregates produced in an upflow anaerobic sludge bed and filter reactor *Appl Environ Microbiol* **56** 1598–1607

McLean RJ Whiteley M Strickler DJ Fuqua WC 1997 Evidence of autoinducer activity in naturally occurring biofilms *FEMS Microbiol Lett* **154** 259–263

Moller S Pedersen AR Poulsen LK Arin E Molin S 1996 Activity and three-dimensional distribution of toluene-degrading *Pseudomonas putida* in a multispecies biofilm assessed by quantitative *in situ* hybridization and scanning confocal laser microscopy *Appl Environ Microbiol* **62** 4632–4640

Moller S Sternberg C Andersen JB Christensen BB Ramos JL Givskov M Molin S 1998 *In situ* gene expression in mixed-culture biofilms: evidence of metabolic interactions between community members *Appl Environ Microbiol* **64** 721–732

Montgomery MT Kirchman DL 1994 Induction of chitin-binding proteins during the specific attachment of the marine bacterium *Vibrio harveyi* to chitin *Appl Environ Microbiol* **60** 4284–4288

Nichols WW Evans MJ Slack MPE Walmsley HL 1989 The penetration of antibiotics into aggregates of mucoid and non-mucoid *Pseudomonas aeruginosa J Gen Microbiol* **135** 1291–1303

Nichols WW Dorrington SM Slack MPE Walmsley HL 1988 Inhibition of tobramycin diffusion by binding to alginate *Antimicrob Agents Chemother* **32** 518–523

Nickel JC Ruseska I Wright JB Costerton JW 1985 Tobramycin resistance of *Pseudomonas aeruginosa* cells growing as a biofilm on urinary tract catheter *Antimicrob Agents Chemother* **27** 619–624

Nickel JC Downey JA Costerton JW 1989 Ultrastructural study of microbiologic colonization of urinary catheters *Urology* **34** 284–291

Okabe S Satoh H Watanabe Y 1999 *In situ* analysis of nitrifying biofilms as determined by in situ hybridization and the use of microelectrodes *Appl Environ Microbiol* **65**:3182–3191

O'Toole GA Kolter R 1998a Flagellar and twitching motility are necessary for *Pseudomonas aeruginosa* biofilm development *Mol Microbiol* **30** 295–304

O'Toole GA Kolter R 1998b Initiation of biofilm formation in *Pseudomonas fluorescens* WCS365 proceeds via multiple convergent signaling pathways: a genetic analysis *Mol Microbiol* **28** 449–461

Pratt LA Kolter R 1998 Genetic analysis of *Escherichia coli* biofilm formation: roles of flagella motility chemotaxis and type I pili *Mol Microbiol* **30** 285–293

Prigent-Combaret C Vidal O Dorel C Lejeune P 1999 Abiotic surface sensing and biofilm-dependent regulation of gene expression in *Escherichia coli J Bacteriol* **181** 5993–6002

Proctor LM Fuhrman JA 1990 Viral mortality of marine bacteria and cyanobacteria *Nature* **343** 60–62

Pruzzo C Crippa A Bertone S Pane L Carli A 1996 Attachment of *Vibrio alginolyticus* to chitin mediated by chitin-binding proteins *Microbiology* **142**:2181–2186

Riley M A 1998 Molecular mechanisms of bacteriocin evolution *Annu Rev Genet* **32**:255–278

Roberts A P J Pratten M Wilson and P Mullany 1999 Transfer of a conjugative transposon Tn5397 in a model oral biofilm *FEMS Microbiol Lett* **177:**63–66

Sayler G S and S Ripp 2000 Field applications of genetically engineered microorganisms for bioremediation processes *Curr Opin Biotechnol* **11**:286– 289

Schink B 1997 Energetics of syntrophic cooperation in methanogenic degradation *Microbiol Mol Biol Rev* **61**:262–280

Shi W Zusman D R 1993 Fatal attraction *Nature* **366**:414–415

Stickler D J Morris N S McLean R J Fuqua C 1998 Biofilms on indwelling urethral catheters produce quorum-sensing signal molecules *in situ* and *in vitro Appl Environ Microbiol* **64**:3486–3490

Stretton S Techkarnjanaruk S McLennan A M Goodman A E 1998 Use of green fluorescent protein to tag and investigate gene expression in marine bacteria *Appl Environ Microbiol* **64**: 2554–2559

Suci P A Mittelman M W Yu F P Geesey G G 1994 Investigation of ciprofloxacin penetration into *Pseudomonas aeruginosa* biofilms Antimicrob Agents *Chemother* **38**:2125–2133

Sutherland J W 1985 Biosynthesis and composition of gram-negative bacterial extracellular and wall polysaccharides *Annu Rev Microbiol* **39**:243–270

Watnick P I Fullner K J Kolter R 1999 A role for the mannose-sensitive hemagglutinin in biofilm formation by *Vibrio cholerae* El Tor *J Bacteriol* **181**:3606–3609

Watnick P I Kolter R 1999 Steps in the development of a *Vibrio cholerae* biofilm *Mol Microbiol* **34**:586–595

Whittaker C J Klier C M Kolenbrander P E 1996 Mechanisms of adhesion by oral bacteria *Annu Rev Microbiol* **50**:513–552

Witte W 1998 Medical consequences of antibiotic use in agriculture *Science* **279**:996–997

Wommack K E and R R Colwell 2000 Virioplankton: viruses in aquatic ecosystems *Microbiol Mol Biol Rev* **64**:69–114

Xu K D Stewart P S Xia F Huang C T McFeters G A 1998 Spatial physiological heterogeneity in *Pseudomonas aeruginosa* biofilm is determined by oxygen availability *Appl Environ Microbiol* **64**:4035–4039

Yildiz F H Schoolnik G K 1999 *Vibrio cholerae* O1 El Tor:identification of a gene cluster required for the rugose colony type exopolysaccharide production chlorine resistance and biofilm formation *Proc Natl Acad Sci* USA **96**:4028–4033

5

GENETIC ENGINEERING OF CROP PLANTS FOR ABIOTIC STRESS TOLERANCE

SHABIR HUSSAIN WANI, J S SANDHU AND S S. GOSAL

Environmental stresses such as drought, salinity, extreme temperatures and oxidative stress represent the most limiting factors for agricultural productivity world wide. Among various abiotic stresses affecting crop productivity, drought is considered most damaging (Borlaug and Dowswell, 2005). Drought and salinity are already widespread in many regions and are expected to cause serious salinization of more than 50% of all arable lands by 2050 (Vincour and Altman, 2005). The world food grain production needs to be doubled by the year 2050 to meet the ever growing demand of the population (Tilman *et al.*, 2002) which is going to reach 9 billion by that time (Virmani and Ilyas-Ahmed, 2007). Abiotic stresses present a major challenge in our quest for sustainable food production as these may reduce the potential yields as high as 70% in crop plants (Katiyar-Agarwal *et al.*, 2006). There is an increasing scarcity of freshwater and plants account for about 65% of global freshwater use (Postel *et al.*, 1996). Soil salinity limits crop production in about 20% of irrigated lands (Flowers and Yeo, 1995). Temperature extremes, drought and salinity stresses are also limiting production even in irrigated conditions (Chinnusamy *et al.*, 2006). Flooding is a major natural disaster affecting half of the 40 MHa of rainfed lowland rice in South and South East Asia (Vanavichit *et al.*, 2006). Likewise, aluminium (Al) toxicity is a severe problem for crop production, therefore, a considerable emphasis is placed on the selection of aluminium tolerant cultivars of crop and fodder plant species (Ernst, 2006). In temperate areas, suboptimal temperatures during spring, decrease productivity and cause poor yield stability. Low temperatures limit the production of about 7 MHa of modern rice cultivars in South and South East Asia. High yielding cultivars cannot be cultivated in many temperate areas because of their susceptibility to cold stress (Revilla *et al.*, 2006).

Breeding for abiotic stress tolerance at whole plant level is very difficult due to the complexity of stresses. Conventional breeding has not been effective in breeding for resistance to abotic stresses because (i) these are time and labour intensive (ii) deleterious genes also get transferred along with the gene of interest (iii) gene sources are not accessible and (iv) lack of efficient screening techniques. Genetic engineering offers a way to transfer gene(s) across crossability barriers (Chahal and Gosal, 2002).

Plants respond and adapt to water deficit, both at the cellular and molecular levels, for instance by the reaccumulation of osmolytes and proteins specifically involved in stress tolerance. A number of genes are switched on, when a plant gets exposed to various abiotic stresses. This results in an increase in the levels of several metabolites and proteins, some of which may be responsible for providing a certain amount of tolerance to these stresses. Therefore it is important to know about the cellular, molecular and biochemical changes occurring in plants in response to stress (Bhatnagar-Mathur *et al.*, 2007). Further after micro array analysis in *Arabidopsis* the products of drought inducible genes have been categorized into two groups. The first group includes that probably function under abiotic stress tolerance. These include molecules such as chaperons, water, LEA proteins, osmotic, antifreeze proteins, mRNA binding proteins, key enzymes for osmotic biosynthesis, water channel proteins, sugar and proline transporters, detoxification enzymes and various proteases. The second group consists of regulatory proteins, which involve in further regulation of signal transduction and stress responsive gene expression. These include various transcriptional factors, protein kinases, protein photospheres, enzymes involved in phospholipid metabolism (Shinozaki and Yamaguchi-Shinozaki, 2007).

Genetic engineering for abiotic stress tolerance involves different kinds of genes such as:

(A) Structural genes for;

(a) Key enzymes for biosynthesizing osmolytes

(1) Proline

(2) Glycinebetaine

(3) Mannitol

(4) Trehalose

(b) Redox proteins and detoxifying enzymes

(c) Stress induced proteins

(1) LEA proteins

(2) Heat-shock proteins

(d) Transporter proteins

(B) Regulatory genes

(1) DREB

(2) Zinc finger proteins

(3) NAC

(A) Structural genes

(a) Key enzymes for biosynthesizing osmolytes

(1) Proline

Amino acid proline is the most commonly distributed compatible osmolyte. From bacteria to maize, there is a strong correlation between increased cellular proline levels and the capacity to survive both drought and salinity stresses (Taylor, 1996). In addition to its role as an osmoprotectant (Christian, 1955) proline also acts as a sink for energy to regulate redox potentials (Blum and Ebercon, 1976 ; Saradhi and Saradhi, 1991), as scavenger of hydroxyl radicals (Smirnoff and Cumbes, 1989), as a solute for avoiding macromolecule denaturation (Schobert and Tschesche, 1978) and as a mean of reducing cell acidity (Venekamp *et al.*, 1989).

Proline synthesis pathway in plants which takes place in cytoplasm is from glutamate, through α-glutamyl phosphate and glutamly-α-semialdehyde. Enzyme P5CS (Pyrroline 5-carboxylate synthetase) in plants and animals catalyzes this reaction in two steps. The glutamyl-α-semialdehyde is naturally cyclised to proline-5-carboxylate, which is converted to proline by pyrroline 5-carboxylate reductase (P5CS) (Taylor, 1996). In many crop plants and forest trees, role of proline in stress tolerance has been proved through the production of transgenic plants overexpressing proline. Overexpression of *Vigna aconitifolia* gene P5CS (Pyrroline 5 carboxylate synthetase) in tobacco resulted in 10-18 fold increase in proline synthesis. This proline acted as an osmoprotectant and its over production resulted in increased tolerance to osmotic stress in plants (Kishore *et al.*, 1995). The P5CS gene from *Vigna aconitifolia* was transformed into rice under the control of ABA-responsive element from barley *HVA22* gene. Transgenic plants showed stress inducible proline accumulation under water stress. The root biomass of the transgenic plants was significantly higher than wild types. After a limited period of salt stress, the transgenic plants showed quick recovery (Zhu *et al.*, 1998)

Agrobacterium mediated gene transfer of *Vigna aconitifolia* gene P5C5 into wheat resulted in increased salt tolerance, thus proving the action of proline as an osmoprotectant in wheat (Sawahel and Hasan, 2002). Enzyme denaturation is caused by many abiotic stress treatments like water, salt, heavy metals and UV radiation. In addition to its role as osmoprotectant, proline has also been shown to reduce this enzyme denaturation (Iyer and Chaplan, 1998). Proline is mainly synthesized in chloroplasts during stress and protects photo system II against photo damage. Pyrroline 5 carboxylate (P5CS) and glutamine which are the two intermediates in proline biosynthesis and catabolism, increase the expression of many osmotically regulated genes in rice, including salt, *dhn4* (Iyer and Chaplan, 1998). There are many experiments done on several crop plants like

Arabidopsis (Nanjo *et al.*, 1999), soybean (De Ronde *et al.*, 2004), tobacco (Yonamine *et al.*, 2004) showing the role of proline in preventing from osmotic stress.

A forest tree *Larex leptoeuropaea* was transformed with P5CS gene from *Vigna aconitifolia* through *Agrobacterium* mediated gene transfer. Transgenics showed 30 fold increase in proline level against non-transformed control. Transgenic tissue lines were significantly more tolerant to cold, salt and freezing stresses and grew under conditions 200 mM NaCl or 4° C that completely inhibit the growth of control cell lines (Gleeson *et al.*, 2005). Further P5CS from *Vigna aconitifolia* was introduced through particle gun method of gene transfer into *Saccharam officinarum* under the action of AIPC promoter. Results indicated that stress-inducible proline accumulation in transgenic sugarcane plants under water-deficit stress acts as a component of antioxidative defense system rather than as an osmotic adjustment mediator (Molinari *et al.*, 2007) (Table 1).

(2) Glycinebetaine

Glycinebetaine is another common compatiable solute in various organisms including higher plants (Csonka and Hanson, 1991; Le Rudulier, 1993; Rhodes and Hanson 1993). Plant species which accumulate betaine grow in saline and arid areas and accumulate the compound in response to drought and salinity (Rhodes and Hanson, 1993). The protection of plants by betaine is through its action as an osmolyte maintaining the water balance between the plant cell and the environment (Robinson and Jones, 1986) and by stabilization of macro molecules during cell dehydration (Incharoensakdi *et al.*, 1986). Betaines also protect photosystem II from salt induced inactivation (Papageorgiou and Murata, 1995). Glycinebetaine is synthesized from choline into betaine aldehyde which is calatysed by choline mono oxygenase (CMO) and the second step involves the conversion of betaine which is catalysed by betaine aldehyde dehydrogenase (BADH) (Rhodes and Hanson, 1993).

Choline is converted into glycinebetaine in a single step by the action of enzyme choline oxidase encoding (COD) (Ikuta *et al.*, 1977). Choline is also converted into glycine betaine via betaine aldehyde, catalysed by choline dehydrogenase (CDH) and BADH encoded by both *bet A* and *bet B* in *E. coli*. Another three step pathway has been found in halophillic microorganisms like *Aphanothece halophytica* and *Ectothiorhodospira halochloris*. In this case, methylation of glycine results in biosynthesis of glycinebetaine and is catalysed by glycine sarcosine methyl transferase and sarcosine dimethyl glycine methyltransferase (Nyyssola *et al.*, 2000). Although many crops synthesize glycine betaine like maize, sorghum, sugarbeet, barley and *Artiplex* (Rhodes and Hanson, 1993), but crops like *Brassica juncea, Arabidopsis,* tobacco and rice do not accumulate glycinebetaine. Among these crop rice is the only important cereal crop which does not accumulate glycinebetaine (Shirasawa *et al.*, 2006).

Therefore efforts have been made to produce transgenic plants which accumulate glycine betaine, by introducing genes cloned from different plant species and micro organisms (Sakamoto and Murata, 2002). Initial attempts for producing transgenic plants through the introduction of CMO (Choline monooxygenase) and betaine aldehyde dehydrogenase (BADH) pathway were made in tobacco. Tobacco plants were transformed with cDNA for BADH from spinach (*Spinacia oleraceae*) and sugar beet (*Beta vulgaris*) under the control of CaMV 35S promoter. The BADH was produced in chloroplasts of tobacco. Betainealdehyde was converted to betaine by BADH, thus conferring resistance to betaine aldehyde. However, transgenic plants were not able to accumulate betaine in the absence of exogenously supplied betaine aldehyde. This was because the enzyme required for oxidation of choline was absent. This showed that expression of *BADH* alone was not sufficient for synthesis of betaine in transgenic plants (Rathinasabapathi *et al.*, 1994). In another attempt cDNA for choline monooxygenase from *Spinacia oleraceae* was introduced into tobacco and the enzyme thus synthesized was transported to its functional place *i.e.*, chloroplasts. But the leaves of tobacco accumulated betaine at a very low *i.e.*, 10-100 folds lower (Nuccio *et al.*, 1998). The reason for insufficient synthesis of betaine most probably was the absence of engineered BADH activity in chloroplasts. Therefore, both CMO and BADH need to be present in the chloroplasts for efficient synthesis of betaine in transgenic plants which do not accumulate glycinebetaine.

Genetic transformation of tobacco with *E. coli* genes, resulted in transgenic plants accumulating glycinebetaine. The plants which produced both CDH (choline dehydrogenase) and BADH (betaine aldehyde dehydrogenase) generally accumulated higher amounts of glycinebetaine, than plants producing *CDH* alone. Increased tolerance to salt stress was visible by biomass production of transgenic tobacco lines under green house conditions. In addition, the transgenic plants showed enhanced recovery from photo inhibition caused by high light, salt stress and under low temperature condition (Holmstrom *et al.*, 2000). Thus the above theory that both CMO and BADH should be present in chloroplasts was again confirmed.

Further, a gene for CMO cloned from spinach (*Spinacia oleracea*) was introduced into rice through *Agrobacterium* mediated transformation. The level of glycinebetaine in rice was low to the expectations. The author has given several reasons for the low productivity of rice and low glycinebetaine accumulation. Firstly, the position of spinach CMO and endogenous BADH might be different and secondly, the catalytic activity of spinach CMO in rice plants might be lower than it was in Spinach (Shirasawa *et al.*, 2006) (Table 1).

Thus it can be concluded that glycinebetaine has role as compatible solute and its engineering into non-accumulations will be a success only if both CMO and BADH pathways are introduced and if the localization of both CMO and BADH is in chloroplasts.

(3) Mannitol

Mannitol is an osmolyte which is normally synthesized in numerous plant species. On exposure to low water potential, plants accumulate mannitol at an increased rate

(Patonnier *et al.*, 1999). Mannitol is synthesized from fructose 6 P. Fructose 6 P is converted to mannitol 1P by mannose-6 P isomerase. Mannitol1-P is then reduced to mannitol by mannitol1-Phosphate dehydrogenase (Chinnusamy *et al.*, 2006). Attempts have been made to produce transgenics in case of plants which do not accumulate mannitol (Tarczynski *et al.*, 1992, 1993; Thomas *et al.*, 1995; Karkas *et al.*, 1997; Shen *et al.*, 1997; Abebe *et al.*, 2003).

. Mannitol accumulated to more than 6 μmol g^{-1} fw in the leaves of transgenic tobacco plants (Tarczynski *et al.*, 1992) transformed with *mtlD* gene for mannitol1-phosphate dehydrogenase from *E. coli*. These transgenic plants showed increased tolerance to high salinity (Tarczynski *et al.*, 1993). Mannitol accumulating transgenic plants were 20-25% smaller than wild plants under non-stressed conditions. The dry weight of the wild type plants was reduced by 44% under 150 mM NaCl stress but the dry weight of transgenic plants remained unchanged. In contrary to wild type plants, transgenic plants showed adjustment against osmotic stress (Karakas *et al.*, 1997). In another experiment, *mtlD gene* construct was transformed into tobacco chloroplasts. Mannitol accumulation ranged from 2.5 – 7.0 μl mol g^{-1} fw. This resulted in increased resistance to methyl viologen induced oxidative stress. Such resistance was due to increased capacity to scavenge hydroxyl radicals (Shen *et al.*, 1997). The concentration of mannitol reached 10 μmol g^{-1} dw in the seeds of *Arabidopsis* plants accumulating mannitol (Thomas *et al.*, 1995). At a salinity stress of 400 mM NaCl, the mannitol accumulating seeds were able to germinate, whereas, the control seeds were unable to germinate at even 100mM NaCl.

Genetic transformation of *Arabidopsis* plants with *M6PR* gene from celery under the control of CaMV35S promoter resulted in transgenic plants accumulating 0.5-6 mM g^{-1} fw mannitol. Transgenic *Arabidopsis* plants were tolerant to salt stress as shown by application of irrigation water with 300 mM NaCl in nutrient solution (Zhifang and Loescher, 2003). In wheat, *mtlD gene* from *E. coli* was transformed through particle gun under the control of maize ubiquitin 1 promoter. Transgenic plants lacked the osmotically significant quantities of mannitol. However, they showed tolerance to water stress and salinity (Abebe *et al.*, 2003) (Table 1). Thus from the above findings it can be concluded that mannitol has a role as compatible osmolyte and osmoprotection through mannitol is one of the ways to produce salt and water stress tolerant transgenic plants.

(4) Trehalose

α, α-Trehalose (α-D-glucopyranosyl-1, 1-α-D glucopyranoside is a non reducing dissacharide that is commonly found in bacteria, fungi and other vertebrates (Elbein, 1974). From the research since a decade now, it is now clear that almost all green plants can synthesize trehalose but the function of trehalose metabolism in plants is still not clear. Resurrection plant *Selaginella lepidophylla* recovers from almost complete dessication, and it accumulates high levels of trehalose during rehydration. Protein and membrances are stabilized *in vitro* by trehalose by displacing water molecules and forming a glass like structure. These properties of *S. lepidophylla* and other resurrection plants help in protecting the intracellular structures from damage during oxidative stress (Lunn, 2007).

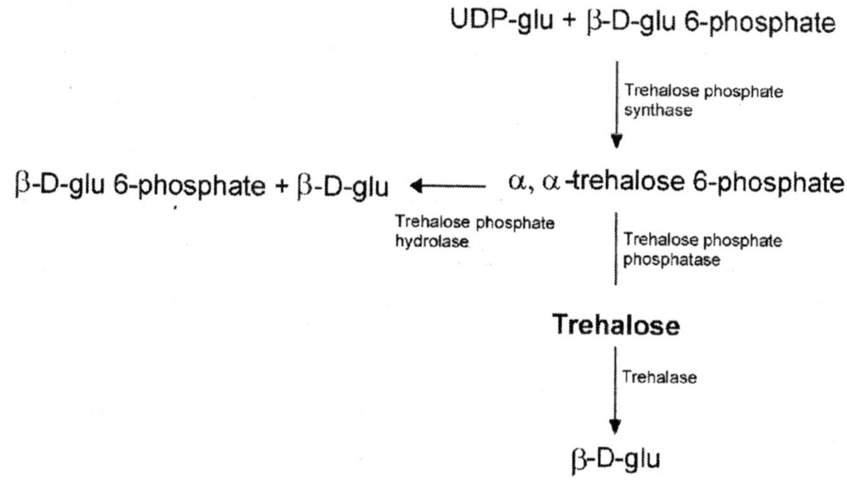

Fig. 2: Trehalose, synthesis and metabolism

Source: Rathinasabapathi and Kaur, (2006)

Biological synthesis of trehalose is a two step pathway. It is formed from glucose-6-phosphate and uridine diphosphoglucose via trehalose-6-phosphate.The first and the second step are mediated by trehalose phosphate synthase (*TPS*) and trehalose-6-phosphate phosphatase (*TPP*) as shown in fig. 2. In case of plants, (*TPS*) level is highly regulated by enzymes that either directly metabolizes it or by trehalase, which breaks down trehalose (Rathinasabapathi and Kaur, 2006).

The genetic engineering of trehalose genes in plants has led to an improvement of stress tolerance. When the TPS coding gene was expressed alone it resulted in striking morphological changes (Pilon-smits *et al.*, 1998; Yeo *et al.*, 2000). To overcome this problem both *TPS1* and *TPS2* homologues were used in a stress inducible expression in rice.

Bacterial genes *ots A* and *ots B* were transformed into rice. Transgenic rice accumulated trehalose 3-10 times that of non-transgenic controls and exhibited tolerance to salt, drought and low-temperature without stunting growth. Thus stress inducible transgene expression is important in recovering trehalose accumulating transgenic without deleterious effects (Garg *et al.*, 2002). Further, *TPS1-TPS2* fusion gene construct was introduced into *Arabidopsis* through *Agrobacterium* mediated gene transfer under the control of CaMV 35S or stress regulated *Cd 29A* promoter. No morphological growth alterations were observed in lines overexpressing the *TPS1-TPS2* construct, while the plants overexpressing the *TPS1* alone under the control of 35S promoter has abnormal growth, colour and shape (Miranda *et al.*, 2007) (Table 1). Thus it can be concluded that engineering trehalose metabolism in plants can substantially increase their capacity to tolerate abiotic stresses.

(b) Redox proteins and detoxifying enzymes

Oxidative stress refers to a serious imbalance between the production and removal of reactive oxygen species (ROS) (Halliwell, 1997). Although oxygen is important for living organisms to survive it makes survival difficult to aerobic living beings through the production of reactive oxygen species (ROS) such as superoxide radicals (O_2^-), hydrogen peroxide (H_2O_2) and hydroxyl radicals (OH). The increased production of reactive oxygen leads to drought and salinity stresses. Oxidative stress is also caused by the development of ROS. The quick build up of toxic products results in oxidative stress (Rathinasabapathi and Kaur, 2006). Plants have evolved various anti-oxidative response system (ARS) against ROS. Anti-oxidants such as ascorbate, reduced glutathione, tocopherols and oxidative enzymes which scavenge ROS also play key roles in oxidative stress response in plants (Asada, 1999; Noctor and Foyer, 1998; Wingsle *et al.*, 1999). In *Arabidopsis*, a network of at least 152 genes is involved in managing the levels of ROS (Mittler *et al.*, 2004). The increasing knowledge on the genetical, molecular and sub molecular basis of plant response to stress factors may be exploited for the development of oxidative stress-tolerant crops (Edreva, 2005). In this context many attempts have been made to produce transgenic plants overexpressing ROS-scavenging enzymes (*CAT, APX, SOD, GR*) and improvement in stress tolerance has been achieved.

SOD overexpressing alfalfa and tobacco plants were tolerant to water deficit and high salt, respectively (Mckersie *et al.*, 1996; Van Camp *et al.*, 1996). Field trials also confirmed the tolerance of transgenic plants to oxidative stress in field drought environment. Similar results were obtained with over production of *SOD* in mitochondria of lucerne (Mckersie *et al.*, 1997) and in cytosol of potato (Perl *et al.*, 1993). In another attempt, *Mn SOD* overexpression in chloroplasts of tobacco resulted in transgenic plants tolerant to Mn deficiency mediated oxidative stress (Yu *et al.*, 1999). The overexpression of cell wall peroxidase (*TPX2*) in tobacco plants improved germination under oxidative stress (Amaya *et al.*, 1999). Ascorbate peroxidase (*APX*) gene from *A. thaliana* was transformed into tobacco chloroplasts following *Agrobacterium* mediated gene transfer. The first generation of transgenic line showed enhanced tolerance to PEG and water stress as determined by net photosynthesis. This demonstrated the overexpression of cytosolic APX in tobacco chloroplasts reduces the toxicity of hydrogen peroxide (H_2O_2) (Badawi *et al.*, 2004).

Ascorbate (AsA) is a major antioxidant and free radical scavenger in plants. Mono dehydro ascorbate reductase MDAR is crucial for AsA regeneration and essential for maintaining a reduced pool of AsA. Further, MDAR gene from *A. thaliana* was over expressed in tobacco cytosol. Transgenic plants exhibited 2.1 fold higher MDAR activity and 2.2 fold higher level of reduced AsA as compared to wild type control plants. Transgenic plants also showed enhanced stress tolerance in terms of significantly higher net photosynthesis rates under ozone, salt and PEG stress. In addition, these transgenic plants showed significantly lower hydrogen peroxide level when tested under salt stress (Eltayeb *et al.*, 2007). Transgenic plants showing tolerance to H_2O_2 and paraquat included oxidative stress have been produced. *Gm TP55* antiquitin homologue gene from soybean has been overexpressed in *A. thaliana* and *N. tabaccum*. Both the transgenic plants showed tolerance to salinity during germination and to water deficit during plant growth. These transgenic lines also exhibited a lower concentration of lipid peroxidation derived reactive aldehydes under oxidative stress (Rodrigues *et al.*, 2006).

Targeting detoxification pathways is an appropriate approach for obtaining plants with multiple stress tolerance traits (Bartels, 2001). Glyoxalase enzymes are important for the glutathione (*GSH*) based detoxification of methylglyoxal (MG) which is formed primarily as a by-product of carbohydrate and lipid metabolism. Therefore many attempts have been made to introduce these enzymes for producing transgenic plants tolerant to environmental stresses including salinity and metal stress (Singla-Pareek *et al.*, 2003). *Gly I* responsible for glyoxalase pathway was overexpressed in tobacco. Transgenic plants showed improved tolerance to methylglyoxal (MG). In addition, the transgenic plants showed tolerance to high salinity (Veena *et al.*, 1999). Similarly introduction of both *Gly I* and *Gly I* genes resulted in double transgenic showing better response than the tobacco transgenic plants overexpressing the either gene singly. Double transgenic, i.e, *Gly II* gene transferred in an already transgenic *Gly I* background (Singla-Pareek *et al.*, 2003). Further, *Gly II* from rice was transformed into rice through *Agrobacterium* mediated gene transfer. Transgenic rice plants showed higher tolerance to toxic concentrations of methylglyoxal (MG) and NaCl (Singla-Pareek *et al.*, 2007). We have standardized transformation procedures in sugarcane (Kaur *et al.*, 2007) and for *indica* rice (Gosal *et al.*, 2001; Grewal *et al.*, 2006). Attempts are now being made to introduce transgene(s) such as *Gly I, Gly II, DREB1A, and ZAT 12* in different combinations into rice, sugarcane and maize for developing resistance to abiotic stresses.

Antioxidant genes are significant in promoting an understanding of the roles of specific antioxidant defenses under stress conditions (Zhao and Zhang, 2007). Transgenic rice overexpressing the *GST* (Glutathione S-transferase) and *CAT I* (Catalase) from *Suaeda salasa* showed increased tolerance to oxidative stress caused by salt and paraqaut (Zhao and Zhang, 2006[a], 2006[b]). Double transgenic showed (*GST + CAT 1*) higher abiotic stress tolerance as compared to single *GST*. Thus again showing the need of gene pyramiding or multigene transfer for a complex trait of abiotic stress.

(c) Stress-induced proteins

Coping with environmental stresses plants activate a large set of genes leading to the accumulation of specific stress-associated proteins (Vierling, 1991; Ingram and Bartels, 1996; Bohnert and Sheveleva, 1998; Thomashow 1999; Hoekstra *et al.*, 2001). Late embryogenesis abundant (LEA)-type proteins and Heat-shock proteins (Hsps) are two major types of stress-induced proteins that accumulate upon water, salinity, and extreme temperature stress. They have been shown to play a role in cellular protection during the stress (Vierling and Kimpel, 1992; Boston *et al.*, 1996; Close 1996; Ingram and Bartels 1996; Waters *et al.*, 1996; Thomashow 1998).

(1) LEA proteins

Late embryogenesis abundant (LEA) proteins are mainly low molecular weight (10–30 kDa) proteins, which are involved in protecting higher plants from damage caused by environmental stresses, especially drought (dehydration) (Hong *et al.*, 2005). LEA genes encode mRNAs and proteins that are late embryogenesis abundant in higher plant seed embryos. Initially LEA genes were characterized in cotton as a set of proteins that are highly accumulated in the embryos at the late stage of seed development (Park *et al.*, 2005[b]). Subsequently many LEA proteins or their genes have been characterized from different plant species. LEA proteins are generally divided into five groups, and they appear to protect cellular structures from dehydration stress; however, the exact functional role of these hydrophilic proteins remains poorly understood. Amongst the several groups of LEA proteins, those belonging to group 3 are predicted to play a role in sequestering ions that are concentrated during cellular dehydration. These proteins have 11-mer amino acid motifs with the consensus sequence TAQAAKEKAGE repeated as many as 13 times (Dure, 1993). The group 1 LEA proteins are predicted to have enhanced water-binding capacity, while the group 5 LEA proteins are thought to sequester ions during water loss. LE25, a group 5 LEA protein from tomato (*Lycopersicum esculentum* L.) improved resistance to high salinity and freezing when expressed in *Saccharomyces cerevisiae* (Imai *et al.*, 1996); similarly, the hiC6 gene encoding a homologue of LEA enhances freezing tolerance of *S. cerevisiae* (Honjoh *et al.*, 1999).

Overexpression of barley (*Hordeum vulgare* L.) group 3 LEA protein HVA1 in rice resulted in transgenic plants constitutively accumulating HVA1 protein both in leaves and roots. Second generation transgenic plants showed increased tolerance to water deficit and salinity stress. Transgenic plants maintained higher growth rates than the non transformed wild type control plants (Xu *et al.*, 1996). Similar attempt resulted in improved biomass productivity and water use efficiency under water deficit conditions in transgenic wheat constitutively expressing the barley HVA1 gene (Sivamani *et al.*, 2000). In this case the gene was transformed into immature embryos of green house grown wheat by particle gun transformation method under the control of maize *ubi1* promoter. After five years, field evaluation of these transgenic plants showed yield increase in the transgenic lines under drought conditions (Bahieldin *et al.*, 2005) (Table 1).

Overexpression of *COR15a*, a group-II LEA protein which was targeted to the chloroplasts, increased freezing tolerance of chloroplasts *in vivo*, and of protoplasts *in vitro* (Artus *et al.*, 1996). This increase most likely resulted from the membrane-stabilizing effect of *COR15a* (Artus *et al.*, 1996; Steponkus *et al.*, 1998). However, the protective effect of *COR15a* was insignificant for the survival of whole plants during freezing (Jaglo-Ottosen *et al.*, 1998).

According to an investigation done on third generation of transgenic oat (*Avena sativa* L.) expressing barley *HVA1* stress tolerance (*uidA*; *gus*) and *bar* herbicide resistance genes. Transgenic plants showed normal 9:7, third generation inheritance for glufosinate ammonium herbicide resistance. Molecular and histochemical studies confirmed the presence and stable expression of all three genes. Compared with the non-transgenic control plants, transgenic R3 plants exhibited greater growth and showed a significant increase in tolerance to salt stress conditions (200 mM NaCl) for traits including number of days to heading, plant height, flag leaf area, root length, panicle length, number of spikelets/panicle, number of tillers/plant, number of kernels/panicle, 1000-kernel weight, and kernel yield/plant (Oraby *et al.*, 2005).

In separate experiments *ME-lea N4* gene from *Brassica napus* was transformed through *Agrobacterium* mediated gene transfer under the control of CaMV 35S promoter into lettuce *Lactuca sativa* L (Park *et al.*, 2005[a]) and chinese cabbage *Brassica campestris* spp. *pekinensis* (Park *et al.*, 2005[b]). In both the cases transgenic plants showed enhanced growth ability as compared to non-transgenic controls under salt and water deficit stress.

Further, *OsLEA 3-1* has been transformed into rice through *Agrobacterium* mediated gene transfer method under the control of different promoters. Constitutive and stress inducible expression of *Os LEA 3-1* under the control of CaMV 35S and HVA1-like promoter respectively resulted in transgenic rice plants showing increased tolerance to drought under field conditions (Xiao *et al.*, 2007). Thus the above findings show that LEA genes have the potential to confer environmental stress protection in various crop plants.

(2) Heat-shock proteins

Since plants are sedentary, they have evolved a range of mechanisms that protect them upon exposure to high temperatures. These mechanisms serve to protect the plant from heat-induced tissue damage. One such mechanism involves the accumulation of various types of heat-shock proteins (HSPs), including *HSP70, HSP90, HSP100/Clp,* and small *HSP* (Yamada *et al.*, 2007). Heat-shock proteins (*HSPs*) protect proteins, membranes, and other cellular components during heat-stress and facilitate repair or degradation of damaged proteins following a stressful event (Parsell and Lindquist, 1994; Wang *et al.*, 2004[b]). The heat-shock response in plants is similar to that of other organisms. The induction of heat-shock genes is known to be regulated mainly by the trimerization and activation of heat-shock factors (*HSFs*). *HSFs* act in the promoter region of the *HSP* genes through a well-defined and highly conserved heat-shock element (HSE) (Prasinos *et al.*, 2005). It is believed that HSPs exhibit chaperone functions by assisting other proteins in proper post-translational folding and maintaining them in a functional state (Rathinasabapathi and Kaur, 2006)

Many attempts have been made to produce transgenic plants to overproduce heat-shock proteins in *Arabidopsis*. Overexpression of *At hsp101* in *A. thaliana* resulted in manipulation of thermotolerance in transgenic plants (Queitsch *et al.*, 2000). Sun *et al.*, (2001) demonstrated that the expression of *Arabidopsis AtHSP17.6A* is regulated by heat-shock and osmotic stress, and is induced during seed maturation. Overexpression of

AtHSP17.6A in *A. thaliana* conferred higher osmotolerance; however, the *AtHSP17.6A* transgene failed to increase heat tolerance. Another gene from *hsp* 90 family, *At hsp 90-3* expressed in cytosol of *Arabidopsis* plants showed the induction of hsp genes (Prasinos *et al.*, 2005). *Hsp 90.2* a gene from same family regulated the heat-shock response in transgenic *Arabidopsis* plants overexpressing it (Yamada *et al.*, 2007) (Table 1). In rice, expression of *Athsp101* resulted in higher survival and better growth performance in the recovery phase following the heat stress (Katiyar-Agarwal *et al.*, 2003).

(d) Transporter proteins

Salinity has three potential effects on plants (1) lowering of the water potential, (2) direct toxicity of any Na and Cl absorbed and (3) interference with the uptake of essential nutrients. Salt tolerance depends upon: morphology; compartmentation and compatible solutes; regulation of transpiration; control of ion movement; membrane characteristics; tolerating high Na/K ratios in the cytoplasm and salt glands. With so many factors involved, it is to be expected that salt tolerance would depend on the action of many genes (Flowers and Flowers, 2005). Electrical conductivity of saturated paste extract (ECe) of 4.0 deci-siemens per meter (dS/m) is considered a lower limit to define saline soils. Rice and most grain crops are glycophytes that show stress symptoms and reduced yield even when the ECe is lower than 4.0 dS/m (~40 mmol/L NaCl). The salinity threshold for rice is 3.0 dS/m with a 12% reduction in yield, per dS/m, beyond this threshold (Maas, 1990).

Sodium does not freely permeate the plasma membrane, as it enters the plant cell through transporters and nonselective cation channels. *HKT* (High-affinity K^+ transporter) is a group of well-studied plant Na^+ transporters identified in several species (Platten *et al.*, 2006). In wheat (Laurie *et al.*, 2002), rice (Garciadeblás *et al.*, 2003) and barley (Haro *et al.*, 2005), it has been shown that *HKT* transporters are involved in root Na^+ uptake. Ion transporters selectively transport ions and maintain them at physiologically relevant concentrations while Na^+/H^+ antiporters also play a crucial role in maintaining cellular ion homeostasis, thus permitting plant survival and growth under saline conditions. The Na^+/H^+ antiporters catalyze the exchange of Na^+ for H^+ across membranes and have a variety of functions, such as regulating cytoplasmic pH, sodium levels and cell turgor (Serrano *et al.*, 1999).

Genetic engineering of ion transporters and their regulators, and of the CBF (C-repeat-binding factor) regulons, holds promise for future development of salt-tolerant crops (Chinnusamy *et al.*, 2005). Overexpression of the vacuolar Na^+/H^+ antiporter *AtNHX1* in *Arabidopsis* plants (Apse *et al.*, 1999) promoted growth and development in potting medium irrigated with up to 200 mM sodium chloride. This salinity tolerance was positively correlated with elevated levels of *AtNHX1* transcript, and with protein and vacuolar Na^+/H^+ antiporter activity. Transgenic *Brassica napus* plants overexpressing *AtNHX1* were able to grow, flower and produce seeds, in the presence of 200 mM sodium chloride, even though they accumulated sodium at a rate of up to 6% dry weight.

Moreover, their seed yields and seed oil quality were not altered by the high soil salinity (Zhang *et al.*, 2001). *AtSOS1* from *Arabidopsis thaliana* has been shown to encode a plasma membrane Na^+/H^+ antiport with significant sequence similarity to plasma membrane Na^+/H^+ antiporters from bacteria and fuñgi (Shi *et al.*, 2000). The overexpression of *SOS1* improved the salt tolerance of *Arabidopsis*, demonstrating that improved salt tolerance can be attained by limiting Na^+ accumulation in plant cells (Shi *et al.*, 2003).

Ohta *et al.*, (2002) engineered salt-sensitive rice cultivar 'Kinhuikari' using the Na^+/H^+ antiporter gene from *Atriplex* spp (*AgNHX1*). The activity of the vacuolar type Na^+/H^+ antiporter in the transgenic rice plants was eight fold higher than in wild type rice plants. Salt tolerance assays followed by non-stress treatments showed that the transgenic plants over-expressing *AgNHX1* could survive under conditions of 300 mM NaCl for 3 days while the wild type plants failed to survive.

In *Arabidopsis thaliana*, the calcium binding protein salt overly sensitive3 (*SOS3*) interacts with and activates the protein kinase *SOS2*, which in turn activates the plasma membrane *Na^+/H^+* antiporter *SOS1* to bring about sodium ion homeostasis and salt tolerance. Expression of some of the activated *SOS2* proteins in wild-type *A. thaliana* conferred increased salt tolerance. These studies demonstrate that the protein kinase activity of *SOS2* is partially sufficient for activation of *SOS1* and for salt tolerance in vivo and *in planta* and that the kinase activity of *SOS2* is limiting for plant salt tolerance. The results also reveal an essential *in planta* role for the *SOS2* C-terminal regulatory domain in salt tolerance (Guo *et al.*, 2004). Likewise, overexpression of wheat Na^+/H^+ antiporter *TNHX1* and H^+-pyrophosphatase *TVP1* through *Agrobacterium* mediated gene transfer under the action of Ca MV 35S promoter improved salt-and drought-stress tolerance in *Arabidopsis thaliana* plants (Brini *et al.*, 2007) (Table 1).

(B) Regulatory genes

Until now, we have seen that genetic engineering for stress tolerance included genes involed in encoding enzymes required for the biosyntheses of various osmo protecants. Other genes that have been selected for transformation include those that encode enzymes for modifying membrane lipids, *LEA* proteins and detoxification enzymes. In all the above experiments, a single gene for a protective protein or an enzyme was constitutively expressed in the transgenic plants. Many of these genes have been shown to function in stress conditions.

The gene regulating protein factors that regulate gene expression and signal transduction and that function in stress responses may be useful for improving the tolerance of plants to abiotic stresses by gene transfer. These genes comprise of regulatory proteins *i.e.*, transcription factors (*bZip, MYC, MYB, DREB and NACs, NAM, ATAF* and *CUC etc.*, protein kinases (*MAP* kinase, CDP kinase, receptor protein kinase, ribosomal protein kinase and transcription regulation protein kinase *etc.*) and proteinases (phosphoesterases, phospholipase *etc.*) (Katiyar-Agarwal *et al.*, 2006).

(1) DREB

There has been a continuous and tremendous research on trancriptional control of abiotic stress in model and crop plants. So to enumerate all that work is out of the scope of this chapter. More emphasis will be given on work done on drought, salinity and cold tolerance in regard to transcription factors especially *DREB* family.

The transcription factors activate cascades of genes that act together in enhancing tolerance towards multiple stresses. It has been revealed that dozens of transcription factors are involved in the plant response to drought stress (Vincour and Altman, 2005; Bartels and Sunkar, 2005). The transcription activation factors interact with cis-elements present in the promoter region of various abiotic stress related genes and thus upregulating the expression of many genes resulting in imparting tolerance to abiotic stresses. Model plant *Arabidopsis* dedicates about 5.9% of its genome coding for more than 1500 transcription factors (Riechmann *et al.*, 2000). The transcription factors which are involved in stress responsive gene expression are traditionally classified in ABA dependent and ABA-independent regulatory pathways (Shinozaki and Yamaguchi Shinozaki, 2000; Thomashow, 1999; Xiong *et al.*, 2002). The DRE (dehydration responsive element) CRT (C-Repeat) was identified as a *cis*-acting element regulating gene expression in response to dehydration (salt, drought and cold stresses) in *Arabidopsis* (Yamaguchi-Shinozaki and Shinozaki, 1994). Initial attempts for genetic transformation using these *DREB* genes started with *Arabidopsis*. Overexpression of *DREB1/CBF* in *Arabidopsis* resulted in the activation of expression of many stress tolerance genes and tolerance of the plant to drought, high salinity and/or freezing was improved (Jaglo-Ottosen *et al.*, 1998, Liu *et al.*, 1998). Transgenic plants overexpressing *DREB1/CBF3* under the control of CaMV 35S promoter also showed increased tolerance to drought, high salinity and freezing stress (Kausga *et al.*, 1999, Gilmour *et al.*, 2000). Constitutive expression of *DREB1A* protein lead to growth retardation under normal growth conditions. Thus the use of stress inducible *rd 29A* promoter for the overexpression of *DREB1A* minimizes negative effects on plant growth (Kasuga *et al.*, 1999).

In their first attempt to transform wheat with *DREB1A* transcription factor, (Pellegrineschi *et al.*, 2004) used particle gun method as the means for gene transfer. *DREB1A* was transformed under the control of stress inducible *rd29A* promoter. Under green house conditions substantial resistance to water deficit was demonstrated. Oh *et al.*, (2005) ectopically expressed *Arabidopsis DREB1A /(CBF3)* in transgenic rice plants under the control of CaMV 35S promoter. *DREB1A* transgenic rice plants showed enhanced tolerance to drought and salinity but to a little extent for low temperature stress without any stunted phenotype despise its constitutive expression.

Four rice *CBF/DREB1A* orthologs, *OsDREB1A, OsDREB1B, OsDREB1C* and *OsDREB1D* have been isolated (Dubouzet *et al.*, 2003). *OsDREB1* transgenic rice plants conferred improved tolerance to drought, salt and low temperatures. On further analysis a large portion of stress inducible genes were identified. Thus giving the confirmation that *DREB1/CBF* cold responsive pathways is conserved in *Arabidopsis* and rice (Ito *et al.*, 2006). In another attempt constitutive or stress inducible expression of *ZmDREB2A* resulted in an improved drought stress tolerance in plants. In addition, transgenic plants showed enhanced thermo tolerance, thus indicating that *ZmDREB2A* has a dual function mediating expression of genes responsive to both water and heat stress (Qin *et al.*, 2007). Similar results were earlier found in *Arabidopsis* plants overexpressing constitutive active *DREB2A* in which transgenic plants showed increased thermo tolerance in addition to tolerance against water stress (Sakuma *et al.*, 2006).

Stress tolerance is a complex trait so unlikely to be under single gene control wise strategy is to use genetic engineering to switch on a transcription factor regulating the expression of several genes related to abiotic stress. Transpiration efficiency is the most important component trait of drought. About 90% of the rooted shoots transformed with *rd 29A:DREB1A* construct survived and appeared to be phenotypically normal. Fifty three *rd29A:DREB1A* and eighteen *35S:DREB1A* plants were successfully transferred to green house and their seeds collected. Transgenic peanut had higher TE than the wild type under well watered conditions. Some transgenic events showed 70% increase over wild type JL 24, RD 2 had 40% more TE than WT JL 24 under water limited conditions (Bhatnagar-Mathur *et al.*, 2007) Transgenic over-expression of *HvCBF4* from barley in rice resulted in an increase in tolerance to drought, high-salinity and low-temperature stresses without stunting growth. Using the 60 K Rice Whole Genome microarrays, 15 rice genes were identified that were activated by *HvCBF4*. When compared with 12 target rice genes of *CBF3/DREB1A*, five genes were common to both *HvCBF4* and *CBF3/DREB1A*, and ten and seven genes were specific to *HvCBF4* and CBF3/*DREB1A*, respectively (Oh *et al.*, 2007) (Table 1).

(2) Zinc finger proteins

Zinc finger proteins (*ZFPs*) play important roles in growth and development in both animals and plants. *Arabidopsis* genes encoding distinct *ZFPs* have been identified. However, the physiological role of their homologues with putative zinc finger motif remains unclear. *ThZF1*, a novel gene was characterized from salt stressed cress (*Thellungiella halophila*, Shan Dong), encoding a functional transcription factor. *ThZF1* contains two conserved C2H2 regions and shares conserved domains, including DNA-binding motif, with *Arabidopsis thaliana ZFP* family members. The transcript of the *ThZF1* gene was induced by salinity and drought. Transient expression analysis of *ThZF1–GFP* fusion protein revealed that *ThZF1* was localized preferentially in nucleus. A gel-shift assay showed that *ThZF1* specially bind to the wild-type (WT) *EP2* element, a *cis*-element present in the promoter regions of several target genes regulated by *ZFPs*. Furthermore, a functional analysis demonstrated that *ThZF1* was able to activate HIS marker gene in yeast. Finally, ectopic expression of *ThZF1* in *Arabidopsis* mutant azf2 suggested that *ThZF1* may have similar roles as *Arabidopsis AZF2* in plant development as well as regulation of downstream gene (Xu *et al.*, 2007)

(3) NAC genes

The NAC gene family encodes one of the largest families of plant specific transcription factors and is absent in other eukaryotes (Gao *et al.*, 2007). Rice and *Arabidopsis* genomes contain 75 and 105 putative *NAC* genes, respectively (Ooka *et al.*, 2003). The role of NAC gene family in abiotic stresses was discovered in *Arabidopsis*. Three NAC genes were induced under salt and/ or drought stress. Overexpression of three genes *ANACO19*, *ANACO55*, *ANACO72*, greatly enhanced drought tolerance in *Arabidopsis* (Tran *et al.*, 2004). Multiple rice transcription factors, including a NAC gene were induced in the early stages of salt stress (Chao *et al.*, 2005) *OsNAC6*, a member of *ATAF* family, was also induced by cold, salt, drought and abscisic acid (ABA) (Ohnishi *et al.*, 2005). Overexpression of stress responsive gene *SNAC1* (stress-responsive *NAC* 1) significantly enhances drought resistance in transgenic rice (22–34% higher seed setting than control) in the field under severe drought stress conditions at the reproductive stage while showing no phenotypic changes or yield penalty. The transgenic rice also shows significantly improved drought resistance and salt tolerance at the vegetative stage. Compared with wild type, the transgenic rice were more sensitive to abscisic acid and lost water more slowly by closing more stomatal pores, yet displayed no significant difference in the rate of photosynthesis. DNA chip analysis revealed that a large number of stress-related genes were up-regulated in the *SNAC1*-overexpressing rice plants. *SNAC1* holds promising utility in improving drought and salinity tolerance in rice (Hu *et al.*, 2006).

CONCLUSION AND FUTURE PERSPECTIVE

In conclusion abiotic stresses, causing up to 70% losses in the crop yield, present a major challenge in our quest for sustainable food production. Due to complex nature of abiotic stresses, the conventional breeding approaches have not been so successful in developing stress tolerant crop varieties. Concerted efforts combining molecular, physiological and metabolic aspects relating to abiotic stresses are required for bridging the knowledge gaps between short-and long-term effects of the genes and their products, and between the molecular or cellular expression of the genes and the whole plant phenotype under stress. With the advent of innovative approaches of biotechnology our understanding of the processes underlying plant responses to abiotic stresses at molecular and whole plant levels have rapidly progressed. Hundreds of genes that are induced under abiotic stresses have been identified and some of these have been cloned. A range of tools from gene expression pattern to transgenic plants have now become available for better understanding. Recent success on laboratory-production of drought stress-tolerant transgenics has been achieved that must be exploited in future. While insect, viral, and herbicide-resistant transgenic plants are being commercially grown, abiotic stress-tolerant transgenics are still under pot experiments or under field evaluation. Monsanto is field testing drought-tolerant GM cotton and maize in USA which are expected to be commercialized in the near future.

Fig. 1

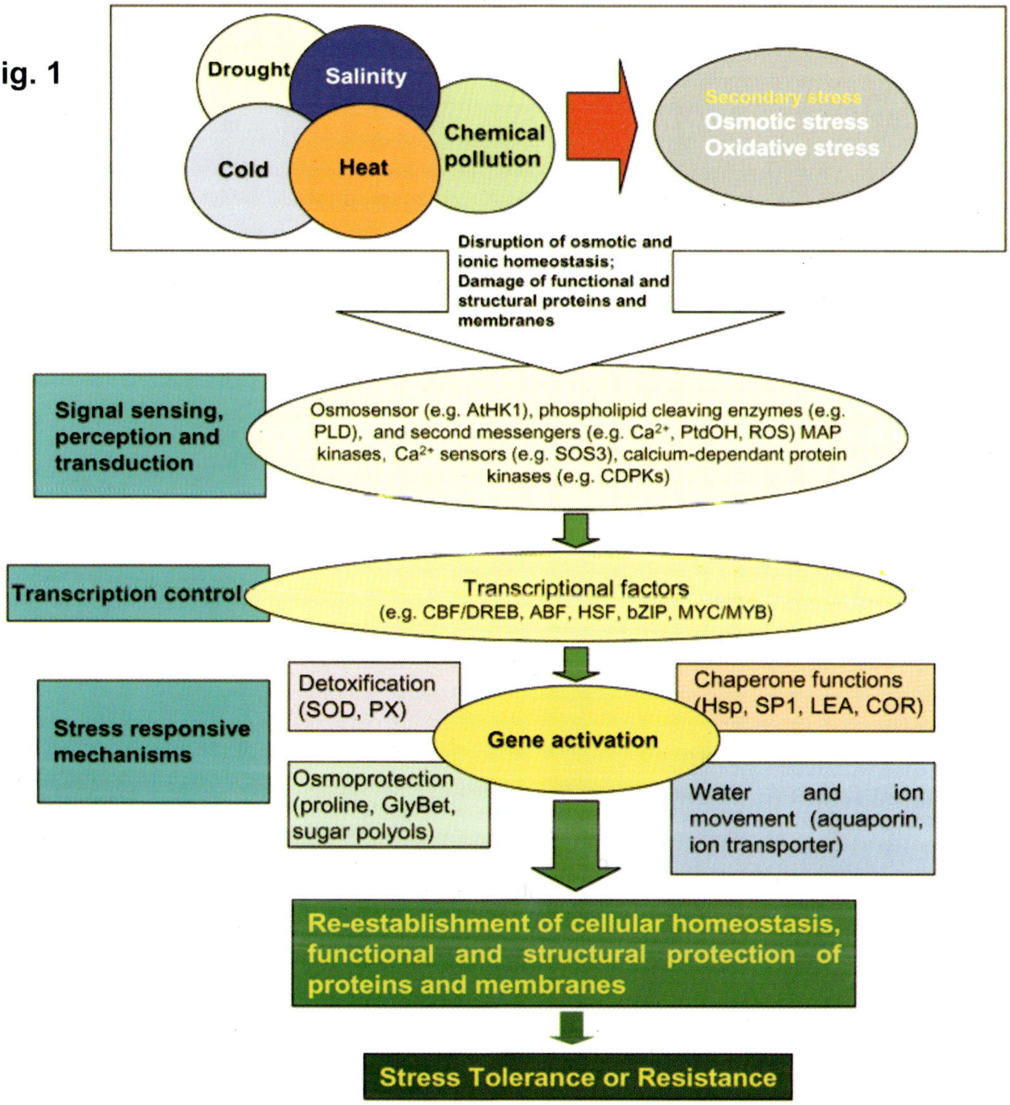

Fig. 1) The complexity of the plant response to abiotic stress. Primary stresses, such as drought, salinity, cold, heat and chemical pollution are often interconnected, and cause cellular damage and secondary stresses, such as osmotic and oxidative stress. The initial stress signals (e.g. osmotic and ionic effects, or temperature, membrane fluidity changes) trigger the downstream signaling process and transcription controls which activate stress-responsive mechanisms to re-establish homeostasis and protect and repair damaged proteins and membranes. Inadequate response at one or several steps in the signaling and gene activation may ultimately result in irreversible changes of cellular homeostasis and in the destruction of functional and structural proteins and membranes, leading to cell death. *(Source: Wang et al. 2003)*

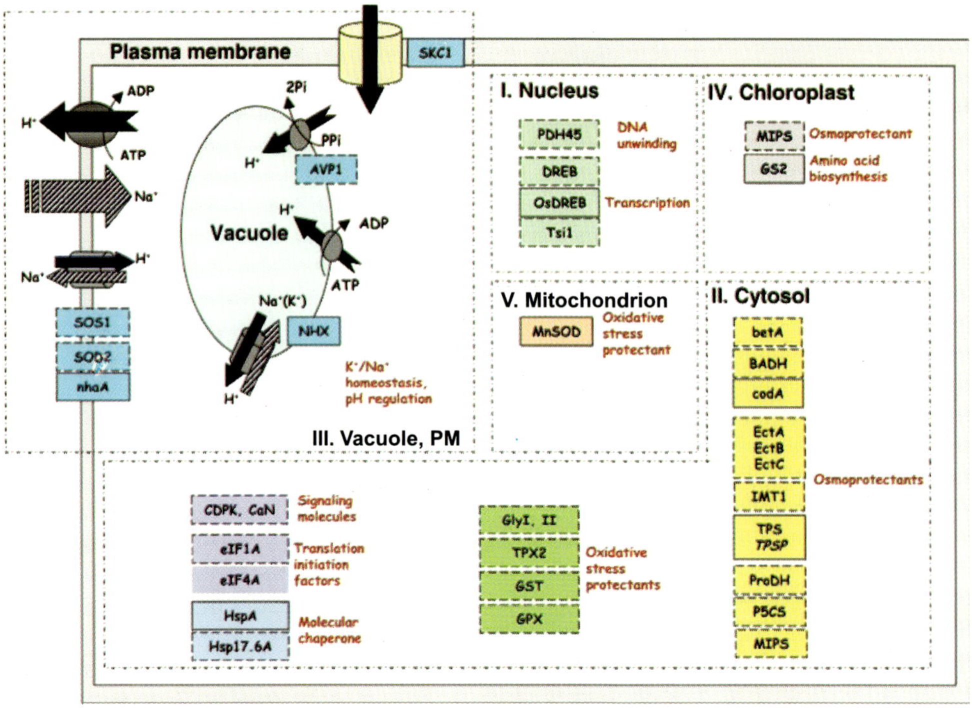

Fig. 3) Schematic representation of mechanisms associated with salt tolerance phenotype in different cellular compartments of the plant cell. For validating the functionality of proteins shown in solid boxes in salt response, transgenic rice plants have been raised. Proteins shown in broken boxes represent examples where the transgenic host employed is other than rice. Evidence for the involvement of proteins not shown in solid or broken boxes in salt response has been obtained using varied other approaches (such as biochemical experiments, transcript/protein expression studies, mutant analysis, or yeast complementation assays). Involvement of SKC1 in salt tolerance was initially suggested on the basis of QTL analysis and, later on, it was functionally validated by producing transgenic rice plants. DREB, dehydration responsive element binding protein (Kasuga et al. 1999; Dubouzet et al. 2003); PDH45, pea DNA helicase 45 (Sanan-Mishra et al. 2005); Tsi1, tobacco stress-induced gene 1 (Park et al. 2001); BADH, betaine aldehyde dehydrogenase (Holmstrom et al. 2000; Jia et al. 2002; Yilmaz and Bulow 2002); CaN, calcineurin (Pardo et al. 1998); codA, choline oxidase (Lilius et al. 1996; Hayashi et al. 1997; Sakamoto et al. 1998; Alia et al. 1999; Prasad et al. 2000a, b); CDPK, calcium-dependent protein kinase (Saijo et al. 2000); ectA, ectB, and ectC, L-2,4- diaminobutyric acid acetyltransferase, L-2, 4-diaminobutyric acid transaminase, and Lectoine synthase (Nakayama et al. 2000); eIF1A, eukaryotic translation initiation factor 1A (Rausell et al. 2003); eIF4A, eukaryotic translation initiation factor 4A (Montero-Lomeli et al. 2002); Gly I and II, glyoxylase I and II (Veena et al. 1999; Singla-Pareek et al. 2003); GST, glutathione-S-transferase (Roxas et al. 1997); GPX, glutathione peroxidase (Roxas et al. 1997); HspA (Asadulghani et al. 2004); Hsp17.6A (Sun et al. 2001); IMT1, myo-inositol O-methyl transferase (Vernon et al. 1993); P5CS, pyrroline-5-carboxylate synthetase (Kishore et al. 1995); ProDH, proline dehydrogenase (Nanjo et al. 1999); SOD2, sodium 2 (Gao et al. 2003); TPS, trehalose-6-phosphate synthase; TPSP, TPS phosphatase (Garg et al. 2002; Jang et al. 2003); TPX2, tomato peroxidase 2 (Amaya et al. 1999); AVP1, vacoular H+ pyrophosphatase (Gaxiola et al. 2001; Park et al. 2005); NHX1, Na+ /H+ antiporter (Apse et al. 1999; Zhang and Blumwald 2001; Zhang et al. 2001; Ohta et al. 2002; Fukuda et al. 2004; Wang et al. 2004; Wu et al. 2004; Xue et al. 2004; Yin et al. 2004; He et al. 2005; Lu et al. 2005; Wu et al. 2005); SKC1, HKT type (Ren et al. 2005); SOS1, salt overly sensitive 1 (Shi et al. 2003); GS2, glutamine synthetase (Hoshida et al. 2000); MIPS, L-myoinositol 1-phosphate synthase (Majee et al. 2004); MnSOD, Mn superoxide dismutase (Tanaka *et al.* 1999) *(Source :Sahi et al. 2006)*

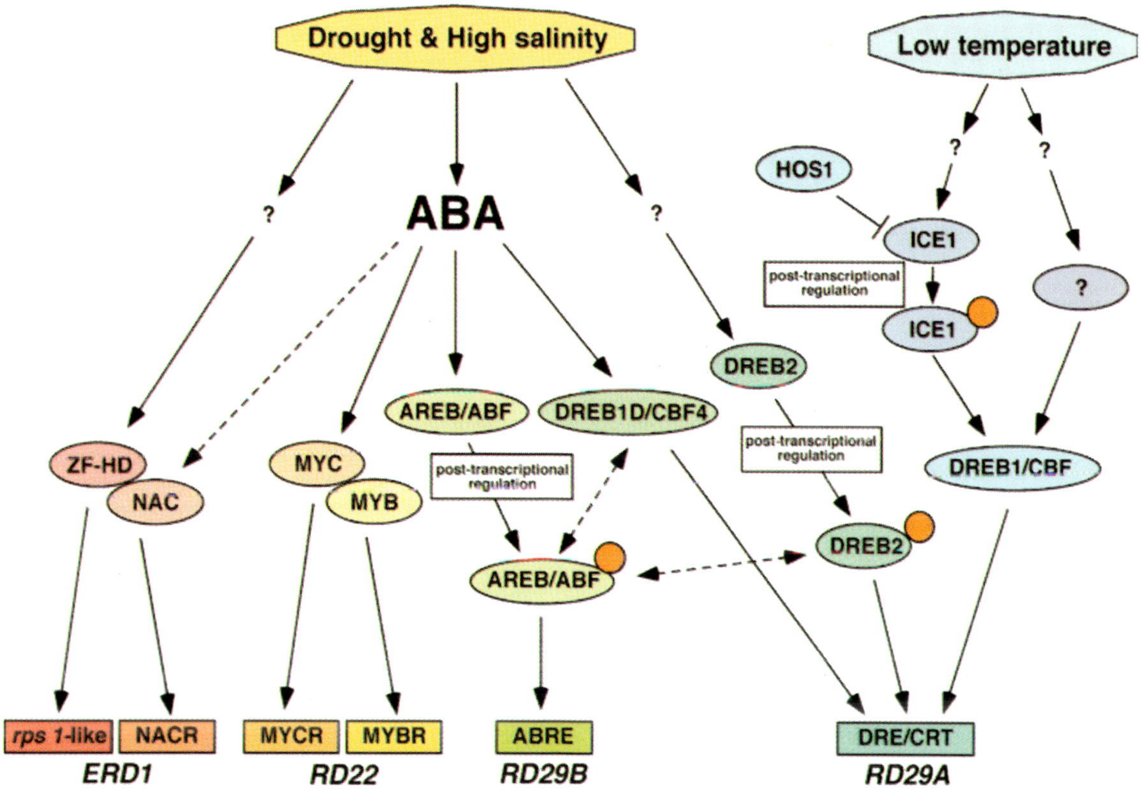

Fig. 4) Regulatory network of gene expression in response to drought, high salinity and cold stresses: specificity and crosstalk of gene networks.

REFERENCES

Abebe T A C Guenzi B Martin and J C Cushman 2003 Tolerance of mannitol-accumulating trangenic wheat to water stress and salinity *Plant Physiol* **131**:1748-1755

Alia K Y A Sakamoto H Nonaka H Hayashi P P Saradhi T H H Chen and N Murata 1999 Enhanced tolerance to light stress of transgenic *Arabidopsis* plants that express the codA gene for a bacterial choline oxidase *Plant Mol Biol* **40**:279–288

Amaya I M A Botella M De La Calle M I Medina A Heredia R A Bressan P M Hasegawa MA Quesada and V Valpuesta 1999 Improved germination under osmotic stress of tobacco plants overexpressing a cell wall peroxidase *FEBS Lett* **457**:80–84

Apse M P G S Aharon W A Snedden E Blumwald 1999 Salt tolerance conferred by overexpression of a vacuolar Na^+/H^+ antiport in *Arabidopsis* Sci **285**: 1256–1258

Artus N N M Uemura P L Steponkus S J Gilmour C Lin and M F Thomashow 1996 Constitutive expression of the cold-regulated *Arabidopsis thaliana* COR15a gene affects both chloroplast and protoplast freezing tolerance *Proc Natl Acad Sci* **93**:13404-13409

Asada K 1999 The water-water cycle in chloroplasts: scavenging of active oxygens and dissipation of excess photons *Annu Rev Plant Physiol Plant Mol Biol* **50**: 601–639

Asadulghani K Nitta Y Kaneko K Kojima H Fukuzawa H Kosaka and H Nakamoto 2004 Comparative analysis of the hspA mutant and wild-type *Synechocysti* sp. Strain PCC 6803 under salt stress: evaluation of the role of hspA in salt-stress management *Arch Microbiol* **182**:487–497

Badawi G H N Kawano Y Yamauchi E Shimada R Sasaki A Kubo and K Tanaka 2004 Overexpression of ascorbate peroxidase in tobacco chloroplasts enhances the tolerance to salt stress and water deficit *Physiol Plant* **121**: 231–238

Bahieldin A H T Mahfouz Eissa O M Saleh A M Ramadan I A Ahmed W E Dyer H A El-Itriby and M A Madkour 2005 Field evaluation of transgenic wheat plants stably expressing the *HVA1* gene for drought tolerance *Physiol Plant* **123**: 421–427

Bartels D 2001 Targeting detoxification pathways: an efficient approach to obtain plants with multiple stress tolerance *Trends Plant Sci* **6**:284–286

Bartels D and R Sunkar 2005 Drought and salt tolerance in plants *Crit Rev Plant Sci* **21**:1–36

Behnam B A Kikuchi F Celebi-Toprak S Yamanaka M Kasuga K Yamaguchi Shinozaki K N Watanabe 2006 The *Arabidopsis* DREB1A gene driven by the stress-inducible rd29A promoter increases salt-stress tolerance in proportion to its copy number in tetrasomic tetraploid potato (*Solanum tuberosum*) *Plant Biotech* **23**:169–177

Bhalla-Sarin N P Bhomkar S Debroy N Sharma M Saxena C P Upadhyaya A Muthusamy M Pooggin and T Hohn 2004 Transformation of *Vigna mungo* (blackgram) for abiotic stress tolerance using marker free approach Proceedings of the 4th International Crop Science Congress Brisbane Australia 26 Sep – 1 Oct

Bhatnagar-Mathur P M J Devi D S Reddy M Lavanya V Vadez R Serraj K Yamaguchi-Shinozaki and K K Sharma 2007 Stress inducible expression of *At DREB1A* in transgenic peanut (*Arachis hypogaea*) increases transpiration efficiency *Pl Cell Rep* (in press)

Blum A and Ebercon A 1976 Genotypic response in sorghum to drought stress III Free praline accumulation and drought resistance *Crop Sci* **16**: 428-431

Bohnert H J and Sheveleva E 1998 Plant stress adaptations – making metabolism move *Curr Opin Plant Biol* 1:267–274

Borlaug NE and Doswell C R 2005 Feeding a world of ten billion people :a 21st century challenge In proceeding s of "In the wake of double helix; From the green revolution to the gene revolution" 27-31st may 2003 Bologna Italy

Boston R S Viitanen PV Vierling E 1996 Molecular chaperones and protein folding in plants *Plant Mol Biol* **32**:191–222

Brini F M Hanin I Mezghani G Berkowitz A and Masmoudi K 2007 Overexpression of wheat Na^+/H^+ antiporter TNHX1 and H^+-pyrophosphatase TVP1 improve salt-and drought-stress tolerance in *Arabidopsis thaliana* plants *J Exp Bot* **58**: 1013-1019

Chahal G S and Gosal S S 2002 Principles and Procedures of Plant Breeding: Biotechnological and Conventional Approaches Narosa Publ House New Delhi

Chen S C Liu H W Lee K T and Yamakawa T 2007 High-efficiency *Agrobacterium rhizogenes*-mediated transformation of heat inducible sHSP18 2-GUS in *Nicotiana tabacum Plant Cell Rep* **26**: 29–37

Chiappetta L Tomes D Xu D Sivasankar S Sanguineti1 M C and Tuberosa R 2005 Dreb1 overexpression improves tolerance to low temperature in maize In: Tuberosa R Phillips R L Gale M (eds) Proceedings of the International Congress "In the Wake of the Double Helix: From the Green Revolution to the Gene Revolution" 27-31 May 2003 Bologna Italy 533-544 Avenue media Bologna Italy

Chinnusamy V Jagendorf A and Zhu J K 2005 Understanding and Improving Salt Tolerance in Plants *Crop Sci* **45**:437–448

Chinnusamy V Xiong L and Zhu J K 2006 In: Abiotic stresses Plant resistance through Breeding and Molecular Approach Ashraf M and PJC Harris (ed) IBDC U P India 47-89

Christian J H 1955 The influence of nutrition on the water relationsof *Salmonella oranienburg Aust J Biol Sci* **8**: 75-82

Close T J 1996 Dehydrins: emergence of a biochemical role of a family of plant dehydration proteins *Physiol Plant* **97**:795–803

Csonka L N and Hanson A D 1991 Prokaryotic osmoregulation:genetics and physiology *Ann Rev Microbiology* **45**:569-606

De Ronde J Laurie A R N Caetano T Greyling M M and Kerepesi I 2004 Comparative study between transgenic and non-transgenic soybean lines proved transgenic lines to be more drought tolerant *Euphytica* **138**:123-132

Dubouzet J Sakuma G Y Ito Y Kasuga M Dubouzet E G and Miura S 2003 OsDREB genes in rice *Oryza sativa* L encode transcription activators that function in drought- high-salt-and cold-responsive gene expression *Plant J* **33**: 751–763

Dure L 1993 III A repeating 11-mer amino acid motif and plant desiccation *Plant J* **3**:363–369

Earnst W H O 2006 In: Abiotic stresses Plant Resistance through Breeding and Molecular Approach Ashraf M and PJC Harris (ed) IBDC U P India 209-226

Edreva A 2005 Generation and scavenging of reactive oxygen species in chloroplasts: a submolecular approach *Agri Ecosy and Env* **106:** 119–133

Elbein AD 1974 The metabolism of α-α-trehalose *Adv in Carbohydrate Chem and Biochem* **30**: 227–257

Eltayeb A Kawano E N Badawi G H Kaminaka H Sanekata T Shibahara T Inanaga S and Tanaka K 2007 Overexpression of monodehydroascorbate reductase in transgenic tobacco confers enhanced tolerance to ozone salt and polyethyleneglycol stresses *Planta* **225**:(5) 1255-1264

Flowers TJ and Yeo AR 1995 Breeding for salinity resistance in crop plants-where next? *Aust J of Plant Physiol* **22**: 875-884

Flowers T J and Flowers S A 2005 Why does salinity pose such a difficult problem for plant breeders? *Agricultural Water Management* **78** : 15–24

Fukuda A Nakamura A Tagiri A Tanaka H Miyao A Hirochika H and Tanaka Y 2004 Function intracellular localization and the importance in salt tolerance of a vacuolar Na^+/H^+antiporter from rice **45**:146–159

Gao X Ren Z Zhao Y and Zhang H 2003 Overexpression of SOD2 increases salt tolerance of *Arabidopsis Plant Physiol* **133**:1873– 1881

Gao J P Chao D Y and Lin H X 2007 Understanding abiotic stress tolerance mechanisms: recent studies on stress response in rice *J of Integ Plant Biol* **49** (6): 742-750

Garciadeblás B Senn M E Banuelos M A and Rodríguez-Navarro A 2003 Sodium transport and HKT transporters: The rice model *Plant J* **34**: 788–801

Garg A K Kim J K Owens T G Ranwala A P Choi Y D Kochian L V and Wu R J 2002 Trehalose accumulation in rice plants confers high tolerance levels to different abiotic stresses *Proc Natl Acad Sci* **99** (25) 15898–15903

Gaxiola R Li A J Undurraga S Dang L M Allen G J Alper S L and Fink G R 2001 Drought-and salt-tolerant plants result from overexpression of the AVP1 H^+-pump *Proc Natl Acad Sci* **98**:11444–11449

Gosal S Gill S R Sindhu A S Kaur D Kaur N and Dhaliwal H S 2001 Transgeic basmati rice carrying genes for stem borer resistance and bacterial leaf blight resistance pp 353-360 In: Rice Research for Food Security and Poverty Alleviation (eds Peng S and B Hardy) IRRI Philippines

Grewal D Gill K R and Gosal S S 2006 Genetic engineering of *Oryza sativa* by particle bombardment *Biol Plant* **50** (2): 311-314

Gilmour S J Sebolt A M Salazar M P Everard J D and Thomashow M F 2000 Overexpression of the *Arabidopsis* CBF3 transcriptional activator mimics multiple biochemical changes associated with cold acclimation *Plant Physiol* **124**: 1854–1865

Gleeson D Walter M A L and Parkinson M 2005 Overproduction of proline in transgenic hybrid larch (*Larix leptocuropaea* (Dengler)) cultures renders them tolerant to cold salt and frost *Mol Breed* **15**: 21–29

Guo Y Qiu Q S Quintero F J Pardo J M Ohta M Zhang C Schumaker K S and Zhua J-K 2004 Transgenic evaluation of activated mutant alleles of sos2 reveals a critical requirement for its kinase activity and c-terminal regulatory domain for salt tolerance in *Arabidopsis thaliana The Plant Cell* **16**: 435–449

Halliwell B 1997 Introduction: free radicals and human disease—trick or treat? In: C E Thomas B Kalyanaraman (Eds) Oxygen Radicals and the Disease Process Harwood Academic Publishers 1–14

Haro R M Bañuelos A Senn M E Barrero-Gil J Rodríguez-Navarro A 2005 HKT1 mediates sodium uniport in roots Pitfalls in the expression of HKT1 in yeast *Plant Physiol* **139**: 1495–1506

Hayashi H Alia M L Deshnium P G Ida M and Murata N 1997 Transformation of *Arabidopsis thaliana* with the codA gene for choline oxidase; accumulation of glycinebetaine and enhanced tolerance to salt and cold stress *Plant J* **12**:133–142

He C J Shen Yan G A S L Holaday Fu Auld D Blumwald E and Zhang H 2005 Expression of an *Arabidopsis* vacuolar sodium/proton antiporter gene in cotton improves photosynthetic performance under salt conditions and increases fiber yield in the field *Plant Cell Physiol* **46** (11): 1848–1854

Hoekstra F Golovina A E A and Buitink J 2001 Mechanisms of plant desiccation tolerance *Trends Plant Sci* **6**:431–438

Holmstrom K Somersalo O S Mandal A Palva T E and Bjorn W 2000 Improved tolerance to salinity and low temperature in transgenic tobacco producing glycine betaine *J Exp Bot* **51**(343):177-85

Hong B S Zong-Suo L and Ming-An S 2005 LEA proteins in higher plants: Structure function gene expression and regulation Colloids and Surfaces B: *Biointerfaces* **45** (3-4) :10 131-135

Honjoh K I Oda Y Takata R Miyamoto T and Hatano S 1999 Introduction of the hiC6 gene which encodes a homologue of a late embryogenesis abundant (LEA) protein enhances freezing tolerance of yeast *J Plant Physiol* **155**: 509–512

Hoshida H Tanaka Y Hibino T Hayashi Y Tanaka A Takabe T and Takabe T 2000 Enhanced tolerance to salt stress in transgenic rice that overexpresses chloroplast glutamine synthetase *Plant Mol Biol* **43**:103–111

Hsieh T H Lee J T Charng Y Y and Chan M T 2002 Tomato plants ectopically expressing *Arabidopsis CBF1*show enhanced resistance to water deficit stress *Plant Physiol* **130**: 618–626

Hu H Dai M Yao J Xiao B Zhang X Li Q and Xiong L 2006 Overexpressing a NAM ATAF and CUC (NAC) transcription factor enhances drought resistance and salt tolerance in rice *Proc Natl Acad Sci* **103** 12 987–12 992

Ikuta S Imamura S Misaki H and Horiuti Y 1977 Purification and characterization of choline oxidase from *Arthrobacter globiformis J Biochem* **82:** 1741–1749

Imai R Chang L Ohta A Bray E A and Takagi M 1996 A lea-class gene of tomato confers salt and freezing tolerance when expressed in *Saccharomyces cerevisiae Gene* **170**:234-248

Incharoensakdi A Takabe T and Akazawa T 1986 Effect of betaine on enzyme activity and subunit interaction of ribulose 1 5-bisphosphate carboxylase/oxygenase from *Aphonothece halophytica Plant Physiol* **81**:1044-1049

Ingram J and Bartels D 1996 The molecular basis of dehydration tolerance in plants *Annu Rev Plant Biol* **47**:377–403

Ito Y Katsura K Maruyama K Taji T Kobayashi M Seki M 2006 Functional analysis of rice DREB1/CBF-type transcription factors involved in cold-responsive gene expression in transgenic rice *Plant Cell Physiol* **47**: 141–153

Iyer S and Caplan A 1998 Products of proline catabolism can induce osmotically regulated genes in rice *Plant Physiol* **116**: 203-211

Jaglo-Ottosen K R Gilmour S J Zarka D G Schabenberger O and Thomashow M F 1998 *Arabidopsis* CBF1 overexpression induces COR genes and enhances freezing tolerance *Sci* **280**:104–106

Jang I C Oh S J Seo J S Choi W B Song S I Kim C H Kim Y S Seo H S Choi Y D Nahm B H and Kim J K 2003 Expression of a bifunctional fusion of the Escherichiacoli genes for trehalose-6-phosphate synthase and trehalose-6-phosphate phosphatase in transgenic rice plants increases trehalose accumulation and abiotic stress tolerance without stunting growth *Plant Physiol* **131**:516–524

Jia GX Z Zhu Q Chang F Q Li Y X 2002 Transformation of tomato with the BADH gene from Atriplex improves salt tolerance *Plant Cell Rep* **21**:141–146

Karakas B Ozias-Akins P Stushnoff C Suefferheld M and Rieger M 1997 Salinity and drought tolerance of mannitol-accumulating transgenic tobacco *Plant Cell Environ* **20**:609-616

Kasuga M Liu Q Miura S Yamaguchi-Shinozaki K Shinozaki K 1999 Improving plant drought salt and freezing tolerance by gene transfer of a single stress inducible transcription factor *Nat Biotechnol* **17**:287–291

Katiyar-Agarwal S Agarwal M and Grover A 2003 Heat-tolerant basmati rice engineered by overexpression of hsp101 *Plant Mol Biol* **51**: 677–686

Katiyar-Agarwal P Agarwal P Reddy M K and Sopory S K 2006 Role of DREB transcription factors in abiotic and biotic stress tolerance in plants *Plant Cell Rep* **25**: 1263–1274

Kaur A Gill M S Gill R and Gosal S S 2007 Standardization of different parameters for 'particle gun' mediated genetic transformation of sugarcane (*Saccharum officinarum*) *Indian J Biotechnology* **6**(1): 31-34

Kishore P B Hong K Z Miao G H Hu C A A Verma D P S 1995 Overexpression of Ä-1-pyrroline-5-carboxylate synthetase increases proline production and confers osmotolerance in transgenic plants *Plant Physiol* **108**:1387–1394

Laurie S K Feeney A Maathuis F J M Heard P J Brown S J and Leigh R A 2002 A role for HKT1 in sodium uptake by wheat roots *Plant J* **32**:139–149

Le Rudulier D Strøm A R Dandekar A M Smith L T and Valentine R C 1984 Molecular biology of osmoregulation *Sci* **224**: 1064–1068

Lilius G Holmberg N and Bulow L 1996 Enhanced NaCl stress tolerance in transgenic tobacco expressing bacterial choline dehydrogenase *Biotechnology* **14**:177–180

Liu Q Kasuga M Sakuma Y Abe H Miura S Yamaguchi-Shinozaki K Shinozaki K 1998 Two transcription factors DREB1 and DREB2 with an EREBP/AP2 DNA binding domain separate two cellular signal transduction pathways in drought-and low temperature-responsive gene expression respectively in *Arabidopsis Plant Cell* **10**:1391–1406

Lu S Y Y Jing X Shen S H Zhao H Y Ma L Q Zhou X J Ren Q and Li Y F 2005 Antiporter gene from *Hordum brevisubulatum* (Trin) Link and its overexpression in transgenic tobaccos *J Integr Plant Biol* **47**:343–349

Lunn J E 2007 Gene families and evolution of trehalose metabolism in plants Functional *Plant Biol* **31**: 550–563

Maas E V 1990 Crop salt tolerance In K K Tanji (ed) Agricultural salinity assessment and management ASCE Manuals and Reports on Engineering No 71 American Society of Civil Engineers New York Chapter 13 : 262–304

Majee M Maitra S Dastidar K G Pattnaik S Chatterjee A Hait N C Das K P and Majumder A L 2004 A novel salt-tolerant L-myoinositol-1-phosphate synthase from *Porteresia coarctata* (Roxb) Tateoka a halophytic wild rice: molecular cloning bacterial overexpression characterization and functional introgression into tobacco-conferring salt tolerance phenotype *J Biol Chem* **279**:28539–28552

McKersie B D Bowley S R Harjanto E and Leprince O 1996 Water deficit tolerance and field performance of trangenic alfalfa overexpressing superoxide dismutase *Plant Physiol* **111**:1177-1181

McKersie B D Murnaghan J and Bowley S R 1997 Manipulating freezing tolerance in transgenicm plants *Acta Physiol Plant* **19**: 485-495

Miranda J Avonce A N Suárez R Thevelein J M Dijck P V and Iturriaga G 2007 A bifunctional TPS-TPP enzyme from yeast confers tolerance to multiple and extreme abiotic-stress conditions in transgenic *Arabidopsis Planta* (in press)

Mittler R Vanderauwera S Gollery M and Breusegem F V 2004 Reactive oxygen gene network of plants *Trends in Plant Sci* **9**: 490–498

Molinaria H B C Marura C J Darosb E Freitas de Camposa M K Portela de Carvalhoa J F R Filhob J C B Pereirac L F P and Vieira L G E 2007 Evaluation of the stress-inducible production of proline in transgenic sugarcane (*Saccharum* spp) Osmotic adjustment chlorophyll fluorescence and oxidative stress *Physiol Planta* **130**: 218–229

Montero-Lomeli M B Morais L B Figueiredo D L Neto D C S Martins J R P and Masuda C A 2002 The initiation factor eIF4A is involved in the response to lithium stress in *Saccharomyces cerevisiae J Biol Chem* **277**:21542-21548

Nakashima K and Yamaguchi-Shinozaki K 2005 Molecular studies on stress-responsive gene expression in *Arabidopsis* and improvement of stress tolerance in crop plants by regulon biotechnology *JARQ* **39** (4): 221-229

Nakayama H Yoshida K Ono H Murooka Y and Shinmyo A 2000 Ectoine the compatible solute of Halomonas elongata confers hyperosmotic tolerance in cultured tobacco cells *Plant Physiol* **122**:1239–1247

Nanjo T K Masatomo Y Yoshiba Y Sanada K Wada H Tsukaya Y Kakubari K Yamaguchi-Shinozaki and K Shinozaki 1999 Biological functions of proline in morphogenesis and osmotolerance revealed in antisense transgenic *Arabidopsis thaliana Plant J* **18**:185-193

Noctor G and Foyer C H 1998 Ascorbate and glutathione: keeping active oxygen under control *Annu Rev Plant Physiol Plant Mol Biol* **49**: 249–279

Nuccio M L Russell B L Nolte K D Rathinasabapathi B D Gage D A and Hanson A D 1998 The endogenous choline supply limits glycine betaine synthesis in transgenic tobacco expressing choline monooxygenase *Plant J* **16**:487-496

Nyyssola A Kerovvuo J Kaukinen P Weymarn N V and Reinikainen T 2000 Extreme halophiles synthesize betaine from glycine by methylation *J Biol Chem* **275**: 22196-22201

Oh S J Song S I Kim Y S Jang H J Kim S Y and Kim M 2005 *Arabidopsis* CBF3/DREB1A and ABF3 in transgenic rice increased tolerance to abiotic stress without stunting growth *Plant Physiol* **138**: 341–351

Oh S J Kwon C W Choi D W Song S I and Kim J K 2007 Expression of barley HvCBF4 enhances tolerance to abiotic stress in transgenic rice *Plant Biotech Journal* **5**: 646–656

Ohnishi T Sugahara S Yamada T Kikuchi K Yoshiba Y and Hirano H Y 2005 OsNAC6 a member of the NAC gene family is induced by various stresses in rice *Genes Genet Syst* **80**: 135– 139

Ohta M Hayashi Y Nakashima A Hamada A Tanaka A Nakamura T and Hayakawa T 2002 Introduction of a Na^+/H^+ antiporter gene from *Atriplex gmelini* confers salt tolerance to rice *FEBS Lett* **532**:279–282

Ooka H Satoh K Doi K Nagata T Otomo Y and Murakami K 2003 Comprehensive analysis of NAC family genes in *Oryza sativa* and *Arabidopsis thaliana DNA Res* **10:**239–247

Oraby H F Ransom C B Kravchenko A N and Sticklen M B 2005 Barley Hva1 gene confers salt tolerance in R_3 transgenic oat *Crop Sci* **45**:2218–2227

Papageorgiou G C and Murata N 1995 The unusually strong stabilizing effects of glycine betaine on the structure andfunction of the oxygen-evolving photosystem II complex *Photosynth Res* **44:** 243–252

Pardo J M Reddy M P Yang S Maggio A Huh G H Matsumoto T Coca M A Paino-D'Urzo M Koiwa H Yun D J Watad A A Bressan R A and Hasegawa P M 1998 Stress signaling through $Ca2^+$/ calmodulin-dependent protein phosphatase calcineurin mediates salt adaptation in plants *Proc Natl Acad Sci* **95**:9681–9686

Park JM Park C J Lee S B Ham B K Shin R and Paek K H 2001 Overexpression of the tobacco Tsi1 gene encoding an EREBP/ AP2-type transcription factor enhances resistance against pathogen attack and osmotic stress in tobacco *Plant Cell* **13**:1035–1046

Park B J Liu Z Kanno A Kameya T 2005[a] Genetic improvement of Chinese cabbage for salt and drought tolerance by constitutive expression of a *B napus* LEA gene a *Plant Sci* **169**: 553–558

Park B J Liu Z Kanno A and Kameya T 2005[b] Increased tolerance to salt and water deficit stress inntransgenic lettuce (*Lactuca sativa* L) by constitutive expression of LEA *Plant growth regulation* **45**: 165-171

Parsell D A and Lindquist S 1994 Heat-shock proteins and stress tolerance In R Morimoto A Tissieres and C Georgopoulos [eds] The biology of heat-shock proteins and molecular chaperones Cold Spring Harbor Laboratory Press Cold Spring Harbor New York USA p 457–494

Pellegrineschi A M Reynolds M Pacheco R M Brito R Almeraya K Yamaguchi-Shinozaki and D Hoisington 2004 Stress-induced expression in wheat of the *Arabidopsis thaliana* DREB1A gene delays water stress symptoms under greenhouse conditions *Genome* **47**:493–500

Perl A Perl-Treves R Galili S Aviv D Shalgi E Malkin S and Galun E 1993 Enhanced oxidative-stress defense in transgenic potato overexpressing tomato Cu Zn superoxide dismutase *Theor Appl Genet* **85**: 568-576

Pilon-Smits EA Terry N Sears T Kim H Zayed A Hwang S Dun K van Voogd E Verwoerd T C Krutwagen R W H H Gooddijn O J M 1998 Trehalose-producing transgenic tobacco plants show improved growth performance under drought stress *J Plant Physiol* **152**:525-532

Platten J D Cotsaftis O Berthomieu P Bohnert H Davenport R J and Fairbairn D J 2006 Nomenclature for HKT transporters key determinants of plant salinity tolerance *Trends Plant Sci* **11** 372–374

Postel S L Daily G C and Ehrlich P R 1996 Human appropriation of renewable fresh water *Sci* **271:**785-788

Prasad K V S Sharmila K P Kumar P A Saradhi P P 2000[a] Transformation of *Brassica juncea* (L) Czern with bacterial codA gene enhances its tolerance to salt stress *Mol Breed* **6**:489–499

Prasad S R Bagali P G Hittalmani S Shashidhar H E 2000[b] Molecular mapping of quantitative trait loci associated with seedling tolerance to salt (*Oryza sativa* L) *Curr Sci* **78**:162–164

Prasinos C Krampis K Samakovli D and Hatzopoulos P 2005 Tight regulation of expression of two *Arabidopsis* cytosolic Hsp90 genes during embryo development *J Exp Bot* **412**: 633–644

Qin F Kakimoto Sakuma M Y Maruyama K Osakabe Y Tran L-S P Shinozaki K and Yamaguchi-Shinozaki K 2007 Regulation and functional analysis of *ZmDREB2A* in response to drought and heat stress in *Zea mays* L the *Plant J* **50:** 54-69

Queitsch C Hong S W Vierling E and Lindquist S 2000 Heat-shock protein 101 plays a crucial role in thermotolerance in *Arabidopsis Plant Cell* **12:** 479-492

Rathinasabapathi B and Ramandeep K 2006 In: Physiology and Molecular Biology of Stress Tolerance in Plants K V Madhava Rao A S Raghavendra and K Janardhan Reddy (eds) Springer Printed in the Netherlands 255–299

Rathinasabapathi B Burnet M Russel B L Gage D A Liao P C Nye G J Scott P Golbeck J H Hanson A D 1997 Choline monooxygenase an unusual iron-sulfur enzyme catalyzing the first step of glycine betaine synthesis in plants: prosthetic group characterization and cDNA cloning *Proc Natl Acad Sci* **94**: 3454-3458

Rausell A Kanhonou R Yenush L Serrano R and Ros R 2003 The translation initiation factor eIF1A is an important determinant in the tolerance to NaCl stress in yeast and plants *Plant J* **34**:257–267

Ren Z H Gao J P Li X L L G Huang Cai W Chao D Y Zhu M Z Wang Z Y Luan S Lin H X 2005 A rice quantitative trait locus for salt tolerance encodes a sodium transporter *Nat Genet* **37**:1141–1146

Revilla P Butron A Cartea M E Maivar R A and Ordas A 2006 In: Abiotic stresses Plant Resistance through Breeding and Molecular Approach Ashraf M and PJC Harris (ed) IBDC U P India 301-349

Rhodes D and Hanson AD 1993 Quaternary ammonium and tertiary sulfonium compounds in higher plants *Annu Rev Plant Physiol Plant Mol Biol* **44**:357-384

Riechmann J L Heard J Martin G Reuber L Jiang C Keddie J Adam L Pineda O Ratcliffe O J and Samaha R R 2000 *Arabidopsis* transcription factors: genome-wide comparative analysis among eukaryotes *Sci* **290**:2105–2110

Robinson S P and Jones G P 1986 Accumulation of glycine betaine in chloroplasts provides osmotic adjustment during salt stress *Australian Journal of Plant Physiol* **13:** 659-668

Rodrigues S M Andrade M O Gomes A P S DaMatta F M Baracat-Pereira M C and Fontes E P B 2006 Arabidopsis and tobacco plants ectopically expressing the soybean antiquitin-like *ALDH7* gene display enhanced tolerance to drought salinity and oxidative stress *J Exp Bot* **55**: 301-308

Roxas V P Smith R K Allen E R and Allen R D 1997 Overexpression of glutathione S-transferase/ glutathione peroxidase enhances the growth of transgenic tobacco seedlings during stress *Nat Biotechnol* **15**:988–991

Sahi C Singh A Kumar K Blumwald E and Grover A 2006 Salt stress response in rice: genetics molecular biology and comparative genomics *Funct Integr Genomics* (in press)

Saijo Y Hata S Kyozuka J Shimamoto K and Izui K 2000 Overexpression of a single $Ca2^+$dependent protein kinase confers both cold and salt/ drought tolerance on rice plants *Plant J* **23**:319–327

Sakamoto A Alia Murata N and Murata A 1998 Metabolic engineering of rice leading to biosynthesis of glycinebetaine and tolerance to salt and cold *Plant Mol Biol* **38**:1011–1019

Sakamoto A and Murata N 2002 The role of glycine betaine in the protection of plants from stress: clues from transgenic plants *Plant Cell Environ* **25** 163-171

Sakuma Y Maruyama K Osakabe Y Qin F Seki M Shinozaki K Yamaguchi-Shinozaki K 2006 Functional analysis of an *Arabidopsis* transcription factor DREB2A involved in drought responsive gene expression *The Plant Cell* **18**:1292–1309

Sanan-Mishra N Pham X H Sopory S K and Tuteja N 2005 Pea DNA helicase 45 overexpression in tobacco confers high salinitytolerance without affecting yield *Proc Natl Acad Sci* **102**:509–514

Saradhi A and Saradhi P P 1991 Proline accumulation under heavy metal stress *J Plant Physiol* **138**: 554-558

Sawahel W A and Hassan A H 2002 Generation of transgenic wheat plants producing high levels of the osmoprotectant proline *Biotech Lett* **24** 721-725

Schobert B and Tschesche H 1978 Unusual solution properties of proline and its interaction with proteins *Biochim Biohys* Acta **541**: 270-277

Serrano R J Mulet M Rios G Marquez J A Larrinoa I F de Leube M P Mendizabal I Pascual-Ahuir A Proft M Ros R Montesinos C 1999 A glimpse of the mechanisms of ion homeostasis during salt stress *J Exp Bot* **50**:1023–1036

Shen B Jensen R G and Bohnert H J 1997 Increased resistance to oxidative stress in transgenic plants by targeting mannitol biosynthesis to chloroplasts *Plant Physiol* **113**: 1177-1183

Shen B Jensen R G Bohnert H J 2001 Increased resistance to oxidative stress in transgenic plants by targeting mannitol biosynthesis to chloroplasts *Plant Physiol* **113**:1177-1183

Shi H Ishitani M Kim C and Zhu J K 2000 The *Arabidopsis thaliana* salt tolerance gene SOS1 encodes a putative Na^+/H^+ antiporter *Proc Natl Acad Sci* **97**:6896–6901

Shi H B Lee H Wu S J and Zhu J K 2003 Overexpression of a plasma membrane Na^+/H^+ antiporter gene improves salt tolerance in *Arabidopsis thaliana Nature Biotech* **21**:81–85

Shinozaki K and Yamaguchi-Shinozaki K 2000 Molecular responses to dehydration and low temperature: differences and cross-talk between two stress signalling pathways *Curr Opin Plant Biol* 3:217–223

Shinozaki K and Yamaguchi-Shinozaki K 2007 Gene networks involved in drought stress response and tolerance *J Exp Bot* **58**(2):221-227

Shirasawa K Takabe T Takabe T and Kishitani S 2006 Accumulation of glycine betaine in rice plants that over express choline monooxygenase from spinach and evaluation of their tolerance to abiotic stress *Annals of Botany* **98** : 565-571

Shou H Bordallo P and Wang K 2004 Expression of the *Nicotiana* protein kinase (*NPK1*) enhanced drought tolerance in transgenic maize *J Exp Bot* **55**: 301-308

Singla-Pareek S L Reddy M K Sopory S K 2003 Genetic engineering of the glyoxylase pathway in tobacco leads to enhanced salinity tolerance *Proc Natl Acad Sci* USA **100**:14672–14677

Singla-Pareek S L Yadav S K Pareek A Reddy M K and Sopory S K 2007 Enhancing salt tolerance in a crop plant by overexpression of *glyoxalase II Transgen Res* (In press)

Sivamani E Bahieldin A Wraith J M Al-Niemi T Dyer W E Ho T-H D and Wu R 2000 Improved biomass productivity and water use efficiency underwater deficit conditions in transgenic wheat constitutively expressing the barley HVA1 gene *Plant Sci* **155**:1-9

Smirnoff N and Cumbes Q J 1989 Hydroxyradical scavenging activity of compatible solutes *Phytochem* **28**: 1057-1060

Steponkus P L Uemura M Joseph R A Gilmour S J and Thomashow M F 1998 Mode of action of the COR15a gene on the freezing tolerance of *Arabidopsis thaliana Proc Nat Acad Sci* **95**: 14570-14575

Sun W Bernard C van de B Cotte Montagu M Van and Verbruggen N 2001 At-HSP17 6A encoding a small heat-shock protein in *Arabidopsis* can enhance osmotolerance upon overexpression *Plant J* **27**:407–415

Suzuki N Rizhsky L Liang H Shuman J Shulaev V and Mittler R 2005 Enhanced tolerance to environmental stress in transgenic plants expressing the transcriptional coactivator multiprotein bridging factor 1 *Plant Physiol* **139**: 1313–1322

Tanaka Y Hibin T Hayashi Y Tanaka A Kishitani S Takabe T Yokota S Takabe T 1999 Salt tolerance of transgenic rice overexpressing yeast mitochondrial Mn-SOD in chloroplasts *Plant Sci* **148**:131–138

Tarczynski M C Jensen R G and Bohnert H J 1992 Expression of a bacterial mt1D gene in transgenic tobacco leads to production and accumulation of mannitol *Proc Natl Acad Sci* **89**:2600-2604

Tarczynski MC Jensen R G and Bohnert H J 1993 Stress protection of transgenic plants by production of the osmolyte mannitol *Sci* **259**:508-510

Taylor C B 1996 Proline and water deficit: ups downs ins and outs *The Plant Cell* **8**: 1221-1224

Thomas J Sepahi C M Arendall B and Bohnert H J 1995 Enhancement of seed germination in high salinity by engineering mannitol expression in *Arabidopsis thaliana Plant Cell Environ* **18**:801-806

Thomashow M F 1998 Role of cold-responsive genes in plant freezing tolerance *Plant Physiol* **118**:1–7

Thomashow M F (1999) Plant cold acclimation: freezing tolerance genes and regulatory mechanisms *Annu Rev Plant Biol* **50**:571–599

Tilman D Cassman KG Matson P A Naylor R and Polasky S 2002 Agricultural sustainability and intensive production practices *Nature* **418**: 671-677

Tran L Nakashima S K Sakuma Y Simpson SD Fujita Y and Maruyama K 2004 Isolation and functional analysis of *Arabidopsis* stress-inducible NAC transcription factors that bind to a drought-responsive cis-element in the early responsive to dehydration stress 1 promoter *Plant Cell* **16**: 2481–2498

Vanavichit A Tragoonrung S and Toojinda T 2006 In: Abiotic stresses Plant resistance through Breeding and Molecular Approach Ashraf M and PJC Harris (ed) IBDC U P India 177-208

Van Camp W Capiau K Montagu M Van Inze D and Slooten L 1996 Enhancement of oxidative stress tolerance in transgenic tobacco plants overproducing Fe-superoxide dismutase in chloroplasts *Plant Physiol* **112**:1703–1714

Veena J Reddy V S Sopory S K 1999 Glyoxalase I from *Brassica juncea*: molecular cloning regulation and its over-expression confer tolerance in transgenic tobacco under stress *Plant J* **17**:385–395

Venkamp J Laurpe H J E M and Koot J T M 1989 Organic acids as sources of drought-induced prolein synthesized in the field bean plants *Vicia faba L J Plant Physiol* **133**: 654-659

Vernon D M Tarczynski M C Jensen R G and Bohnert H J 1993 Cyclitol production in transgenic tobacco *Plant J* **4**:199–205

Vierling E 1991 The roles of heat-shock proteins in plants *Annu Rev Plant Biol* **42**:579–620

Vierling E J Kimpel A 1992 Plant responses to environmental stress *Curr Opin Biotech* **3**:164–170

Vinocur B and Altman A 2005 Recent advances in engineering plant tolerance to abiotic stress: achievements and limitations *Curr Opin Biotechnol* **16**:123–132

Vrimani S S and Ilyas Ahmed M 2007 In : Breeding major food staples M S Kang and P M Priyadarshan (ed) Blackwell publishing USA 141-192

Wang W Vinocur B and Altman A 2003 Plant responses to drought salinity and extreme temperatures: towards genetic engineering for stress tolerance *Planta* **218**: 1–14

Wang J Zuo K Wu W Song J Sun X Lin J Li X and Tang K 2004[a] Expression of a novel antiporter gene from *Brassica napus* resulted in enhanced salt tolerance in transgenic tobacco plants *Biol Plant* **48**:509–515

Wang W Vinocur B Shoseyov O and Altman A 2004[b] Role of plant heat-shock proteins and molecular chaperones in the abiotic stress response *Trends in Plant Sci* **9**: 244–252

Waters E Lee R G J Vierling E 1996 Evolution structure and function of the small heat-shock proteins in plants *J Exp Bot* **47**:325–338

Wingsle G Karpinski S and Hällgren J E 1999 Low temperature high light stress and antioxidant defence mechanisms in higher plants *Phyton* **39:** 253–268

Wu C A Yang G D Meng Q W and Zheng C C 2004 The cotton GhNHX1 gene encoding a novel putative tonoplast Na^+/K^+ antiporter plays an important role in salt stress *Plant Cell Physiol* **45**:600–607

Wu L Fan Z Guo L Li Y Chen Z L and Qu L J 2005 Overexpression of the bacterial nhaA gene in rice enhances salt and drought tolerance *Plant Sci* **168**:297–302

Xiao B Huang Y Tang N and Xiong L 2007 Over expression of *LEA* gene in rice improves drought resistance under field conditions *Theor Appl Genet* **115**:35-46

Xiong L Lee H Ishitani M and Zhu J K 2002 Regulation of osmotic stress responsive gene expression by LOS6/ABA1 locus in *Arabidopsis J Biol Chem* **277**: 8588–8596

Xu D Duan X Wang B Hong B T-H D Ho and R Wu 1996 Expression of late embryogenesis abundant protein gene *HVA1* from Barley confers tolerance to water deficit and salt stress in transgenic rice *Plant physiol* **110**:249-57

Xu C Y Jing R Mao X Jia X Y and Chang X 2007[a] A Wheat (*Triticum aestivum*) protein phosphatase 2a catalytic subunit gene provides enhanced drought tolerance in tobacco *Ann of Bot* **99**: 439–450

Xu S M WangX C and Chen J 2007[b] Zinc finger protein 1 (*ThZF1*) from salt cress (*Thellungiella halophila*) is a Cys-2/His-2-type transcription factor involved in drought and salt stress *Plant Cell Rep* **26**(4): 497-506

Xue ZY Zhi D Y Xue G P Zhang H Zhao Y X Xia G M 2004 Enhanced salt tolerance of transgenic wheat (*Tritivum aestivum* L) expressing a vacuolar Na^+/H^+ antiporter gene with improved grain yields in saline soils in the field and a reduced level of leaf Na^+ *Plant Sci* **167**:849–859

Yamada K Fukao Y Hayashi M Fukazawa M Suzuki I and Nishimura M 2007 Cytosolic HSP90 Regulates the Heat-shock Response That Is Responsible for Heat Acclimation in *Arabidopsis thaliana The J of Biol Chem* **282**(52): 37794–37804

Yamaguchi-Shinozaki K and Shinozaki K 1994 A novel *cis*-acting element in an *Arabidopsis* gene is involved in responsiveness to drought low-temperature or high-salt stress *Plant Cell* **6**: 251-264

Yeo E T Kwon H B Han S E Lee J T Ryu J C and Byu MO 2000 Genetic engineering of drought resistant potato plants by introduction of the trehalose-6-phosphate synthase TPS1 gene from *Saccharomyces cerevisiae Mol Cell* **30**: 263-268

Yilmaz J L and Bulow L 2002 Enhanced stress tolerance in Escherichia coli and Nicotiana tabacum expressing a betaine aldehyde dehydrogenase/choline dehydrogenase fusion protein *Biotechnol Prog* **18**:1176–1182

Yin X Y Yang A F Zhang K W and Zhang J R 2004 Production and analysis of transgenic maize with improved salt tolerance by the introduction of AtNHX1gene *Acta Bot Sin* **46**:854–861

Yonamine I Yoshida K Kido K Nakagawa A Nakayama H and Shinmyo A 2004 Overexpression of NtHAL3 genes confers increased levels of proline biosynthesis and the enhancement of salt tolerance in cultured tobacco cells *J Exp Bot* **55**: 387-395

Yu Q and Rengel Z 1999 Micronutrient defciency infuences plant growth and activities of superoxide dismutases in narrow-leafed lupins *Ann Bot* **183** 175-182

Zhao F Y and Zhang H 2006[a] Salt and paraquat stress tolerance results from co-expression of the *Suaeda salsa* glutathione S-transferace and catalase in transgenic rice *Plant Cell Tissue Organ Cult* **86**(3):349-358

Zhao F Y and Zhang H 2006[b] Expression of *Suaeda salsa* glutathione S-transferacein transgenic rice resulted in a different level of abiotic stress tolerance *J Agri Sci* **144**:1-8

Zhao F Y and Zhang H 2007 Transgenic rice breeding for abiotic stress tolerance –Present and future *Chinese J of Biotech* **23**(1): 1-6

Zhang H X and Blumwald E 2001 Transgenic salt-tolerant tomato plants accumulate salt in foliage but not in fruit *Nat Biotechnol* **19**:765–768

Zhang H X Hodson J N Williams J P and Blumwald E 2001 Engineering salt-tolerant Brassica plants: characterization of yield and seed oil quality in transgenic plants with increased vacuolar sodium accumulation *Proc Natl Acad Sci* **98**:12832–12836

Zhifang G and Loescher W H 2003 Expression of a celery mannose 6-phosphate reductase in *Arabidopsis thaliana* enhances salt tolerance and induces biosynthesis of both mannitol and a glucosyl-mannitol dimmer *Plant Cell Environ* **26**: 275–283

Zhu B J Su Chang M Verma D P S Fan Y L and Wu R 1998 Overexpression of a delta 1-pyrroline-5-carboxylate synthetase gene and analysis of tolerance to water-and salt-stress in transgenic rice *Plant Sci* **139**: 41-48

Table 1: Genetic engineering of crop plants for abiotic stress tolerance

Mechanism	Transgene (s)	Plants Sepcies	Transformation method	Promoter	Remarks	Reference
Compatible Solutes						
Proline	*P5CS* (Pyrroline -5-carboxylate) from *Vigna aconitifolia*	*N. tabacum*	*Agrobacterium*	CaMV 35S	Transgenic plants produced 10-18 fold more proline than control plants. Over production of proline also enhanced root biomass and flower development in transgenic plants. Thus proline acts as an osmo protestant and over production of proline results in increased tolerance to osmotic stress in plants.	Kishor *et al.* 1995
	P5CS from *Vigna aconitifolia*	*Larix leptoeuropaea*	*Agrobacterium*		The integration of the gene into the plant genome was confirmed by Southern blot and by proline content analysis. There was an approximately 30-fold increase in proline level in transgenic tissue compared to non-transformed controls. The transgenic tissue lines were significantly more resistant to cold, salt, and freezing stresses and grew under conditions (200mM NaCl or 4 °C) that completely inhibited the growth of control cell lines.	Gleeson *et al.* 2005
	P5CS from Vigna aconitifolia	*Saccharam officinarum*	Particle gun	AIPC	Stress-inducible proline accumulation in transgenic sugarcane plants under water-deficit stress acts as a component of antioxidative defense system rather than as an osmotic adjustment meciator.	Molinari et at 2007
Glycinebetaine	*ct-codA*, cy-codA from *Arthrobacter* globiformis	*Oryza sativa*		CaMV 35S	Transgenic plants had high levels of *glycine* betaine and grew faster as compared to wild types on removal of stress.	Sakamoto et al. 1998

	bet A and bet B from E. coli	*N. tabaccum*	*Agrobacterium*	rbc SIA	Transgenic plants showed increased tolerance to salt stress as measured by biomass production of green house grown plants.	Holmstrom *et al.* 2000
	CMO (choline monooxygenase) from Spinacia *olearceae*	*Oryza sativa*	*Agrobacterium*	Maize ubi	Transgenic plants were tolerant to salt and temperature stress at seeding *stage*. CMO expressing rice plants were not effective for accumulation of GB and improvement of productivity.	Shirasawa *et al.* 2006
Mannitol	mtlD from E. coli	*Triticum aestivum*	Particle gun	Maize ubi-1	Ectopic expression of the mtlD gene for the biosynthesis of mannitol in wheat improves tolerance to water stress and salinity.	Abebe *et al.* 2003
Trehalose	TPSP from E. *coli*	*Oryza sativa*	*Agrobacterium*	rbcS, and ABA	Transgenic rice plants accumulate trehalose at levels 3-10 times that of the nontransgenic controls. Compared with nontransgenic rice, several independent transgenic lines exhibited sustained plant growth, less photo-oxidative damage, and more favorable mineral balance under salt, drought, and low temperature stress conditions.	Garg *et al.* 2002
	TPS1-TPS-2 from E. coli	*A. thaliana*	*Agrobacterium*	CaMV 35S	No morphological growth alterations were observed in lines over expressing the TPS1-TPS2 construct, while the plants over expressing the TPS1 alone under the control of 35S promoter had abnormal growth, colour and shape.	Miranda *et al.* 2007
Detoxification						
	Mn-SOD Superoxide dismutase from *N. plumbaginifolia*	*Medicago sativa.*	*Agrobacterium*	CaMV 35S	Transgenic plants had reduced injury from water deficit stress. A three year field trial indicated that yield and survival of transgenic plants were significantly improved.	Mckersie *et al.* 1996
	Gly I from *Brassica juncea*	*N. tabacum*	*Agrobacterium*	CaMV 35S	Transgenic plants showed tolerance to methylglyoxal and high salt.	Veena *et al.* 1999
	Gly I from Brassica juncea and Gly II from Oryza sativa	*N. tabacum*	*Agrobacterium*	*CaMV 35S*	Double transgenics showed tolerance to salinity without yield penalty.	Singla-Pareek *et al.* 2003

APX from *A. thaliana*	*N. tabacum*	*Agrobacterium*		The first generation of transgenic line chl-APX5 line showed enhanced tolerance to salt, PEG and water stresses, as determined by net photosynthesis indicating that overproducing the cytosolic APX in tobacco chloroplasts reduces the toxicity of H_2O_2.	Badawi *et al.* 2004
Gly I from B. juncea	*Vigna mungo*	*Agrobacterium*	*csm, rd29A*	Transgenic plants showed tolerance to methylglyoxal (10mM) and high salt (600mM).	Bhalla-Sarin *et al.* 2004
GmTP55 from *Glycine max*	*A. thaliana and N. tabaccum*	*Agrobacterium*	CaMV 35S	Ectopic expression of GmTP55 in both *Arabidopsis* and tobacco conferred tolerance to salinity during germination and to water deficit during plant growth. Antiquitin may be involved in adaptive responses mediated by a physiologically relevant detoxification pathway in plants.	Rodrigues *et al.* 2006
GST (Glulathione S-transfierase) and CAT I (Catalase) from Suaeda *salasa*	*Oryza sativa*	*Agrobactreium*		Transgenic rice over expressing the GST (Glutathione S-transferase) and CA T I (Catalase) from *Suaeda* salasa showed increased tolerance to oxidative stress caused by salt and paraqaut. Double transgenic; showed (GST + CAT 1) higher abiotic stress tolerance as compared to single GST.	Zhao and Zhang, 2006a, 2006'
MDAR from *A. thaliana*	*N tabaccum*	*Agrobacterium*		Transgenic plants exhibited 2.1 fold higher MDAR activity and 2.2 fold higher level of reduced ASA compared to wild type control plants which increased photosynthesis rates under ozone, salt and PEG stress. In addition, these transgenic plants showed significantly lower hydrogen peroxide level when tested under salt stress.	Eltayeb *et al.* 2007
Gly II from *Oryza sativa*	*Oryza sativa*	*Agrobacterium*		Transgenic rice plants showed higher tolerance to toxic concentrations of methylglyoxal (MG) and NaCl.	Singla-Pareek *et al.* 2007

Stress induced proteins						
LEA-type proteins						
	HVA1 from *Hordeum* vulgare	*Oryza sativa*	Particle gun	Rice actin I	Second generation transgenic plans showed significantly increased tolerance to water deficit and salinity.	Xu *et al.* 1996
	HVA1 from Hordeum vulgare	*Tritium aestivum*	Particle gun	Maize ubi I	Transgenic plants showed improved growth characteristics in response to soil water deficits. Field trials showed that HV1I gene has the potential to confer drought stress protection in transgenic spring wheat.	Sivamani *et al.* 2000 Bahieldin *et al.* 2005
	ME.LEA N4 Brassica napus	*Lactuca sativa*	*Agrobacterium*	CaMV 35S	Transgenic plants showed improved growth characteristics under salt and water deficit stress.	Park *et al.* 2005
	ME.LEA N4 Brassica napus	*Brassica campestris*	*Agrobacterium*	CaMV 35S	Transgenic plants showed increased growth ability under salt and water deficit.	Park *et al.* 2005a
	Os LEA-3-1 from *Oryza sativa*	*Oryza sativa*	*Agrobacterium*	*CaMV* 35S, HVA I -like	Transgenic rice plants showed increased growth ability under salt stress and water deficit.	Xiao, *et al.* 2007
Heat-shock proteins						
	Athsp101 from *A. thaliana*	*A. thaliana*	*Agrobacterium*		Manipulated thermotolerance, in transgenic plants.	Queitsch *et al.* 2000
	At hsp I 17.6A from *A. thaliana*	*A. thaliana*	*Agrobacterium*	Ca MV 35S	Transgenic plants showed increased osmotolerance, however they failed to increase heat tolerance	Sun *et al.* 2001
	Athsp101 from *A. thaliana*	*Oryza sativa*	Agrobactrium	Maize ubil	The transgenic rice lines showed significantly better growth performance in the recovery phase following the heat stress.	Katiyar-Agarwal *et al.* 2003
	Athsp 90-3 and At hsp 90-1 from *A. thaliana*	*A. thaliana*	*Agrobacterium*		*Different* regulatory and developmental mechanisms control and differentiate the expression of the two cytosolic members of the *Arabidopsis* AtHsp90 gene family under normal conditions.	Prasinos *et al.* 2005

	sHSP 18.2 from A. thaliana	*N. tabacum*	*Agrobacterium rhizogenes*		The sHSP 18.2 promoter exhibited an extraordinary strong *thermo-induced* activity in A. rhizogenes transformed roots.	*Chen et al.* 2007
	HSP90.2 from A. thaliana	*A. thaliana*	*Agrobacterium*		*Cytosolic HSP90 regulates the* heat *shock* response that *is* responsible *for heat acclimation in transgenic* plants	*Yamada et al.* 2007
Ion transport						
	At NHX1 from *A.thaliana*	*A.thaliana*			Over expression of vacuolar Na+/H+ in Arabidopsis promoted *growth* in soil watered with 200mM of NaCl	Apse *et al.* 1999
	At NHX1 from *A. thaliana*	*Brassica napus*	*Agrobacterium*		Transgenic plants, *were* able to grow, flower, and produce seeds in the presence of 200mM NaCl	Zhang *et al.* 2001
	SOD2 from Schizosaccharomyces pombe	*A. thaliana*	*Agrobacterium*	CaMV 35S	Results from different tests indicated that the expression of the SOD2 gene promoted a higher level of salt tolerance in vivo in transgenic *Arabidopsis* plants.	Gao *et al.* 2003
	SOS I from *A. thaliana*	*A.thaliana*	*Agrobacterium*	CaMV 35S	The overexpression of SOS] improved the salt tolerance of *Arabidopsis,* demonstrating that improved salt tolerance can be attained by limiting Na+ accumulation in plant cells	*Shi et al.* 2000, 2003
	SOS2 from *A. thaliana*	*A. thaliana*			Expression of some of the activated SOS2 proteins in wild-type *A. thaliana* conferred increased salt tolerance	Guo *et al.* 2004
	At NHX1 from *A. thaliana*	*Gossypium hirsutum*	*Agrobacterium*	CaMV 35S	Field-grown *At NHX1* -expressing cotton plants produced more fibers with better quality, indicating that *At NHX1* can indeed be used for improving salt stress tolerance in cotton.	He *et al. 2005*
	TNHX1 and TVP1 from *Triticum aestivum*	*A. thaliana*	*Agrobacterium*	CaMV 35S	Transgenic plants showed increased tolerance to salt and drought stress	Brini *et al* 2007

Transcriptional control					
CBF1 from A. thaliana	*Lycopersicon esculentum*	*Agrobacterium*	CaMV 35S	Transgenic tomato plants were more resistant to water deficit stress than the wild-type plants.	Hsieh *et al.* 2002
DREB 1A from *A. thaliana*	*Triticum aestivum*	Particle gun	rd29A	Transgenic plants expressing *DREB 1A* gene demonstrated substantial resistance to water stress under green house conditions.	Pellegrineschi *et al.* 2004
NPK1 from *N. tabacum*	*Zea mays*	*Agrobacterium Particle gun*	35SC4PPDK	Transgenic maize plants showed enhanced drought tolerance The *Agrobacterium*-derived events contained fewer than five copies of the NPK1 transgene, whereas, the bombardment derived events carried more than 20 copies of the transgene.	Shou *et al.* 2004
DREB I from *A. thaliana*	*Zea mays*	*Agrobacterium*, Particle gun	UB11ZM *ubi*	Two *DREB I* positive families showed significant increase *in* resistance to freezing survival test applied at 3 leaf stage.	Chiappetta *et al.* 2005
CBF31 DREB]A from A. thaliana	*Oryza sativa*		Rice Ubi I	Tolerance to drought and high salinity without growth retardation or any phenotypic alteration.	Oh *et al.* 2005
MBFlc from *A. thaliana*	*A. thaliana*	*Agrobacterium*	CaMV 35S	Constitutive expression of the stress-response transcriptional coactivator multiprotein bridging factor I c (MBF I c) in *Arabidopsis* enhances the tolerance of transgenic plants to bacterial infection, heat, and osmotic stress. Moreover, the enhanced tolerance of transgenic plants to osmotic and heat stress was maintained even when these two stresses were combined.	Suzuki *el al.* 2005
DREB1A from A. thaliana	*Solanum tuberosum*	*Agrobacterium*	*rd29A*	Transgenic plants showed significant tolerance against salinity stress (1Med NaCl)	Benham *et al.* 2006

SNAC I from *Oryza sativa*	*Oryza sativa*	*Agrobacterium*	CaMV 35S	Transgenic plants showed improved drought resistance under favourable conditions and strong tolerance to salt stress.	Hu *et al.* 2006
OS DREB 1A/ *OS DREB* 1B from *Oryza sativa*	*Oryza sativa*	*Agrobacterium*	Ca MV 35S	*DREB I* type genes are quite useful for improvement of stress tolerance in crop plants including rice.	Ito *et al.* 2006
DREB2A from *A. thaliana*	*A. thaliana*	*Agrobacterium*	CaMV 35S	Significant drought stress tolerance but slight freezing tolerance in transgenic *Arabidopsis* plants.	Sakuma *et al.* 2006
DREB IA from *A. thaliana*	*Arachis hypogaea*	*Agrobacterium*	CaMV *35S,rd 29 A*	Transgenic plants showed increased transpiration efficiency, an important feature of drought tolerance.	Bhatnagar Mathur *et al.* 2007
OS NAC 6 from *Oryza sativa*	*Oryza sativa*	*Agrobacterium*		Transgenic plants showed tolerance to dehydration and high salt stress.	Nakashima *et al.* 2007
HvCBF4 from *Hordeum vulgare*	*Oryza sativa*	*Agrobacterium*	Ubi1	Transgenic rice resulted in an increase in tolerance to drought, high-salinity and low-temperature stresses without stunting growth.	Oh *et al.* 2007
Zm DREB 2A from *Zea mays*	*A. thaliana*	*Agrobacterium*	CaMV 35S	Drought stress tolerance in plants was improved.	Qin *et al.* 2007

6

PRODUCTION AND UTILIZATION OF HAPLOIDS IN RICE (*ORYZA SATIVA*)

ANURADHA GAUTAM, C P MALIK AND BHAVNEET KAUR

Rice (*Oryza sativa* L.) is one of the most important cereal crops, which supplies food for more than half of the world's population (Tyagi *et al.*, 2004) that is; it feeds more people than any other single plant species on earth. According to Simpson and Ogarzaly, it has been estimated that around 2.7 billion people depend on rice. Cultivated in submerged rice paddies in Asia for thousands of years, this grain is also grown in many other parts of the world, including the Sacramental Delta flood plain in California's Central Valley. Like other cereal grains, there are different varieties of *Oryza sativa,* and different ways to mill and prepare the grains. Several different grass species in different genera are also called rice. Wild rice (*Zizania aquatica*) is native to North America and was harvested for food by native Americans.

Scientific classification of *Oryza sativa*

Kingdom	:	Plantae
Division	:	Magnoliophyta
Class	:	Liliopsida
Order	:	Poales
Family	:	Poaceae
Genus	:	*Oryza*

Species

- *Oryza glaberrima*
- *Oryza sativa*

Domesticated rice comprises two species in the Poaceae ("true grass") family, *Oryza sativa* and *Oryza glaberrima*. These plants are native to tropical and subtropical southern

Asia and southeastern Africa (Crawford and Shen, 1998). Rice provides more than one fifth of the calories consumed worldwide by humans (Smith, 1998). The term "wild rice" can refer to the wild species of *Oryza*, but conventionally refers to species of the related genus *Zizania*, both wild and domesticated. Rice is a monocarpic annual plant, growing to 1–1.8 m tall, occasionally more depending on the variety and soil fertility. The grass has long, slender leaves 50–100 cm long and 2–2.5 cm broad. The small wind-pollinated flowers are produced in a branched arching to pendulous inflorescence 30–50 cm long. The seed is a grain (caryopsis) 5–12 mm long and 2–3 mm thick.

Rice is a staple for a large part of the world's human population, especially in East, South and Southeast Asia, making it the most consumed cereal grain. Rice cultivation is well-suited to countries and regions with low labour costs and high rainfall, as it is very labour-intensive to cultivate and requires plenty of water for cultivation.

Rice can be grown practically anywhere, even on steep hillsides. Although its species are native to South Asia and certain parts of Africa, centuries of trade and exportation have made it commonplace in many cultures.

Traditional method for cultivating rice is flooding the fields with or after setting the young seedlings. This simple method requires sound planning and servicing of the water barring and ducting but reduces the growth of lesser robust weed and pest plants and reduces vermin that has no submerged growth state.

The seeds of the rice plant are first milled using a rice huller to remove the chaff (the outer husks of the grain). At this point in the process the product is called brown rice. This process may be continued, removing the germ and the rest of the husk, called the bran at this point, creating white rice.

Whereas brown rice contains all ingredients of a healthy cereal, white rice, without the nutrients of rice germ and rice bran, is a standard in industrialized countries for commercial offerings. The former Beri-Beri disease was related to the stripping off of all ingredients of the bran, however the impact of aflatoxins and other mycotoxins contributed to the problem. Today, *parboiling* is a first method to move some of the nutrients from the bran to the rice corn before stripping the bran, however the energy requirements are high compared to dry processing technologies.

White rice may be also buffed with glucose or talc powder (often called *polished rice*, though this term may also refer to white rice in general), parboiled, or processed into flour. The white rice may also be enriched by adding nutrients, especially those lost during the milling process. While the cheapest method of enriching involves adding a powdered blend of nutrients that will easily wash off, more sophisticated methods apply nutrients directly to the grain, coating the grain with a water insoluble substance which is resistant to washing.

Rice bran, called *nuka* in Japan, is a valuable commodity in Asia and is used for many daily needs. It is a moist, oily inner layer which is heated to produce an oil. It is also used in making a kind of pickled vegetable. The raw rice may be ground into flour for many uses, including making many kinds of beverages such as amazake, horchata, rice milk, and sake. Rice flour is generally safe for people on a gluten-free diet. Rice may also be made into various types of noodles.

Table 1: Nutritional value of rice
Rice, raw Nutritional value per 100 g (3.5 oz)

Energy 360 kcal 1510 kJ	
Carbohydrates	79 g
Fat	0.6 g.
Protein	7 g
Vitamin B6	0.15 mg
Water	13 g

In some countries, rice is commonly consumed as parboiled rice, also known as easy-cook rice. Parboiled rice is subjected to a steaming or parboiling process while still a brown rice. This causes nutrients from the outer husk to move into the grain itself. The parboil process causes a gelatinization of the starch in the grains. The grains become less brittle, and the color of the milled grain changes from white to yellow. The rice is then dried, and can then be milled as usual or consumed as brown rice. Milled parboil rice is nutritionally superior to standard milled rice. Parboiled rice has an additional benefit in that it does not stick to the pan during cooking as happens when cooking regular white rice.

A nutritionally superior method of preparing brown rice known as **GABA Rice** or GBR (Germinated Brown Rice) may also be used. This involves soaking washed brown rice for 20 hours in warm water (38 °C or 100 °F) prior to cooking it. This process stimulates germination, which activates various enzymes in the rice. By this method, it is possible to obtain a more complete amino acid profile, including GABA (a result of research carried out for the United Nations Year of Rice).

Genetic history

Two species of rice were domesticated, Asian rice (*O. sativa*) and African rice (*O. glaberrima*). According to Londo and Chiang, *O. sativa* appears to have been domesticated from wild (Asian) rice, *Oryza rufipogon* around the foothills of the Himalayas, with *O. sativa* var. *indica* on the Indian side and *O. sativa* var. *japonica* on the Chinese and Japanese side (Londo *et al.*, 2006). The different histories have led to different ecological niches for the two main types of rice. *Indica* are mainly lowland rices, grown mostly submerged, throughout tropical Asia, while *japonica* are usually cultivated in dry fields, in temperate East Asia, upland areas of Southeast Asia and high elevations in South Asia. (Oka, 1988) Current genetic analysis suggests that *O. sativa* would be best divided into five groups, labeled *indica, aus, aromatic, temperate japonica* and *tropical japonica*. The same analysis suggests that *indica* and *aus* are closely related, as are *tropical japonica, temperate japonica and aromatic* (Garris *et al.*, 2004). Further analysis of the genetic material of various types of rice indicates that *japonica* was the first cultivar to emerge, followed by the *indica, aus,* and *aromatic* groups, whose genome did show significant differences in age. Within the *japonica* group, there is some genetic evidence that *temperate japonica* is derived from *tropical japonica* (Garris *et al.*, 2004).

Other studies have suggested that there are three groups of *Oryza sativa* cultivars: the short-grained "japonica" or "sinica" varieties, exemplified by Japanese rice; the long-grained "indica" varieties, exemplified by Basmati rice; and the broad-grained "javonica" varieties, which thrive under tropical conditions (Zohary and Hopf, 2000). The earliest find site for the javonica variety, dated to the fifth millennium BC, was in the earliest phases of the Hemudu culture on the south side of Hangzhou Bay in China, but was found along with japonica types.

Rice history in South-East Asia

Wild rice appeared in the Belan and Ganges valley regions of northern India as early as 4530 BC and 5440 BC respectively. Agricultural activity during the second millennium BC included rice cultivation in the Kashmir and mature Harrappan-Pakistan regions. Mixed farming was the basis of Indus valley economy. Farmers planted their crops in integrated fields. Rice, grown on the west coast, was cultivated in the Indus valley (Kahn *et al.*, 2005). Rice, along with barley, meat, dairy products and fish constituted the dietary staple of the ancient Dravidian people.

There is mention of ApUpa, Puro-das and Odana (rice-gruel) in the Rig Veda, terms that refer to rice dishes. The Rigvedic commentator Sayana refers to *"tandula"* when commenting on RV 1.16.2., which means rice. The Rigvedic term *dhana* (dhanaa, dhanya) means rice. Both Charaka and Sushruta mention rice in detail. The Arthasastra discusses aspects of rice cultivation. The *Kashyapiyakrishisukti* by Kashyapa is the most detailed ancient Sanskrit text on rice cultivation (Nene, 2005).

High Yielding Varieties

The High Yielding Varieties are a group of crops created intentionally during the Green Revolution to increase global food production. Rice, like corn and wheat, was genetically manipulated to increase its yield. This project enabled labor markets in Asia to shift away from agriculture, and into industrial sectors. The first 'modern rice', IR8 was produced in 1966 at the International Rice Research Institute, Philippines. IR8 was created through a cross between an Indonesian variety named "Peta" and a Chinese variety named "Dee Geo Woo Gen." (Source: IRRI Knowledge Bank)

With advances in molecular genetics, the mutant genes responsible for reduced height(rht), gibberellins insensitive (gai1) and slender rice (slr1) in *Arabidopsis* and rice were identified as cellular signaling components gibberellic acid (a phytohormone involved in regulating stem growth via its effect on cell division) and subsequently cloned. Stem growth in the mutant background is significantly reduced leading to the dwarf phenotype. Photosynthetic investment in the stem is reduced dramatically as the shorter plants are inherently more stable mechanically. Assimilates become redirected to grain production, amplifying in particular the effect of chemical fertilizers on commercial yield. In the presence of nitrogen fertilizers, and intensive crop management, these varieties increase their yield 2 to 3 times.

Golden Rice

German and Swiss researchers have engineered rice to produce Beta-carotene, with the intent to treat vitamin A deficiency. Additional efforts are being made to improve the quantity and quality of other nutrients in golden rice.

Expression of human proteins

Ventria Bioscience has genetically modified rice to express lactoferrin, lysozyme, and human serum albumin which are proteins usually found in breast milk. These proteins have antiviral, antibacterial, and antifungal effects.

Rice containing these added proteins can be used as a component in oral rehydration solutions which are used to treat diarrhoeal diseases, thereby shortening their duration and reducing recurrence. Such supplements may also help reverse anemia (Bethell *et al.*, 2004).

World production and trade

World production of rice has risen steadily from about 200 million tons of paddy rice in 1960 to 600 million tons in 2006. Milled rice is about 68% of paddy rice by weight. In the year 2006, the top three producers were China (26% of world production), India (20%), and Indonesia (9%).

World trade figures are very different, as only about 5–6% of rice produced is traded internationally. The largest three exporting countries are Thailand (26% of world exports), Vietnam (15%), and the United States (11%), while the largest three importers are Indonesia (14%), Bangladesh (4%), and Brazil (3%).

Rice is the most important crop in Asia. In Cambodia, for example, 90% of the total agricultural area is used for rice production.

HAPLOID PRODUCTION

Haploids are saprophytes with gametophyte chromosome number. The first report of the haploid plant was published by Blakeslee *et al.*, (1922) in *Datura stramonium*. Subsequently, haploids were reported in many other species. Guha and Maheswari (1964) developed an anther culture technique for the production of haploids in the laboratory. Haploid production by wide crossing was reported in barley (Kasha and Kao, 1970) and tobacco (Burk *et al.*, 1979). Tobacco, rapeseed, and barley are the most responsive species for doubled haploid production. Doubled haploid methodologies have now been applied to over 250 species (Maluszynski *et al.*, 2003).

Haploids can be produced following delayed pollination, irradiation of pollen, temperature shocks, colchicines treatment and distant hybridization, the most important methods currently being utilized under biotechnology programmes include:

1. Anther or pollen culture
2. Ovule culture
3. Chromosome elimination following interspecific hybridization (bulbosum technique)

Anther culture has emerged as an adjunct to plant breeding with the distinct advantage of rapid fixation of homozygosis, as a promising target for genetic manipulation, and for producing doubled-haploid populations for molecular and gene tagging (Jähne and Lörz, 1995; Gosal *et al.*, 1997). Techniques to induce microspore embryogenesis, which allows generation of plants from single cells, have been widely used for breeding purposes in cereal crops including rice (Shih-Wei and Zhi-Hong, 1991; Raina and Zapata, 1997). Previous reports of anther culture in rice mainly pertain to japonicas and hybrids involving japonica parent(s) (Guiderdoni *et al.*, 1992; Faruque *et al.*, 1998). Early anther necrosis, poor androgenesis frequencies and albino plant regeneration have been major problems for the successful application of anther culture technique(s) in *indica* rice breeding programmes (Chen *et al.*, 1991; Gosal *et al.*, 1997).

Efforts have been made to optimize the culture conditions, media requirements for androgenesis and green plant regeneration in *indica* rice (Lentini *et al.*, 1995; Raina and Zapata, 1997). The basic anther culture protocol for recalcitrant *indica* rice varieties still needs considerable improvement before it can be routinely exploited in breeding programmes.

Anther culture is used to develop homozygous, or true-breeding, lines from segregating populations by doubling the chromosomes of the haploid pollen and regenerating double haploid plants, all in a single cycle of tissue culture. In contrast, with the standard pedigree method, it normally takes six generations of selfing to produce completely homozygous rice lines. Thus, by accelerating the development of breeding populations, anther culture can cut several years off the approximately 15-year process of developing and releasing a new commercial rice variety.

Scientists are currently employing this technique in gene pool improvement, with particular emphasis on identifying genes for cold tolerance and disease resistance. The technique is also being used to fix enhanced traits in the backcrossed populations resulting from crosses between cultivated rice and wild species. Anther culture should prove useful as well for accelerating the introgression of QTLs for higher yield from wild species into cultivated rice varieties.

A recent advance in plant tissue culture and experimental embryology is the successful induction of haploid plants by the culture of unpollinated ovaries or ovules (Yang and Zhou, 1982). It means that not only the microspore but also the megaspore or female gametophyte of angiosperms can be triggered *in vitro* to saprophytic development, thus opening a new way to genetically research and haploid breeding.

Another method of producing the haploids is wide crossing. In barley haploids can be produced by wide crossing with the related species *H. bulbosum,* fertilization is effected, but during the early stages of seed development the *H.bulbosum* chromosomes are eliminated leaving a haploid embryo. In tobacco, wide crossing with *N. africana* is widely used. When *N. africana* is used to pollinate *N. tabacum,* 0.25 to 1.42 per cent of the progeny survives and can readily be identified as either F1 hybrids or maternal haploids. Although, these percentages appear small, the vast yield of tiny seeds and the early death of most seedlings provide significant numbers of viable hybrids and haploids in relatively small soil containers. This method of interspecific pollination serves as a practical way of producing seed-derived haploids of *N. tabacum,* either as an alternative method or complementary method to anther culture.

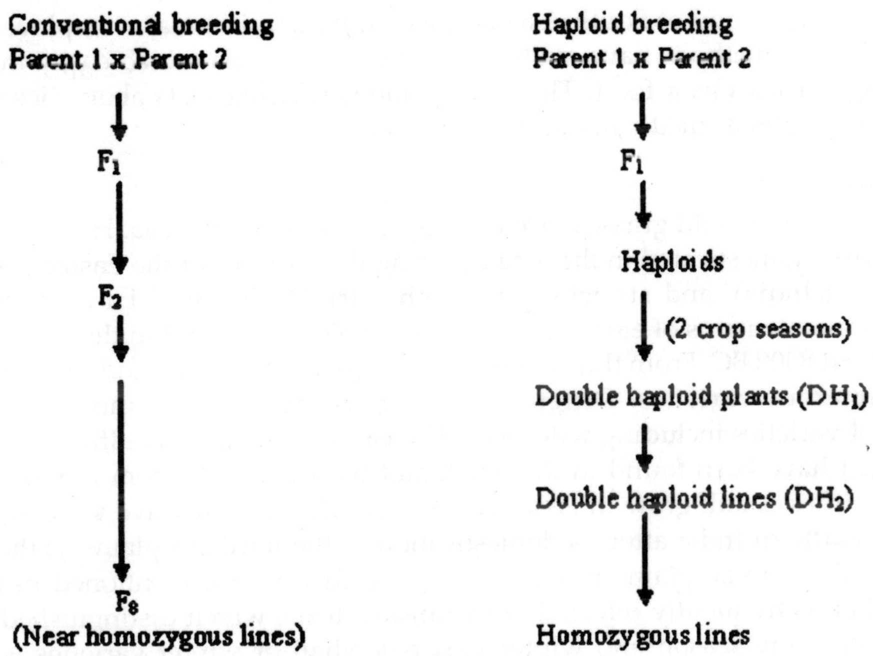

Fig. 1: Period required developing homozygous lines through conventional and haploid breeding

PRODUCTION OF DOUBLED HAPLOIDS

Doubled haploids can be produced via *in vivo* and *in vitro* systems. Haploid embryos are produced *in vivo* by parthenogenesis, pseudo gamy, or chromosome elimination after wide crossing. The haploid embryo is rescued, cultured, and chromosome-doubling produces doubled haploids. The *in vitro* methods include gynogenesis (ovary and flower culture) and androgensis (anther and microspore culture), (Barnabas *et al.*, 1999).

In DH method only two types of genotypes occur for a pair of alleles, A and a, with the frequency of ½ AA and ½ aa, while in diploid method three genotypes occur with the frequency of ¼ AA, ½ Aa, ¼ aa. Thus, if AA is desirable genotype, the probability of obtaining this genotype is higher in haploid method than in diploid method. If *n* loci are segregating, the probability of getting the desirable genotype is (1/2)n in haploid method and (1/4)n in diploid method. Hence the efficiency of haploid method is obviously high when the number of genes concerned is large.

Studies were conducted comparing DH method and other conventional breeding methods and it was concluded that adoption of doubled haploid does not lead to any bias of genotypes in populations, and random DHs were even found to be compatible to selected line produced by conventional pedigree method (Winzeler *et al.*, 1987).

RICE

Rice is a staple food of South Asia and its cultivation is a major employer. A wide range of growing methods have evolved all over South Asia which easily matches its diverse range of uses as a food. There are some surprising facts about rice too, from cosmetics and crafts to medicines and mythology.

Early history

Descended from wild grasses, rice is a staple food in South Asia. Historians believe that it was first domesticated in the area covering the foothills of the Eastern Himalayas (north-eastern India), and stretching through Burma, Thailand, Laos, Vietnam and Southern China. Remains of early cultivated rice have been found in the Yangtze valley dating to about 8500 BC. From this region, it spread in all directions and human selection created numerous varieties. Different rices cross-breed easily and there are now thousands of varieties including wild rices. The earliest remains of cultivated rice in the sub-continent have been found in the north and west and date from around 2000 BC. Perennial wild rices still grow in Assam and Nepal. It seems to have appeared around 1400 BC in southern India after its domestication in the northern plains. It then spread to all the fertile alluvial plains watered by rivers. Rice is first mentioned in the Yajur Veda and then is frequently referred to in Sanskrit texts, which distinguished summer varieties from rainy season and winter varieties. Shali or winter varieties were most highly regarded. About 2000 years ago, rice was well-established as the main cereal of the sub-continent, with barley second and wheat a barely mentioned winter food.

Greek visitors noted the popularity of rice amongst Indians. The Greek emissary Megasthenes, visiting Pataliputra (modern Patna) in 315 BC, wrote that they ate it ceremonially, boiled, placed in a bowl and then various other dishes added to it. Hundreds of years later, the Portuguese in the 15th century observed cooked rice being eaten in much the same way. The 17th century traveler Francois Bernier described fields of rice in Kashmir and Bengal, irrigated by endless channels. The Muslim rulers of India created famous rice and meat dishes such as pilafs and biriyanis. The number of dishes made with rice was by this time legion.

Rice types

Over the centuries, three main types of rice had developed in Asia, depending on the amylose content of the grain. They were called *indica* (high in amylose and cooking to fluffy grains to be eaten with the fingers), japonica (low in amylase and cooking to sticky masses suitable for eating as clumps with chopsticks), and javanica (intermediate amylose content and stickiness). Rice is further divided into long, medium and short-grained varieties, and in the sub-continent different regions grows and consumes different varieties. Basmati rice is probably the best-known variety of rice from the sub-continent. Basmati denotes 'queen of fragrance' and this fragrant rice is chiefly grown and exported by Pakistan and India. The dominant food crop in Bangladesh is rice which accounts for at least 70% of the land under agriculture. The best varieties are rice grown in shallow, slowly moving water and so irrigation is crucial to the success of rice-growing in the sub-continent. Slash and burn methods of creating rice fields in some parts of India have placed tremendous pressure on the environment.

Rice finds applications in the arts and crafts of India. Rice paste is employed in the resist-dyeing techniques of creating patterns on cloth. In traditional homes, decorative features frequently consist of wall paintings and floor patterns. It is usually women who paint renditions of folklore and mythology on domestic spaces, passed on from mother to daughter. In many parts of India, as part of daily ritual, ephemeral, abstract designs created by women are traced in rice powder or paste on domestic thresholds and floors. These are known by various names, alpana in Bengal, mandana in Rajasthan and kolam in South India, for instance. The designs are meant to bring good luck to the home.

RICE-PLANT PROFILE

Names

Rice, paddy rice, chowdhury rice (English)

Dhanya, vrihi, nivara, syali (Sanskrit)

Dhan, chaval (Hindu)

Chal (Bengal)

Dangar, choka (Gujarat)

Nellu, arisi (Tamil)

Botanical name: *Oryza sativa*

Family: Poaceae, the grass family, also known as the Gramineae.

The plant

The rice plant is a grass which can grow to over 1 m tall or to 5 m long in deep water.

Stem - upright and composed of a series of joint-like nodes. A leaf grows from each node.

Seed (or grain) - grow on branch-like spikes which arch over. The grain is the most economically important part of the rice plant, and its endosperm is the final product consumed.

Rice should not be confused with 'wild rice' which is the North American *Zizania aquatica*, also in the grass family but not closely related.

Nutritional Benefits of Rice

Rice is an excellent food to help keep the body healthy. Rice has a number of nutritional benefits like it is an excellent source of carbohydrates, good energy source, low fat, low salt content, no cholesterol, low sugar, no gluten, no additives, no preservatives etc.

Rice is an excellent source of carbohydrates and energy

Good nutrition is vital to everyday performance, and eating more carbohydrate rich foods means having more energy. Rice is an excellent source of carbohydrates. Carbohydrates provide energy to fuel working muscles and feed the brain to enhance concentration.

Carbohydrates are not normally converted to body fat, as they are the body's preferred source of energy. Carbohydrates are broken down to glucose, most of which is used as energy for exercise and as essential fuel for the brain.

Rice is low in fat, salt and has no cholesterol

Dieticians recommend eating plenty of breads and cereals (preferably wholegrain). Rice is an excellent food to include in a balanced diet. It is low in total fat and saturated fat, is cholesterol free, and contains negligible amounts of sodium.

Being low in fat, rice is suitable to include in a diet for those watching their weight. Rice is also cholesterol free, therefore being an excellent food to include in a cholesterol lowering diet. Brown rice contains a small amount of rice bran oil.

Rice is gluten free

Some people are unable to tolerate the proteins found in wheat, barley, rye, oats and triticale. These people choose foods that are gluten free. All rice is gluten free, making rice the essential choice for people with gluten free dietary requirements.

Rice contains no additives or preservatives, making it an excellent inclusion in a healthy and balanced diet. Rice also contains resistant starch, which is the starch that reaches the bowel undigested. This encourages the growth of beneficial bacteria, keeping the bowel healthy.

Rice and the Glycaemic Index

Worldwide research has shown that different carbohydrate foods have dramatically different effects on blood sugar levels. The glycaemic index (GI) was developed as a means of ranking foods based on their immediate effect on blood sugar levels.

Carbohydrate foods that break down quickly during digestion have the highest GI factors. Carbohydrates which break down slowly, releasing glucose gradually into the bloodstream have low GI factors.

Health benefits of Low GI Foods :

- Result in lower insulin levels which makes fat easier to burn and less likely to be stored
- Are more satisfying and reduce appetite
- Help to sustain endurance exercise for longer

Rice nutrition facts

With high nutrients, rice is a good source of insoluble fiber, which is also found in whole wheat, bran and nuts. Insoluble fiber reduces the risk of bowel disorders and fights constipation. Among other nutrients, rice is rich in carbohydrates, the main sources of energy, low in fat, contains some protein and plenty of vitamins B.

Table 2: Nutrition value of different rice types

Nutrition Facts Serving: 100 g	White Rice	Jasmine	Brown	Glutinous
Calories, kcal	361	355	362	355
Moisture (water), g	10.2	11.9	11.2	11.7
Total Fat, g	0.8	0.7	2.4	0.6
Dietary Fibre, g	0.6	0.8	2.8	0
Calcium, mg	8	5	12	7
Phosphorus, mg	87	65	255	63
Potassium, mg	111	113	326	0
Sodium, mg	31	34	12	0
Vitamin B1, mg	0.07	0.12	0.26	0.08
Vitamin B2, mg	0.02	0.02	0.04	0.03
Niacin, g	1.8	1.5	5.5	1.8
Protein, g	6	6.1	7.4	6.3
Carbohydrates, g	82.0	81.1	77.7	81

Source: Thai Food Composition Table (1999), Institute of Nutrition, Mahidol University

World's 'oldest' rice found !

Scientists have found the oldest known domesticated rice in the world. The handful of 15,000-year-old burnt grains was discovered by archaeologists in Korea (BBC News Science-Nature).

Oldest domesticated rice

Their age challenges the accepted view that rice cultivation originated in China about 12,000 years ago. Rice is especially important in Asia, where it is responsible for almost a third of all calorific intake.

Tracer of evolution

The oldest known rice was discovered by Lee Yung-jo and Woo Jong-yoon of Chungbuk National University in South Korea. They found the ancient grains during excavations in the village of Sorori in the Chungbuk Province.

Radioactive dating of the 59 grains of carbonized rice has pushed back the date for the earliest known cultivation of the plant.

DNA analysis shows the early rice sample to be different from the modern intensively farmed varieties, thereby offering scientists the opportunity to study the evolution of one of the world's principal food sources.

The region in central Korea where the grains were found is one of the most important sites for understanding the development of Stone Age man in Asia.

Rice pests

Rice pests are any organisms or microbes with the potential to reduce the yield or value of the rice crop (or of rice seeds) (Jahn *et al.*, 2005, 2007). Rice pests include weeds, pathogens, insects, rodents, and birds. A variety of factors can contribute to pest outbreaks, including the overuse of pesticides and high rates of nitrogen fertilizer application (Jahn *et al.*, 2005). Weather conditions also contribute to pest outbreaks. For example, rice gall midge and army worm outbreaks tend to follow high rainfall early in the wet season, while thrips outbreaks are associated with drought (Douangboupha *et al.*, 2006).

One of the challenges facing crop protection specialists is to develop rice pest management techniques which are sustainable. In other words, to manage crop pests in such a manner that future crop production is not threatened (Jahn *et al.*, 2001). Rice pests are managed by cultural techniques, pest-resistant rice varieties, and pesticides (which include insecticide). Increasingly, there is evidence that farmers' pesticide applications are often unnecessary (Jahn *et al.*, 1996, 2004a,b). By reducing the populations of natural enemies of rice pests (Jahn 1992), misuse of insecticides can actually lead to pest outbreaks (Cohen *et al.*, 1994). Botanicals, so-called "natural pesticides", are used by some farmers in an attempt to control rice pests, but in general the practice is not common. Upland rice is grown without standing water in the field. Some upland rice farmers in Cambodia spread chopped leaves of the bitter bush (*Chromolaena odorata* (L.)) over the surface of fields after planting. The practice probably helps the soil retain moisture and thereby facilitates seed germination. Farmers also claim the leaves are a natural fertilizer and helps suppress weed and insect infestations (Jahn *et al.*, 1999).

Among rice cultivars there are differences in the responses to, and recovery from, pest damage (Jahn *et al.*, 2004c; Khiev *et al.*, 2000). Therefore, particular cultivars are recommended for areas prone to certain pest problems. Major rice pests include the brown planthopper (Preap *et al.*, 2006), armyworms, the green leafhopper, the rice gall midge (Jahn and Khiev 2004), the rice bug (Jahn *et al.*, 2004c), hispa (Murphy *et al.*, 2006), the rice leaf folder, stem borer, rats (Leung *et al.*, 2002), and the weed *Echinochloa crusgali* (Pheng *et al.*, 2001). Major rice diseases include Rice Ragged Stunt, Sheath Blight and Tungro. Rice blast, caused by the fungus *Magnaporthe grisea*, is the most significant disease affecting rice cultivation.

Cultivars

The largest collection of rice cultivars is at the International Rice Research Institute (IRRI), with over 100,000 rice accessions held in the International Rice Genebank. Rice cultivars are often classified by their grain shapes and texture. For example, Thai Jasmine rice is long-grain and relatively less sticky, as long-grain rice contains less amylopectin than short-grain cultivars.

Indian rice cultivars include long-grained and aromatic Basmati (grown in the North), long and medium-grained Patna rice and short-grained Masoori. In South India the most prized cultivar is 'ponni' which is primarily grown in the delta regions of Kaveri River. Kaveri is also referred to as ponni in the South and the name reflects the geographic region where it is grown. Rice in East India and South India, is usually prepared by boiling the rice in large pans immediately after harvesting and before removing the husk; this is referred to in English as *parboiled rice*. It is then dried, and the husk removed later. It often displays small red speckles, and has a smoky flavor from the fires. Usually coarser rice is used for this procedure. It helps to retain the natural vitamins and kill any fungi or other contaminants, but leads to an odor which some find peculiar. In South India, it is also used to make idlis, dosas and several breakfast and tiffin items. In the Western Indian state of Maharashtra, a short grain variety called Ambemohar is very popular. This rice has a characteristic fragrance of Mango blossom.

Aromatic rices have definite aromas and flavors; the most noted cultivars are Thai fragrant rice, Basmati, Patna rice, and a hybrid cultivar from America sold under the trade name, Texmati. Both Basmati and Texmati have a mild popcorn-like aroma and flavour. In Indonesia there are also *red* and *black* cultivars.

Draft genomes for the two most common rice cultivars, *indica* and *japonica*, were published in April 2002. Rice was chosen as a model organism for the biology of grasses because of its relatively small genome (~430 megabase pairs). Rice was the first crop with a complete genome sequence (Gillis, 2005). Basmati rice is the oldest, common progenitor for most types.

HISTORY AND CONCEPT OF HYBRID RICE

The first artificial hybrid was produced by Thomas Fairchild during 1717 by crossing between sweet William (*Dianthus barbatus*) and carnation (*Dianthus caryophyllus*). Thereafter, several scientists used hybridization for scientific studies as well as for crop improvement in the breeding programme. Joseph Koelreuter, one of the famous scientists made several crosses in Tobacco during 1760-1766 and reported hybrid vigor in F1. At the end of nineteenth century, hybridization was widely used for different crop improvement programme. In the modern era hybridization is the most common method is applied in crop improvement and large number of varieties have been developed in various crops by hybridization.

For development of hybrid, intensive research was started in India during 1952 in the Maize Coordinated Project and first hybrid was released during 1960. Subsequently three more maize hybrids were also released during 1960. Later on, research work was initiated on Bajra and Pearl millet and Sorghum (Jowar) crops during 1961 and first hybrid in Sorghum and in Pearl millet was released during 1964.

Success in hybrids of maize, pearl millet, sorghum, cotton, sunflower and pigeonpea breeding programme encouraged researchers to explore commercial exploitation of heterosis in rice. In mid seventies, rice hybrids were also developed and cultivated in China.

Efforts to develop and use of hybrid rice technology in India was initiated during 1970. But the research works were systematized and intensified since 1989 with a mission mode project. With the concerted research work, the country developed half a dozen rice hybrids each from public and private sectors within a short span of five years. The first four rice hybrids were released in the country viz. APHR-1, APHR-2, MGR-1 and KRH-1 during 1994. Subsequently, two more hybrids viz. CNRH-3 and DRRH-1 were also released. The total number of 18 rice hybrids have been released till December 2001.

In the mid seventies rice hybrids were also developed and cultivated in China. The China cultivated rice hybrids in an area of about 55% of its total area under rice during 1991. The result indicated that the hybrid rice have greater potential to increase productivity by 15-20% than the high yielding varieties.

Golden rice

Golden rice is a variety of rice (*Oryza sativa*) produced through genetic engineering to biosynthesize the precursors of beta-carotene (pro-vitamin A) in the edible parts of rice. The scientific details of the rice were first published in *Science* in 2000 (Ye *et al.*, 2000). Golden rice was developed as a fortified food to be used in areas where there is a shortage of dietary vitamin A (Frist, 2006). In 2005 a new variety called *Golden Rice* 2 was announced which produces up to 23 times more beta-carotene than the original variety of golden rice (Paine *et al.*, 2005). Neither variety is currently available for human consumption. Although golden rice was developed as a humanitarian tool, it has met with significant opposition from environmental and anti-globalization activists.

Golden rice was created by Ingo Potrykus of the Institute of Plant Sciences at the Swiss Federal Institute of Technology, working with Peter Beyer of the University of Freiburg. The project started in 1992 and at the time of publication in 2000, golden rice was considered a significant breakthrough in biotechnology as the researchers had engineered an entire biosynthetic pathway.

Golden rice was designed to produce beta-carotene, a precursor of Vitamin A, in the part of rice that people eat, the endosperm. The rice plant can naturally produce beta-carotene, which is a carotenoid pigment that occurs in the leaves and is involved in photosynthesis. However, the plant does not normally produce the pigment in the endosperm since photosynthesis does not occur in the endosperm.

Golden rice was created by transforming rice with two beta-carotene biosynthesis genes:

1. *psy* (phytoene synthase) from daffodil (*Narcissus pseudonarcissus*)
2. *crt1* from the soil bacterium *Erwinia uredovora*

(The insertion of a *lyc* (lycopene cyclase) gene was thought to be needed but further research showed that it is already being produced in wild-type rice endosperm.)

The *psy* and *crt1* genes were transformed into the rice nuclear genome and placed under the control of an endosperm specific promoter, so that they are only expressed in the endosperm. The exogenous *lyc* gene has a transit peptide sequence attached so that it is targeted to the plastid, where geranyl geranyl diphosphate formation occurs. The bacterial *crt1* gene was an important inclusion to complete the pathway, since it can catalyze multiple steps in the synthesis of carotenoids, while these steps require more than one enzyme in plants (Hirschberg, 2001). The end product of the engineered pathway is lycopene, but if the plant accumulated lycopene the rice would be red. Recent analysis has shown that the plant's endogenous enzymes process the lycopene to beta-carotene in the endosperm, giving the rice the distinctive yellow color for which it is named (Schaub, 2005). The original Golden rice was called SGR1, and under greenhouse conditions it produced 1.6nµg/g of carotenoids.

Subsequent development

Golden rice has been bred with local rice cultivars in the Philippines, Taiwan and with the American rice variety Cocodrie. The first field trials of these golden rice cultivars were conducted by Louisiana State University AgCenter in 2004. Field testing allows a more accurate measurement of the nutritional value of golden rice and enables feeding tests to be performed. Preliminary results from the field tests have shown that field grown Golden rice produces 3 to 4 times more beta-carotene than Golden rice grown under greenhouse conditions.

In 2005, a team of researchers at biotechnology company Syngenta produced a variety of golden rice called "Golden Rice 2". They combined the phytoene synthase gene from maize with *crt1* from the original golden rice. Golden rice 2 produces 23 times more carotenoids than golden rice (up to 37nμg/g), and preferentially accumulates beta-carotene (up to 31nμg/g of the 37nμg/g of carotenoids) (Paine *et al.*, 2005). To receive the Recommended Dietary Allowance (RDA), it is estimated that 144 g of this rice would have to be eaten. Bioavailiability of the carotene from either variety has not been tested in any model (Datta *et al.*, 2007).

In June 2005, researcher Peter Beyer received funding from the Bill and Melinda Gates Foundation to further improve Golden rice by increasing the levels of or the bioavailability of pro-vitamin A, vitamin E, iron, and zinc, and to improve protein quality through genetic modification.

Because many children in countries where there is a dietary deficiency in Vitamin A rely on rice as a staple food, the genetic modification to make rice produce pro-vitamin A (beta-carotene) can prove to be a simple and less expensive alternative to vitamin supplements or an increase in the consumption of green vegetables or animal products. It can be considered as the genetically engineered equivalent of fluoridated water or iodized salt.

Theoretical analyses of the potential nutritional benefits of golden rice show that consumption of golden rice would not eliminate the problems of blindness and increased mortality, but should be seen as a complement to other methods of Vitamin A supplementation (Dawe *et al.*).

New Technologies

A considerable improvement has been made by exploiting the natural variation through conventional breeding. Albeit the success made in the last century, traditional breeding efforts alone cannot meet the increasing demand of rice consumers in the coming century. One of the possible options for further increase in rice production is through the application of new technologies such as biotechnology and improved agronomy, complimented by traditional farming systems that need to be adapted to changing environment and needs of local people. At present, various tissue culture techniques are being used for the genetic improvement of rice plant throughout the world (Raina, 1989). Among techniques, anther culture, protoplast fusion and culture, leaf culture, root culture, immature embryo culture and mature seed culture are important in rice to create additional variation and novel rice varieties (Lutts *et al.*, 2001; Sathish *et al.*, 1995).

Rice improvement in many parts of the world, especially Africa and Latin America has benefited from some interventions of biotechnology such as embryo rescue and anther culture achieving yield and quality improvements. Genetic engineering, supported by gene mapping, gene tagging and gene transfer techniques are emerging as a potential area in rice research. However it is true that biotechnology is certainly moving the pace of rice improvement both in qualitative as well as quantitative terms.

Use of anther culture has contributed more than 100 rice cultivars in China (Meifang 1992), and over 42 Japonica rice cultivars in Korea. Exploitation of gametoclonal variation has resulted in the development and release of cv. Dama in Hungary (Heszkey *et al.*, 1991). At IRRI (International Rice Reseach Institute, Phillipines), the anther culture technique is being employed to obtain DH lines from many crosses for different rice breeding objectives (Khush and Brar, 1998).

ANTHER AND MICROSPORE CULTURE

One of the very popular methods for production of haploids is through culturing anthers or microspores on artificial culture medium. This leads to the growth of microspores into saprophytes.

After the initial reports of successful production of haploids from anther culture in Datura (Guha and Maheshwari, 1964, 1966), haploids have been obtained in more than 150 species belonging to 23 families of angiosperms (Maheshwari *et a1.*, 1980). These include a wide variety of economically important species.

More often, anthers rather than microspores are cultured, since the extraction and culture methods for microspores differ and have been successful only in a few species (*Datura inoxia, Nicotiana sylvestris, N. tabacum, Oryza sativa,* etc.).

The following parameters have been recognized as particularly important for successful anther and microscope culture;

(i) Conditions of growth of donor plant;

(ii) Genotype of donor plant:

(iii) The pretreatment;

(iv) The developmental stage of anther/microspore;

(v) The culture medium and the conditions during culture growth.

In either case, flower buds are brought to a laminar flow chamber and sterilized using appropriate chemical treatment before their dissection. While removing anthers from the flower buds, care is exercised to avoid injury, because injury leads to development of callus, giving a mixture of diploids, haploids and aneuploids.

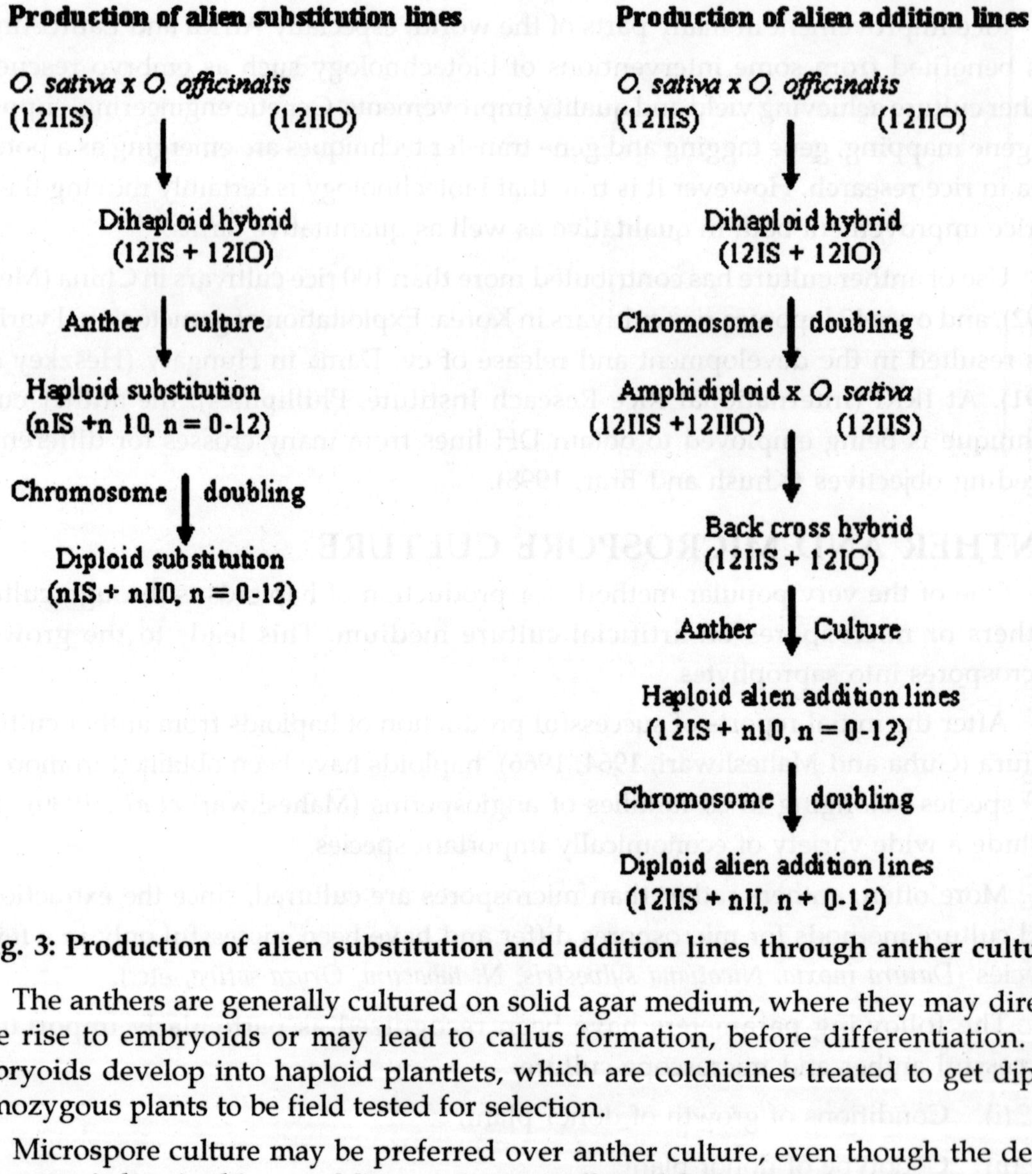

Fig. 3: Production of alien substitution and addition lines through anther culture

The anthers are generally cultured on solid agar medium, where they may directly give rise to embryoids or may lead to callus formation, before differentiation. The embryoids develop into haploid plantlets, which are colchicines treated to get diploid homozygous plants to be field tested for selection.

Microspore culture may be preferred over anther culture, even though the degree of success is low in this case. Microspores are collected first using the following steps:

(i) Anthers, dissected as above, are taken in a sterile beaker containing liquid medium, and are pressed with a glass rod or syringe piston to allow the microspores squeeze out;

(ii) The suspension with anthers and microspores is filtered through a nylon sieve, which allows only the microspores to pass through.

(iii) The filtrate is centrifuged thrice, for 5 minutes each, at 500-800 rpm (the pellet resuspended in a fresh medium each time);

Table 3: Anther and microspore culture for production of haploids/doubled haploids in rice

Plant material	*Explant/stage*	*Medium*	*Remarks*	*References*
O.sativa	Anthers	Blaydes 1966 medium – NAA (9.4 M) +kin (10 M)	obtained 0.57% regeneration after 4-8 wks in culture.	Niizeki & Oono (1986)
20 Indica cultivars	Anthers with uninucleate	medium supplemented with IAA, 2,4-D, YE & CW	observed variation between cultivars for androgenic response. Highest Callus seen in "Assam 5".	Guha *et al.* (1970)
Keng, Keng x Shien F_1 hybrids	Anthers	N_6 or SK-8 liquid medium	Callus induction from unreleased pollen in both media	Chen and Chen (1979)
O.sativa	Anthers with Binucleate to Trinucleate Microspores	Miller's macro+micro + Morel's vit. + NAA (3 *mg*/l) +YE (2000 *mg*/l) + sucrose (6%w/v)	Obtained green haploid & mixoploid plant	de Be- uville (1980)
O.sativa	Anthers with late uni- to binucleate microspores. Uninucleate to mature ovary	1.N_6 + MCPA (0.125 *mg*/l) +sucrose (3%w/v) 2.N_6 or MS+IAA (0.5 *mg*/l) + kin (1-2 *mg*/l) +NAA (0.25 *mg*/l) + sucrose (3%w/v) or N_6 +MCPA (0.33 *mg*/l) + sucrose (3%w/v)	green as well as albino haploids were obtained; float culture of unhusked flowers was used.	Zhou and Yang (1981)
O.sativa	Anthers with the uni to early binucleate microspores. Uninucleate to mature ovary	1.N_6 +2,4-D(2 *mg*/l) + CH (500 *mg*/l)+ sucrose (4%w/v) 2. N_6 + CH(500 *mg*/l)+ kin (2 *mg*/l) + NAA (0.55 *mg*/l) + sucrose (3%w/v)	Green haploid, diploid & some albino plants	Kuo (1981)
Minekari (Japonica)	Anthers with mid to late uninucleate uninucleate microspores	R-3(modification of LS) + various combinations of NAA, IAA & 2,4-D + agarose (0.2%w/v)	Sucrose conc. of 4%w/v was found to be beneficial for callus induction in both 0.4%w/v & 0.2%w/v agarose; Addition of 50 M mannitol increased callus induction frequency.	Chaleff & Stolarz (1981)

Plant material	Explant/stage	Medium	Remarks	References
Calrose-76, Labelle, Mars, Nortal, Starbonnet	Anthers with mid to late uninucleate uninucleate microspores	1.Blaydes salts + YE (1g *mg*/l)+ CM (150 m/l) + IAA (2 *mg*/l)+ sucrose (3%w/v)2. Blaydes salts + CM (500 *mg*/l)+ YE 250 *mg*/l) 2,4-D (1 *mg*/l)+ NA(0.5 *mg*/l) + kin(1 *mg*/l) + sucrose(3%w/v) 3.MS + glutamine (10^{-3}M)+ inositol(100 *mg*/l)+ IAA (1 *mg*/l) + kin(1 *mg*/l)+ sucrose (3%w/v)	Obtained plants in Calrose 76 and observed variability in regenerated plants for height, seeds, flowering time, awn length, seed weight, etc.	Schaeffer (1982)
Giza 170, Taipei 390	Anthers with mid uninucleate to early binucleate microspores	1.E10 + sucrose (2% w/v)+ glucose (0.5% w/v)+ BA (0.5 *mg*/l)+ 2,4-D (1 *mg*/l) 2.MS+ NAA(1 *mg*/l) + kin(1 *mg*/l)	Used conditioned 1:8 dilution and obtained 1.6 times higher callus than control for Taipei 309 while conditioned medium reduced callus induction as well as green plant regeneration in Giza 170.	Zapata *et al*., (1985)
Indica cultivars ARC1557, Teep, Mohini, Pankaj, Govindabhoj, Radhunipagal, Meghna	Anthers with uninucleate microspores	1.He-2+ 2,4-D(9 M) or He-5 + sucrose (5%w/v) or N_6 or R_3 2. He-2+ different conc.& combinations of 2,4-D, NAA, picloram, zeatin and kin+ glucose(5%w/v) 3.He-2+ 2,4-D(9 M) + sucrose (3,5,7,9 or 11%)4.MS +NAA (1.5 M) + kin (9.5 M)	Different cultivars respond differently to different media He-2 & He-5 media induced callus at higher frequency than R-3 & N_6. Medium with 2,4-D (9 M) picloram(0.3 M) & zeatin (0.5 M) was most effective for callus induction & green plant regeneration. Best results obtained when anthers were treated at 35°C for 5 min. followed by 10°C for 7 days. An increase in sucrose conc.from 3 to 5% resulted in an increase in percentage of callus induced.	Reddy *et al*., (1985)
3 Indica, Indica x Japonica 4 F_2 lines	Anthers	N_6, modified N_6H_5, modified H_5, SK_8	The response of different cultivars was different in 5 media indicating existence of genotypic variability Green	Shahjahan *et al*., (1985)

Plant material	*Explant/stage*	*Medium*	*Remarks*	*References*
			plant regeneration was higher in calli induced on H_5 medium.	
Indica cultivars IR-42	Anthers with mid uninucleate to early binucleate microspores	B5 + 2,4-D(1 *mg*/l) + BA(0.5 *mg*/l)+IAA (0.5 *mg*/l) or B5 + zeatin + 2,-D(2 *mg*/l) + picloram(0.01 *mg*/l)	Heat treatment at 35°C for 15 min. increased significantly the callus induction in both media.	Zapata and Torrizo (1986)
Basmati 370	Anthers with mid uninucleate to binucleate microspores	1.B5 + 2,4-D(1 *mg*/l) NAA (0.5 *mg*/l)+BA (0.5 *mg*/l)+ sucrose (2%w/v) with or without glucose (0.5%w/v) liquid medium. 2.MS+ kin (1 *mg*/l)+ NAA(1 *mg*/l)+ ABA (20 *mg*/l) for 4weeks medium(semi-solid) 3.ABA free medium	20K rad treatment of seeds of donor plants reduced the callus induction but also reduced the percentage of albino plant regeneration. No green plants were regenerated in control addition of glucose (5g/l) improved the callus induction frequency.	Zapata *et al.*, (1986)
Taipei-309 (Japonica), IR-5, Basmati 370, Pankaj	Anthers	1.E_{10}+ 2,4-D(1 *mg*/l) BA (0.5 *mg*/l) + sucrose (2%w/v) + glucose(5g/l) 2.MS +NAA(1 *mg*/l) + kin(1 *mg*/l) + sucrose (3%w/v)	60% of the callus forming anthers were plated on edge & 40% were plated flat. Most of the plated flat were callused on both lobes but callusing on one lobe was also observed. Most of the anthers plated on edge produced callus on the upper lobe.	Mercy & Zapata (1987)
Taipei-309 (Japonica)	Uninucleate microspores	$E_{10}E_{10}$ +alanine(1mM)E_{10} +glutamine(1mM)E_{10} + proline(1mM)	Glutamine & proline had stimulatory effects on pollen culture. On large 50-58 M diam with thin pink colored cell walls produced pollen embryos. Division of 40 M diam thick-walled pollen was not observed.	Cho & Zapata (1988)
PR106 *X jaya* *O.sativa* x *O.rufipogon*	Anthers	Potato-2 or N_6 +NAA or 2,4-D(2 *mg*/l) NAA (2.5 *mg*/l)+ kin(1 *mg*/l) sucrose(6%w/v)+agar	1. Potato-2 medium was superior to N_6 2. Early callusing when NAA was present instead of 2,4-D	Rout *et al.*, (1989)

Plant material	Explant/stage	Medium	Remarks	References
O.sativa x *O.logistaminata*		(0.75%w/v)	3. *O.sativa* x *O.logistaminata* gave no response except their F2 progeny formed callus from anthers.	
17 Japonica 3 indica, 2 indica x Japonica and one non classified cv.	Anthers	1. N_6 + NAA (1 *mg*/l) + agar (.8% w/v) 2. MS + Kin (3 *mg*/l) + NAA (.5 *mg*/l)	Japonica cultivars were more responsive than indicas; 27 % of the plants obtained were albinos.	Reiffers and Freire (1990)
F_1 heterotic hybrids	Anthers	MSN or SK1 with 2,4-D (.7 *mg*/l) + kin (.7 *mg*/l) + sucrose (4% w/v)	MSN medium was better in all the F_1 hybrids while regeneration frequencies were higher in calli raised on SK-1medium.	Raina *et al.*, (1989)
Indica cv.: PTB-33, Rasi, Crossa, Pokkali, Tellahemasa, Yerregaluvadlu, Mahsuri, Indica x Japonica; TCA-2 x YVSD, Basmati D_5 x Basmati, Mahsuri x Bala, TN-1 x Carreon	Anthers with early to mid uninucleate microspores	1. N_6, He_2, He_5, LS or R_2 with different concentration and combination of 2,4-D, NAA, kin and sucrose (2, 3, 4 or 5 % w/v) 2. R_2 + NAA(2 *mg*/l) + kin (2 *mg*/l) 3. LS + NAA (1 *mg*/l) + kin (4 *mg*/l)	1. Variation for callus induction and green plant regeneration was found. 2. R_2 with NAA (2 *mg*/l) + kin (.5 *mg*/l) was found suitable for callus induction3. 3% w/v sucrose gave better results.	Kavi Kishore *et al.*, (1989)
WU IOB	Anthers	1. N_6 + 2,4-D (2 *mg*/l) + sucrose (6% w/v) + gourd stem exudates (0,10,15, 20 or 25% w/v) 2. MS + kin (2 *mg*/l) + NAA (.5 *mg*/l) + BA (.5 *mg*/l)	Callus response was highest from main tiller. However, regeneration frequencies on calli bases did not differ except for tertiary tiller.	Hadi and Raina (1989)
Indica and Japonica	Anthers	1. N_6 + 2,4-D (2 *mg*/l) + sucrose (6% w/v) + gourd stem exudates (0,10,15,20 or 25% w/v)	Addition of gourd stem exudates (15-20% w/v) lead to early callus induction and also increased the callus	Chengmei and Zhenghua (1990)

Plant material	Explant/stage	Medium	Remarks	References
		2. MS + kin (2 *mg*/l) + NAA (.5 *mg*/l) + IAA (.5 *mg*/l)	induction frequencies in both Japonica and Indica rices and also had increased plant regeneration.	
Taipei-309 (Japonica) IR43 (Indica)	Anthers with late uninucleate microspores	E_{10} + 2,4-D (1 *mg*/l) + BA (.5 *mg*/l) + IAA (.5 *mg*/l) + sucrose (20 g/l) + glucose (5 gm/l) + alanine (1mM) or glutamine (1mM) or proline (1mM)	Addition of proline and glutamine reduced the albinism frequency; 72% of regenerated plants were homozygous diploids.	Zapata *et al.*, (1990)
Japonica x Japonica, Japonica x Indica crosses	Anthers	1. N_6 + NAA (1 *mg*/l) + sucrose (6% w/v) 2. MS +NAA (.5 *mg*/l) +kin (3 *mg*/l)	Variability for callus induction and plant regeneration was observed.	Reiffers and Freire (1990)
Nagdongbyea (Japonica) So mgaughbyeo (Tongil types)	Anther	One step method N_6 + NAA (1 *mg*/l) + kin (4 *mg*/l) + agar (1.2% w/v)Two step method 1. N_6 Y-1 + NAA (2 *mg*/l) + kin (1 *mg*/l) + agar (.8% w/v) 2. N_6 Y-1 + IAA (2 *mg*/l) + kin (1 *mg*/l) + agar (.8% w/v)	Single step anther culture can be effectively employed which can save time energy and money in obtaining haploids directly from anther.	Chung *et al.*, (1991)
Taipei-309	Anthers with mid uninucleate to early	1. B_5 + 2,4-D (1 *mg*/l) + BA (.5 *mg*/l) + IAA (.5 *mg*/l) + sucrose (2% w/v) 2. M_{10} + NAA (1.5 *mg*/l) BA (.5 *mg*/l) + IBA (.5 *mg*/l) + kin (.75 *mg*/l) + sucrose (6% w/v) 3. P_2 + NAA (2 *mg*/l) + kin (.5 *mg*/l) + sucrose (6% w/v)	The procedure saved time (10-15 days); direct plantlets were produced on medium M-10 and P-2'% albino plantlets was greater on medium P-2.	Karim and Zapata (1990)
Indica and Japonica	Microspores	MS + Glutamine + CH + Ficoll (10% w/v)	Green plants obtained in both Indica and Japonica cv.	Dutta *et al.*, (1990a)

Plant material	Explant/stage	Medium	Remarks	References
Indica cv. IR-24	Mechanically isolated pollen grains (50 μM)	R_2 (1/4 strength) + B_5 vitamins + 2,4-D (.5 *mg*/l) + sucrose (5mM) + mannitol (.4M)	Green and albino plants were regenerated; a cold treatment of 25 days gave the highest colony formation.	Ogawa *et al*., (1992)
Three F_1 hybrids of Japonica cv.	Anthers	1. N_6 medium with hormones in disposable Petri dishes or in glass jam jars 2. MS + kin (2 *mg*/l) + NAA (.5 *mg*/l) + sucrose (4% w/v)	Regeneration of green plants was better in jam jars in all genotypes.	Courtois (1992)
Japonica x Japonica, Japonica x Indica	Anthers with uninucleate microspores	1. N_6M or N_6AK supplemented with MCPA, kin, NAA with sucrose (4.5 % or 5.5% w/v) 2. for Indica R_2 + kin (2 *mg*/l) +NAA (.5 *mg*/l)+sucrose (6% w/v) or for Japonica R_6 + kin (1 *mg*/l) + NAA (.1 *mg*/l) + sucrose (4% w/v)	With an increase in time in culture, the ratio of green/ albino plantless did not change; however, the overall plant regeneration frequency was affected with the time in culture.	Guiderdoni *et al*., (1992)
Indica rice cv. F_1 and F_2 resulting from high yielding Indica rice x basmati cv.	Anthers with uninucleate microspores	N_6 + 2,4-D (2 *mg*/l) + sucrose (3% w/v)	Addition of tryptophan (25 *mg*/l) use of mannitol (1% w/v) along with sucrose (3% w/v), cold pre-treatment (8°C for 7 days) and culturing anthers from primary tillers enhanced anther callusing.	Gosal *et al*., (1991, 1993) Sandhu *et al*., (1993)
Indica cv. CR666-34-S, SR26-B, Basmati Sd-10, Ptb-33	Anthers	1. R_3, N_6 or MS with different combination of 2,4-D and NAA. 2. R_2, N_6 or MS with different conditions of IAA, NAA, Ba and kin with sucrose (3-6% w/v)	CR666-34-3, and SR-26-B were poorer responding than basmati Sd-10 and Ptb-33. Early callus induction was observed in media having NAA:R_2 and N_6 were better callus induction media; MS media supplemented with 6% w/v sucrose gave higher green plant regeneration.	Lenka and Reddy (1994)

Plant material	Explant/stage	Medium	Remarks	References
Mediterranean Japonica rice cv. Miara	Anthers with uninucleate microspores	1. R_2-control 2. R_2 + modified KNO_3 + modified $(NH_4)_2SO_4$ 3. Different combinations of Glutamine and Alanine	Better callus haploid induction and plant regeneration was observed by using R_2 modified medium. Alanine was best supplement.	Ogawa *et al.*, (1995)
Pankaj x FR 13 A	Uninucleate microspores, binucleate microspores	N_6 + 2,4-D (2 *mg*/l) +kin (1 *mg*/l) + sucrose (5% w/v)	More embryogenic calli from uninucleate microspores than binucleate pollen grains.	Mandal and Gupta (1996)
IR 43 (Indica)	Anthers with uninucleate microspores	M-019 supplemented with casein hydrolysate	Casein hydrolysate (50 *mg*/l) significantly better for pollen callusing as well as frequency of green plant regeneration.	Raina and Zapata (1997)
Japonica cv. F_1 hybrids with Japonica cv	Anthers	1. N_6 + 2,4-D (4 *mg*/l) + kin (.5 *mg*/l) 2. LS + sorbitol (3 g/l) – CH (2 g/l) + basal media + MES (1 g/l) + NAA (2 *mg*/l) + kin (.5 *mg*/l)	N_6 medium with 2,4-D (4 *mg*/l) effectively induced calli from microspores. Green spots were produced effectively from 2.5 basal medium supplemented with plant growth substances.	Shimada *et al.*, (1999)
Taipei 309 (Japonica)	Anthers with uninucleate Pollen	N_6 medium	The anthers from the basal part of the spikelet showed High callus induction compared to top parts of spikelet	Afza *et al.*, (2000)
Taipei 309 (Japonica)	Pollen at all uninucleate stages – early, middle and late	1. N_6 + ABA (0, 2, 5 and 10 *mg*/l) + sucrose (6 % w/v) + 2,4-D (2 *mg*/l) 2. MS + BAP (2 *mg*/l) + NAA (1 *mg*/l) + kin (2 *mg*/l) +agarose (0.45%)	N_6 media with 5 and 10 *mg*/l ABA gave highest callus induction. Green plant regeneration frequency raised from 30% to 70% by using rationed donor plants.	Guzman and Zapata (2000)
Pusa basmati-1, basmati 370, basmati 386	Anthers containing uninucleate to early binucleate microspores	1. N_6 + sucrose (3% w/v) + cystein (40 *mg*/l) + 2,4-D (2.5 *mg*/l) + kin (0.5 *mg*/l) 2. N_6 + sucrose (3% w/v) + proline (560 *mg*/l) + 2,4-D	Plating of microspores on medium containing proline gave maximum calli followed by cystein and tryptophan by cystein and tryptophan.	Grewal *et al.*, (2001)

Plant material	Explant/stage	Medium	Remarks	References
		(2.5 *mg*/l) + kin (0.5 *mg*/l) 3. N_6 + sucrose (3% w/v) + tryptophan (560 *mg*/l) + 2,4-D (2.5 *mg*/l) + kin (0.5 *mg*/l)		
10 cv. of Japonica rice and 9 cv. of Indica rice	Anthers with uninucleate to mid binucleate microspores	N_6 + colchicine (10-20 *mg*/l) with different doses (10-50 Gy) of gamma rays	The dose of 20 Gy gamma rays and 30 *mg*/l colchicines have significant stimulation on green plant regeneration.	Chen *et al.*, (2001)
Indica cv., Japonica cv., Indica x Japonica cv.	Anthers with uninucleate microspores	1. N_6 + 2,4-D (2 *mg*/l) + kin (1 *mg*/l) + CM (10% w/v) 2. MS + kin (0.5 *mg*/l) + BA (2 *mg*/l) + NAA (1 *mg*/l) – 1:4:2	Manipulation of 2,4-D, kin and BA ratio in the regeneration medium of 1:4:2 gave excellent regeneration. Organic additives – yeast extract, casein hydrolysate and coconut milk enhanced no regeneration further.	Rout and Sharma (2001)
HKR 120, HKR 86-3, HKR 86-217, PR 106, Gobind, CH2 (Indica) Bas-370, Taraori Bas. Indica x Basmati hybrids	Anthers with uninucleate to early binucleate microspores	N_6M, Heh 5M - RZM	Best callusing frequencies (2.6-7.8%) obtained in RZM medium with 4% w/v maltose. F_2 plants compared to F_1 hybrids and parental cv. were more responsive.	Bishnoi *et al.*, (2000)
Japonica x Indica hybrid	Anthers with uninucleate to mid binucleate microspores	N_6 medium	Large scale deletions in the plastid genome of albino/ green plants was detected through PCR and competitive PCR techniques.	Yamagishi (2002)
Indica *Oryza sativa*	Anthers with uninucleate stage	N_6 and R_2 medium	Marker assisted programme was developed to improve plant regeneration of Indica rice. Two significant QTLs associated with green	Sohn *et al.*, (2002)

Plant material	Explant/stage	Medium	Remarks	References
			plant regeneration on chromosome 10 were detected.	
BR-3, BR-29 and BR-10 (Indica)	Anthers with late uninucleate stage	Z medium with various combinations of auxins and cytokinins	The best callusing from cultured anthers was obtained in medium containing 2,4-D (2 *mg*/l), NAA (2.5 *mg*/l), kin (0.5 *mg*/l), BR-3 produced green plants only.	Shahnewaz *et al*., (2003)
BR-5, BR-31, BR-34, BR-37, BR-38, Indica	Anthers with uninucleate microspores	N_6, Z_2, with NAA (2.5 *mg*/l), kin(0.5 *mg*/l), 2,4-D (0.5 *mg*/l)	Z_2 medium performed best for callus induction BNR-38 produced highest percentage of androgenic calli. Both albino and green plantlets appeared in five Indica cv.	Asaduzzman *et al*., (2003)
Pankaj FR-11A Mahsuri x *O. rufipogon*	Anthers with uninucleate to early binucleate stage	Z_2	The dihaploid lines derived from the crosses displayed better survivability under submergence conditions.	Mandal *et al*., (2004)
Japonica x Indica, Indica x Japonica	Anthers with early binucleate to mid binucleate stage	N_6 with different concentrations of auxins (2,4-D, NAA, IAA) and carbon source (maltose and sucrose)	The results obtained were genotype dependent. A cold pretreatment of 10°C for 8 days was essential. The greatest callus induction was obtained in the hybrid A92 x Koshihikari using 10.74 µM NAA and 30 g/l maltose.	Gabriela *et al*., (2005)

(iv) the microspores are inoculated on a solid or in a liquid medium maintained at 25°C and 16/8 hr photoperiod.

The microspores may develop directly into embryoids within 15 days or follow one of the several indirect paths to produce haploid plantlets.

In anther culture as well as in microspore culture, spontaneous doubled haploids (SDH) are also obtained, so that no colchicines treatment is needed in such cases. Efforts are being made to increase the frequency of SDH, so that these can be multiplied and directly field tested.

Protocol for Anther Culture in Laboratory

A simple stepwise procedure to isolate haploid plants from cultured anthers is given below:

1. Flower buds are collected from the plants after all preparations for culturing in the laboratory have been made.
2. Buds are collected in a no sterile petri dish and the length of each bud measured using a cm scale. Buds of corolla length 21-23 mm are chosen as the pollen in them would have usually completed the first mitosis.
3. The buds are chilled for 12 days (7 to 8°C) in a refrigeration unit.
4. They are then surface sterilized in a petri dish containing a suitable sterilizing agent, for instance 0.01% solution of mercuric chloride.
5. Using sterile, double distilled water; the buds are rinsed 3-4 times in a sterile air cabinet.
6. Using forceps and a dissecting needle, the buds are carefully teased open and the anthers removed.
7. The anthers dissected out from each bud are grouped separately on removal.
8. One anther from each group is then removed and squashed in acetocarmine stain in order to determine the stage of pollen development (The pollen should be in the first pollen division).
9. Filaments are cut from their anthers to avoid callusing from the cut ends in vitro.
10. Anthers from the group showing pollen in the first division are placed on a suitable culture medium and the cultures are incubated at 25°C in darkness.
11. After about 2 weeks in culture, young embryos emerging from cultured anthers can be detected by gently bursting the anthers on a microscope slide in a drop of acetocarmine stain and observed under a microscope.
12. The cultures are shifted to light (intensity of about 300 lux) after the specified period of approximately 2 weeks. Complete plantlets are observed after 4-5 weeks of culture.

13. Plantlets emerging out of the anthers are separated using a pair of forceps and the remaining anther tissues are discarded.
14. For quicker development of the plantlets, they may be transferred to a root inducing medium after they are about 15 mm long.

During this period the plants are placed on a 12-h day length provided with 500 lux illumination from white fluorescent tube lights. The plantlets can be transferred to soil after they attain a height of about 50 mm.

Protocol for Pollen Culture

1. Flower buds are collected from the plants after all preparations for culturing in the laboratory have been made.
2. Buds are collected in a non sterile petri dish and the length of each bud measured using a cm scale. Buds of corolla length 21-23 mm are chosen as the pollen in them would have usually completed the first mitosis.
3. The buds are chilled for 12 days (7° to 8°C) in a refrigeration unit.
4. They are then surface sterilized in a petri dish containing a suitable sterilizing agent, for instance 0.01% solution of mercuric chloride.
5. Using sterile, double distilled water, the buds are rinsed 3-4 times in a sterile air cabinet.
6. Using forceps and a dissecting needle, the buds are carefully teased open and the anthers removed.
7. The anthers dissected out from each bud are grouped separately on removal.
8. One anther from each group is then removed and squashed in acetocarmine stain in order to determine the stage of pollen development. The pollen should be in the first pollen division.
9. For each culture, anthers from tobacco buds are placed in 5 ml of liquid medium in a petri dish.
10. After regular intervals of 6, 10 and 14 days of culture, the anthers are removed from the culture and discarded. That ensures the dehiscence of all anthers and release of pollen into the culture medium.
11. The dishes are then sealed with parafilm and incubated at 28°C in darkness for the first 14 days of culture.
12. After 14 days, the cultures are transferred to an illuminated growth chamber (500 lux, 12th day length, 25°C).
13. The growth of haploid embryos to plantlets developed from released pollen should be observed.

14. Plantlets emerging are separated using a pair of forceps and the remaining tissues are discarded.

15. For quicker development of the plantlets, they may be transferred to a root inducing medium after they are about 15 mm long.

Sometimes the nurse culture technique is used for raising haploid tissue clones from isolated pollen grains. Intact microspores are mounted on a filter paper disk which is then placed over an intact anther or callus derived from the floral parts.

The extract of cultured anthers has equal nursing effects on the growth and development of microspores and may be used in place of a nurse callus or anther. A purely synthetic medium developed for raising haploid plants from isolated pollen grains of tobacco and Datura may also be suitable for other systems.

The efficiency of isolated pollen cultures for haploid production may be increased by the gradient centrifugation technique. Anthers at the proper stage of development are gently macerated to obtain a microspore suspension.

After removing the debris either by sieving or filtration, the suspension is layered onto a sucrose solution (30% or 55% percoll and 4% sucrose) and centrifuged at 1200 g for 5 min. The pollen (vacuolated and starch filled) with embryogenic potential forms a band at the top of the sucrose solution while non embryogenic pollen (starch-filled) settles to the bottom.

The upper band containing embryogenic or saprophytic pollen is now pipetted out and suspended in the washing medium, centrifuged again, and the supernatant removed. Finally, the pellet containing these grains is resuspended in the culture medium. Such a pollen suspension gives a better response than control in terms of induction of cell division and continued growth of the callus.

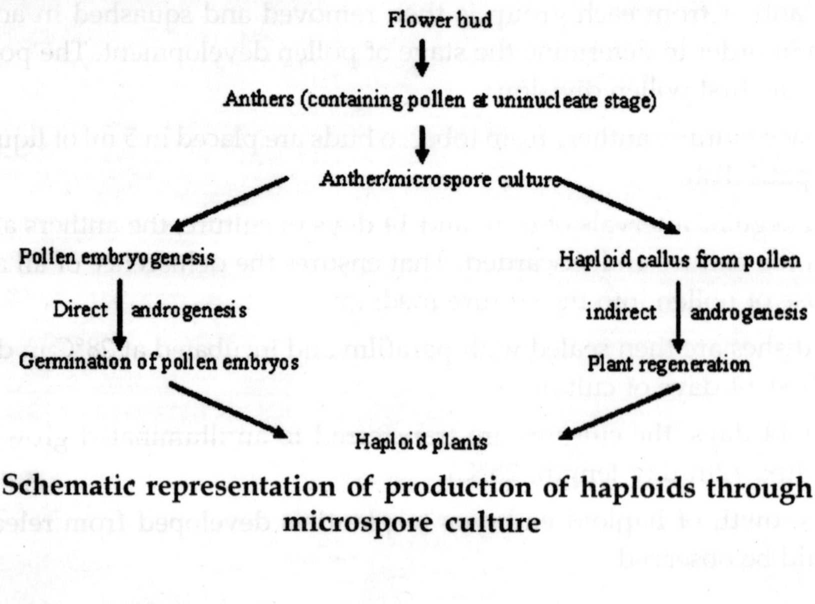

Fig. 5: Schematic representation of production of haploids through anther/ microspore culture

The stages of anther culture from anther to haploid plantlet can be described as follows: a) an unopened flower bud, 1b) anthers, 1c) the anthers in culture, 1d) and 1e) proliferating anther, 1f) haploid callus, 1g) differentiating callus, h) haploid plantlet. Isolated pollen culture is as follows: a) an unopened flower bud, 3b) isolated pollen from a cultured anther, 3c) pollen culture, 3d) multinucleate pollen, 3e) and 3f) pollen embryo. Homozygous plants can be obtained by treating the haploid plantlets with colchicine.

Diploidization of Haploids Plants

Haploid plants obtained either from anther or ovule culture may grow normally under *in vitro* conditions up to the flowering stage but viable gametes are not formed due to the absence of one set of homologous chromosomes and, consequently, there is no seed set.

The only mechanism for perpetuating the haploids is by duplicating the chromosome complement in order to obtain homozygous diploids. In pollen derived plants duplication of chromosomes may occur spontaneously in cultures but due to the small percentage of such double-haploids it is necessary to diploidize the haploids by chemical means.

A simple procedure designed to achieve diploidization involves immersion of very young haploids in a filter sterilized solution of colchicine (0.4%) for 2-4 days, followed by their transfer to the culture medium for further growth. In this procedure chromosome or gene instabilities are minimal compared to other methods of colchicine or chemical treatment.

Detection and Identification of Haploids

A number of methods have been developed for the detection of haploids in a population of diploids. Although the most effective method is the use of genetic markers, phenotypic effects can also be used for identification of haploids.

The vegetative and floral parts and the cell size in haploids are reduced relative to diploids.

In maize, Greenblatt *et al.*, (1967) selected 45 seedlings out of 10,000 seedlings, on the basis of size (i.e. thin coleoptiles, etc.) and five of them were found to be monoploids showing some success in screening for haploidy.

Stomata size, pollen size and pollen abortion can be other criteria to help initial screening, although in each case cytological confirmation will be essential.

Factors Affecting Haploid Production

A number of factors influence androgenesis *in vitro*. The genotype of the donor plant plays a significant role in determining the frequency of pollen plant production. Anther wall factors also support pollen embryo development.

As induction of the pollen into embryoids occurs most easily within the confines of an anther, the anther wall seems to provide a nursing effect. However, there are two schools of thought regarding the role of the anther wall. One is that it may have a stimulatory effect on the growth of pollen embryos (probably due to the presence of enhanced levels of some amino acids such as glutamine and serine); the other view holds that it may emanate some inhibitory substances into the culture medium thereby blocking the growth of more pollen into embryos.

The culture medium also plays a vital role since the requirements vary with the genotype and probably the age of the anther as well as conditions under which donor plants are grown. The medium should contain the correct amount and proportion of inorganic nutrients to satisfy the nutritional as well as physiological needs of the many plant cells in culture.

In addition to basal salts and vitamins, hormones in the medium are critical factors for embryo or callus formation. Cytokinins (e.g. kinetin) are necessary for induction of pollen embryos in many species of Solanaceae except tobacco. Auxins, in particular 2,4-D, greatly promote the formation of pollen callus in cereals. For regeneration of plants from pollen calli, a cytokinin and lower concentration of auxin are often necessary.

Sucrose has been considered the most effective carbohydrate source which cannot be substituted by other disaccharides. Glucose can be used in anther culture in some cases but fructose is far less effective. The concentration of sucrose also plays an important role in induction of pollen plants. Activated charcoal is also added to the culture medium.

It helps in the removal of inhibitors from the agar used for gelling the medium. Another role assigned to activated charcoal is the adsorption of 5-hydroxymethylfurfural, a product of sucrose dehydration during autoclaving, assumed to be an inhibitor of growth in anther cultures.

Certain organic supplements added to the culture medium often enhance the growth of anther cultures. Some of these include the hydrolyzed products of proteins such as casein (found in milk), nucleic acids, and others. Coconut milk obtained from tender coconuts is often added to tissue culture media. It contains a complex mixture of nucleic acids, sugars, growth hormones and some vitamins.

The physiological state of the parent plant plays a role in haploid production. Success in haploid induction is in part dependent on knowledge of the physiology of the pollen yielding plant. In various plant species it has been shown that the frequency of androgenesis is higher in anthers harvested at the beginning of the flowering period and declines with plant age.

This may be due to deterioration in the general condition of the plants, especially during seed set. The lower frequency of induction of haploids in anthers taken from older plants may also be associated with a decline in pollen viability. Seasonal variations, physical treatment, and application of hormones and salts to the plant also alter its physiological status, which is reflected in a change in anther response.

Temperature and light are two physical factors which play an important role in culture of anthers. Higher temperatures (30°C) yield better results. Temperature shocks also enhance the induction frequency of microspore androgenesis. Frequency of haploid formation and growth of plantlets are generally better in light.

Certain physical and chemical treatments given to flower, buds or anthers prior to culture, can be highly conducive to the development of pollen into plants. The most significant is cold treatment.

The developmental stage of pollen greatly influences the fate of the microspore, Androgenesis occurs when a microspore or pollen is induced to shift from a gametophyte pathway to a saprophytic pathway of embryo formation.

Anthers of some species (*Datura,* tobacco) give the best response if pollen is cultured at first mitosis or later stages (postmitotic), whereas in most others (barley, wheat, rice) anthers are most productive when cultured at the uninucleate microspore stage (premitotic). Anthers at a very young stage (containing microspore mother cells m tetrads) or a late stage (containing binucleate, starch filled pollen) of development are generally ineffective, albeit some exceptions are known.

OVULE CULTURE

Haploid have also been successfully produced from cultured female gametophytes. Success in this connection was first achieved with some Gymnosperms like Zamia, Ephedra and some Cycads.

Although initial attempts for the production of haploids from cultured ovules in angiosperms were unsuccessful, recently haploids could be obtained from cultured ovaries in a few crop species, like barley, wheat and tobacco. With more efforts in the future, haploids from ovule culture will be produced in a number of plant species. Further research is, therefore, necessary to improve the understanding about the following;

(i) The events leading to induction of haploidy in female gametophytes,

(ii) The factors which control the *in vitro* development of proembryo into the fully organized plants and

(iii) The differences in the growth patterns of *in vitro* development of unfertilized female gametophytes (in ovules) and the male gametophytes (in pollen).

Haploids from Unpollinated Ovaries

Production of a haploid individual by development of an unfertilized egg cell as a result of delayed pollination is referred to as gynogenesis. Gynogenesis is found in the interspecific cross *Solanum tuberosum* (2n = 4x) x *S. phureja* (2n = 2x), resulting in the production of a dihaploid (2n = 2x) potato.

San Noeum (1976) was the first to demonstrate that gynogenesis, an essentially *in vivo* phenomenon, can be induced under *in vitro* conditions. She obtained gynogenic haploids using an ovary culture of *Hordeum vulgare.*

Subsequently, gynogenic haploids have been obtained from unpollinated ovaries and ovules. Induction of haploids from female gametophytes may not be inaccessible.

Haploid Plants Raised Through *in vitro* Gynogenesis

Table 4: Haploid plants raised through *in vitro* gynogenesis

Family	Plant species
Chenopodiaceae	*Beta vulgaris*
Asteraceae	*Helianthus annuus*
Poaceae	*Hordeum vulgare; Oryza sativa; Triticum aestivum*
Zea mays	
Solanaceae	*Nicotiana tabacum; Petunia axillaries*
Solanum tuberosum	

Comparison of Haploid Breeding *Vs* Conventional Breeding

Table 5: Comparison of conventional and haploid breeding methods

Years	Pedigree Method	Bulk Method	Haploid Method
1	Cross parents	Same as in pedigree method	Cross parents
2	Grow Fl	Same as in pedigree method	Grow F1: derive haploids and DHs (anther culture)
3	Grow F2 (2000-1000 seeds space planted); select for qualitative traits; harvest seed on single plant basis	Harvest seed in bulk in F2 - F5; no selection; only natural selection	Dihaploids are used for intial trials
4 - 5	Grow F3, F4 (plant to progeny rows from selected plants); select rows, plants	Harvest seed in bulk in F2 - F5; no selection; only natural selection	Trials to verify performance; release of variety
6	GrowF5; initial trial (3-4 rows of each plant progeny); compare and select	Harvest seed in bulk in F2 - F5; no selection; only natural selection	
7 - 9	F6 - F8; trials to; verify performance	Space planting; select single plants; plant to progenies; yield trials	
10	Final trial and release of variety	Final trial and release of variety	

Hybrid Sorting in Haploid Breeding

One of the essential steps in haploid breeding involves selection of superior plants among haploids derived from F1 hybrids through anther culture. It is popularly described as hybrid sorting and virtually means selection of recombinant superior gametes.

The haploid method of breeding involving hybrid sorting is considered superior over pedigree and bulk methods, firstly because the frequency of superior gametes is higher than the frequency of corresponding superior plants in F2 generation, and secondly because haploid breeding reduces significantly the time required for development of a new variety.

For instance, if the frequency of superior F1 gametes is one in one hundred, then the frequency of homozygous F2 plants derived from the fusion of two such superior similar gametes would be one in ten thousand. Therefore, smaller populations of doubled haploids derived from haploids will need to be handled.

The technique of hybrid sorting has been successfully utilized, particularly in China and Japan, for development of new varieties in several crops.

In Japan, an excellent tobacco cultivar, 'F-211' bred through anther culture, is resistant to bacterial wilt and has a mild smoking quality. In China, more than 100 rice varieties developed using the same technique, gave an increased yield and, therefore, occupy significant areas of rice cultivation.

Haploids have also been successfully utilized for breeding of barley, maize, sugarcane, oilseed rape or canola (*B. napus*), bell peppers and some other crops.

In barley, at least two varieties (Mingo, Gwylan) have been produced in Canada using haploids derived through *bulbosum* technique of chromosome elimination.

The varieties were superior for characters like disease resistance (against powdery mildew and barley yellow mosaic virus) and other yield contributing traits. Similarly, in maize superior inbred lines were developed through screening of haploids derived from anther culture in a hybrid. In sugarcane, selection among anther-culture derived haploids led to the development of superior lines with tall stem and higher sugar content.

In *Brassica napus* also, anther derived doubled haploids, produced in Canada, had low erucic acid and glucosinolates content. In another case, a haploid derived from twin seedlings in *B. napus* led to the development of a superior cultivar 'Maris Haplona'. In bell peppers, DH lines produced from anther derived haploids, exhibited all shades of color ranging from dark green found in one parent to light green found in the other parent of the hybrid employed.

Despite the above conspicuous achievements made through haploid breeding, the acreage occupied by varieties developed through haploid breeding is negligible on a world wide level.

Production of Spontaneously Double Haploids for Breeding

The utility of haploids in plant breeding eventually requires production of doubled haploids (DH). For this purpose, often efficient colchicine treatments are applied and colchicine doubled haploids (CDH) are obtained. In a system where haploids are detected through genetic markers, the spontaneously doubled haploids (SDH) can also be identified, if doubling does occur spontaneously. This will avoid unnecessary colchicine treatment and can save time and plant material. Although there are several reports of SDHs produced during haploid production, but at Svalov in Sweden, hap gene is being utilized, not only for the production of haploids, but also for the production of SDHs (Hagberg and Hagberg, 1987). It utilizes a marker gene system for mildew resistance due to an allele (ml-o), which is recessive to susceptibility allele (Ml-o).

The breeding scheme used to produce spontaneously doubled haploids, gave spontaneously doubled haploids (SOH) which were three times as many as the normal haploids, which seem to be really promising.

These SOH plants were found to be completely homozygous thus proving their maternal origin due to doubling of chromosome number in the original egg cell.

Use of Haploids in Production of Disomic Inheritance of Homozygous Lines

The most important use of haploids, is their use in the production of homozygous lines, which may be directly used as cultivars or may be used in breeding programmes.

For instance, a doubled haploid, 'Marglobe tomato' has been used commercially. Similarly, doubled haploids were used as inbred lines in hybrid maize programme. However, the chief limitation in the use of haploids for production of homozygous lines had been the low frequency, with which they were available.

Limitations of Haploid Breeding

Haploid breeding has not given the desired and expected dividends, despite substantial investments made on haploid research during the last two decades. Following are some of the widely recognized limitations:

(i) Haploids can not be obtained in high frequency required for selection;

(ii) The cost benefit ratio in haploid breeding is often not favorable, thus discouraging the use of haploid breeding despite its obvious advantages;

(iii) Haploids will express recessive deleterious traits and deleterious mutations may arise during anther culture;

(iv) Different ploidy levels may be available so that haploid status may need to be confirmed cytologically; alternatively pollen culture may need to be used, which is more expensive and gives relatively low success rate.

Problems Concerned with Haploid Production

Although haploids have been raised mainly from anther culture of a large number of species, this technique has not proved successful in respect of all genotypes of crop species. A number of problems are encountered with the induction of haploids.

1. Anthers often fail to grow *in vitro* or the initial growth is followed by abortion of the embryos.
2. The tissue or callus developing from the anther generally comprises a chimera of diploid, tetraploid, and haploid cells.
3. Selective cell division must take place in the haploid microspores, concomitantly restricting proliferation of unwanted diploid and polyploid tissues. Selective cell division is often impossible.
4. Formation of albinos in anther cultures, especially with cereals can neither be avoided nor the loss of plants due to albinism reduced.
5. The technique of inducing haploids *in vitro* is not economically viable due to low success rate.
6. Callus derived from anther or pollen in a medium supplemented with growth regulators is usually detrimental for haploid production.
7. It is difficult to isolate a haploid from a mixture of haploids and higher ploidy levels since the polyploids outgrow the haploids.
8. Doubling of haploids is time consuming and may not always result in the production of a homozygote. Double haploids sometimes exhibit segregation in their progeny.
9. The anther culture technique works well on plants belonging to only a few families, of which Solanaceae is one.
10. The frequency of embryoids obtained is often very low.

Progress in rice breeding system and techniques

The conventional breeding system such as pedigree and bulk method mainly used in handling the japonica rice materials was modified to the general breeding system accompanying the rapid generation advancement (RGA) scheme utilizing the greenhouse or IRRI field during winter season. This started to change the real rice breeding works to introduce the useful characters from tropical indica rice from the 2nd half of 1960s. Some mutation rice breeding by irradiation were conducted during this time, but it was interrupted with almost no practical outcomes. The RGA rice breeding system and rapid extension of Tongil-type rice cultivars through seed multiplication in IRRI field during winter were actively carried out during 1970s since the development of the first Tongil-type rice cultivar 'Tongil' in 1971. Also, the breeding efficiency was largely enhanced through the continuous improvement in testing techniques for evaluating the grain quality and resistance to disease and insect pests during this time. **The haploid rice breeding utilizing the anther culture technique was practically established in japonica rice during 1980s since the first varietals development of 'Hwaseongbyeo' by this system in 1986.** The rice breeding period was shortened to 5-6 years by this technique.

A total of 15 rice cultivars were developed by anther-culture breeding system during 1980s-1990s. Although the hybrid rice breeding using the cytoplasmic genic male sterility (CGMS) was started from the early 1970s, the real breeding work was actively conducted during 1980s and the first development of hybrid rice 'Suweonjapjong No.1 and No.2' was realized with a milled rice yield of 7.3-8.3t/ha in 1989.

The establishment of testing system and effective selection of breeding kines for cold tolerance was carried out with the construction of cold-tolerance test nursery. Rice varieties were subjected to the cold-water irrigation drained from Soyang dam in Chuncheon from 1979. This cold-tolerance testing facility has been utilized by IRRI as the international rice cold-tolerance test nursery since 1975. The establishment of embryo rescue technique was realized to get the interspecific hybrid between the different genomic wild species and cultivars and to introduce resistant genes for disease and insect pests from wild rice species to cultivars by recurrent backcrossing in 1990s. Also, the basic techniques to ensure the practical utilization of biotechnologies in rice breeding were developed, and studies on improvement of evaluation techniques and breeding systems were actively driven forward for rice breeding of direct seeding and high-quality or diversification of food-processing utility in recent years (Table 5).

Table 6: Phases of rice breeding and improvement

Period	Major progress in rice breeding technology & system
Before 1970	Settlement of conventional breeding system of pedigree and bulk methods. Mutation breeding by irradiation.
1970s	Development of 'Tongil' cultivar from a three-way remote cross between semidwarf indica and japonica rices. Introduction of rapid generation advancement (RGA) scheme in conventional rice breeding system. Establishment of effective testing & evaluation technologies for resistance to pests and grain quality.
1980s	Practical settlement of anther culture technique in japonica rice breeding. Development of Tongil-type hybrid rices using cytoplasmic- genic male sterility (CGMS). Improvement of cold & salinity tolerance testing system. Development and operation of recurrent population improvement breeding scheme using genetic male sterility (GMS).
1990s	Establishment of embryo rescue technique for interspecific hybridization between wild rice species and cultivars. Development of basic technique for molecular breeding. Driving of effective rice breeding system for direct seeding. Mutation breeding for high grain quality and diversification of food-processing utility.

Plant regeneration by pollen culture of rice

Pollen culture is useful for producing haploid plants (Chen *et al.*, 1980). It also offers an efficient experimental system for mutation induction and genetic manipulation. Haploid plants of rice were obtained by isolated pollen culture.

Three Japonica varieties, Sasanishiki, Nipponbare and Reimei, were used. Panicles were sampled at the mid-uninucleate microspore stage, and the anthers were kept at 10°C for 10 days. They were then cultured on the N6 medium supplemented with NAA (2-4 mg/1), kinetin (0.3 mg/1), casein hydrolysate (1 g/1), and 5% sucrose at pH 5.7, in a 6.0 X 1.5 cm petri dish. Pollen grains were released from the anthers by pressing them gently, and the remaining somatic tissues were removed using a nylon mesh with a pore size of 40μm. The pollen suspension thus obtained was centrifuged at 80 G. The pellets were washed with culture medium, precipitated again by centrifuging and finally resuspended in a medium supplemented with NAA or 2,4-D (2 mg/1), kinetin (0.3 mg/1), casein hydrolysate (1 g/1), and sucrose (3 or 5%). Pollen density was adjusted at 5 X 10^4 per ml. The pollen suspension (2-3 ml) was poured directy into a petri dish or on a 1 mm depth of agarose N6 culture medium. These were incubated under a low light intensity (ca. 500 lux).

Pollen grains precultured for 10 days in the anther were divisible into two types, one with rich cytoplasm and the other poor (Fig. A). The former type grew larger and performed nuclear division (Fig. B). In three weeks after the nuclear division, a cell mass consisting of 8-16 cells appeared out of the exine (Fig. C). One month after this, the cell cluster developed to form a cell colony which was visible (Fig. D). Many calli with diameters ranging from 0.5 cm to 2 cm were formed after two months (Fig. E).

These calli were transferred to 100 ml flasks containing an N6 regeneration medium supplemented with kinetin (1 mg/1), casein hydrolysate (1 g/1) and 3% sucrose. Several plantlets (Sasanishiki 7, Reimei 3, and Nipponbare 1) were obtained from the calli in petri dishes (Fig. F). The Reimi calli subcultured for about one year on the AA medium (Muller and Grafe,1978) showed competence for regeneration on the medium containing kinetin (1 mg/1). The regeneration process would be due to somatic embryogenesis. Chromosome numbers in the root tips of green regenerated plants were 12, indicating that they were haploids (Fig. G). This method of pollen culture is effective in obtaining haploid plants.

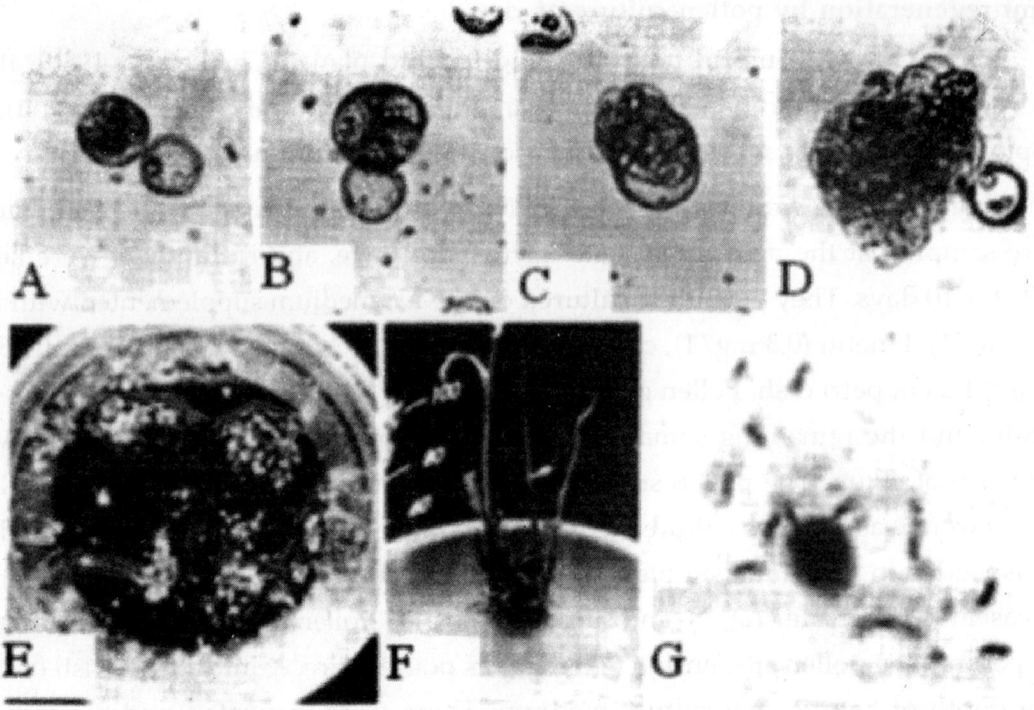

Fig. A.	Isolated pollen grains released after preculture of the anther for 7 days. One of the two contains rich cytoplasm, and the other is poor.
Fig. B.	An enlarged pollen grain containing two or more nuclei within the exine, cultured for 10 days.
Fig. C.	An androgenic mass of 8-16 cells extruding out of the broken exine, after 10 days in culture.
Fig. D.	An androgenic colony after one month in culture.
Fig. E.	Calli formed in a petri dish after two months in culture.
Fig. F.	A plantlet emerging from the pollen callus, two months after transfer to the regeneration medium.
Fig. G.	Giemsa-stained chromosomes in the root tip of a regenerated plantlet.

Factors Affecting Androgenesis in *Indica* Rice

For the effective utilization of haploids in breeding programmes, the production technique should allow genotype-independent production of large numbers of haploids. Although anther culture has been successfully used to hasten the breeding programmes in several crop species, including rice, there still remain problems to realize its full potential. Unlike the highly responsive model systems, most of the indica cultivars respond rather poorly in anther cultures. There is also concern regarding the gametic spectrum representation by the anther culture-derived doubled haploids. If there is a biased representation of recombinants possessing better cultivability, it is important to analyze how seriously it effects the objectives of the breeder.

Whereas in maize androgenesis occurs via embryogenesis, in barley, wheat and rice the microspores divide to form a callus which later differentiate plants. Both the steps require different culture conditions and are affected by the genetic make-up of the plants. Recently, the physiology of the donor plants has been identified as a critical factor in achieving better anther culture efficiency.

Genotype

Most of the *in vitro* morphogenesis responses are genotype-dependent (Bhojwani and Razdan, 1996). In general, indica cultivars of rice exhibit poorer androgenic response than the japonica cultivars (Hu, 1985, Raina, 1997). Miah *et al.*, (1985) reported that anther culture response varied from 41% for a japonica cultivar to 0% for an indica cultivar. Even among the indica cultivars a considerable variation for pollen callusing and plant regeneration has been observed. Guha-Mukherjee (1973) reported that only 5 out of 18 indica cultivars showed pollen callusing and in only four cases did the calli differentiate plants. Similarly, Lentini *et al.*, (1995) reported that only one out of 35 *indica* cultivars exhibited pollen callusing on N6 medium. The purple pigmented, coarse-grained *indica* cv Crossa performed significantly better (40%) than the fine grained, aromatic cv Basmati-370 (<10%; Raina, 1989).

Physiology of Donor Plants

The physiological state of the donor plants, which is affected by several factors, has a profound effect on the androgenic response of their microspores. In *Brassica juncea*, a recalcitrant species, the androgenic response was improved from 3% to 16% only by late sowing of the donor plants, which probably acted as a stress on the plants (Agarwal and Bhojwani, 1993). Recent experience with this oilseed crop shows that under controlled growth conditions, in a phytotron or growth chamber, the donor plants can be maintained for a long time without affecting the androgenic potential of their anthers, and the difference between the anthers from main branch and lateral branches of different order is considerably reduced. There is no similar study with rice but there is sufficient evidence to suggest that the growth conditions of the donor plants have significant effect on the yield of androgenic pollen in this crop.

A remarkable effect on the androgenic response of an indica rice cultivar by light and day/night temperature regime was observed by Raina and Zapata (see Raina, 1997). The plants of cv IR43 that reached the panicle emergence stage under long days (>12 h), high solar radiation (>18Mj m-2) and sunshine (>8 h) and day/night temperature (34°C /24°C) showed highest anther culture response; the response declined with lowering values of these parameters. They also observed that the plants grown in the field were significantly superior in this respect to those grown in the glasshouse or in pots near the field. Superiority of field grown plants over glasshouse-grown plants has also been reported for other cereals, including maize (Petolino and Thompson, 1987) and wheat (Lu *et al.*, 1991).

The physiology of the donor plants can also be altered, and the androgenic response modified, by treating the donor plants with certain chemicals, such as etherel (Zhao *et al.*, 1991, *cited in* Sun, 1999).

Pre-Treatment

Application of a variety of stresses, such as temperature pre-treatment, osmotic shock and sugar starvation, during the labile developmental period of pollen grains is known to be promotory or essential for the induction of androgenesis in several plants, including cereals (Bhojwani and Razdan, 1996; Bhojwani *et al.*, 1997). However, the type, duration and the time of application of these pre-treatments may vary with the species or even variety (Datta, 2001).

Low Temperature

The most widely used pre-treatment for androgenesis is the low temperature shock. In most of the published works on androgenesis in rice, panicles were given a cold pre-treatment but the temperature and duration varied. Matsushima *et al.*, (1988) had reported that a pre-treatment at 10°C for 10-30 days was necessary to induce sporophytic divisions in microspores of the japonica cv Nipponbare. This was subsequently confirmed by several workers for japonica and indica cultivars (Ogawa *et al.*, 1992, Datta *et al.*, 1990, Raina and Irfan, 1998). Ogawa *et al.*, (1995) observed that 28 days of pre-treatment at 10°C was optimum for the indica cv IR24. Gupta and Borthakur (1987) had selected pre-treatment at 10°C for 11 days for anther culture of the *indica* cultivar khonorullo. Although the frequency of anthers showing pollen callusing after cold-treatment for 25 days was fairly high, most of the plants regenerated from the calli formed after such a long cold pre-treatment were albinos. Similarly, Pande (1997) observed that cold pre-treatment was essential for androgenesis in anther cultures of the *indica* cv IR43, and 10°C for 10 days was most suitable. Pre-treatments longer than 11 days resulted in albino production. Reddy *et al.*, (1985) reported that a brief (10 min) exposure to high temperature (35°C) before cold-treatment was better for pollen callusing but it adversely affected green plant production.

Osmotic Stress

Another pre-treatment, which can substitute cold treatment for the induction of androgenesis, is osmotic shock. Wei *et al.*, (1986), who first time reported *ab initio* microspore culture of barley, followed a protocol involving isolation of microspores in 0.3 M mannitol and treating them in this solution for 3 days. According to them, mannitol pre-treatment for 3 days was essential and for 7 days optimum to induce androgenesis in this system. Roberts-Oehlschlagerand Dunwell (1990) had reported that 4 day incubation of barley anthers in a medium containing 3.2% mannitol raised the pollen callusing response from 23% to 78% which was more than cold treatment alone or in combination with mannitol. According to them, mannitol stress improves sugar uptake causing build-up of glucose pool in the anther tissue.

Similar results were reported by Ziauddin *et al.*, (1990). Hoekstra *et al.*, (1992) and Cistue *et al.* (1994) confirmed the importance of mannitol pre-treatment. In the latter case, application of 0.7 M mannitol for 3-5 days improved green plant regeneration significantly. Increase in mannitol concentration showed linear increase in the number of dividing microspores and an increase in the proportion of green plants regenerated. The optimum duration of treatment varied with the genotype. Hoekstra *et al.*, (1993), who gave mannitol stress (440 mOs kg-1) for 4 days and cultured the microspores on a medium of the osmolarity of 350 mOs kg-1 at adensity of 2x104 microspores per ml, obtained 320 embryo-like structures per 1x104 microspores. The yield of green plants increased from 50% to 97%.

Recently, Raina and Irfan (1998) have reported that treatment of anthers in 0.4 M mannitol solution was essential to induce androgenesis in microspore cultures of indica and japonica cultivars. For indica cv, pre-treatment at 33°C was better than at 25°C, with respect to the number of embryo-like structures or calli formation. In the absence of cold treatment, mannitol treatment promoted androgenesis in anther cultures of cv IR43 from 3% to 33.4%; with cold treatment it had no promotory effect (Pande, 1997). For wheat microspore culture, Mejza *et al.*, (1993) isolated and fractionated microspores in 20% maltose solution and cultured them on a medium with 9% maltose. This allowed genotype non-specific regeneration in *ab initio* microspore cultures. Cold treatment in combination with the osmotic treatment was detrimental. Therefore, maltose treatment was given at 25°C.

Ogawa *et al.*, (1995) reported that sugar starvation of microspores for 3 days could, to some extent, substitute for cold treatment for the induction of androgenesis in microspore cultures of indica rice. Sugar starvation of the anthers of IR43 for 2 days in the beginning of culture caused 12-fold increase in the androgenenic response (39%) of cv IR43 (Pande, 1997). However, cold treatment was superior to sugar starvation. Sugar starvation has been earlier shown to promote androgenesis in tobacco (Aruga *et al.*, 1985) and barley (Wei *et al.*, 1986).

Culture Medium

Chu (1975) demonstrated that the level of ammonium nitrogen in the culture medium is critical for androgenesis in rice. On this basis he developed the N_6 medium which has been most widely used for rice anther culture (Raina, 1989, 1997). However, this medium has proved significantly better than most other media for japonica rice. The indica cultivars require even lower NH_4^+ ions. Reddy *et al.*, (1985), who studied 8 *indica* cultivars, found He_2 medium to be better than N_6 medium. He_2 medium is derived from N_6 medium by reducing NH_4^+ to half strength and $MgSO_4$ to 1/50th level, and doubling the concentration of KH_2PO_4.

In Korea N6-Y1 medium was recommended for indica-japonica hybrids (Chung, 1987). This medium is essentially the same as N_6 medium except that $(NH_4)_2SO_4$ has been reduced from 3.5 mM to 1.5 mM. In 1989, Raina reported that a medium with high KNO_3 and NH_4^+ ions completely replaced by 50 mg/l casein hydrolysate (CH) was significantly better than the original MSN medium for *indica x indica* F1 hybrids, with regard to the frequency of green plant regeneration. This SK-1 medium was half as effective as N_6 medium so far as the frequency of pollen callusing is concerned but the calli formed on this medium produced twice as many green plants as those on N_6 medium. More recently, Raina and Zapata (1997), based on their detailed study on the medium requirement of the indica cultivar IR43 evolved a new medium called M-019 (Raina and Irfan, 1998). This medium differs from SK-1 medium in substitution of CH by a small amount of $(NH_4)_2SO_4$ and the level of KH_2PO_4 rose from 170 mg/l to 540 mg/l.

Ogawa *et al.*, (1995) studied the effect of nitrogen source on androgenesis in another indica cultivar IR24, using R-2 medium as the control. R-2 has 40 mM KNO_3 and 2.5 mM $(NH_4)_2SO_4$. When 20 mM KNO_3 was combined with a small amount of $(NH_4)_2SO_4$, glutamine or alanine all treatments induced pollen callusing but alanine was the best supplement. It not only induced high frequency androgenesis but also showed maximum regeneration of green plants. Amino acids have been used as nitrogen source in *in vitro* culture of various tissues, including isolated microspores. Cho and Zapata (1988) reported that proline and glutamine promoted callus formation in microspore cultures of a japonica cultivar. These amino acids also induced a higher degree of plant regeneration and green plant production than the medium containing alanine or no amino acid. Ogawa *et al.*, (1995) also found glutamine to be promotory for pollen callusing in microspore cultures of an indica cultivar but alanine was far better than glutamine for plant regeneration and green plant production.

Since 1983, maltose has been shown to be a superior source of carbohydrate than sucrose for androgenesis in several species, including cereals (Finnie *et al.*, 1989; Last and Brettell, 1990; Pande and Bhojwani 1999). A significant increase in anther culture efficiency and green plant formation in otherwise highly recalcitrant indica rice cultivar occurred when sucrose was replaced by maltose (Raina, 1997).

Lentini *et al.*, (1995) reported that on N6 medium with 146 mM sucrose only one out of 23 indica cultivars exhibited pollen callusing and green plant production. Substitution of sucrose by equimolar amount of maltose enhanced pollen callusing from 6.3% to 10.1% and green plant regeneration from 0.6% to 1%. Glucose was inhibitory. Earlier, Mejza *et al.*, (1993) had observed that isolation and fractionation of microspores in 20% maltose solution, and their culture on a medium containing 9% maltose allowed genotype-independent plant regeneration in *ab initio* microspore cultures of wheat. Maltose promoted the direct embryogenic pathway of plant formation from pollen in wheat (Navarro-Alvarez *et al.*, 1994), maize (McDonald, 1992) and barley (Powell *et al.*, 1992).

Fig. 2

Fig. 4

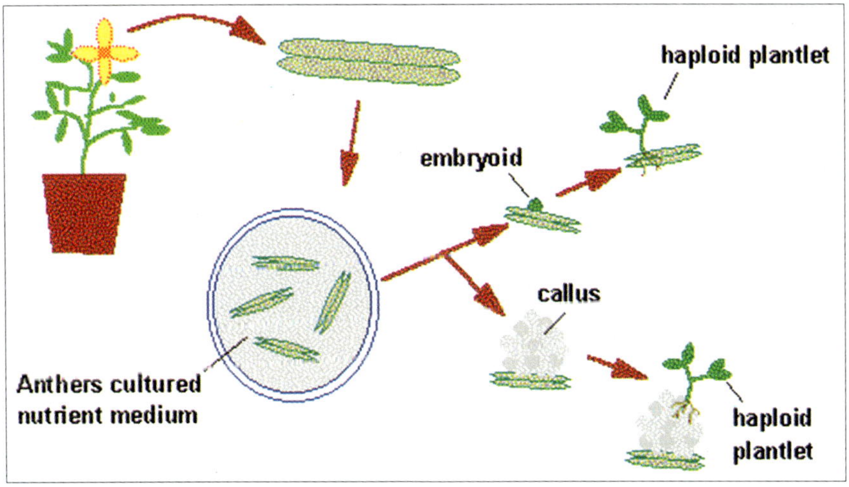

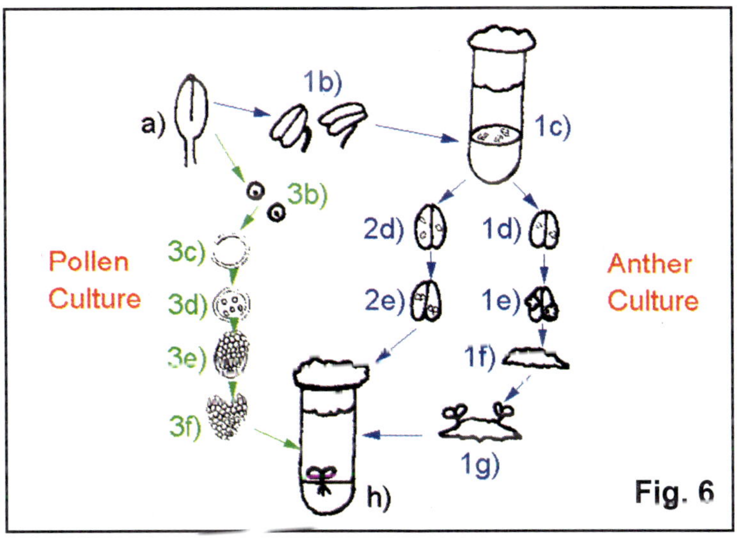

Fig. 6

Fig. 2) World's oldest rice grain. *(Source: BBC News Online)*

Fig. 4) Anther culture technique for haploid production.

Fig. 6) Various stages of anther and isolated pollen culture.

Table 7: Composition of the culture media used for rice anther/microspore culture

Constituents	Media (mg /l)			
	He_2*	MO19**	N_6***	C****
KNO_3	3181.5	3101.0	2830.0	3134.0
KH_2PO_4	800.0	540.0	400.0	540.0
$(NH_4)_2SO_4$	231.0	264.0	463.0	231.5
$MgSO_4.7H_2O$	3.5	370.0	185.0	185.0
$CaCl_2.2H_2O$	166.0	440.0	166.0	150.0
KI	0.8	0.8	0.8	0.83
H_3BO_3	1.6	6.2	1.6	6.2
$MnSO_4.4H_2O$	4.4	22.3	4.4	22.3
$ZnSO_4.7H_2O$	1.5	8.6	1.5	8.6
$Na_2MoO_4.2H_2O$	-	0.25	-	0.25
$CuSO_4.5H_2O$	-	0.025	-	0.025
$CoCl_2.6H_2O$	-	0.025	-	0.025
$FeSO_4.7H_2O$	167.1	27.8	27.8	27.8
$Na_2EDTA.2H_2O$	223.5	37.3	37.3	37.3
Inositol	-	100.0	-	-
Nicotinic acid	2.5	2.5	0.5	2.5
Pyridoxine HCl	0.5	2.5	0.5	2.5
Thiamine HCl	0.5	1.0	1.0	2.5
Glycine	2.0	2.0	2.0	2.5
$AgNO_3$	-	-	-	10.0
Casein hydrolysate	250.0	-	-	-
Yeast extract	1000.0	-	-	-
Sucrose	30000.0	40000.0	40000.0	-
Maltose	-	-	-	50.0

* Huang *et al.*, (1978)
** Raina & Irfan (1998)
*** Chu *et al.*, (1975)
**** Letini *et al.*, (1995)

In cultures, sucrose rapidly breaks down to glucose and fructose, so that after 3 weeks of barley anther culture the medium did not contain any sucrose (Last and Brettell, 1990). On the other hand, the hydrolysis of maltose during the same period was below the detectable level. The toxicity of sucrose for androgenesis is due to the sensitivity of microspores to fructose but not to glucose.

Calcium in the medium is known to stimulate ethylene production in many plant tissues. Addition of the calcium ionophore A23187 (0.5 mM) along with $CaCl_2$ (1 mM) enhanced pollen callusing over $CaCl_2$ alone. Addition of 10 mg L-1 of $AgNO_3$, an anti-ethylene compound, to the callus induction medium enhanced pollen callusing frequency in indica cultivars from 10.1% to 20.6%. With $AgNO_3$ the frequency of green plant differentiation doubled (Lentini *et al.*, 1995). It is also known to promote pollen embryo production in anther cultures of Brussels sprouts (Biddington *et al.*, 1988).

Microspore Culture

There are many problems associated with raising androgenic haploids by anther culture.

The pollen grains within an anther being genetically heterogeneous, the plants arising from an anther would constitute a heterogeneous population. In cereals, where androgenesis occurs *via* callusing, mixing of calli of different pollen origin within an anther lobe enhances the chances of chimeric plant formation. This limitation can be overcome by isolated microspore culture, which offers many other advantages: (a) it provides an efficient system of isolated, haploid, single cells for biochemical and molecular analysis of androgenesis/embryogenesis, (b)enrichment of androgenic microspore population by gradient centrifugation becomes possible, (c) isolated microspores can be genetically manipulated before culture to obtain solid mutants or transformants, and selection of new genotypes can be made at an early stage. Unlike the earlier belief, isolated microspore culture is less tedious and time consuming than anther culture. Considering these advantages, during the last decade considerable progress has been made in isolated microspore culture of cereals (Jahne and Lorz 1995; Raina 1997). Plant regeneration from microspore culture has been achieved in barley (Wei *et al.*, 1986; Hunter 1987), maize (Coumans *et al.*, 1989; Pescitelli *et al.*, 1989), wheat (Mejza *et al.*, 1993) and rice (Raina and Irfan, 1998).

Most of the work on rice microspore culture involved an anther pre-culture step (Jia *et al.*, 1987; Cho and Zapata 1988, 1990; Xie *et al.*, 1995). Datta *et al.*, (1990) following the float culture method to raise microspore cultures of indica rice cultivars (Basmati-370, Chinsurah Boro II, Upland-13, and a local West Bengal cultivar). Anthers from florets exposed to 4°C /10°C for 10-18 days were cultured in liquid medium. Within 3-4 days the anthers burst and released the microspores in the medium. The released microspores and those within the anthers were collected and plated in fresh liquid medium. A total of 1300 anthers of the indica cultivars yielded 42 green and 13 albino plants. Against this the Japonica cultivars (Nipponbare, Yamabiko, Yamahoushi) yielded 140 green and 63 albino plants from 3380 anthers cultured.

Raina and Irfan (1998) reported high frequency androgenesis and plant regeneration in *ab initio* microspore cultures of two indica cultivars (IR43, IR54) and one japonica cultivar(T-309). They observed that co-cultivation of microspores with young ovaries of rice (10 ovaries/ml of medium) significantly increased the androgenic response. However, pretreatment of anthers in 0.4 M mannitol solution for 4 days at 25/30°C was essential to induce androgenesis. Another critical factor for sporophytic division in *ab initio* cultured microspores was substitution of sucrose by 9% maltose. The japonica cv T-309 showed up to70% division of the microspores. The response of indica cultivars was comparatively poor but some green plants were obtained for all the cultivars (T-309 9%; IR43 7% and IR52 2%). Some of the indica microspore calli turned brown soon after transfer to semi-solid medium for regeneration. Occasionally, browning was also observed in liquid medium. This problem in barley was overcome by adding Ficoll, a high molecular weight polymer, which keeps the anthers and microcalli afloat on the medium. Ogawa *et al.*, (1992, 1994, 1995) achieved androgenic plant formation in *ab initio* microspore cultures of the indica cv IR24 without a nurse tissue. The critical factor was cold pretreatment of the panicles at 10°C for 28 days before isolation of the micropsores (Ogawa *et al.*, 1995). In this study the microspores were purified by filtration and enrichment of androgenic grains by using 0/35% Percoll gradient. They obtained 56 colonies per dish containing 5x104 pollen. Only 6 calli exhibited regeneration and 1 out of 37 plants was green.

Albinism

The occurrence of a large proportion of albinos in the pollen plant population is probably the most frustrating feature of androgenesis in its application to rice breeding. The frequency of albinos may vary from 5% to 100%. Indica rice cultivars are more prone to this problem than japonica rice.

Several factors, including pre-treatment, culture medium and the protocol, affect the frequency of albinos. However albinism can be considerably reduced by shortening the culture period. In most of the cereals androgenesis occurs via pollen callusing and the callus needs to be transferred from the induction medium to regeneration medium for obtaining plants. In rice cv IR43, the pollen calli left on the induction medium longer than 30 days after release from the anthers lost their regeneration potential (Pande and Bhojwani, unpublished). Therefore, the microcalli was routinely transferred to regeneration medium 10 days after their release from the anther. A more effective approach to reduce the culture period is to alter the mode of androgenesis from pollen callusing to direct pollen embryogenesis. In barley, by reducing the culture period to 6 weeks by osmotic shock pretreatment and low density microspore plating, Hoekstra *et al.* (1993) could change the proportion of green to albino plants from 1:1 to 34:1.

Generally, liquid medium is used for anther/microspore culture of barley. Consequently, the anthers and large pollen calli released by them sink in the culture medium, and the tissues are exposed to anaerobic conditions. This results in the production of lactate and alcohol dehydrogenase in the tissues which damages the internal structure and DNA of the plastids, leading to albino plant regeneration (Kao *et al.*, 1991). Proper aeration of the tissue by addition of Ficoll (40%) to the induction medium, supplementing the regeneration medium with Ficoll and high sucrose (4.5%) and gelling the embryo maturation and germination medium with starch and agar, Kao *et al.*, induced direct embryogenesis in barley, which cut down the period from anther culture to plant regeneration to 6 weeks. It raised the proportion of green plants from 0% to 50%. The authors have remarked that the calli should be transferred from induction medium to regeneration medium within 70 days of culture. High Ficoll/high sucrose kept the anthers and calli afloat on the medium and delayed anther dehiscence, maintaining sugar starvation-like condition around the microspores. The plants regenerated through this protocol showed very little or no genetic variation in the field or with 100 RFLP markers.

The stage of pollen at the time of culture may also contribute towards albinism. In rice the cytoplasm of microspores at the tetrad stage is rich in plastids. With further development simplification of the cytoplasm occurs and the granular matrix of the plastids is replaced by starch grains, which has been positively correlated with deletion of pDNA (Kawata *et al.*, 1995). Consequently, calli from anthers cultured at the early uninucleate stage of pollen produce only green plants, and the frequency of albinism increased with advancing stage of the pollen (Pandey, 1997). Kawata *et al.*, (1995) are of the opinion that more than the process of androgenesis, the culture period is responsible for pDNA deletions. The plants regenerated from one-month-old pollen or seed callus did not show pDNA deletion but those from 11-year-old callus showed substantial pDNA deletion. This theory is further supported by the fact that with the passage of time the calli initially producing only green plants may start producing albinos but it never happens the other way round.

Also, the *in vitro* selection of microspores is important. The phenomenon of pollen dimorphism has been observed in several species, including rice. For example, in IR43 15% of the grains in an anther are large and thin-walled. These grains are normally incapable of forming viable pollen, which may be due to minor genetic abnormalities. However, these so called androgenic grains are most capable of forming pollen plants in cultures. Albinism may be one of the expressions of this genetic deficiency.

Adapting rice anther culture to gene transformation and RNA interference

Anther culture offers a rapid method of generating homozygous lines for breeding program and genetic analysis. To produce homozygous transgenic lines of rice (*Oryza sativa* L.) in one step, an efficient protocol of anther-callus-based transformation mediated by *Agrobacterium* has been developed after optimizing several factors influencing efficient transformation, including callus induction and *Agrobacterium* density for co-cultivation. Using this protocol, 145 independent green transform ants were obtained from five cultivars of japonica rice by transformation with a binary vector pCXK1301 bearing the rice gene, Xa21 for resistance to bacterial blight, of which 140 were further confirmed by PCR and Southern hybridization analysis, including haploids (32.1%), diploids (62.1%) and mixoploids (7.5%). Fifteen diploids were found to be doubled haploids, which accounted for 10.7% of the total positive lines. Finally, by including 28 from colchicine induced or spontaneous diploidization of haploids later after transformation, a total of 43 doubled haploids (30.7%) of Xa21 transgenic lines were obtained. Two RNAi transgenic haploids of the rice OsMADS2 gene (a putative redundant gene of OsMADS4), were also generated based on their sequence similarity, to investigate its possible roles in rice flower development by this method. Flowers from the two OsMADS2 RNAi transgenic haploids displayed obvious homeotic alternations, in which lodicules were transformed into palea/lemma-like tissues, whereas identities of other floral organs were maintained. The phenotypic alternations were proved to result from specific transcriptional suppression of OsMADS2 gene by the introduced RNAi transgene. The results confirmed that OsMADS2 is involved in lodicule development of rice flower and functionally redundant with OsMADS4 gene. The results demonstrate that rice anther culture could be adapted to gene transformation and RNAi analysis in rice.

Mapping AFLP markers associated with subspecific differentiation of *Oryza sativa* (rice) and an investigation of segregation distortion

Asian rice has undergone ecogeographical differentiation, and the fundamental division is between indica and japonica types (Chang, 1976). Crosses between indica and japonica types result in heterosis, which is exploited by breeders in the development of new varieties (Zhang *et al.*, 1994). Efforts to enhance the use of crop genetic resources and the exploitation of conserved biodiversity more effectively by using molecular markers have led to address some of the practical issues related to the maintenance of plant germplasm in a gene bank. This has included the designation of duplicate accessions (Virk *et al.*, 1995a) and taxonomic identification of rice species (Martin *et al.*, 1997).

Discrimination between indica and japonica types using morphological characters is not always easy, and material is frequently classified using isozyme data (Glaszmann, 1987). It is also possible to distinguish between the two types using RFLP (Zhang *et al.*, 1992), RAPD (Virk *et al.*, 1995b), AFLP (Mackill *et al.*, 1996) and ISSR (Parsons *et al.*, 1997) data.

AFLP marker data have been generated for a doubled haploid rice mapping population developed from a cross between IR64 (indica) and Azucena (japonica) varieties (Huang *et al.*, 1994; Maheswaran *et al.*, 1997), as well as for a diverse set of rice germplasm (Virk *et al.*, 1996a). A total of 151 AFLP markers were incorporated into an existing restriction fragment length polymorphism framework map and compared with a published map. Markers associated with the indica-japonica subspecific differentiation in Asian rice were also mapped and were found to be distributed throughout the genome. This was the first example of mapped markers associated with subspecific differentiation in a plant species, namely indica and japonica types of Asian rice.

Sources of mapping errors were examined and the possibility of such errors being caused by co-migration of bands and segregation distortion was explored. Segregation distortion is the deviation of segregation ratios from the expected Mendelian fractions (Zamir and Tadmor, 1986; Lyttle, 1991) and has been reported in a wide range of plant species (Xu *et al.*, 1997). There have been several previous reports of segregation distortion in rice (e.g. Lin *et al.*, 1992; Harushima *et al.*, 1996; Yamagishi *et al.*, 1996). In the case of markers that show distorted segregation, it was concluded that one-third of these may be affected by gametophyte/sterility genes and a further one-third may arise because of association with the indica-japonica differentiation (Virk *et al.*, 1998).

RFLP mapping of isozymes, RAPD and QTLs for grain shape, brown plant hopper resistance in a doubled haploid rice population

An RFLP framework map with 146 RFLP markers based on a doubled haploid population derived from a cross between an indica variety IR64 and a japonica variety Azucena has been developed. The population carries 50.2% of IR64 loci and 49.8% of Azucena loci, indicating an equal amount of genetic materials from each parent has been transmitted to the progenies through anther culture. However, some markers show segregation distortion. These distorted marker loci are located on 10 chromosomal segments. Using this map 8 isozymes, 14 RAPDs, 12 cloned genes, 1 gene for brown planthopper (BPH) resistance, and 12 QTLs for grain length, grain width and length/width ratio were placed onto rice chromosomes. The major gene for BPH resistance was mapped on chromosome 12 near RG463 and isozyme Sdh-1. Most of the QTLs identified for the three grain characters were closely linked on chromosomes 1, 2, 3 and 10 (Huang *et al.*, 1997). It was concluded that the RFLP framework map presented will be useful for mapping other genes segregating in this doubled haploid population. Thus rapid generation of doubled haploid lines and their unbiased segregation makes it very attractive for gene mapping.

Performance of Backcrosses Involving Transgress and Doubled Haploid Lines in Rice under Contrasting Moisture Regimes

Improvement of grain yield in the rain fed lowland rice (*Oryza sativa* L.) ecosystem is important because that ecosystem covers a considerable area. Studies were carried out with the objective of finding relationship(s) among grain yield and its components, second to assess the influence of root morphological characters on grain yield, and finally to find the relationship of molecular marker heterozygosis with heterosis–performance of yield related characters in rain fed lowland rice. Nine backcrosses involving transgressants for maximum root length in rice from a doubled haploid mapping population of IR64/Azucena along with parents were evaluated simultaneously for grain yield and related characters as well as root traits under contrasting moisture regimes. Grain yield showed maximum reduction under severe moisture stress conditions. Significant negative correlation between drought tolerance index and grain yield was observed in the well-watered conditions only. Multiple linear regression analysis revealed total dry weight (root + shoot dry weight) as the most significant variable under well-watered conditions since it could explain as much as 30% of variability in grain yield, followed by root number. Under severe stress conditions, root dry weight explained 20% of variability in grain yield. The relationships between yield and root related characters were determined. General heterozygosis, based on molecular marker data, was found to be correlated with hybrid performance and heterosis of yield related characters. Among the nine backcrosses, P331 x IR64 and P124 x IR64 were selected to map quantitative trait loci for root morphological characters and grain yield (Toorchi *et al.*, 2003).

Agro-morphological characterization of ovary culture-derived plants of rice (*Oryza sativa* L.)

Plants derived from unpollinated ovary culture of ten rice genotypes showed significant variability in agro-morphological characteristics. The ovary-derived plant (H1) populations were found to be completely haploid, doubled haploid or haploid-doubled haploid mixture. Haploids showed drastic reduction in plant height, panicle length, grain length, breadth and number and spike let fertility (0.0%–2.1%). Doubled haploids from the hybrid of UPRI 95–121 × UPRI 95–165 were normal with fertility ranging between 69.6% and 97.7%. A genetic segregation in ratio of 1:1 was observed for five pigmentation characters in the H1 population derived from hybrid UPRI 95–122 × UPRI 95–165. Plant height showed the largest coefficient of variability (28.5%) followed by the number of spike lets per panicle (24.2%), number of grains per panicle (22.0%), per cent seed set (9.2%) and panicle length (9.0%). The range of variation in the H1 population from fully fertile hybrid PMS 2A (CMS) × IR 31802 (restorer) was similar to its corresponding F2 population for plant height, spike let fertility and number of grains/panicle. A single clone of plants from the cultivar BG 1321 exhibited complete male sterility but normal female fertility when pollinated with other varieties. Ovary-derived plants from the CMS lines PMS 2A and IR 58025A showed stable male sterility and those from thermo sensitive genetic male sterile line UPRI 95–140 showed thermo sensitive genetic male sterility. These lines have potential in the hybrid breeding program and are being currently exploited.

Characterization of the genetic basis for yield and its component traits of rice revealed by doubled haploid population

Main-effect QTL, epistatic effects and their interactions with environment are important genetic components of quantitative traits. The QTLs, epistatic effects and QTL by environment interactions (QE) underlying nine traits of yield and yield-component were analyzed, using a doubled-haploid (DH) population consisted of 190 lines from the cross between an indica parent Zhenshan 97 and a japonica parent WYJ 2, and tested in two-year replicated field trials. A genetic linkage map with 179 SSR (simple sequence repeat) marker loci was constructed. A mixed linear model approach was applied to detect QTLs, digenic interactions and QEs for the nine traits. In total, 57 QTLs of main effects, 41 digenic interactions, eight QEs and seven interactions of epistasis by environment were detected. Each of the main-effect QTLs individually explained

1.3% to 25.8% of the phenotypic variations. And they collectively explained 11.5% to 66.8% of the phenotypic variations for these traits. Most of the traits (except seed setting) had the QTLs simultaneously detected in two years. Many of the traits shared same QTLs with each other, which is consistent with their significant phenotypic correlations. The pleiotropism or tight linkage of QTLs for different traits might be the important genetic base for trait correlations.

High frequency androgenesis in indica - Basmati rice hybrids using liquid culture media

Breeding efforts to improve the yield potential of Basmati rice have resulted in limited success (Khush and de la Cruz, 1997). Crosses of Basmati rice with high yielding dwarf rice cultivars show considerable sterility and do not yield a full spectrum of recombinants in the segregating generations. Furthermore, aroma and grain elongation both are quantitative traits and it is difficult to combine all the polygene for these traits into the same line. Scientists have been interested in the introgression of desirable genes/traits such as dwarfism, photoperiod insensitivity, higher yield, etc. from indica rice cultivars into Basmati rice using anther culture as a breeding tool. The most common practice has been to use F1 hybrids, thereby fixing the products of recombination between the two parental genomes at the earliest opportunity (Raina *et al.*, 1989). Anther donors have also been effectively selected from F2 populations (Shi-Wei and Zhi-Hong, 1991) as F2 plants have been shown better for anther culture especially for characters such as disease resistance and grain quality.

An improved procedure has been developed for high frequency androgenesis in indica Basmati rice hybrids using a liquid culture medium (Bishnoi *et al.*, 2000). Anthers from fourteen genotypes comprising of indica Basmati rice F_1 hybrids, F2 plants and the parental rice cultivars, were floated in liquid RZM, N6M, and Heh5M media. Anther culture frequencies (percentage of anthers forming calluses) in most of the genotypes were significantly higher in RZM medium (16–75%) compared to those obtained in N6M (7–29%) and Heh5M (7–41%) media. Agarose (1.0% w/v)-solidified MSR1 medium containing 3.0% (w/v) maltose, 1 mg l^{-1} kinetin, 1 mg l^{-1} 6-benzyladenine (BA) and 0.5 mg $l^{-1}\alpha$-naphthalene acetic acid (NAA) induced green shoot regeneration at high frequencies compared to the medium (MSR2) lacking BA. In all the genotypes, microspore calluses initiated in RZM medium regenerated green shoots with over tenfold higher frequencies compared to the calluses initiated in other two media.

High plant regeneration frequencies (up to 270 green plants/1000 anthers) were obtained from microspore-derived calluses of some of the F_1 hybrids (Gobind Basmati 370, Gobind Taraori Basmati) and F2 plants (Gobind Basmati 370, Gobind Taraori Basmati, HKR86-3 Taraori Basmati) as compared to their actual parents. Cytological analysis of the root tips of the progeny seedlings of the microspore-derived plants revealed haploids at a frequency of about 50%; 22% of the microspore-derived plants had more than 5% spikelet fertility and were diploid. Use of RZM liquid and MSR1 media, respectively for anther culture and plant regeneration resulted in several fold increase in the recovery of green plants from recalcitrant indica Basmati rice F_1 hybrids/F_2 plants which were comparable to those reported for japonica rice varieties/hybrids leading to the improved feasibility of using doubled haploids in genetic, breeding and mapping research with indica rice.

Utilization of anther culture technique for rice improvement in the Philippines

Elite breeding lines have been developed at PhilRice through anther culture (Lapitan *et al.*). Anther culture technique offers great opportunities for accelerating breeding progress and generating as well as directing variation to increase genetic diversity useful for rice improvement.

Through anther culture, 10 doubled haploid lines (DHLs) from F1-506, a cross between C4044-2B-2-2 and IR13540-56-3-2-1 have been developed. Further evaluations of these lines both under greenhouse and field conditions resulted to selection of three outstanding DHLs. Aside from giving stable performance; these lines have been selected for their high phenotypic acceptability, high yield, and uniformity. Likewise, results of grain quality evaluation revealed that the selected DHLs were better than the parents. On the succeeding replicated yield trials (RYT), from R4 to R6 generations, these elite DHLs consistently outyielded the check varieties, IR72 and PSBRc 18. The results connote that anther culture technique shortened the breeding cycle as and developed elite breeding lines within 3 years while conventionally, it takes 6 to 7 years for selected inbred to reach RYT.

Also, variants have been generated from the popular IR64 variety utilizing the technique. Three outstanding DHLs have been produced with resistance to different soil-related stresses such as P, K, Zn, S, and Cu deficiencies. Evaluation of these variants showed improvements on their root system in terms of root length and weight as compared to the seed-derived IR64 variety.

ANTHER CULTURE: TODAY AND TOMORROW

As early as 1968, Niizeki and Oono reported the production of haploids from anther culture of rice. Since then, the anther culture technique has been refined greatly; it is now possible to produce haploids from the anther culture of many *japonica* and *indica* rice, although the frequency of plant regeneration is lower in *indica* varieties. Anther culture is important for the development of true breeding lines in the immediate generation from any segregating population, producing a shorter breeding cycle in new varieties. Selection efficiency in doubled haploid (DH) lines is higher, especially when dominance variation is significant. DH lines are also useful for developing mapping populations for molecular analysis. Seeds of such populations can be distributed to workers in other laboratories and populations can be grown repeatedly in many different environments, greatly facilitating the additional mapping of both DNA markers and genetic loci controlling traits of agronomic importance. In rice, mapping populations (DH lines) produced through anther culture of IR 64 x Azucena (*indica/japonica* cross) and CT9993 x IR 62266 are being used in the molecular mapping of genes with quantitative trait loci (QTL).

A number of varieties and improved breeding lines have been developed through anther culture in China, the Republic of Korea, Japan and the United States. Most of the anther-culture-derived varieties are *japonica*. *Indica* rices are generally regarded as recalcitrant for anther culture. At IRRI, the anther culture technique is being employed to obtain DH lines from many crosses for different rice breeding objectives. In the Philippines, one of the anther-culture-derived lines, IR 51500-AC-11-1, has been named as variety PSBRc50. Many DH lines produced through anther culture are now being used as parents in breeding programmes. DH lines are being produced to overcome sterility in the progenies of *O. sativa* x *O. glaberrima* crosses (Enriquez *et al.*, 2000).

Potential impact

Recent advances in cellular and molecular biology of rice offer new opportunities to enhance the efficiency of both the evolutionary and the evaluation phase of rice breeding. Biotechnology is becoming an important component of rice breeding. Anther culture has become an important technique for use by plant breeders to shorten the breeding cycle for the development of rice varieties, to fix recombinants and in overcoming sterility in distant crosses. Several rice varieties have been developed through anther culture. Doubled haploid populations are important for mapping genes governing agronomic traits including QTL.

Many rice cultivars have been developed through the application of biotechnology tools. Many more will be forthcoming. Future food and nutritional security will depend upon the availability of rice cultivars with higher yield potential, durable resistance to diseases and insects, tolerance to abiotic stresses and higher levels of micronutrients in the grain. Conventional breeding methods and biotechnology tools will help meet these challenges.

REFERENCES

Agarwal P K and Bhojwani S S 1993 Enhanced microspore embryogenesis and plant regeneration in anther cultures of *Brassica juncea* cv PR-45 *Euphytica* **70** 191-196

Aruga K Nakajima T and Yamamoto K 1985 Embryogenic induction of pollen grains of *Nicotiana tabacum* L *Japan J Breed* **35** 50-58

Barnabás B Obert B Kovács G Colchicine an efficient genome-doubling agent for maize (Zea mays L) microspores cultured in anthero Journal *Plant Cell* Reports Publisher Springer Berlin/Heidelberg ISSN 0721-7714 (Print) 1432-203X (Online) Issue Volume **18** Number 10/June 1999 DOI 10 1007/s002990050674 Pgs 858-862

Bhojwani S S and Razdan M K 1996 *Plant Tissue Culture: Theory and Practice* A Revised Edition Elsevier Amsterdam

S S Bhojwani and W Y Soh eds Kluwer Academic Publishers Dordrecht 471-488

Bhojwani S S Soh W Y and Pande H 1999 Morphogenesis in haploid cell cultures In W Y Soh and S S Bhojwani (eds) Morphogenesis in Plant Tissue Cultures Kluwer Dordrecht

Biddington N L Sutherland R S and Robinson H T 1988 Silver nitrate increases embryo production in anther culture of Brussels sprouts *Ann Bot* **62** 181-185

Bishnoi U S Jain R K Gupta K R Chowdhury V K and Chowdhury J B 2000 High frequency androgenesis in indica Basmati rice hybrids using liquid culture media *Plant Cell Tissue and Organ Culture* **61** 153–159

Blakelsee A F Belling J Farhnam M E and Bergner A D 1922 A haploid mutant in the Jimson weed *Datura stramonium Science* **55** 646-647

Burk G Gerstel D U and Wernsman E A 1979 Maternal haploids of *Nicotiana tabacum* L from seed *Science* **206** 585

Chen C Xiao H Zhang W Wang A Xia Z Li X Zhai W Cheng Z Zhu L State Key Laboratory of Plant Genomics and National Center for Plant Gene Research Institute of Genetics and Developmental Biology Chinese Academy of Sciences Beijing 100101 China

Chen C C Tsay H S and Huang C R 1991 Factors affecting androgenesis in rice (*Oryza sativa* L) In Bajaj YPS (ed) Rice Biotechnology in Agriculture and Forestry Springer Verlag Berlin **14** : 192–215

Chen Y Wang R Tian W Zuo O Zeng S Lu D and Zheng S 1980 Studies on pollen culture *in vitro* and induction of plantlets in *Oryza sativa* subsp *Keng Acta Genetica Sinica* **7** 46-54 Chinese/English

Cho M S and Zapata F J 1988 Callus formation and plant regeneration in isolated pollen culture of rice *Oryza sativa* L cv Taipei 309 *Plant Sci* **58** 239-244

Chu C C Wang C C Sun C S Hsu C Yin K C Chu C Y and Bi F Y 1975 Establishment of an efficient medium for anther culture of rice through comparative experiments on the nitrogen sources *Sci Sin* **18** 659-668

Chu Q Zhang Z and Gao Y 1985 Cytogenetical analysis on aneuploids obtained from pollen clones of rice (*Oryza sativa* L) *Theor Appl Genet* **71** 506-512

Chung G S 1987 Application of anther culture technique for rice (*Oryza sativa* L) improvement Res Rep Yeongnam Crop Exp Sta (RDA) Milyang Korea 36-55

Cohen J E Schoenly K Heong K L H Justo G Arida A T Barrion J A Litsinger 1994 A Food Web Approach to Evaluating the Effect of Insecticide Spraying on Insect Pest Population Dynamics in a Philippine Irrigated Rice Ecosystem *Journal of Applied Ecology* **31** No 4 747–763

Coumans M P S Sohota and E B Swanson 1989 Plant development from isolated microspores of *Zea mays* L *Plant Cell Rep* **7** 618-621

Crawford G W and C Shen 1998 The Origins of Rice Agriculture: Recent Progress in East Asia *Antiquity* **72** 858–866

Datta S K 2001 Androgenesis in cereals In: Current Trends in Embryology of Angiosperms

Datta S K *et al* 2007 Golden rice: introgression breeding and field evaluation *Euphytica* **154** (3) 271-278

Datta S K K Datta and I Potrykus 1990 Embryogenesis and plant regeneration from microspores of both 'indica' and 'japonica' rice (*Oryza sativa*) *Plant Sci* **67** 83-88

Dawe D Robertson R and Unnevehr L 2002 Golden rice: what role could it play in alleviation of vitamin A deficiency? *Food Policy* **27** 541-560

Douangboupha B K Khamphoukeo S Inthavong J Schiller and G Jahn 2006 Pests and diseases of the rice production systems of Laos 265–281

Enriquez E Brar D S Rosario T L Jones M and Khush G S 2000 Production and characterization of doubled haploids from anther culture of the F_1s of *Oryza sativa* L x *O. glaberrima* Steud *Rice Genet Newsl* **17** 67-69

Faruque M O Farzana T Seraj Z I Sarker R H and Khatun A A 1998 Variations in green plant regeneration response from anthers of indica rice and their hybrids with japonica cv Taipei 309 *Plant Cell Tiss Org Cult* **54** 191–195

Finnie S J B P Forster K J Chalmers A F Dyer R Wough and W Powell 1991 Genetic stability of microspore-derived doubled haploids of barley: A cytological bio-chemical and molecular study *Genome* **34** 923-928

Frist B 2006 One existing crop genetically engineered "golden rice" that produces vitamin A already holds enormous promise for reducing blindness and dwarfism that result from a vitamin-A deficient diet in a *Washington Times* commentary - November 21 2006

Garris Amanda Tai Thomas Coburn Jason Kresovich Steve McCouch Susan 2004 "Genetic Structure and Diversity in *Oryza sativa* L"

Gillis Justing "Rice Genome Fully Mapped" Washington post com August 11 2005

Gosal S S Sindhu A S Sandhu J S Sandhu-Gill R Singh B Khera G S Sidhu G S & Dhaliwal H S 1997) Haploidy in rice In: Jain S M Sopory S K & Veilleux R E (eds) Cereals *In vitro Haploid Production in Higher Plants* **4** pp 1–35 Kluwer Academic Publishers Dordrecht 159

Guha S and Maheswari S C 1964 *In vitro* production of embryos from anthers of Datura *Nature* **204** 497

Guha-Mukherjee S 1973 Genotypic differences in the *in vitro* formation of embryoids from rice pollen *Exp Bot* **24** 139-144

Guiderdoni E Galinanto E Luistro J & Vergara G 1992 Anther culture of *tropical Japonica Indica* hybrids of rice *Oryza sativa* L *Euphytica* **62** 219–224

Gupta H S and D N Borthakur 1987 Improved rate of callus induction from rice anther culture following microscopic staging of microspores in iron alum-haemotoxylin *Theor Appl Genet* **74** 95-99

H Y Yang and C Zhou 1982 In vitro induction of haploid plants from unpollinated ovaries and ovules *TAG Theoretical and Applied Sciences* **63** 97-104

Heszky L E Nam L S Simon I K Kiss E Lokos K and Birth D 1991 *In vitro* studies on rice in Hungary In Y P S Bajaj Ed *Biotechnology in Agricultural Forestry* **14** pp 619-641 Springer-Verlag Berlin

Hirschberg J 2001 Carotenoid biosynthesis in flowering plants *Current Opinion in Plant Biology* **4** 210-218

Hoekstra S M H van Zijderveld F Heidekamp and F van der Mark 1993 Microspore culture of *Hordeum vulgare*: L the influence of density and osmolality *Plant Cell Rep* **12** 661-665

Hoekstra S M H van Zijderveld J D Louwerse F Heidekamp and F van der Mark 1992 Anther and microspore culture of *Hordeum vulgare* L cv Igri *Plant Sci* **86** 89-96

Hu H 1985 Use of haploids for crop improvement in China *Genet Manipulation Crops Newslett* **1** 11-23

Huang N Khush G S Parco A 1997 RFLP mapping of isozymes RAPD and QTLs for grain shape brown planthopper resistance in a doubled haploid rice population *Molecular Breeding* **3** 105-113

Hunter C P 1987 Plant generation method Australian Patent Application No AU-A-7271 Shell Int Res

J P Londo Y Chiang *et al* 2006 "Phylogeography of Asian wild rice *Oryza rufipogon* reveals multiple independent domestications of cultivated rice *Oryza sativa*" *PNAS* **103**(25) 9578–83

Jahn G C 1992 Rice pest control and effects on predators in Thailand *Insecticide & Acaricide Tests* **17** 252–253

Jahn G C C Pol B Khiev S Pheng and N Chhorn 1999 Farmer's pest management and rice production practices in Cambodian upland and deepwater rice Cambodia-IRRI-Australia Project Baseline Survey Report No 7

Jahn G C S Pheng B Khiev and C Pol 2000 Ecological characterization of biotic constraints to rice in Cambodia *International Rice Research Notes* (IRRN) **25** (3) 23–24

Jahn G C S Pheng B Khiev and C Pol 1996 Farmers' pest management and rice production practices in Cambodian lowland rice Cambodia-IRRI-Australia Project CIAP Baseline Survey Report No 6 CIAP Phnom Penh Cambodia 28 pages

Jahn G C B Khiev C Pol N Chhorn S Pheng and V Preap 2001 Developing sustainable pest management for rice in Cambodia 243–258

Jahn G C I Domingo L P Almazan and J Pacia 2004c Effect of rice bugs (Alydidae: Leptocorisa oratorius Fabricius)) on rice yield grain quality and seed viability *Journal of Economic Entomology* **97**(6) 1923–1927

Jahn G C J A Litsinger Y Chen and A Barrion 2007 Integrated Pest Management of Rice: Ecological Concepts In Ecologically Based Integrated Pest Management eds O Koul and G W Cuperus CAB International 315–366

Jahn G C L P Almazan and J Pacia 2005 Effect of nitrogen fertilizer on the intrinsic rate of increase of the rusty plum aphid *Hysteroneura setariae* (Thomas) (Homoptera: Aphididae) on rice (*Oryza sativa* L) *Environmental Entomology* **34** (4) 938–943

Jahn G C N Q Kamal S Rokeya A K Azad N I Dulu J B Orsini A Barrion and L Almazan 2004a Completion Report on Livelihood Improvement Through Ecology LITE PETRRA IPM Subproject SP 27 02 Poverty Elimination Through Rice Research Assistance PETRRA IRRI Dhaka 20 pages text plus 20 pages appendices

Jahn G C N Q Kamal S Rokeya A K Azad N I Dulu J B Orsini M Morshed NMS Dhar N A Kohinur 2004b Evaluation Report on Livelihood Improvement Through Ecology LITE PETRRA IPM Subproject SP 27 02 Poverty Elimination Through Rice Research Assistance PETRRA IRRI Dhaka 42 pages plus 40 pages of annexes

Jahne A and Lorz H 1995 Cereal microspore culture *Plant Sci* **109** 1-12

Kahn C *et al* 2005 World History: Societies of the Past Portage & Main Press ISBN 1553790456 pg 92

Kao K N M Saleem S Abrams M Pedras D Horn and C Mallard 1991 Culture conditions for induction of green plants from barley microspores by anther culture methods *Plant Cell* Rep **9** 595-601

Kasha K J and Kao K N 1970 High frequency haploid production in barley *Hordeum vulgare* L *Nature* **225** 874-876

Kawata M A Ohmiya Y Shimamoto K Oono and F Takaiwa 1995 Structural changes in the plastid DNA of rice *Oryza sativa* L during tissue culture *Theor Appl Genet* **90** 364-371

Khiev B G C Jahn C Pol and N Chhorn 2000 Effects of simulated pest damage on rice yields *IRRN* **25** (3) 27–28

Khush G and de la Cruz N 1997 Developing Basmati rices with high yield potential In: Proceedings of International Symposium on Rice Quality Nottingham UK November 24–27

Khush G S Brar D S 1998 The application of biotechnology in rice In Ives C and Bedford B eds *Agricultural biotechnology in international development* 92-121 Wallingford U K CAB International

Last D J and R I S Brettell 1990 Embryo yield in wheat anther culture is influenced by the choice of sugar in the culture medium *Plant Cell Rep* **9** 14-16

Lentini Z Reyes Z P Martinez C P and Roca W M 1995) Androgenesis of highly recalcitrant rice genotypes with maltose and silver nitrate *Plant Sci* **110** 127–138

Lentini Z P Reyes C P Martinez and W M Roca 1995 Androgenesis in highly recalcitrant rice genotypes with maltose and silver nitrate *Plant Sci* **110** 127-138

Leung L K P Peter G Cox Gary C Jahn and Robert Nugent 2002 Evaluating rodent management with Cambodian rice farmers *Cambodian Journal of Agriculture* **5** 21–26

Lu C S H C Sharma and H W Ohm 1991 Wheat anther culture: effect of genotype and environmental conditions *Plant Cell Tiss Org Cult* **24** 233-236

Lutts S J N Kinet and J Bouharmont 2001 Somaclonal variation in rice after two successive cycles of mature embryo derived callus culture in presence of NaCl *Biol Plant* **44** (4) 489-495

Macdonald M V 1992 Donor plant growth factors affecting anther culture of maize and sweet corn *Zea mays Ann Bot* **70** 357-363

Maluszynski M Kasha K J Forster B P and Szarejko I 2003 Doubled haploid production in crop plants: A manual Kluwer Academic Publ Dordrecht Boston London

Matsushima T S Kikuchi F Takaiwa and K Oono 1988 Regeneration of plants by pollen culture in rice *Oryza sativa* L *Plant Tissue Cult Lett* **5** 78-81

Meifang L I 1992 Anther culture breeding of rice at the CAAS In K Zeng and T Murashige eds *Anther Culture for Rice Breeders* 75-85 Hangzhou China

Mejza S J V Morgant D E DiBona and J R Wong 1993 Plant regeneration from isolated microspores of *Triticum aestivum Plant Cell Rep* **12** 149-153

Miah M A A E D Earle and G S Khush 1985 Inheritance of callus formation ability in anther cultures of rice *Oryza sativa* L *Theor Appl Genet* **70** 113-116

Muller A J and R Grafe 1978 Isolation and characterization of cell lines of *Nicotiana tabacum* lacking nitrate reductase *Mol Gen Genet* **161** 67-76

Murphy S J Stonehouse J Holt J Venn N Q Kamal M F Rabbi M H Haque G Jahn B Barrion 2006 Ecology and management of rice hispa Dicladispa armigera in Bangladesh pp 162––164 In Perspectives on Pests II Achievements of research under UK Department for International Development Crop Protection Programme 2000–05 Natural Resources International Limited 206 pages

Navarro-Alvarez W P S Baenzigar K M Eskridge D R Shelton V D Gustafson and M Hugo 1994 Effect of sugars in wheat anther culture media *Plant Breed* **112** 53-62

Nene Y L 2005 Rice Research in South Asia through Ages *Asian Agri-History* **9** No 2 85–106

Ogawa T H Fukuoka and Y Okhawa 1994 Induction of cell division of isolated pollen grains by sugar starvation in rice *Breed Sci* **44** 75-77

Ogawa T H Fukuwa and Y Ohkawa 1995 Plant regeneration through direct culture of isolated pollen grains in rice *Breed Sci* **45** 301-307

Ogawa T T Hagio and Y Ohkawa 1992 Plant regeneration from isolated pollen grains in indica type rice *Japan J Breed* **42** 675-679

Oka H I 1988 Indica/Japonica differentiation of rice cultivars In *Origin of Cultivated Rice* pp141-179 Japan Sciences Social Press Tokyo/Elsevier Amsterdam

Paine *et al* 2005 Improving the nutritional value of Golden Rice through increased pro-vitamin A content *Nature Biotechnology* doi:10 1038/nbt1082

Pande H 1997 Androgenesis in anther cultures of an indica cultivar of *Oryza sativa* L Ph D Thesis University of Delhi

Pande H and S S Bhojwani 1999 Promotion of androgenesis in rice anther cultures by substitution of sucrose with maltose and mannitol *Biol Plant* **42** 125-128

Pescitelli S M J C Mitchell A M Jones D R Pareddy and J F Petolino 1989 High frequency androgenesis from isolated microspores of maize *Plant Cell* Rep **7** 673-676

Petolino J F and S A Thompson 1987 Genetic analysis of anther culture response in maize *Theor Appl Genet* **74** 157-159

Pheng S B Khiev C Pol and G C Jahn 2001 Response of two rice cultivars to the competition of *Echinochloa crus-gali* L P Beauv *International Rice Research Institute Notes* IRRN **26** (2) 36–37

Powell W W T B Thomas and D M Thompson 1992 The agronomic performance of anther culture derived palnts of barley produced via pollen embryogenesis *Ann Appl Biol* **120** 137-150

Preap V M P Zalucki and G C Jahn 2006 Brown planthopper outbreaks and management *Cambodian Journal of Agriculture* 7(1) 17–25

Raina S K & Zapata F J 1997 Enhanced anther culture efficiency of indica rice *Oryza sativa* L through modification of the culture media *Plant Breeding* **116** 305–315

Raina S K Balachandra S M Virmani S S and Zapata F J 1989 Improved medium for efficient anther culture of some indica rice haploids *Int Rice Res Newsl* **14** 4

Raina S K 1989 Tissue culture in rice improvement: status and potential *Adv Agron* **42** 339-397

Raina S K 1997 Doubled haploid breeding in cereals *Plant Breed* Rev **15** 141-186

Raina S K and F J Zapata 1997 Enhanced anther culture efficiency of indica rice *Oryza sativa* L through modification of the culture media *Plant Breed* **116** 305-315

Raina S K and Irfan S T 1998 High-frequency embryogenesis and plantlet regeneration from isolated microspores of indica rice *Plant Cell Rep* **17** 957-962

Reddy V S Leelavathi S and Sen S K 1985 Influence of genotype and culture medium on microspore callus induction and green plant regeneration in anthers of *Oryza sativa Physiol Plant* **63** 309-314

Roberts-Oehlschlager S L and Dunwell J D 1990 Barley anther culture: Pretreatment on mannitol stimulates production of microspore-derived embryos *Plant Cell Tiss Organ Cult* **20** 235-240

Sathish P Gamborg O L and Nabors N W 1995 Rice anther culture: callus initiation and androclonal variation in progeny of regenerated plants *Plant Cell Rep* **14** 432-436

Schaub P *et al* 2005 Why Is Golden Rice Golden (Yellow) Instead of Red? *Plant Physiology* **138** 441–450

Shih-Wei L and Zhi-Hong X 1991 Anther culture for rice improvement in China In Bajaj Y P S ed *Rice Biotechnology in Agriculture and Forestry* **14** pp 151–179 Springer Verlag Berlin

Smith Bruce D 1998 *The Emergence of Agriculture* Scientific American Library A Division of HPHLP New York

Sun Z 1999 Methods to increasing frequency of rice anther culture In Regional RCA Training Course on Production and Utilization of Doubled Haploid Lines in Rice Republic of Korea 18-30 October 1999

Toorchi M Shashidhar H E T M Gireesha T M and Hittalmani S 2003 Performance of Backcrosses Involving Transgressant Doubled Haploid Lines in Rice under Contrasting Moisture Regimes *Crop Science* **43** 1448-1456

Tyagi A K J P Khurana P Khurana S Raghuvanshi A Gaur A Kapur V Gupta D Kumar V Ravi S Vij P Khurana and S Sharma 2004 Structural and functional analysis of rice genome *Genet* **83** 79-99

Virk P S Ford-Lloyd B V and Newbury H J 1998 Mapping AFLP markers associated with subspecific differentiation of *Oryza sativa* rice and an investigation of segregation distortion *School of Biological Science* **81** 6 pp 613-620 *University of Birmingham Edgbaston Birmingham B15 2TT U K*

Wei Z M M Kyo and H Harada 1986 Callus formation and plant regeneration through direct culture of isolated pollen of *Hordeum vulgare* cv Sabarlis *Theor Appl Genet* **72** 252-255

Winzeler H Schmid J and Fried P M 1987 Field performance of androgenetic doubled haploid spring wheat line in comparison with line selected by the pedigree system *Plant breeding* **99** 41-48

Ye *et al* 2000 Engineering the provitamin A (beta-carotene) biosynthetic pathway into (carotenoid-free) rice endosperm *Science* **287** (5451) 303-305

Ziauddin A Simion E and Kasha K J 1990 Improved plant regeneration from shed microspore culture in barley (*Hordeum vulgare* L) cultivar Igri *Plant Cell* Rep **9** 69-72

Zohary D and Hopf M 2000 *Domestication of plants in the Old World* third edition Oxford University Press

7

MORTALITY OF AGROFORESTRY TREES AND ITS MANAGEMENT

J C KAUSHIK AND O P TOKY

In the recent past, it has been noticed that mortality of trees in plantations or agroforestry systems is high in several parts of the country. In the present paper, the interaction between trees, environment and other organisms is expressed and various methods to control important diseases have been suggested. The changing environment and spread of diseases have caused serious economic and ecological losses. Among the disease agents of forest plants, fungal pathogens caused the maximum damage. Damping-off, root rot, cankers, vascular wilts and foliar diseases are serious in agroforestry plantations. Damping-off in nurseries can be effectively managed by treating the seeds, propagating materials and soil with fungicides and solarization. Root rots in plantations can be managed by proper sanitation, isolation of diseased trees, adequate spacing and by the use of biological control agents. Canker diseases can be controlled by the use of fungicides, proper silvicultural practices, avoiding injury and keeping the trees in more vigorous conditions. Mortality of important tree such as *shisham, kikar, khejri,* etc. cannot be attributed to single causing factor but a combination of factors, which has been discussed. Quarantine regulations, choice of species, selection of planting site and use of resistant plant material are some of the other measures employed for the management of tree diseases.

INTRODUCTION

Population growth and industrialization have enhanced the demand for wood products in India accentuating the pressure on forest resources which further widened the gap between demand and supply. During the past two decades, several strategies have been adopted to bring more area under tree cover. This unprecedented expansion of forest plantations in the arid and semi-arid regions led to increasing incidence of pest and diseases and evolving of new problems. The recent large scale mortality of sissoo (*Dalbergia sissoo* Roxb.) and *kikar (Acacia nilotica)* in northern Gangatic plains in India and neighbouring countries, are indications of what can happen if timely attention is not paid to the management of pests and diseases. The tree species are affected by a variety of biotic and abiotic factors including environmental, edaphic, hydrological, pathological, entomological and biotic. Among the biotic factors fungi, prokaryotes, viruses, nematodes and phanerogamic parasites are predominant.

Fungal pathogens are responsible for causing maximum damage to forest trees and therefore, have been studied in greater detail. Biotic plant diseases are the product of three interacting factors namely, host, pathogen and environment. These diseases are generally host specific but sometime may occur on a limited number of related and unrelated hosts. They affect the productivity of trees and sometimes threaten the very survival of the species. This chapter elucidates a brief account of important nursery and plantation diseases of some Indian forest trees and suggests the management practices to minimize the losses. Mass mortality of *Dalbergia sissoo*, an important agroforestry tree species in northern India, reported recently, has been discussed in detail.

NURSERY DISEASES

Seeds and subsequently, the seedlings are attacked by a number of seed and soil borne pathogens. These pathogens affect the planting stock qualitatively as well as quantitatively and thus, disturb the planting programmes. A brief account of various nursery diseases, factors affecting them and control measures are given below.

(i) Damping-off

Damping-off in forest seedlings is widely distributed all over the world and is probably, the most destructive disease. Almost all the tree species except *Cedrus deodara* (Cedar) and its relatives are susceptible to this disease. Species of *Fusarium, Pythium, Rhizoctonia, Cylindrocladium, Phytophthora, Botrytis, Alternaria and Phoma* have been reported to cause this disease (Boyce, 1961; Sharma, 1989; Kaushik, 1991). The importance of pathogen varies from region to region and the host species. In acidic soils, *Pythium* spp. appear to be most serious in conifers. In India, *Rhizoctonia solani* is an important cause of damping-off disease among broad-leaved tree species. *Fusarium* species are sometimes more important than other species, e.g. *Fusarium solani, F. moniliforme, F. roseum, F. oxysporum and F. tricinctum* causing damping-off in different host species. *Cylindrocladium* species is sometime very destructive in eucalypts nurseries causing damping-off disease (Gupta and Kaushik, 1988). *Phytophthora* is more important in plantations and natural forests than in the nurseries. Sehgal *et al.*, (1989) observed heavy mortality in fir and spruce seedlings due to damping-off in India.

Damping-off fungi cause two types of symptoms i.e. pre-emergence and post-emergence; in the former case, seedlings are attacked and killed by the pathogen before they emerge above the ground, which may be wrongly attributed to poor viability of seeds; while in post-emergence mortality, the seedlings are attacked on the hypocotyls or on roots or both while still in cotyledon stage. The fungi spread rapidly in the tissues and the seedlings either wilt completely or fall down on ground due to rotting of the tissue at the base.

Damping-off may also be clarified as normal, late and top damping-off (Tattar, 1989). Normal damping-off refers to infection on the lower part of the stem, the root or both. Late damping-off occurs after lignification of stems and roots is initiated. This type of infection is not fatal. Top damping-off occurs when pathogens infect cotyledons that are kept for abnormally extended periods after seedling emergence.

(ii) Root rot

The most common fungi that cause root rot in nurseries belong to genera *Phytophthora, Rosellinia, Macrophomina and Cylindrocladium.* Affected seedlings exhibit yellowing, stunted growth or they grow slowly or wilt easily on a warm day. Foliage may turn yellow to brown and fall prematurely. The severity depends upon fungal pathogen, the host plant susceptibility and the soil conditions. Diseased roots show various degree of water soaking and usually some shade of brown or black lesions/spots. Charcoal or black root rot caused by *Macrophomina phaseolina* renders serious losses in eucalypts, mulberry, conifers, acacias, poplars and casuarinas. The pathogens enter through fine roots and spread slowly upwards. Roots become blackened and small black sclerotia are formed on the dead seedlings. Species of *Cylindrocladium* produce serious root rot symptoms on eucalypts, oaks and black walnut.

(iii) Web blight

It is caused by *Rhizoctonia solani* in different host species viz., eucalypts, *neem, sissoo,* etc. The disease is characterized by profuse mycelial growth entangling the affected seedlings giving the appearance of a cobweb. The infected seedlings wilt and die. The disease spread rapidly under high moisture conditions, especially when seedlings are crowded.

(iv) Seedling blight

This disease results in heavy mortality of eucalypts seedlings and is caused by *Cylindrocladium quinqueseptatum* (Kaushik, 1988). The disease has been characterized by the appearance of a water soaked lesion on the stem near the soil level just above the root collar region. The affected seedlings turn brown and die (Anonymous, 1992).

(v) Grey mould

Botrytis cineraria, though seriously affects conifer nurseries, but the seedlings of *Rhododendron, Eucalyptus and Ailanthus* are also seriously affected. Seedlings show top die back symptoms with yellowing to browning of the leaves which get covered with fungal conidiophores and conidia. The pathogen becomes more serious because of its ability to grow and survive under diverse environmental and edaphic conditions.

(vi) Bacterial wilt

The disease is commonly observed on the seedlings of *Casuarina equisetifolia* and *Tectona grandis*. The bacterium (*Pseudomonas solanacearum)* causes heavy losses in these crops. Initially, the small patches of brown color appear on leaves between the veins. It is followed by wilting of leaves and then seedling death. The dark discoloration of vascular tissues confirms the disease.

Besides, there are other minor diseases such as leaf spots, rusts, mildews, snow blight, etc. (Table 1), which cause mild to severe losses in the nursery depending upon the prevailing environmental conditions.

Table 1 : Some other nursery diseases of forest trees

Disease 1	Casual organism 2	Symptoms 3	Occurrence on 4	Host 5
Sissoo leaf rust	*Maravalia achroa*	—	Leaves and young branches	Sissoo
Crown gall	*Agrobacterium tumefaciens*	Tumor like structure	—	Several trees
Seedling blight of wattles	*Glomerella cingulata*	Light brown to dark brown lesions on above ground part of seedlings	Leaves and stem	Acacias
Snow blight of conifers	*Phacidium infestans*	Foliage become soft and brown and white mycelium on affected foliage and later stage	Foliage	Conifers
Needle caste of pines	*Lophodermium* spp.	Pale spots on needles, these gradually enlarge and merge giving brown colour to needles	Foliar	Pines
Powdery mildew	*Microsphaera alphitoides*	Cinnamon coloured spots on leaf and shoot both. On severe attack of disease, leaves shrivel and fall down	Foliage and shoot	Oak, chestnut and beech

FACTORS AFFECTING NURSERY DISEASES

Several factors such as location, soil characteristics, climate, water table, tree species, plant density and soil treatment play an important role in determining the disease occurrence. One or combination of these factors influence the kind and amount of losses caused to the host.

Location of nursery is one of the most determining factors for disease development (Sutherland *et al.*, 1989; Aldhons and Mason, 1994). Climate of elevation zones and specific localities within a zone dictate the cause of a particular disease, e.g. *Sirococcus* blight of pines predominates in nurseries located in coastal areas.

Soils deficient in organic matter and rich in clay content favor damping-off disease whereas light textured soils, having good organic matter deter the damping-off caused by *Fusarium solani* in *Dalbergia sissoo* (Kaushik and Singh, 1996). Kaushik *et al.* (1990) reported high mortality of eucalypts seedlings due to blight disease caused by *Cylindrocladium quinqueseptatum* under high soil temperature and moisture conditions. *Rhizoctonia solani* is also favored by high temperature and high moisture conditions. Saline or alkaline pH favors damping-off disease caused by most of the fungal pathogens except *Pythium* which prefers acidic soils. Adverse climatic conditions affect seedlings and make them susceptible to several diseases. Cool and wet weather during spring and early summer often favors damping-off and shoot blight such as *Sirococcus* in pines.

High water table or poor drainage favors certain water borne pathogens such as *Phytophthora*. Different tree species have resistance or susceptibility to different diseases. Variability is also prevalent within a species based on their genotypes. Deep and dense seed sowing favor damping-off pathogen in all species and snow blight pathogen in conifers. Pre-sowing insecticidal treatment also play some role in occurrence and development of certain nursery diseases.

MANAGEMENT OF NURSERY DISEASES

Damping-off and root-rot diseases in the nurseries are caused by numerous unrelated pathogens and it is not possible to give exact measure applicable under all conditions. An integrated disease management approach is therefore, suggested comprising appropriate nursery practices, seed and soil sterilization before sowing and timely use of fungicides, bactericides and other chemicals (Kaushik and Bisla, 1993).

Following are the important methods to control nursery diseases:

(i) Healthy nursery practices

Nursery should be located on well-drained, light textured and fertile soils. Heavy and alkaline soils can cause disease severity by delaying seedling emergence, thereby giving more exposure by young succulent seedlings to pathogens. Generally, the pathogen survival is enhanced and effectiveness of chemical control is reduced in heavy soils. Such soils should be avoided or mixed with sand and organic matter. Deep and dense sowing increases disease incidence of damping-off.

In most cases, shallow seed sowing to a depth of about 1½ times of the diameter of seed and removal of the covering soil from the crowns of seedlings reduces the disease incidence.

A medium density of about 200 seedlings per 30 cm² of area of eucalypts has been recommended (Sharma and Mohanan, 1984). Regular raising of seedlings of same species on the same piece of land flare-up the disease. If possible, nursery site should be shifted regularly.

(ii) Solarization

Energy from sunlight trapped in the soil elevates temperature to a high level enough to inactivate soil borne pathogens. The temperature by solarization treatments ranges between 25 to 52°C which is high enough to reduce the potential of soil borne pathogens (Kaṭan, 1981). The nursery plots covered with polythene sheets for solar heating up to 55 days had 64 to 75 per cent less incidence of damping-off and root-rot caused by *Pythium, Fusarium* and *Rhizoctonia*. This indicates a great potential of this source of energy for management of nursery diseases (Hildebrand, 1985; Kaushik *et al.*, 2002).

(iii) Biological control

Management of some pathogens, through biological means has been proved very effective. Nesme *et al.* (1987) reported that crown gall bacterium of poplar was successfully controlled by *Agrobacterium* radiobacter strain K-84 which produced a bacteriocin "agracin 84" which was very effective against the pathogen. Species of *Trichoderma, Epicoccum, Bacillus* and *Entrobacter* were found effective against *Cylindrocladium, Fusarium, Rhizoctonia, Pythium* and *Phytophthora* (Mukhopadhaya *et al.*, 1988; Kaushik *et al.*, 1996, 2002). Mycorrhizae roots are well known, both in agricultural and forest crops to provide resistance to infection by root pathogens (Lakhanpal, 1988).

(iv) Chemical control

Chemical control is generally expensive. They are applied in several ways as seed and soil treatments and foliar sprays and dusts. A vast group of fungicides available in the market having broad spectrum status as well as specific to a group of pathogens are recommended to apply at proper time and locality keeping in view the economic threshold level.

PLANTATION DISEASES

A. Root diseases

The causal factors for root diseases may be fungi, bacteria, nematodes and/or insects. However, fungi form an important group of organisms causing root rots and wilt. Many fungi that cause root diseases are facultative parasites and are commonly found growing saprophytically in the soil. Though some fungi penetrate directly into healthy roots, but most enter through wounds or the dead roots in the lower trunk or buttress. Most root disease fungi produce specialized structures for survival or movement in the soil. They may be reproductive structures such as thick walled or motive spores or vegetative structures that enable translocation or storage of materials.

(i) Root rots

Majority of the root rots of mature trees are caused by soil and root inhabiting basidiomycotina fungi. Initially, the symptoms produced by root rots look like nutrient and water deficiencies where the foliage becomes sparce and yellow. In some cases, the leaves remain small, yellow and fall down prematurely. The extremely important root pathogens of forest trees are, *Armillariella mellea (Armillaria mellea), Heterobasidion annosus, Ganoderma lucidum (Polyporus lucidus), Phytophthora cinnamoni, Phymatotrichum omnivorum* and *Inonotus weirrii* (Bakshi, 1976; Kaushik and Bisla, 1993).

Armillariella mellea is one of the most widespread of all root diseases causing disease of broad leaved and coniferous trees and shrubs. Once the fungus has succeeded in entering and girdling a tree root, it spreads distally from the point of entry along the cambial zone between bark and wood and produces its characteristics sheets of mycelium (rhizomorphs). It survives on dead stumps and dying trees. Fruiting bodies are recognized by their white spores and ring of veil tissue remaining on the stripe. *Heterobasidion annosus* is another important root rot pathogen mainly of *Pinus* and *Juniperus* trees, but it may also attack hard woods such as birch, beach, oak and alder. The spread of *H. annosus* in stumps on ex-hardwood sites may be restricted by *Armillaria mellea* invading either from infected roots or by means of rhizomorphs at the root collar.

Ganoderma lucidum (Polyporus lucidus) causes root and butt rot in a wide range of broad-leaved trees such as *Dalbergia, Acacia, Tectona, Azadirachta, Shorea, Quercus, Prosopis, Eucalyptus, Populus, Albizia,* palms, willows and maple (Bakshi, 1976; Phillips and Burdekin, 1982; Kaushik and Bangarwa, 1992). Bakshi *et al.* (1976) described in detail the symptoms, epidemiology and management of this disease in Acacia *catechu.* Trees, both young and old, are killed as a result of attack by *Ganoderma lucidum* and infection could be traced back to root contacts with stumps and roots of the previous forest cover. Heavy mortality of sissoo (*Dalbergia sissoo*) due to *Ganoderma lucidum* at Hisar, India has been recorded.

The feeder root disease caused by *Phytophthora cinnamomi* is wide spread in the natural forests throughout the world. *Phytophthora* root disease occurs on a range of broad-leaved and coniferous hosts in India (Gupta *et al.*, 1988; Negi, 1996). The crown symptoms of *phytophthora* root diseases are typical of root diseases. In broad-leaved species, leaves may be abnormally small and yellow, which may be followed by ultimate death of the tree but in some cases, trees may recover. Infection of *phytophthora* root diseases is particularly favoured by wet and warm soil conditions. Long distance transmission of *phytophthora* root diseases can occur through drainage or irrigation water.

(ii) Wilt diseases

The xylem of the host tree is plugged with enzymatically hydrolyzed constituents of affected cell walls or secondary metabolites of the pathogen. Toxic metabolic products from the pathogen often contribute to the syndrome. The fungi which cause wilt diseases are typically either ascomycetes or deutromycetes. The woody plants are usually invaded through wounds in the stem or roots. Insects are usually the agents of transmission of vascular wilt pathogens (Tattar, 1989). Vascular wilt diseases induced by either *Fusarium* spp. or *Vetricillium albo-atrum* occur in many crop plants throughout the world. *Dalbergia sissoo* suffers heavily to *Fusarium* wilt throughout the Asian continent.

The disease incidence attain its maximum when the plantations are raised on stiff and clayey soils with poor drainage, as on such sites finer roots get killed due to asphyxiation and are subsequently colonized by wilt pathogen. As the disease progresses, the permanent wilting point is reached and turgor is never regained. In mimosa wilt of silk tree *(Bombax ceiba)* caused by *Fusarium oxysporum* fsp. *permiciosum*, the foliage on affected branches becomes yellow and drops off.

Another important wilt pathogen, *Trichosporium vesculosum* causes stem wilt in *Casuarina equisetifolia* in almost all coastal tracts of Orissa, West Bengal, Andhra Pradesh, Tamil Nadu, Kerala, Maharashtra and Gujarat (Bakshi, 1976). The air borne spores of the pathogen invade through the wounds or injuries. The disease of trees may be found in groups as the secondary spread of this pathogen in by root contacts. Dutch elm disease is another vascular wilt caused by *Ceratocystis ulmi*. This disease is one of the most serious tree disease known and has caused extensive losses in Europe, North America and Western Asia. In India, it has been recorded in Kashmir (Gibbs, 1978). The fungi are carried by bark beetles from diseased to healthy plants. Some of the important wilt diseases of trees are listed in Table 2.

Table 2 : Some important wilt diseases of forest trees

Disease	Pathogen/ Casual organism	Symptoms	Host	Management
1	2	3	4	5
Maple wilt	*Verticillium albo-atrum*	Necrosis of one or more limbs with subsequent die back, greenish discoloration in outer rings of infected sapwood	Maple	Proper selection of site
Dutch elm disease tion	*Ceratocystis ulmi*	Yellowing and/or wilting of leaves on a single branch followed by browning and die back known typically as "flag"	Elm	Insect vector control, removing all infected branches, planta- of resistant hybrids
Oak wilt	*C. fugaceurum*	Yellowing and wilting of leaves. It spreads through root graft	Red and black oak	Use of resistant varieties
Verticillum wilt	*Verticillium albo atrum, V. dahliae*	Cup shaped leaves and sudden wilting of one or more branches of tree	Ash, elm, maple, poplar	Planting resistant species, removing of diseased roots from soil and fumigation of soil
Mimosa wilt	*Fusarium oxysporium* fsp. *Permiciosum*	Wilting of leaves in one/ two branches in upper crown. Symptomatic branches contain brown discoloration in streaks or in rigs rust under the bark in outer wood	Mimosa or silk tree	Use of resistance var. and avoiding plating trees in areas having past history
Sissoo wilt	*Fusarium solani*	Yellowing, browning and ultimate drying of leaves from the top of branch, resulting in drying of whole tree	Sissoo	Proper selection of site

B. Stem diseases

Canker is regions of necrosis in the bark of trees that occur in gymnosperms as well as in angiosperms group. Canker diseases are mostly induced by fungi belonging to class ascomycetes, deutromycetes and a few from basidiomycetes. Canker symptoms are grouped as diffuse, perennial and annual.

(i) Diffuse canker types

The casual pathogen has the capacity to grow rapidly through host bark and not allowing the host to produce callus defense wall and then girdles the stem and results in host mortality. Hypoxylan canker (*Hypoxylon mammatum*) is very serious. Diseased bark and wood tissue become black and crack in checkerboard like fashion. Chestnut blight (*Endothia parasitica*), pink disease (*Corticium salmonicolor*) of poplar, teak, eucalypts and semul (Bombax ceiba). Canker of conifers (*Gremmeniiellea abietina*) is other example of diffuse canker.

(ii) Perennial cankers

These cankers characteristically develop more slowly in host tissue than diffuse cankers as the casual fungi grow more slowly in their hosts. Perennial canker diseases are nevertheless, more important. The presence of red perithecia on the margins of the canker is a characteristic identify feature of disease, for example Nectria canker (*Nectria galligena*) of hardwood species. Ceratocystis canker caused by *Ceratocystis fimbriata* is diagnosed based on the target like series of concentric rings. This pathogen has a very wide host range including poplars (Kaushik and Toky, 1992).

(iii) Annual cankers

These are formed by fungi which are normally saprophytic but become temporarily pathogenic. This disease is most prevalent on trees exposed to adverse environmental conditions. The fungus is checked by callus development after first year invasion. Irregularly roughened bark caused by callusing is a typical symptom. *Fusarium solani* infestation on sugar and red maple is a classical example.

C. Bacterial diseases

Literature available on bacterial diseases of forest trees is scanty. Bacteria don't have any structure to allow them to penetrate directly into plant tissues except natural openings in the plants such as stomata, lenticels, openings at the ends of leaf veins and glands in flowers. Wounds are major site of infection by bacteria. Wounding from pruning, grafting, mechanical injury and insects can commonly result in infection of trees with bacteria. The pathogen from infected plant can easily be spread to many nearby healthy plants through routine operations. Some of the bacterial diseases of forest trees are enlisted in Table 3. Bacterial pathogens of leaves and shoots produce water soaking lesions initially, followed by necrosis and blightening symptoms.

Table 3 : Important bacterial diseases of forest trees

Disease	Pathogen	Host
1	2	3
Crown gall	*Agrobacterium tumefaciens*	Woody plants
Canker	*Alpanobacter populi*	Poplars
Bacterial Gall	*Pseudomonas sevastanoi pv. nerii*	Oleander
Shoot blight	*Pseudomonas syringae*	Woody plants
Wet wood decay	*Corynebacterium humiferan*	Poplar
Fire blight	*Erwinia amylovora*	Woody plants
Woody decay	*Erwinia nimipressuralis*	Elm
Bacterial wilt	*Pseudomonas solanacearum*	Teak, Casuarina

The bacterium *Agrobacterium tumefaciens* induces crown galls in a large number of host plants. Uncontrolled proliferation of host tissue (crown galls) is associated with the transfer of tumor inducing plasmid from the bacterium to the host plant cells. This uncontrolled cell multiplication continues in the absence of the bacterium.

D. Virus diseases

Virus replicates within host and cause hindrance in normal metabolic functions like chlorophyll production of cells. Most viruses spread from infected trees to healthy trees by insects.

The symptoms caused by each virus disease of a tree or shrub are unique. A virus disease may cause distinctive symptoms, quite subtle symptoms, or it may produce no symptoms at all. The symptoms of most virus diseases may be lack of chlorophyll formation in normally green organs, dwarfing, distorted growth of a part or the entire plant and necrotic areas of lesions.

The virus diseases of forest trees (Table 4) can be prevented by using virus free tools during grafting, budding and pruning. Use of insecticides to control insect vector population can also be useful in reducing transmission of viruses. Mycoplasma is another group of plant pathogens, the diseases caused by mycoplasma in trees and shrubs have been reviewed by Maramorosch and Raychaudhuri (1981).

Table 4 : Common viral diseases of forest trees

Host	Virus/Disease
1	2
Ash	Ash ring spot (tobacco ring spot virus)
	Ash line pattern (tobacco mosaic virus)
Birch	Birch line pattern (apple mosaic virus)
	Cherry leaf roll virus on birch
Elder	Tobacco ring spot virus
Elm	Elm mosaic virus
	Elm ring spot virus
	Elm scorch virus
	Elm zonate canker virus
Maple	Maple mosaic (maple mottle)
	Peach rosette virus
	Tobacco ring spot virus
Mulberry	Mulberry mosaic virus
	Mulberry ring spot virus
	Yellow net vein virus
Oak	Oak mosaic virus
	Oak ring spot virus
Poplar	Arabis mosaic
	Poplar mosaic
	Potato virus Y
	Tobacco necrotic ring spot virus
Sandal	Leaf curl mosaic

E. Foliar diseases

Foliar diseases are of minor importance in angiosperms. These diseases are usually expressed only in partial defoliation which results in ɪeduced growth and reduced resistance to other stress factors. Leaf spots, leaf blotch, leaf blister, anthracnose, shoot holes, mildews and rusts are some of the leaf diseases induced by different pathogens in different host species (Table 5).

Table 5 : Important foliar diseases and their management

Disease	Pathogen	Host	Management
Anthracenose	*Gnomonia spp.* *Gleosporium spp.*	Ash, maple, oak, Sycamore	1)Avoid highly susceptible species 2)Application of fungicides in plantation 3)Application of balanced fertilizers and irrigation during dry period
Scab	*Venturia inqequalis* *V. pyrina* *V. saliciperda*	Hawthorns, Ornamental apple willows	No control measure but fungicide application in plantation
Powdery mildew to	*Erysiphae sp.* *Microsphaera sp.* *Phyllactinia sp.* *Podosphaera sp.* *Sphaerotheca sp.* *Uncinula sp.*	Sissoo, teak, khair, juglans, salix	Optimum plant density to permit sunlight penetrate
Oak leaf blister	*Tapharina Eaerulescens*	Oak, elm, maple, poplar, willows	
Tar spot	*Rhytisma acerinum* *R. punctatum* *R. salicinum*	Maples and Willows	
Leaf spot	*Cercospora sissoo*	Sissoo	
Vein necrosis and leaf spot	*Cercosporidium bolleanum* *Mycosphaerella bolleanum*	*Ficus auriculata*	Minute, linear, brackish brown necrotic area develops on the veinlets, veins, mid-ribs and petioles of the leaves
Leaf spot/ Anthracenose	*Glomerella cingulata*	*Ficus auriculata*	Irregular brown spot with fant chlorotic halo develop on leaves
Leaf tar spot	*Phyllachora repens*	*Ficus infectoria*	Symptoms on young leaves, leaf lesions become elliptical, at first yellow tith irregular border, amphigenous, the stroma forms a dark crust on towards the center
Powdery mildew	*Oidium caesalpinia-cearum*	*Bauhinia purpurea*	Symptoms on older leaves first as small, circular, powdery patches of different size
Leaf spot/mould	*Terminalia bellerica*	*Pseudocerco-spora terminalia*	Leaf spots are first indistinct, becoming large, irregular blotches, up to 15 mm diameter, pale brown, a slightly darker line margin

Source : Srivastava and Srivastava, 2000.

F. Rust diseases

Rust diseases are caused by a specialized group of fungi belonging to order of basidiomycotina. These are commonly associated with leaves, cones (fruit) and stems of the trees. Most rust fungi that infect trees have different spore stages on two completely unrelated hosts. However, some tree rust pathogens complete their entire life cycle on one host. Those fungi which cause stem galls or canker are generally most devastating. Important rust diseases occurring on forest trees are enlisted in Table 6.

Table 6 : Rust diseases of forest trees

Disease 1	Host 2	Pathogen(s) 3
Albizia leaf rust	*Albizia*	*Ravenelia japonica* *R. indica* *R. sessilis*
Poplar rust	Poplars	*Melampsora ciliata*
Bamboo leaf rust	Bamboo	*Puccinia* spp.
Khair leaf rust	Khair	*Ravenelia tandonii*
Mulberry leaf rust	Mulberry	*Aecidium mori*
Semul leaf rust	Semul	*Chrysomyxa bombacis*
Sissoo leaf rust	Sissoo	*Maravalia achroa*
Spruce needle rust	Spruce	*Chrysomyxa deformans*
Teak leaf rust	Teak	*Olivea tectonae*
Pine needle rust	Blue Pine	*Coleosporium barclayense*
Stem blister rust	Pines	*Cronartium ribicola*

Source : Gupta et al., 1988; Sehgal et al., 1989.

MASS MORTALITY OF SISSOO (*DALBERGIA SISSOO*)

A. Extent of damage

Sissoo is an extremely valuable tree species on farmlands in north India with nearly 10-15% of total forest cover. In the recent years, mortality of sissoo occurred on large scale in northern Gangatic plains including the states of Haryana, Punjab, Delhi, Himachal Pradesh, Uttar Pradesh and Bihar. Heavy mortality in sissoo plantations has also been recorded in the neighbouring countries like Pakistan, Nepal, Bangladesh, Mayanmar and Sri Lanka. The exact mortality could not be recorded, though it was significantly high in some pockets. In north Bihar, an area of 8400 sq. km. stretching from the fertile Gangatic plains to the tarai grasslands bordering Nepal has been severely affected by sissoo mortality. More than 60 per cent mortality was recorded on farmer's field in Bettiah district of Bihar. In Haryana, an alarming 30% mortality in the major plantations of Sirsa, Hisar, Rota and Gurgaon districts have been reported. However, the mortality was as high as 80 to 100 per cent in some pockets having prolonged stagnation of water and or along the canal sides in Sirsa district. In Yamunanagar, Jind, Panipat, Sonepat and parts of Bhiwani, mortality was attributed to hard kankar pan in the sub-soil. However, in such soils the reason was attributed to low moisture in soils, especially in low rainfall years. According to the survey conducted by the forest research Institute, Dehradun, "sissoo" plantations along canal and road sides had high mortality. It might be due to seepage of water from canals or stagnation of water along the roadsides for a longer period caused anaerobic conditions in the root zone resulting into the death of roots.

In Himachal Pradesh, 2000 trees of *sissoo* were recorded as withered away within 3 years on 35 km Jawalamukhi – Kangra highway. In Dehra Gopipura area, according to State Forest Department, 656 dead sissoo trees were noticed in 1996, followed by 350 trees in 1997. In Kangra, Dehradun and Nadaun sub-division along more than 2000 trees have been dried up. Most of the trees were between 30-35 years old. Possible reasons for mass mortality are discussed as below:

(i) Pathological

High incidence of Fusarium wilt disease has been reported in plantations from Uttar Pradesh, Bihar, Haryana and Punjab which were raised on stiff and clayey soils and had poor drainage system. However, the mortality was significantly less in natural forests and plantations raised on suitable sites.

The disease manifests during humid months from July to September, the characteristic symptoms of the disease are yellowing and drooping of leaves in acropetal succession up to the main stem of the tree. Eventually, the entire tree presents a chlorotic appearance. The affected trees die within a few months. Trees of all the ages are usually susceptible to the disease. The pathogen is mostly restricted to roots. The fungal hyphae and jelly like substances plug the conducting vessels resulting in wilt symptoms.

Another serious disease recorded was Ganoderma root rot caused by *Ganoderma lucidum*. It is very common particularly when this tree is raised as a reforested pure crop without removing the stumps and roots of the original crop. In light soils, the pathogen spreads rapidly on roots and the trees are killed in a short period, while in heavy soils, the fungus spreads slowly. The pathogen is a root inhabitant and infects the roots through root contact as well as injured surfaces. It causes white spongy rot in the sapwood. Lateral spread of the disease is through root contact. The infected trees start drying from the top giving stag head appearance in early days. It may take 2-3 year or more to kill a tree depending upon the prevailing environment and vigor of the tree.

The fungus is primarily a wood parasite and is known to attack the tree in plantations. Sometimes, it occurs in association with *Fusarium solani* which enhance the death of the tree. Recently, an association of *Ganoderma lucidum* and root borer (*Celosterna scabrator)* has also been recorded in sandy soils causing mortality in sissoo as well as in *Prosopis cineraria.*

(ii) Hydrological

Water is a critical component of plant health. Most key plant functions such as photosynthesis and transpiration cannot occur unless the necessary water balance is available throughout the plant. However, if there is insufficient water absorbed from soil to replace that lost during transpiration, a water deficiency occurs. If the balance is not restored, wilting of foliage will occur within a few days. On the other hand, too much moisture can also be detrimental and can cause the roots to suffocate and die. The roots of trees are obligatory aerobic and need gaseous exchange of oxygen and carbon dioxide between the air and the soil for them to survive. Each tree species has different capabilities to tolerate various types of soil moisture conditions.

Recently, sudden rise of water table has been recorded in western and central part of Haryana state, might be due to erratic rainfall, seepage from canal network and un-judicious irrigation. Rise of water table has also been observed in Meerut, Aligarh, Bulandshahar, Moradabad and Bareilly areas of Uttar Pradesh. In Piplee block of Bareilly forest division, the water table remained at 2-3 m. (Shukla, 2002). Drying of sissoo in these areas may be attributed to the rise in water table when tap roots get submerged in water causing asphyxiation to the respiratory root hairs.

Dalbergia sissoo develops a strong root system both vertically as well as horizontally and allocates a significant amount of biomass (about 20%) into roots (Toky and Bisht, 1992). When the tap root comes in contact with excess water, the roots die due to asphyxiation because this species has been adapted to semi-arid conditions. Their dead roots serve as infection court, and it becomes more susceptible for *Fusarium solani* and ultimately killing the entire plant.

In Ginnaur, a pure stand of sissoo along the canal bank recorded very high mortality, as 400 trees in a block were found suffering from wilt disease. The area was waterlogged and the site was silty. But in a nearby block having a mixed stand of *Azadirachta indica, Pongamia pinnata, Dalbergia sissoo, Albizia lebbek* and *Eucalyptus tereticornis,* there was mortality of only one tree of sissoo.

The north-western and southern part of the Haryana state witnessed receding water table. The southern part of the state has sandy or sandy loam soils with poor quality underground water. In this part, decline in sissoo was also observed but not in severe form. The trees exhibit yellowing of foliage, reduction in leaf size and interveinal necrosis. The leaf tissues around the major veins remain green but the tissue between the veins turn brown. There is pre-mature leaf fall but most of the trees sprout again in spring. The process goes on 2-3 or more years till the complete drying of whole tree (Fig. 3).

Fig. 1) Ganoderma root rot of Sissoo.

Fig. 3) Drought stressed trees of Sissoo.

Fig. 2) Wilt disease in Sissoo.

(iii) Soil texture

It was observed that the soil texture has a significant correlation with the disease incidence. The disease is very less to moderate in sandy and sandy-loam soils. The stiff and clayey soil cause root mortality of feeding roots as a result of asphyxiation and subsequently colonized by wilt fungus *Fusarium solani.* The fungus grows over a wide range of pH and its growth is optimum at pH 4.6. The pH of the root of sissoo varies from 4.6 to 5.2 which are ideal for the infection of the pathogen. The plant tissues maintain an acidic pH in the range of 4.6 to 5.2 even in highly alkaline soils, indicating that the roots possess a high buffering capacity.

B. Nutrient deficiency

Stress sites exhibit shortage of phosphorus and potassium in foliage. Further, higher pH (7.4 - 8.5) in sites also creates hindrance for uptake of phosphorus and potassium. Potassium is known to be responsible for controlling stomatal transpiration and its deficiency disturbs the water balance of the plant thereby causing stress. The growing period of *Dalbergia sissoo* is about 195 days and this species completes its shoot extension and radial growth within 120 days during the dry period (March to June) of the year (Bisht and Toky, 1993) following a conservative growth strategy. Water stress conditions during dry period cause heavy damage to new growth. From March to May, transpiration rate is high on these stress sites. But as a result of reduced surface area, plants are unable to translocate the requisite amount of water to the transpiring leaves which ultimately causes stomatal closure resulting in the rise of internal leaf temperature and leading to desiccation and finally leaves become scorched. This process is repeated continuously for subsequent years and finally the entire tree dries up within a period of 2-3 years.

It is suggested that monoculture of sissoo should be strictly avoided to prevent the root grafting and root contact in case of infection of *Ganoderma lucidum*. The new plantations must be raised on suitable sites. Uprooting of infected trees should be immediately done. For long-term management, periodic monitoring and assessment of disease is desirable.

DISEASE MANAGEMENT STRATEGIES

Chemical control in forest plantations is neither economical nor feasible. Thus an approach towards Integrated Disease Management (IDM) based on the principles of exclusion, avoidance, eradication and protection of trees from disease attack is best suited to minimize the disease losses. These strategies are given below.

A. Exclusion of pathogens

Exclusion of exotic pathogens can best be achieved by strictly following quarantine practices. The majority of quarantine regulations are imposed by importing countries. Within a country, spread of diseases can be checked only if healthy planting stock is moved from one region to another. For example, *Populus ciliata* in hill areas of Uttar Pradesh and Himachal Pradesh are infected with *Melampsora ciliata* (Singh *et al.*, 1983), and cutting from should not be planted in plain areas unless it is ensured that the material is from diseased free trees. Similarly, planting of exotic pines should be avoided in the natural zone of chir pine to avoid cercoseptoria needle blight in chir pine (Singh, 1982).

In seeds of *Hevea brasiliensis* from Malaya, *Lasidiplodia theobromae* (die-back or collar rot) and *Phomopsis heveae* (Stem die-back) were intercepted at Madras Quarantine Station in India.

Table 7 : Common diseases of important Indian forest trees

Host 1	Name of disease 2	Pathogen 3
Shorea robusta	Polyporus root-rot, Root rot, Stem canker, Leaf spot, Heart rot	*Polyporus shoreae, Ganoderma lucidum Macrophoma shoreae, Cercospora* spp. *Fomes caryophlli*
Tectona grandis Helicobasidium	Bacterial wilt, Armillaria root-rot Peniophora root rot, Helicobasidium root rot, Pink disease, Nectria stem canker Powdery mildew	*Pseudomonas solanacearu Armillaria mellea, Peniophora rhizomorpho-sulphecea, compactum, Corticium salmonicolor, Nectria haematococca, Phyllactinia corylea*
Acacia catechu	*Ganoderma* root-rot, Polyporus gilvus, Root-rot, Powdery mildew	*Ganoderma lucidum, Polyporus gilvus, Microstroma acaciae*
Dalbergia sissoo	Wilt, Root-rot, Leaf spot, Powdery mildew, Leaf rust, Pod fungi	*Fusarium solani, Ganoderma lucidum, Cercospora sissoo,Phyllactina dalbergia, Marvalia achroa, Glomerella cingulata*
Albizia lebbek	Root-rot, Pit canker, Nectris canker, Heart rot, Leaf spot	*Clitocybe tabescens, Fusarium solani, Nectria distissima, Fomes fastuosus, Cercospora albiziae*
Heavea brassiliensis	Red root-rot, White root, Black stripe, Die back, Bird's eye spot	*Ganoderma pseudoferreum,Fomes lignosus, Phytophthora palmivora,Diplodia* spp.*, Drechslera heveae*
Acacia spp.	Seedling blight, Black butt, Heart rot	*Glomerella cingulata, Phytophthora nicotianae, Fomes mastoporus*
Eucalyptus spp.	Seedling blight, *Ganoderma* root-rot, Pink disease	*Cylindrocladium scoparium, Ganoderma lucidum, Corticium salmonicolor*
Alm	Root and butt rot fomes	*Fomes annosus*
Fir and spruce	Root-rot, Heart rot, Needle rust	*Armillaria mellea, Fomes fomentarius. F. robustus,Peridermium thomsoni*
Pinus spp.	Root rots, Swertia felt rust, Heart rot, Caleosporium needle rust	*Fomes annosus,Ganoderma lucidum, Cronartium himalayense,Fomes pini, Coleosporium companufae*

(i) Avoidance

Raising of susceptible tree species may be avoided on sites where there is a potential risk of any disease development in epidemic form. For example, stiff and clayey soils with poor drainage support wilt disease in *Dalbergia sissoo,* and thus, for raising sissoo, soil should be sandy or sandy loam. In high water table areas, sissoo plantations should be raised through root-shoot cuttings to encourage the plants to develop shallow root system so that killing of finer roots by repeated inundation in sub-soil water can be avoided (Singh, 1980). Susceptible species of *Eucalyptus* to pink disease should not be planted in areas having more than 250 cm of rainfall per annum and high humidity prevailing for prolonged periods (Singh *et al.*, 1983). Plantation of susceptible species such as khair, eucalypts and poplars should be avoided on sites where *Ganoderma* root-rot develops in epidemic form on residual roots and stumps. Resistant species such as *Bombax ceiba, Tectona grandis, Ailanthus excelsa,* etc. should be raised on *Ganoderma lucidum* infected soils in the first rotation.

(ii) Eradication

In selection fellings, diseased trees should be marked and removed for instance; heartwood in sal *(Shorea robusta)* continues to decay in affected trees if left in the forest. Control of heteroceous, macrocyclic rusts on pines is best achieved by eradication of alternate, uneconomic host from affected areas.

Collection and burning of the fruiting bodies of *Fomes badius* and *Ganoderma lucidum* helps to check further spread of disease in *Acacia catechu* (Bakshi *et al.*, 1976). The principle of eradication of pathogens is best utilized in forest nurseries. This is done by treating the nursery soil with fumigants (formalin, chloropicrin), fungicides (captan, thiram, blitox) and acidifying chemicals (Sulphur, Sulphuric acid). Nursery soil can also be sterilized by physical agencies like steaming, burning of wood and flooding etc. Likewise, rust and stem pathogens can also be checked by use of therapeutants. Raychaudhuri *et al.* (1972) reported control of spike disease of sandal by treatment of diseased plants with tetracycline antibiotic.

(iii) Protection

Most of the tree diseases establish through wounds or injuries caused to trees during lopping and logging. In agroforestry systems, agricultural operations by tractor are one of the major causes of injuries. Adverse environmental conditions also cause injuries that serve as infection court, which must be avoided to check the disease progress.

The diseased trees can be isolated by digging trenches (1.5 m long, 0.7 m deep and 0.3 m wide) around them to check further spread of the disease through root contacts.

Protection of trees from diseases by biological agents is gaining importance due to dangers of pollution inherent in use of chemicals. This approach is very effective in the management of a number of soil borne diseases. For example, the fungus *Peniophora gigantea* is an effective competitor of *Heterobasidium annosum* which cause root-rot in conifers. *Trichoderma* spp. and *Epicoccum* spp. have also show inhibitory influence on the growth of *Rhizoctonia, Cylindrocladium* and *Fusarium* spp. in forest nurseries. Mycorrhizal roots are more resistant to infection by various root-rot pathogens than non-mycorrhizal roots. *Pisolithus tinctorius* or *Thelephora terrestris* make the plant resistant to *Phytophthora cinnamon* infection. Another mechanism of using microorganisms in biological control is the use of hypovirulent (non-virulent) strain of the same pathogen. Hypovirulent strain of *Endothia parasitica* has been shown to induce hypovirulence in the virulent strains of the same pathogen. In fact, some of the most successful biological control agents are aberrant types of mutants of the pathogen itself. A very interesting approach to the control of crown gall *(Agrobacterium tumefaciens)* is already in practice. Another strain of this pathogen, i.e. *A. radiobacter Pv.* Radiobacter strain-84, produces a bacteriocin molecule (Agrocin 84) which inhibits *a tumefaciens* tumor inducing plasmid (TIP) from transferring to the host plant and causing disease.

CONCLUSION

Agroforestry has picked up as an important landscape practice in green revaluation belt in India, and this was already prevalent in traditional farming belts of arid and semi-arid region of the country. However, in recent past several strategies have been made to bring more area under tree cover through social forestry programmes. The expansion of forest plantations, injudicious use of irrigation, use of chemical fertilizers and pesticides and changes in land use pattern also increased the pest and disease and other problems in various area species. Timely attention to the management of these problems is needed to avoid enormous losses of times. Management of these problems in trees through chemicals is neither feasible nor economical thus integrated management strategies through eco-friendly means i.e. silvicultural, physical and biological methods should be followed.

REFERENCES

Aldhons J R and Mason W L 1994 Forest nursery practices Forestry Commission Bulletin 111 London

Anonymous 1992 Nursery diseases of eucalypts in Kerala and their control KFRI Information Bulletin No 6 India

Bakshi B K 1976 forest pathology : Principles and practice in forestry Controller of Publications Delhi 400 pp

Bakshi B K Reddy M A R and Singh S 1976 Ganoderma root-rot mortality in khair(*Acacia catechu*) in reforested stands *Eur J For Path* **6**: 30-38

Bisht R P and toky O P 1993 Growth pattern and architectural analysis of nine important multipurpose trees in an arid region of India *Can J For Res* **23** : 722-730

Boyce J C 1961 Forest Pathology Third ed 572 pp McGrew Hill Book Co Inc New York

Gibbs J N 1978 Development of Dutch elm disease epidemic in southern England *Ann Appl Biol* **88** : 219-28

Gupta V K and Kaushik J C 1988 Cylindrocladium spp Associated with eucalypts in Himachal Pradesh *Journal of Tree Sci* **7** : 17-19

Gupta V K Sharma R C Garg R C and Kaushik J C 1988 Forest plant diseases and their management Proc of national seminar on tree protection (eds Gupta V K and Sharma N K) held at Solan India Pp 103-121

Hildebrand D M 1985 Soil Solar heating for control of damping-off fungi and weeds at the Colorado state forest service nursery *Tree Planter Notes* **36** : 28-34

Katan J 1981 Solar heating (solarization) of soil for control of soil borne pests *Ann Rev Phytopathol* **19** : 211-236

Kaushik J C 1988 Studies on seedlings disease of eucalypts Ph D thesis submitted to Dr Y S Parmar University of Hort & forestry Nauni Solan (HP) 135 pp

Kaushik J C 1991 Epidemiology and management of forest diseases Paper presented in symposium on "Role of nursery practices in forest renewal" held at HAU Hisar pp 189-194

Kaushik J C and Bangarwa K S 1992 *Ganoderma* root-rot a threat to leguminous trees Nitrogen Fixing tree Res Reports **12** : 1-2

Kaushik J C and Bisla S S 1993 Pathogen management under agroforestry systems In : Agroforestry-Indian perspective (eds Jha L K and Sen Sarma P K) pp 207-272

Kaushik J C and Singh A 1996 Effect of soil texture and moisture on seedling root-rot of *Dalbergia sissoo* and its control In : Proc IUFRO symposium on "Impact of disease and insect pests in tropical forests" pp 209-216

Kaushik J C and Toky O P 1992 *Ceratocystis fimbriata* Ell and Halst A New record on poplar (*Populus deltoids* Marsh) from India *Myforest* **28** : 79-80

Kaushik J C Arya S and Gupta P P 2002 Management of damping-off disease in forest nurseries by soil solrization and biocontrol agents *Forage Research* **28** : 29-31

Kaushik J C Gupta V K and Sharma S 1996 Biological control of damping-off of *Eucalyptus tereticornis* *Annl of Agribios* Res **1** : 155-158

Kaushik J C Gupta V K Saharan G S and Andotra P S 1990 Influence of soil temperature and moisture on the severity of *Cylindrocladium* blight in eucalypts and its control Plant Pathological Research (ed) Srivastava M P and Saharan G S pp 60-63

Lakhanpal T N 1988 Role of mycorrhiza in tree health In : Tree protection (ed Gupta V K and Sharma N K pp 122-129

Maramorosch Karl and Raychaudhuri S P 1981 Mycoplasma Diseases of trees and shrubs Academic Press pp 362

Mukhopadhaya A N Upadhyay J P and Kaur N P 1988 Biological control research on fruit and forest trees In : tree protection(ed) V K Gupta and N K Sharma pp 143-155

Negi S S 1996 An introduction to forest pathology International Book Distributors Dehradun India 236 pp

Nesme X Michel M F and Digat B 1987 Population heterogeneity of *Agrobacterium tumefaciens* in galls of *Populus* from a single nursery *Appl Environ Microbiol* **53** : 655-659

Phillips D H and Burdekin D A 1982 Diseases of forest and ornamental trees MacMillan Pren Ltd London 411 pp

Raychaudhuri S P Chenelu V V Ghos S K Verma A Rao P S Srimathi R A and Nag K C 1972 Chemical control of spike disease of sandal *Curr Sci* **41** : 72-73

Sehgal H S Tivari R K Khan S N and Misra B M 1989 Diseases of forest trees in Himachal Pradesh and their control *Indian Forester* **115** : 228-234

Sharma J K and Mohanan 1984 Management of seedling diseases of eucalypts in nurseries In : J K Sharma C T S Nair S Kedharnath S Kundu (Eds) Proceedings : *Eucalyptus* in India Past Present and Future Pp 377-383 Kerala India

Sharma R C 1989 Forest nursery diseases and their management Tech Bull Dr Y S Parmar Uni Pub Solan India

Shukla A N 2002 Mortality of *Dalbergia sissoo* in India Indian Forester **128(9)** : 1209-1215

Singh S Khan S M and Misra B M 1983 Status of melampsora rusts on poplar in *India Indian Forester* **109** : 745-747

Singh S 1980 Mortality in sissoo (*Dalbergia sissoo Roxb*) plantations Contributed to second forestry conference Jan 16-19 1980 FRI and College Dehradun

Singh S 1982 *Cercoseptoria* neeble blight : A potential threat to indigenous pines in India *Ind For* **108** : 314-318

Sutherland J R Shrimpton G M and Sturroch R N 1989 Diseases and Insects in British Columbia forest seedling nurseries Forestry Canada and D C Ministry of Forests FRDA Report 065 Victoria BC Canada

Tattar T A 1989 Diseases of Shade Trees Academic Press Inc New York

Toky, O P and Bisht, R P 1992 Observations on the rooting patterns of some agro-forestry trees in an arid region of north-western India *Agro-forestry Systems* **18** : 245-263

8

TOXIC EFFECTS OF MERCURY ON PLANTS: CURRENT DEVELOPMENTS AND REMEDIAL MEASURES

SHILPA GOEL, JAHID ALI MALIK AND HARSH NAYYAR

INTRODUCTION

Mercury (Hg) is a rare, shiny, silver white, odorless liquid, which is found naturally in several forms like a gas or as an inorganic compound with oxygen, sulfur or chlorine. Mercury may also combine with carbon to form organic mercury compounds. The most common organic compound is methyl mercury, which is also released into the atmosphere by small organisms like bacteria (Meagher and Rugh, 1997). Mercury toxicity has become a major problem over several parts of the world, especially in the developing countries as a result of pollution on a large scale (Patra and Sharma, 2000). Considerable amount of mercury may be added to agricultural land with sludge, fertilizers, lime and manure that has the potential to pollute the plants and their harvestable parts. Seed coat dressing with mercury, which is used as an antifungal treatment, is one of the major causes of mercury toxicity in plants (Frank *et al.*, 1976a, 1976b). This mercury from plants moves to higher trophic levels. Methyl mercury builds up in the tissues of fish and other organisms (Risher, 2004). Its levels in tissues of animals increase as we go up the food chain. Hg is a very hazardous substance because it is a charged species that has great affinity for the thiol group on cysteine residues of proteins and other important biological molecules (Wagner-Dobbler, 2003). Natural levels of mercury in a plant must not exceed $0.1 \mu g\ g^{-1}$ (Markert, 1994). Being a ubiquitous element it is present in trace amounts throughout lithosphere, hydrosphere, atmosphere and biosphere as well as in rocks (Goldwater, 1971). Permissible levels of soil mercury are 17 mg/kg on calcareous and 6 mg/kg on acid soil (Wang *et al.*, 1982). The critical value of mercury in soil is 0.5 ppm and normal soils contain 20-150 ppb of mercury (Patra and Sharma, 2000). We present here the toxic effects of Hg on plants, the adaptive mechanisms acquired by the plants for detoxification and remedial measures to minimize Hg toxicity.

Sources of mercury

Mercury is a naturally occurring element which is generally present in little quantities in rocks, soils and the oceans. As mercury is an element, it is not converted into a less dangerous form. Instead, it persists in the environment and can cycle for decades between land, air and water until eventually it is covered by sediments. Although there are some natural sources of mercury, humans are responsible for nearly two-third of the mercury currently circulating in our environment.

Alkali and metal processing, incineration of coal, medical and other waste, and mining of gold and mercury contribute greatly to mercury concentrations in some areas, but atmospheric deposition is the dominant source of mercury over most of the landscape. Once in the atmosphere, mercury is widely disseminated and can circulate for years, accounting for its wide-spread distribution. Natural sources of atmospheric mercury include volcanoes, geologic deposits of mercury, and volatilization from the ocean. Although all rocks, sediments, water, and soils naturally contain small but varying amounts of mercury, scientists have found some local mineral occurrences and thermal springs that are naturally high in mercury. The normal soils may have 20-150 ppb mercury; it can reach to 80% at sites close to the known deposits (WHO, 1976). Usually, Hg is strongly bound to the organic components of soil, consequently its mobility gets restricted. The contamination of ground water occurs when mercury leaches from the municipal landfills (USEPA, 1984).

Mercury has been extracted and used in the past. Industrial development has led to significant increase in Hg content in the environment. The incineration of waste, combustion of coal, smelting of base metals are the major source of Hg release. Its use in production of fluorescent lights and dental amalgam has additionally elevated its levels in water, soil, sediments and inhabitants of these environments.

Natural emissions of Hg account for two-third of the inputs, man-made releases form one-third (NRC, 1978), man-made emissions include mining ore deposits, burning coal and medical waste, gold mining (Veiga and Hinton, 2002), bleaching of paper and textiles, bleach production, use of agricultural pesticides and manufacturing plants. Natural emissions take place from soil erosion and volcanic activities. These sources release inorganic mercury (metallic mercury and inorganic mercury compounds) into the air. In water and soil, it enters from natural deposits, disposal of wastes, and the use of mercury-containing fungicides (Patra and Sharma, 2000). It is also circulated in earth's atmosphere by natural degassing of earth's crust and oceans (Patra and Sharma, 2000). The most important sources of contaminating agricultural soil have been the use of organic mercurials as a seed coat dressing to prevent fungal diseases in seeds, of mercury sulphate as root dip (Frank *et al.*, 1976a). Earthworms are considered to be indicators of mercury-polluted soils (Hinton, 2002). Another major source of mercury waste comes from mercury cells in the chlor-alkali industry and from sludge deposits in landfills or discharged into waterways in industrialized countries (Hynninen and Lodenius, 1986). Metallic mercury is used to produce chlorine gas, caustic soda, in thermometers, amalgams (dental fillings) and batteries. Mercury salts are used in skin-lightening creams, antiseptic creams and ointments. Mercury is used in scientific and electrical equipment, in the electrolytic production of chlorine and sodium hydroxide; and as a catalyst in polyurethane foam production.

Health Hazards

Mercury enters humans through contaminated salt water or freshwater fish (Tchounwox *et al.*, 2003). Inhalation of mercury vapor is the most important route of uptake of elemental mercury. Elemental mercury vaporizes at room temperature, and is readily absorbed by the lungs. It is then rapidly absorbed and distributed by the blood; some of which is deposited in the brain where it is retained for a long time, and the rest is transported to the liver and kidneys where it is excreted through bile and urine (Clarkson, 2000). Inhalation of low level of mercury causes "pins and needles" feelings, numbness usually in the hands, feet and sometimes around the mouth, lack of coordination of movements, such as writing, impairment of speech, hearing, walking, muscle weakness, skin rashes, mood swings, memory loss and mental disturbance. Inhalation of high concentrations of mercury causes severe respiratory irritation, digestive disturbances, marked kidney damage, acute pulmonary edema and interstitial pneumonitis, which may be fatal. In non-fatal cases coughing may persist. Contact dermatitis is a disease reported to occur due to continuous exposure to mercury. Repeated or prolonged exposure to mercury vapor is highly toxic to the central nervous system. The nervous system is very sensitive to all forms of mercury; some forms are more harmful than others because of how the body handles them. Exposures to high levels of metallic, inorganic, or organic mercury can permanently damage the brain, kidneys, and developing fetus. Irritability, shyness, tremors, changes in vision or hearing, and memory problems are the common effects of exposure to mercury. Exposure to methyl mercury is more damaging in case of young children than adults, because it interferes with normal development of brain. Also, mercury vapor may cause 'Kawasaki' disease, which seems to be immunologically mediated and is similar to Pink disease. In 1950's and 1960's, mercury poisoning due to exposure to industrial effluents caused 'Minamata' disease in Japan (Wagner Dobler, 2003). Mercury intoxication also causes reproductive effects like complications in pregnancy and childbirth, low birth weight, menstrual disturbances, spontaneous abortion, premature deliveries, and malformations and can even cause cancer. However, the World Health Organization (WHO) reported no evidence that inorganic mercury is carcinogenic. Regarding mutagenicity, organic mercury induced sex-linked recessive lethals in *Drosophila melanogaster* (Mathews and Al-Doori,1976).

Transport

Mercury ions are thought to enter the cells by means of the same uptake processes that move essential micronutrient metal ions. Since mercury is a class-B metal, it preferentially binds with sulfur and nitrogen rich ligands in the soils for uptake processes (Patra and Sharma, 2000).

Metals have different affinities for different elements, thus influencing complex formation and binding to different macromolecules. According to Pearson (1968), Hg^{2+} is a soft acceptor characterized by high polarizabilty, high electro negativity and smaller charge density. Hg^{2+} forms strong complexes with P, S, Cl, Br and I and it also forms organic metal complexes. Mercury is found as R-Hg^{+} and R_2-Hg where R can be methyl, ethyl or benzyl group (Clarkson, 1972). Mercury can also get converted into Hg^0 vapor (Patra and Sharma, 2000). Mercury is taken up more easily by plants when supplied as organic mercury compounds rather than inorganic forms (Ribeyere and Boudou, 1994) and in soils; Hg is retained as slightly mobile organo-complexes (Kabata-Pendias and Pendias, 1992).

It is observed that levels of Hg increase in plant tissues with increasing external levels and also the proportion retained in the roots is constant as compared to shoots (Beauford *et al.*, 1977). In soils, mercury remains retained in solid phase by adsorption onto sulphides, clay particles and organic matter. These adsorbed Hg forms are insoluble and immobile so its availability to plants is less but it becomes available to the plants by reacting with various ions like potassium iodide (KI), etc. and others, which are present in the oxygenated, soils e.g. Cl^-, OH^-. After reacting with these ions, Hg becomes mobile and is available to plants in the form of $HgCl_2$, Hg, (OH), Cl, Hg, $(OH)_2$ (Moreno *et al.*, 2005). Mercury also has a strong affinity for thiol groups (Sulphide and Bisulphide complexes.). Humic substances (HS), which are present in soils contain a very high proportion of S-functional groups and is generally present in soils in abundance. The soluble fraction of HS contains Fulvic and Humic Acid (HA), which are known as mercury ligands. Therefore, Hg-HA complexes are mobile in soils and HA enhances the uptake of mercury (Hinton, 2002).

The uptake of mercury in a plant is dependent upon the plant species and the interaction of plant roots with other components, which affect nutrient uptake and transport to root surface (Marschner, 1991). The factors affecting metal uptake include high pH value, abundant lime and accumulated salt in soil (Xuexun and Linhai, 1991). In a specific plant, characters like root morphology, root surface area, root length density, etc. can affect the uptake *i.e.* diffusion rate *e.g.* plant species with finer roots (like *Allium* sp.) and greater surface area show high mercury concentration in their roots whereas it is opposite for the plants with larger root size. It is observed that plants have a tendency to accumulate mercury in their roots, which have a higher tolerance than shoots for high internal metal concentrations (Gothberg *et al.*, 2004); thus roots act as a barrier to Hg uptake. Major portion of mercury is found bound to the cell walls of the plants (Beauford *et al.*, 1977; Wierzbicka, 1998), only 0.5-0.6 % of mercury is transported to upper harvestable parts. When mercury is in the form of a gas it enters the leaves through stomata, while as ions it mainly enters through leaf cuticle (Martin and Juniper, 1970). Gaseous Hg (0) is taken up via stomata (Cavallini *et al.*, 1999) and uptake seems to be higher in C_3 plants rather than C_4 plants and it is justified on the basis that it is due to internal resistance to Hg vapor binding (Browne and Fang, 1983).

John (1972) studied the mercury uptake from soil by various plant species in a growth chamber experiment involving 3 rates of $HgCl_2$ applied to soil. Mercury concentrations, on dry weight basis, in various parts of leaf lettuce, spinach, broccoli, cauliflower, peas, oats, radishes, and carrots were higher in root portions than the above-ground sections. The content of roots ranged from 0.387 µg Hg/g^{-1} of lettuce roots to 2.447 µg Hg/g^{-1} of cauliflower roots when 20 µg Hg/g^{-1} of soil were applied. At the same treatment rate and among the edible plant parts, pea seeds and oat grain accumulated the least mercury while spinach leaves and radish tubers contained the highest concentrations, averaging 0.695 and 0.663 µg Hg/g^{-1} of plant material. In a recent study, the uptake and accumulation of mercury was found more in *Oryza sativa* than *Brassica, Zea mays* and *Triticum* (Patra and Sharma, 2000).

Transportation of mercury to upper harvestable parts of a plant is a very selective process. It is found that in presence of $(NH_4)_2S_2O_3$, the plants have high levels of Hg accumulated in their shoots (Moreno *et al.*, 2005). So it may be possible that $Hg\text{-}S_2O_3$ complexes are selected for shoot transport over other dissolved Hg species present in substrate. It is found that concentration of mercury in the above ground parts is also dependent upon the foliar uptake of volatilized Hg from soil. So we can say that atmospheric or air borne Hg has a very strong effect on mercury accumulation in plants above ground parts.

Aquatic plants are bioaccumulators of Hg (Samecka and Kempers, 1996). In aquatic plants, the uptake is dependent upon plant species, growth rate changes and on the metal ions being absorbed (Patra and Sharma, 2000).

Hence, it is evident that Hg availability and uptake are very much interrelated with each other and is mainly controlled by plant species, sulphur containing ligands, substrate Hg concentration and humic acid content (Moreno *et al.*, 2005).

Toxic Effects

The toxicity of Hg may vary depending upon the a) type or form available in the soil b) soil properties affecting Hg availability to the plants c) the plant species d) the accumulation of Hg in the roots thus preventing its transfer to the shoots e) the metabolic status of the leaves and f) other environmental factors. The effects occurring at various organizational levels are manifested on inhibition of growth depending upon the stage of development and its sensitivity. The effects at various levels are described as follows:

Growth: The growth and seedling development are adversely affected by mercury in many crops. The effects depend upon the form of Hg applied, duration and its concentration, the type of the plant species and its stage sensitivity. Accordingly, the effects may show large variation among different plant species examined for the effects of Hg.

Seed injury is caused by high concentrations of mercury leading to abnormal germination and characteristic hypertrophy of the roots and coleoptiles (Prudy, 1956). The germination was reported to be inhibited in the seeds growing in the presence of Hg as noted in case of *Phaseolus vulgaris* and *Brassica campestris* but at lower concentration the germination was slightly stimulated, the reasons were not reported (Gupta, 1991). The inhibition in germination may possibly occur due to direct injury to the embryo itself where the meristematic tips may get damaged. In case of maize, the primary roots were reduced in growth and also lost their gravitropic response, again emphasizing the effects of Hg on growing tips (Pilet and Versel, 1981). The root and shoot length as well as dry matter accumulation was inhibited with increase in concentration in *Abelmoschus esculentus* (Ganesan and Manoharam, 1983). In case of *Mentha sylvestris* and *Lactuca sativa*, the radicle emergence was decreased by 50%. A reduction in seedling growth was observed in Hg-treated seedlings of *Lactuca sativa* and *Pisum sativum* (Siegel *et al.*, 1984). In rice, the germination and seedling growth was inhibited; root growth was affected largely than shoot growth (Varshney, 1991). In wheat, Hg application to germinating seeds inhibited the germination, root differentiation, growth of the roots and shoots and reduced the dry weight of the seedlings (Setia *et al.*, 1994). In *Vigna radiata*, Hg inhibited the germination and hypocotyls length (Varshney, 1990). In Hg treated seeds of *Stylosanthes guianensis* seeds, the length of root cells showed decrease, the possible reason was suggested to be increased ethylene production (Delachiave *et al.*, 1990). In contrast, a stimulation of germination was observed in *Pinus halepensis* with successive application of mercury and chloride ions (Thalouran, 1976). Mercury inhibits root elongation (Goldbold, 1991). It has been observed that lower period of exposure or exposure to lower concentrations of mercury showed significant inhibition of root growth, however shoot growth showed stimulation but higher exposure period or exposure to higher concentrations brought inhibition of both root and shoot growth (Padhi *et al.*, 2004). At concentrations inhibiting root growth by 50%, root apices were thinner and longer compared to control, and root hairs developed close to root apex in case of maize cv. Diamant. Moreover mercury exerts non-selective root growth inhibition due to general toxicity rather than selective inhibition of any particular process or processes (Ivanov *et al.*, 2003). Mercury toxicity results in inhibition of hypocotyl elongation growth in case of *Phaseolus vulgaris* seedlings (Mor *et al.*, 2002).

The main effect of Hg is abnormal wall organization in cells (Sawidis and Reiss, 1995). Phenyl-Hg compounds hasten plant senescence (Patra and Sharma, 2000). Various aquaporins in plant cell membrane and tonoplast are mercury sensitive (Daniells *et al.*, 1996). Pollen tube growth and pollen germination also decreased in presence of mercury

(Tuna *et al.*, 2002). These two parameters have been used as indicators of pollution with heavy metals (Gur and Topdemir, 2005). Mercury also affects embryo and endosperm development (Patra and Sharma, 2000).

Physiological effects: The rate of respiration (Nag *et al.*, 1989), photosynthesis, transpiration and activity of nitrate reductase were reported to be affected by mercury treatment in spruce seedlings (*Picea abies* (L.) Karst.) (Godbold and Huttermann, 1998). The CO_2 uptake decreased as the concentration of mercury in the soil increased. Decreased CO_2 uptake was explained by lower level of chlorophyll and closed stomata (Godbold and Huttermann, 1998). Hg also increased cuticular resistance to gas diffusion in leaves of plants growing in dry soils by closing their stomata (Moreschet, 1975). Various cell metabolic activities like dissoluble protein concentration, chlorophyll content, chlorophyll a/b ratio in *Elodea nuttalii* (Wu *et al.*, 2005), soluble protein content, carbohydrate content, dry weight and viability of cells are greatly inhibited due to toxicity of mercury.

The uptake of macro and micro elements in plants are inhibited by mercury toxicity (Osvalde, 2001). It also inhibited the mobilization of nitrogen and phosphorus in plants (Varshney,1991). It reduced K, Mg, Mn in roots and accumulated Fe, Ca in root tips (Godbold, 1991). A possible effect of mercury is suggested to promote ethylene production, which would inhibit root growth (Delachiave *et al.*, 1990). Root damage along with decrease in water content and nutrient levels is also observed in the plants treated with mercury (Godbold and Huttermann, 1998). Also, free proline accumulation was observed in roots (Osvalde, 2001). It was also observed that accumulation of proline helps the plant to survive under stress situation (Roy and Bera, 2002), but it also has a direct effect on productivity of plants (Osvalde, 2001). $HgCl_2$ treatment inhibited the uptake of ^{14}C-3-o-methyl-D-glucose and ^{14}C-L-alanine by binding irreversibly to essential –SH groups and can change the cell permeability (Golle and Luttge, 1983).

Effects on Photosynthesis: The toxic effects of Hg occur both on light and dark reactions and the severity of the effects varies depending upon the concentration and type of plant species used (Krupa and Baszynski, 1995; Kupper *et al.*, 1996). In developing seeds of wheat, the chlorophyll content is decreased by Hg (Zhang *et al.*, 1989).

The Hill activity and chlorophyll content were inhibited by Hg in leaves of *Cuscuta reflexa* (Jana and Bhattacharjee, 1988). These authors also reported inhibition in biosynthesis of porphyrin and permeation of membranes. The chlorophyll content was decreased by mercury in *Phaseolus* seedlings (Parmar *et al.*, 2002). Jain & Gadre, (2004) reported inhibitions in chl biosynthesis in the presence of Hg in etiolated cotyledons of maize. By displacing Mg^{2+} in the chlorophyll molecule, mercury prevents photosynthetic light harvesting in these chlorophyll molecules, resulting in the breakdown of photosynthesis (Kupper *et al.*, 1996; Patra and Sharma, 2000; Neculita *et al.*, 2005). The oxygen evolution and chlorophyll fluorescence were inhibited by Hg in the leaves of Barley (Bernier *et al.*, 1992).

In Barley, Hg caused inhibition in electron transport, evolution of oxygen and quenching of chlorophyll fluorescence in photosystem II. It also influences the electron transport on donor side of PSII, the ATP: NADPH ratio was reported to be changed, leading to changes in accumulation of starch and anthocyanins (Gebhard *et al.*, 1990). It was proposed that Hg disturbs the chloride binding or functioning (Bernier *et al.*, 1993). Both Hill reaction and photophosphorylation as well as the dark reduction of cytochrome C_{554} were found to be inhibited by Hg in case of *Chlamydomonas reinhardi*. Mercury also reportedly inhibited the ferredoxin or ferrodoxin –NADP oxidoreductase (Hiyama *et al.*, 1970). Hg was reported to bind with light harvesting chlorophyll (LHC) proteins of the chloroplast even at very low concentrations in leaves of Spinach (Ahmed and Tajmir-Riahi, 1993; Nahar and Tajmir-Riahi 1995). It was found to bind with protein carbonyl group, nitrogen group or both. Hg also bound with the sulfur donor sites for Hg-protein complexation (Ahmed and Tajmir-Riahi, 1993). The activity of enzyme, ribulose-5-phosphate kinase and 3-phosphoglyceric acid kinase was inhibited in *Euglena gracilis* (De Fillippis and Ziegler, 1993).

Murthy and Mohanty (1993) reported inhibition of electron transport at multiple sites in cyanobacterium *Synechococcus* 6301. Inhibition by $HgCl_2$ is dependent on the concentration of Hg ions. The inhibition of whole chain electron transport activity occurs at low concentration of Hg (6 µl). This inhibition occurs mostly due to interaction of Hg with plastocyanin. At an elevated concentration (24 µl), Hg induces inhibition chiefly in photosystem II catalyzed electron trasnport. At this concentration, it also alters both the absorption and emission characteristics of phycocyanin. The photosystem I catalyzed electron transport was inhibited by 50% only at high concentrations (36 µl) of $HgCl_2$. However, electron transport catalyzed by photosystems I and II from reduced duroquinone to methylviologen which involves intersystem electron transport is extremely sensitive to Hg (low concentration of 6–9 µl) like that of whole chain assay indicating that the observed inhibition in whole chain electron transport at low concentrations is mostly contributed by the damage involving other intersystem electron transport carrier(s) like plastocyanin.

Nitrogen metabolism

The uptake of NO_3^- and NH_4^+ was observed to be inhibited by Hg in *Chlorella vulgaris* (Rai *et al.*, 1991). Hg (0.8 µg/ml) inhibited the activity of nitrogenase, ^{14}C uptake and pigment diversity in algal communities (Singh and Rai, 1991). For yeast proteinase yscJ, Hg negatively influenced the activity (Wagner and Wolf, 1992). In *Nostoc calcicola*, Hg (0.05-230 µM) inhibited the activity of glutamine synthetase, nitrate reductase and nitrogenase (Singh and Singh, 1992).

In leaf segments of *Echinocola colona*, 10-5000 µM of Hg proved to be inhibitory for nitrate reductase activity (Rani *et al.*, 1988). In Maize, Hg (0.1-10 mM) inhibited the nitrate reductase (Pandey and Srivastava, 1993). Similar findings were reported by Vyas and Puranik (1993) in case of bean seedlings. On the other hand, Hg (30 μM) was reported to stimulate the activity of glutathione-S-transferase in shoots and roots of maize (Aioub *et al.*, 1993). In contrast, Komives *et al.*, (1994) found 30 mM $HgCl_2$ to be inhibitory for glutathione transferase in roots and stimulatory for the same enzyme in shoots. Vijay *et al.*, (1988) observed inhibition in the activity of amino transferase in leaves of seedling of barley with 50 ppm Hg. Varshney *et al.*, (1990) reported decrease in mobilization and protease activity in *Phaseolus aureus* seeds in the presence of mercuric acetate ($5x10^{-5}$ M).

Oxidative stress

Mercury like other toxic metals is reported to cause oxidative stress that has been quantified by malondialdehyde content and carbonyl accumulation, by-products of lipid peroxidation and protein oxidation. It was observed that the activity of cytosolic ascorbate peroxidase (APX) and superoxide dismutase (SOD) increased (Rellan-alvarez *et al.*, 2006). The glutathione (GSH) is consumed i.e. its content is decreased in roots under stress conditions. It is observed that in *Zea mays* with higher concentration of mercury, the glutathione transferase activity increased in shoots and slightly decreased in roots (Komives *et al.*, 1994). There is also induction of HSP70 under stress conditions. Induction of HSP70 suggests that anti-oxidative response and the mechanism in which GSH is consumed are insufficient for protecting against the harmful affects of severe mercury and other heavy metal exposure (Collin-Hansen *et al.*, 2005).

The inhibitory effects of Hg on chlorophyll biosynthesis in excised greening leaf segments of maize were reported to be prevented in the presence of exogenously applied reduced glutathione (GSH) and other thiols. Moreover, the chlorophyll content was increased with GSH (Jain and Puranik, 1993). In the plants of *Bacopa monnieri* (Brahmi), there was increase in cysteine, total –SH, reduced GSH by mercury as part of the defense mechanism to exogenous application of Hg.

Hernández *et al.*, (2005) while working on Alfalfa (*Medicago sativa*) plantlets exposed them to Hg to study the kinetics of diverse stress indices. In the hydroponic system, plantlets were grown in 30 μM Hg for 7 days. Oxidative stress took place and increased over time without any linear response. To improve the sensitivity of the stress assays used, a micro-assay system, in which seedlings were exposed for 24 h, was developed. Phytotoxicity of metals, quantified as growth inhibition, was observed well before there was any change in the non-protein thiol tissue concentration. Oxidative stress and cell death increased after exposure for 6-24 h and levels of GSH/hGSH decreased acutely. This diminution was suggested to be the result of direct interaction of GSH/hGSH with Hg.

Cargnelutti *et al.*, (2006) studied the effects of exogenous mercury ($HgCl_2$) on time-dependent changes in the activities of antioxidant enzymes (catalase and ascorbate peroxidase), lipid peroxidation, chlorophyll content and protein oxidation in cucumber seedlings (*Cucumis sativus* L.) Cucumber seedlings were exposed to from 0 to 500 μM of $HgCl_2$ during 10 and 15 days. Hg was readily absorbed by growing seedlings, and its content was greater in the roots than the in shoot. Hg-treated seedlings showed elevated levels of lipid peroxides with a concomitant increase in protein oxidation levels, and decreased chlorophyll content when exposed to between 250 and 500 μM of $HgCl_2$. At 10 days, catalase activity increased in seedlings at a moderately toxic level of Hg, whereas at the higher concentration (500 μM), there was a marked inhibition.

Zhou *et al.*, (2006) reported that the Alfalfa (*Medicago sativa*) roots treated with mercuric ions in a concentration- and time-dependent manner caused the peroxidation of membrane lipids and loss of plasma membrane integrity in Hg-treated roots occurred in the meristem and the elongation zone. The histochemical observations were supported by the quantitative determinations of thiobarbituric acid reactive substances (TBARS). However, under the mercuric ions stress, the alfalfa plants showed no significant alteration of hydrogen peroxide in roots. Analysis of lipoxygenase activity showed that there were two isoforms in the root of alfalfa plants, but they showed quite different patterns under the Hg exposure. The activities of SOD and peroxidase (POD) increased in roots after Hg treatment of roots. Activity of APX was stimulated at relatively high concentration of Hg (40 μM), and after prolonged Hg exposure (20 μM, 24 h). In contrast, glutathione reductase activity was depressed at higher concentrations of Hg (10–20 μM). Treatments of seedlings with 10–40 μM Hg decreased the ascorbate and glutathione amounts but increased total non-protein thiols. The above results indicated that Hg exerted its toxic effect on the root growth of alfalfa by induction of oxidative stress.

Biochemical effects: The toxicity of mercury strongly depends on its redox state (Clarkson, 1992). The most toxic form is the highly reactive inorganic mercury ion Hg^{2+}, which can bind to the amino acid cysteine of proteins thus inactivating various enzymes like enzymes of glutathione metabolism (Wagner-Dobler, 2003). Hg^{2+} can also bind to N-atoms in nucleic acids thus resulting in defective metabolism of plants (Wagner-Dobler, 2003). The difference is also observed in lipid solubility, non-specific inhibition of a variety of intracellular enzymes and several specific thiol-containing respiratory enzymes *in vitro*. The causal mechanism of mercury toxicity also includes the changes in the permeability of cell membranes, affinity of Hg to react with phosphate groups and active groups of ADP or ATP and the replacement of many major cations by Hg cations, reactions of –SH groups with cations (Kabata-Pendias and Pendias, 1984). Due to formation of bonds with sulphur in –SH groups or disulphide bridges, any function in which proteins are involved may get disturbed, it may also lead to precipitation of proteins at high concentration (Clarkson, 1972). Mercury compounds can bind to RNA also and can inhibit electron transport activity, oxygen evolution (Lee *et al.*,1992). Increase in cytoplasmic and wall bound peroxidase activities were studied in Mercury treated plants as compared to control (Parmar *et al.*, 2002). Decreased activity of catalase and protease and increase in the content of free amino acids and peroxidase activity occur as a result of mercury poisoning (De *et al.*, 1985).

Genetic alterations: Various genetic defects caused due to mercury include induction of C-mitosis through disturbance of spindle activity; increase in frequency of chromosomal aberrations, reduced mitotic index (Fiskejso, 1969). C-mitosis and chromosomal doubling were observed after exposure to mercuric fungicide of the germinating grains of *Secale cereale, Triticum durum, T. persicum, T. polonicum, T. aegilopoides* and to some extent of *Pisum sativum* but no such observations are found in *Crepis capillaries*. However, inactivation of spindle fiber and polyploidy were recorded in *Crepis capillaries* and *Avena sativa* (Patra and Sharma, 2000). In plant material, organic and inorganic mercury compounds inactivated the spindle-fiber mechanism at cell division causing aneuploidy and/or polyploidy and C-tumours (Fiskejso,1969). Even disorganization of microtubules did occur on exposure of cell suspensions to mercury in *Ducus carota* (Czuba, 1987). Micronuclei were increased in roots of *Eichhornia cressipes* in presence of mercury stress (Panda *et al.*, 1988).

Cellular and ultracellular effects: Mercury induced the secretion of mucilage from epidermal cells, suppressed root differentiation, and decreased the number of vascular bundles and cell size in the stems (Setia *et al.*, 1994). The plants also induced wall thickenings in epidermal and hypodermal cells as a result of Hg toxicity (Setia *et al.*, 1994). Ultrastructural studies conducted under TEM and SEM has shown toxic effects on cell nucleus like nuclear membranes were split and nucleolous disappeared. Membrane bound organelles like chloroplast and mitochondria were injured and thylakoid membranes and mitochondrial cristae were damaged (Wang *et al.*, 2004). Agglutination of chromatin, disintegration of membrane systems and disappearance of ribosomes after the leaf cells were poisoned with mercury were also observed (Wei *et al.*, 2001)

DETOXIFICATION STRATEGIES

Plants expressing a modified bacterial mercury reductase merA, are highly resistant to Hg(II) toxicity as a result of enzymatically catalyzed electrochemical reduction of Hg(II)to much less toxic and volatile Hg(0) (Heaton *et al.*, 2005).

Microorganisms are known to mediate four types of enzymatic transformations of mercury (Barkay, 2000).

a. Reduction of Hg^{2+} to Hg^0.

b. Breakdown of organic mercuric compounds (including $MeHg^+$), resulting in formation of Hg^0.

c. Methylation of Hg^{2+}.

d. Oxidation of Hg^0 to Hg^{2+}.

Microbial mercury resistance operon *i.e.* mer enzymes perform the first two steps, and the result is production of Hg^0. The detoxification process of Hg is based on unique properties of this metal.

The electrochemical potential of Hg^{2+}/Hg^0 at pH 7, which is +430 V means that living cells are able to reduce Hg^{2+} to elemental form, non-toxic to bacteria. So metallic mercury can leave cells either by passive diffusion and is then either volatilized into the air or precipitate due to low solubility.

Plasmids of both gram +ve and gram –ve bacteria can confer resistance to organic and inorganic compounds by converting mercury into non-toxic form by reaction with NADPH with help of enzyme mercuric reductase (mer A) (Hamlett *et al.*, 1992). The best-studied determinants of metal resistance are those of mercuric ion resistance in gram –ve bacteria. The DNA sequence of mer operons (Tn 501 and Tn 21) shows that there are several open reading frames in addition to mer A, which can encode mercuric reductase. Two more genes have been identified for their regulatory functions- mer R and mer D. The cellular machinery encoded by microbial *mer* operon also provides specific uptake proteins (mer T, mer P, mer C) that transport Hg^{2+} into the cytoplasm and so prevent damage to the cell. The mer C and mer T gene products are cytoplasmic and 50% of mer P gene products are located in periplasm. The mer P product initially binds Hg (II) and transfers it to mer T, which in turn transfers Hg (II) to cytoplasm or directly to reductase thus, helping in transport and detoxification of Hg (II). The role of mer C gene is however not very clear, where mer A gene is functional, intracellular Hg (II) concentrations are greatly reduced due to volatilization of Hg (II) to Hg (0) by decreasing toxicity and increasing concentration gradient. It has also been suggested that mer P gene acts as a sponge (Summers, 1986), scavenging Hg (II) from periplasmic space and mer T might have somewhat same function at the cytoplasm level.

Certain higher plants have metal protection mechanisms including sequestration of metal by producing organic compounds, sequestration in certain cell compartments, metal ion efflux and organic ligands exudation. Plants can also form phytochelatins (PC) in response to mercury toxicity (Oritz, 1994). Free PC peptides reactivate metal poisoned nitrate reductase *in vitro* (Kneer and Zenk, 1992). Metals in form of gases Hg (0) are also able to disappear through the stomata, if they are opened, into the air (Siegel *et al.*, 1974; Kozuchowski and Johnson, 1978). This has been found in genetically modified plants, which were able to transform Hg^{2+} into Hg^0 (Rugh *et al.*, 2000). Mercury release in some studies on unmodified plants probably depends on bacterial activity and not on Hg transpiration by the plants (Prasad, 2004).

BIOREMEDIATION

Mercury remediation by conventional methods is very expensive, and thus many areas polluted by mercury are presently left unclaimed. Plant tolerance to mercury is quite low and if it is accumulated in plant tissues, it can be toxic to wildlife. It is found that controlling the electrochemical state of Hg or increasing the thiol pool of a plant as a sink for Hg will increase the resistance to and hyperaccumulation of Hg in plants. A plant based phytoremediation approach uses plant roots to extract, the vascular system to transport, and the leaves as a sink to concentrate the metal above ground for harvest (Che *et al.*, 2003). Several hyper-accumulator plant species are being utilized in the bioremediation of heavy metals from soils and water environments. These plant species can absorb heavy metals like Hg, translocate them through tonoplast and accumulate in vacuoles, thereby, protecting cell metabolism from toxic effects of heavy metals (Maiti *et al.*, 2004). Hg (II), formed by protonolysis of methyl mercury, is thiol reactive, and favors trapping of Hg (II) in above ground tissues as thiol complexes whereas electrochemical reduction of Hg (II) to Hg (0) helps in volatilization of mercury from plants. The plants can be engineered to take up and tolerate several times the level of Hg, which can otherwise be toxic for normal plants. These modified plants can detoxify methyl mercury or ionic mercury by storing it in below or above ground portions of a plant or even by volatilizing Hg (0) as part of transpiration process, thus keeping toxic Hg out of food chain (Che *et al.*, 2003).

Transgenic mercury remediation strategies have worked well with various cultivated and wild plants including *Canola*, rice, yellow and cottonwood and brassica (Che *et al.*, 2003). The transgenic wetland plants can be regenerated by the insertion of *mer* genes in plants such as cordgrass (*Spartina Schreber* spp.), cat-tail (*Typha* L. spp.) and bulrush (*Scirpus* L. spp.) as well as the water-tolerant trees poplar (*Populus* L. spp.) and willow (*Salix* L. spp.) (Memon *et al.*, 2001). It is also suggested that there are thousands of such genes in bacteria, plants and animals, which can be transferred to other plants thus helping in phytoremediation (Memon *et al.*, 2001). Two bacterial genes of *mer* operon – *mer* A and *mer* B are used for phytoremediation and for mercury transportation in plants (Bizily *et al.*, 1999). *Mer* B gene encodes an enzyme organomercurylyase that is responsible for protonolysis of methyl mercury into ionic Hg (II) and methane (an organic compound). It is also found that plants grown in methyl mercury can accumulate more Hg than ones grown in inorganic mercury. *Mer* A gene encodes mercuric reductase which confer resistance to ionic mercury [Hg (II)] in *Arabidopsis thaliana* and tobacco by reducing it to Hg (0) using NADPH as the electron donor. Hg (0) is less toxic and is released into environment through transpiration. Release of Hg (0) into air by plants is a major concern when using phytoremediation in practice. However, no such emissions are found in case of shoots of willow, garden-pea, spring wheat, sugar beet, oil-seed rape and white clover. Thus, it can be concluded that atmospheric burden of Hg via phytoremediation is not increased and *mer* A genes can be used to produce transgenics. Plants expressing these two genes are resistant to extremely high concentrations of mercury in the environment. These two genes *mer A* and *mer B* are now used to create mercury volatilizing plants in other species. Some transgenics of tobacco and yellow poplar have already developed (Rugh *et al.*, 1998; Bizily *et al.*, 2000). These transgenic plants cleave mercury ions and finally evolve less toxic elemental mercury at concentration well below given standards, thus providing us with cost effective, permanent *in situ* solution.

Mercury is a unique metal as there is no known biological function of this metal (Neiss, 1999) and it has low vapour pressure in case of elemental mercury, which is a liquid at room temperature and is highly toxic so, steps should be taken for its complete detoxification by reduction of Hg^{2+} to Hg^0 and also by diffusional loss of Hg^0 from plant cells (Wagner-Dobler, 2003). Volatilization as a detoxification mechanism was first discovered in bacterial cells as a resistance mechanism for mercury contaminated soils (Summers and Silver, 1972).

Another method of phytoremediation involves removal of mercury from water with the help of hyperaccumulator plants such as *Typha* sp. (Krishnan *et al.*, 1988), *Eichhornia crassipes* and *Ixora coccinea* (Tikku *et al.*, 1990). A number of transporters such as members of ZIP family have been identified as possible method in metal exclusion, uptake and detoxification (Maiti *et al.*, 2004) which can be exploited to facilitate bioremediation of mercury from the environment.

CONCLUSION

Mercury is a very toxic metal whose concentration in the environment has increased in the recent past, leading to environmental pollution. It causes various disorders in plants and animals. It is observed that it is more toxic in its ionic form and also in form of organic compounds. Various plants, both terrestrial and aquatic are known which can accumulate mercury in their leaves. The physiological and biochemical mechanisms of plants may provide alternative strategies to adopt measures for purification of soil contaminated with heavy metals. *Mer* operon of bacterial genes is also known to confer resistance and help in hyperaccumulation of mercury. Hence, it is advisable to develop modified or transgenic plants that have bacterial genes in them to combat mercury toxicity.

REFERENCES

Ahmed A Tajmir-Riahi HA 1993 Interaction of toxic metal ions Cd^{2+} Hg^{2+} and Pb^{2+} with light harvesting proteins of chloroplast thylakoid membranes: An FTIR spectroscopic study *J Inorg Biochem* **50**: 235-243

Aioub AAA Gullner G Komives T 1993 Peroxidation of lipid in corn plants exposed to heavy metals and herbicide stress Pp 57-60 *in* G Mozsik I Emerit J Feher B Matkovics A Vincze(eds) Congress of oxygen free radicals and scavengers in the biological and medical sciences *Akademiai Kiado Budapest*

Barkay T 2000 Mercury cycle *Encyclopedia Microbiol* **3** 171-181

Beauford W Barber J Barringer AR 1977 Uptake and distribution of mercury within higher plants *Physiol Plant* **39** 261-267

Bernier Carpentier MR and Murata N 1992 Reversal of mercury inhibition in photosystem II by chloride Pp 2 97-100 *in* Photosynth Res Proceedings of IXth International Congress on Photosynthesis Nagoya Japan August 30- September 4 1992

Bernier M Popovic R Carpentier R 1993 Mercury inhibition at the donor side of the photosystem II is reversed by chloride *FEBS Letters* **321** 19-23

Bizily SP Rugh CL and Meagher RB 2000 Phytodetoxification of hazardous organomercuricals by genetically engineered plants *Nat Biotechnol* **18** 213-217

Bizily SP Rugh CL Summers AO and Meagher RB 1999 Phytoremediation of methylmercury pollution mer B expression in *Arabidopsis thaliana* confers resistance to organomercurials *Proc Natl Acad Sci* USA **96** 6808-6813

Browne CL Fang SL 1983 Differential uptake of mercury vapor by graminaceous C_3 and C_4 plants *Plant Physiol* **72** 1040-1042

Cargnelutti D Tabaldi LA Spanevello RM Jucoski GDO Battisti V Redin M Linares CEB Dressler VL de Moraes Flores ÉM Nicoloso FT Morsch VM Schetinger MRC 2006 Mercury toxicity induces oxidative stress in growing cucumber seedlings *Chemosphere* **65** 999-1006

Cavallini A Natali L Durante M and Maserti B 1999 Mercury uptake Distribution and DNA affinity in Durum wheat (*Triticum Durum* Desf) plants *Sci Tot Environ* **243/244** 119-127

Che DS Meagher RB Heaton ACP Lima A Rugh CL Merkle SA 2003 Expression of mercuric ion reductase in Eastern cottonwood (*Populus deltoides*) confers mercuric ion reduction and resistance *Plant Biotech* **1** 311-319

Clarkson TW 1972 The pharmacology of mercury compounds *Ann Rev Pharmacol* **12** 375-406

Clarkson TW 1992 Mercury major issues in environmental health *Environ Health Perspects* **100** 31-38

Clarkson TW 1997 Mercury poisoning Pp 189-200 *in* Brown S S (ed) Clinical chemistry and chemical toxicology of metals Elsevier/ North Holland Biomedical Press Amsterdam

Clarkson TW 2000 The three important faces of mercury *Environ Health perspect* **110** 11-23

Collin-Hansen C Hansen RA Steinnes E 2005 Molecular defense systems are expressed in the king Bolete (*Boletus edulis*) growing near metal smelters *Mycolo* **97(5)** 973-983

Czuba M 1987 Methylmercury toxicity in plant cultures Modification of resistance and demethylation by light and / or 2 4-dichlorophenoxyacetic acid *Ecotoxicol Environ Saf* **14** 64-72

Daniells MJ Chaumont F Mirkov TE Chrispeels MJ 1996 Pl Cell **8** 587-599

De Fillippis LF Zeigler H 1993 Effect of sublethal concentration of zinc cadmium and mercury on the photosynthetic carbon reduction cycle of Euglena *J Pl Physiol* **142** 167-172

De AK Sen AK Modakand DP and Jana S 1985 Studies of toxic effects of Hg(II) on *Pistia Stratiotes Water Air Soil Pollut* **24** 351-360

Delachiave MEA Pedras JF Rodrigues JD Moraes JAPV Rodrigues SD Boaro CSF and Moraes MA 1990 Seed germination of *Stylosanthes guianensis* III Fungicide concentration in relation to germination and cell size at different water potentials *Cientifica Jaboticabal* **18** 27-34

Fiskejso G 1969 Some results from *Allium* test with organic mercury halogenides *Hereditas* **62** 314-322

Frank R Ishida K Suda P 1976 a Metals in agricultural soil of Ontario *Canad J Soil Sci* **56** 181-196

Frank R Braun HE Ishida K Suda P 1976 b Persistent organic and inorganic pesticide residue in orchard soils and vineyards of southern Ontario *Canad J Soil Sci* **56** 463-484

Ganeshan R and Manoharan 1983 Effect of cadmium and mercury on germination growth and dry matter production of *Abelmoschus esculentus Geobios* **10** 9-12

Gebhard AAC Chetverikov Gerasimenko VV and Tsoglin LN 1990 Effect of mercury ions on duckweedplants (*Spirodela polyrhiza*) Soviet Pl Physiol **37** 262-267

Godbold DL Huttermann A 1998 Inhibition of photosynthesis and transpiration in relation to mercury induced root damage in spruce seedlings *Physiol Plant* **74 (2)** 270

Godbold DL 1991 Mercury induced root damage in spruce seedlings *Water Air Soil Pollut* **56** 823-831

Goldwater LJ 1971 Mercury in the environment *Sci Amer* **224** 15-21

Golle B and Luttge U 1983 Inhibition of glucose and Aminoacid carriers of *Lemma gibba* by pretreatment with mercuric chloride *Physiol Plant* **57(1)** 62

Gothberg A Greger M Holm K Bengtsson BE 2004 Influence of nutrient levels on uptake and effects of mercury cadmium and lead in water spinach *J Enval Qual* **33(4)** 1247-1255

Gupta R 1991 Toxic effects of mercury on seed germination of bean and mustard *Comp Physiol Ecol* **16** 43-45

Gur N and Topdemir A 2005 Effect of heavy metals (Cd, Cu, Pb and Hg) on pollen germination and tube growth of quince (*Cydonia oblonga* M) and plum (*Prunus domestica*) *Fressenius Env Bull* **14(1)** 36-39

Hamlett NV Landale ED Davis BH and Summers AO 1992 Roles of *Tn 21 mer T mer P* and *mer* C gene products in mercury resistance and binding *J Bacteriol* **46** 6377-6385

Heaton ACP Rugh CL Wang NJ and Meagher RB 2005 Physiological responses of transgenic mer A-tobacco (*Nicotiana tobacum*) to foliar and root mercury exposure *Water Air and Soil Pollut* **161 (1/4)** 137-155

Hernández LE Villasante CO Rellán-Álvarez R Del Campo FF and Carpena-Ruiz RO 2005 Cellular damage induced by cadmium and mercury in *Medicago sativa J Exp Bot* **56(418)** 2239-2251

Hinton JJ 2002 Earthworms as bioindicators of mercury pollution in an artisanal gold mining community Cahoeira do Piria Brazil Master Thesis University of British Columbia Canada

Hiyama T Nishimura M Chance B 1970 Energy and electron transfer systems of *Chlamydomonas reinhardi Pl Physiol* **46** 163-168

Hynninen V Lodenius M 1986 Mercury pollution near an industrial source in southwest Finland *Bull Environ Contamination Toxicol* **36** 294-298

Ivanov VB Bystrova EI and Seregin IV 2003 Comparative impacts of heavy metals on root growth as related to their specificity and selectivity *Russian J Plant Physiol* **50(3)** 398-406

Jain M Gadre R 2004 Inhibition of chlorophyll biosynthesis by mercury in excised etiolated maize leaf segments during greening effect of 2-oxoglutarate *Indian J Exp Biol* **42(4)** 419-23

Jain M and Puranik RM 1993 Protective effect of reduced glutathione on inhibition of chlorophyll biosynthesis by mercury in excised greening maize leaf segments *Indian J Exp Biol* **31** 708-710

Jana S Bhattacharjee A 1988 Effects of combinations of heavy metal pollutants on *Cuscuta reflexa Water Air Soil Pollut* **42** 303-310

John MK 1972 Mercury uptake from soil by various plant species *Bull Env Cont Toxicol* **8(2)** 77-80

Kabata-Pendias A Pendias H 1984 Trace elements in soils and plants CRC press Boca Raton FL

Kabata-Pendias A Pendias H 1992 Trace elements in soils and plants 2nd edn CRC Press Boca Raton FL

Kneer R Zenk MH 1992 Phytochelatins protect plant enzymes from heavy metal poisoning *Phytochemistry* **31** 2663-2667

Komives T Aioub AAA Gullner G and Brunold C 1994 Effects of mercuric chloride on the glutathione transferase enzyme activity in corn (*Zea mays* L) plants *Cereal Res Communications* **22** 99-104

Kozuchowski J Johnson DL 1978 Gaseous emission of mercury from an aquatic vascular plant *Nature* **274** 468-469

Krishnan SS Cancilla A Jervis RE 1988 Waste water treatment for heavy metal toxins using plant and hair as adsorbents *Sci Tot Environ* **68** 267-273

Krupa Z Baszynski T 1995 Some aspects of heavy metal toxicity towards photosynthetic apparatus Direct and indirect effects on light and dark reactions *Acta Physiol Pl* **17** 177-190

Kupper H Kupper F Spiller M 1996 Environmental relevance of heavy metal substituted chlorophylls using the example of water plants *J Exp Bot* **47** 259-266

Lee C Chang HH Ha SB Moon BY Lee CB and Murata N 1992 Mercury induced light dependent alterations of chlorophyll Fluorscence kinetics in Barley leaf slices **4** 623-626 in Photosynth Res Proceedings of the IXth International Congress on Photosynthesis Nagoya Japan August 30-September 4 1992

Maiti RK Hernanadez-Pinero JL Gonzalez-Oreja JA Lopez-Santiago D 2004 Plant based bioremediation and mechanisms of heavy metal tolerance of plants a review *Proc Ind Natl Sci Acad* **70(1)** 1-12

Markert B 1994 Plants as biomonitors- potential advantages and problems In Adriano DC Chen ZS Yang SS eds Biogeochemistry of trace elements Science and Technology letters Northwood NY 601-613

Marschner H 1991 Mineral nutrition in higher plants New York USA Academic Press 412-424

Martin TJ Juniper EB 1970 The cuticles of plants Arnold Edinburgh

Mathews C Al-Doori Z 1976 The mutagenic effects of the mercury fungicide Ceresan M in *Drosophila melanogaster Mutat Res* **40** 31

Meagher RB Rugh CL 1997 Phytoremediation of heavy metal pollution Ionic and methylmercury Pp 305-321 in Organization for Economic Cooperation and Development (OCED) document Biotechnology for water use and conservation The Mexico '96 workshop OCED Paris

Memon AR Aktoprakligi D Ozdemir A Vertii A 2001 Heavy metal accumulation and detoxification mechanism in plants *Turk J Bot* **25** 111-121

Mor IR Gokani SJ Chanda SV 2002 Effect of mercuric toxicity on hypocotyl elongation and cell wall loosening in *Phaseolus* seedlings *J Plant Nutr* **25(4)** 843-860

Moreno FN Anderson CWN Stewart RB Robinson BH Ghomshei M and Meech JA 2005 Induced Plant uptake and transport of mercury in the presence of sulphur- containing ligands and humic acid *New Phytol* **166** 445-454

Moreshet S 1975 Effects of phenyl mercuric acetate on stomatal and cuticular resistance to transpiration *New Phytol* **75** 47-52

Nag P Paul AK and Mukherjee S 1989 The effects of heavy metals zinc and mercury on the growth and biochemical constituents of mung bean(*Vigna radiata*) seedlings *Bot Bull Acad Sin* **30** 241-250

Nahar S and Tajmir-Riahi HA 1995 Do metal ions alter the protein secondary structure of a light harvesting complex of thylakoid membranes? *J Inorg Biochem* **58** 223-234

Neculita CM Gerald JZ Louis D 2005 Mercury speciation in highly contaminated soils from Chlor-Alkali plants using chemical extractions *J Environ* **34** 255-262

Neiss DH 1999 Microbial heavy-metal resistance *Appl Microbiol Biotechnol* **51** 730-750

NRC (US National Research Council panel on mercury) 1978 Assessment of mercury in the environment A report National Academy of Sciences Washington DC

Oritz DF 1994 An MDR like vacuolar transport protein implicated in heavy metal tolerance *Fedrip Database* National Technical Information Service(NTIS)

Osvalde A 2001 Plant responses to mercury and arsenic stress possible adaptive role of increased proline accumulation *Proc Lat Acad Sci* **55(2/3)** 100-106

Padhi NP Padhi P Panigrahy AK 2004 Assessment of deposition of MEMC (seed dressing fungicide) in exposed rice grain through study of seedling growth parameters *J Env Biol* **25(3)** 279-285

Panda KK Das BL Lenka M and Panda BB 1988 Water Hyacinth (*Eichhornia crassipes*) to biomonitor genotoxicity of low levels of mercury in aquatic environment *Mutation Res* **206** 275-279

Pandey M and Srivastava HS 1993 Inhibition of Nitrate reductase activity and nitrate accumulation by mercury in maize leaf segments *Indian J Environ Health* **35** 110-114

Parmar NG Vithalani SD and Chanda SV 2002 Alteration in growth and peroxidase activity by heavy metals in *Phaseolus* seedlings *Acta Physiol Plant* **24(1)** 89-95

Patra M Sharma A 2000 Mercury toxicity in plants *The Bot Rev* **66(3)** 379-422

Pearson R 1968 Hard and soft acids and bases HSAB part I Fundamental Principles *J Chem Educ* **45** 581-587

Pilet PE Versel JM 1981 Effect of mercury on growth and gravireaction of maize roots (*Zea mays*) Z *PFlanphysiol* **104** 193-198

Prasad MNV 2004 Metal uptake transport and accumulation Pp 11 *in* Heavy Metal Stress in Plants From Biomolecules to Ecosystem Springer-Verlag

Purdy LH 1956 The nature and extent of mercury injury and associated anatomical responses in wheat seedlings *Rev Appl Mycol* **36** 391

Rai LC Singh AK Mallick N 1991 Studies on photosynthesis the associated electron transport system and some physiological variables of *Chlorella vulgaris* under heavy metal stress *J Pl Physiol* **137** 419-424

Rani SMV Muthuchelian K Pilwal K 1988 Effect of short term exposure of heavy metals on *in vivo* nitrate reductase (EC 1 6 6 1) activity of *Echinochloa colona* Link *Indian J Range Management* **9** 119-123

Rellan-Alvarez R Ortega-Villasante C Alvarez-Fernanadez A Campo FFD Hernanadez LE 2006 Stress responses of *Zea Mays* to cadmium and mercury *Plant soil* **279(1/2)** 41-50

Ribeyre F Boudou A 1994 Experimental study of inorganic and methylmercury bioaccumulation by four species of fresh water rooted macrophytes from water and sediment contaminated sources *Ecotoxicol Environ Saf* **28** 270-286

Risher JF 2004 Too much of good thing(Fish) methylmercury case study *J Environ Health* 67

Roy SB and Bera AK 2002 Individual and combined effect of mercury and manganese on phenol and proline content in leaf and stem of mung bean seedlings *J Env Biol* **23(4)** 433-435

Rugh CL Bizily SP Meagher RB 2000 Phytoreduction of environmental mercury pollution In Ensley BD Raskin I(eds) Phytoremediation of toxic metals using plants to clean the environment Wiley NY 151-169

Samecka CA and Kempers AJ 1996 Bioaccumulation of heavy metals by aquatic macrophytes around Wroclaw Poland *Ecotoxicol Environ Saf* **35** 242-247

Sawidis T Reiss HD 1995 Effects of heavy metal on pollen tube growth and ultrastructure *Protoplasmotologia* **185** 113-122

Seigel AM Puerner NJ Speitel TW 1974 Release of volatile mercury from vascular plants *Physiol Plant* **32** 174-176

Seigel BZ Lasconia M Yeager E Seigel SM 1984 The phytotoxicity of mercury vapour *Water Air Soil Pollut* **23** 15-24

Setia RC Rajna B Bala R 1994 Anatomical changes in root and stem of wheat (*Triticum aestivum* L) in response to different heavy metals *Phytomorphology* **44** 95-104

Singh AK Rai LC 1991 Cr and Hg toxicity assessed *in situ* using the structural and functional characteristics of algal communities *Environ Toxicol Water Qual* **6** 97-107

Singh CB Singh SP 1992 Assessment of Hg^{2+} toxicity to N_2 fixing cyanobacterium in long and short term experiments *Biometals* **5** 149-156

Summers AO and Silver S 1972 Mercury resistance in a plasmid bearing strain of *Escherichia coli J Bacteriol* **112** 1228-1236

Summers AO 1986 Organization expression and evolution of genes for mercury resistance *Annu Rev Microbiol* **40** 607-634

Tchnouwou PB Ayensu WK Ninashuili N Sutton D 2003 Environmental exposure to mercury and its toxopathological implications for public health *Environ Toxicol* **18(3)** 149-175

Thalouarn P 1976 Stimulated seed germination of *Pinus halepensis* Mill by successive applications of mercury and chloride ions Search for a mechanism *Comptes Rendus Hebdomadaires des Seances de l' Acadmie des Sciences Serie D* **282** 1857-1860

Tikku S Sindhu RS and Bhartiya RK 1990 The removal of lead (II) cadmium (II) and mercury (II) from water by *Ixora coccinea* Linn plant substrate *Indian J For* **13** 112-114

Tuna AL Burun B Yokas I Coban E 2002 The effect of heavy metals on pollen germination and pollen tube length in the tobacco plant *Turk J Biol* **26(2)** 109-113

USEPA (US Environment Protection Agency) 1984 Health effects assessment of mercury Environmental Criteria and assessment office Cincinnati OH

Varshney AK 1990 Differential response of *Phaseolus aureus* cultivars to mercury pretreatment on mobilization of N and protease activity during germination and seedling growth *Geobios* **17** 82-87

Varshney AK 1991 Effect of mercuric acetate on mobilization of nitrogen and phosphorus during seedling growth of *Oryza sativa* cvs IR36 and Saket 4 *J Indian Bot Soc* **70** 99-102

Veiga MM Hinton JJ 2002 Abandoned artisanal gold mines in the Brazilian Amazon a legacy of mercury pollution *Natur Res For* **26** 15-26

Vijay SS Kant Bohra SP 1988 Ecophysiological studies on Indian desert plants II Physiological response to heavy metals in *Sorghum vulgare* Pers Trans *Indian Soc Desert Technol* 59-67

Vyas J Puranik RM 1993 Inhibition of nitrate reductase activity by mercury in bean leaf segments *Indian J Plant Physiol* **36** 57-59

Wagner JC Wolf DH 1992 Purification and Characterization of proteinase yscJ a new yeast peptidase *Eur J Biochem* **203** 571-575

Wagner-Dobler I 2003 Pilot plant for bioremediation of mercury-containing industrial wastewater *Appl Microbiol Biotechnol* **62** 124-133

Wang HK Liv BY Feng GZ Wang QM 1982 A study of the permissible level of mercury due to utilization of sludge on land *Acta Agric univ Pekinensis* **8** 69-75

Wang YQ Zheng JG Chen WL Chen LY 2004 Ultrastructure studies of mesophyll cells damaged from Hg^{2+} and Cd^{2+} pollution in rice *J Fujian Agri For Univ Nat Sci Ed* **33(4)** 409-413

Wei G Xin SG Hui HC He DK Li QX 2001 The toxicity damage effect of Hg^{2+} and Cd^{2+} pollution on Myriophyllum verticillatum Linn *China Env Sci* **21(4)** 371-375

WHO 1976 Mercury Pp 1-31 *in* Environmental Health criteria1 World health organization Geneva

Wierzbicka M 1998 Lead in the apoplast of *Allium cepa* L root tips- ultra structural studies *Plant Sci* **133** 105-119

Wu ZB Min MJ Qiang Z Feng H and Ping CS 2005 Toxic effects of Hg^{2+} and Cd^{2+} and their combined stress on *Elodea nuttallii China Environ Sci* **25(3)** 262-266

Xuexun Z Linhai G 1991 Studies on heavy metals pollution of soil and plants in Tianjin waste water irrigated areas *Ekologia* **10** 87-97

Zhang Z Lu Q and Fang F 1989 Effect of mercury on growth and physiological function of wheat seedlings *Huanjing Kexue* **10** 10-13

Zhou ZS Huang SQ Guo K Mehta S K Zhang PC Yang ZM 2006 Metabolic adaptations to mercury-induced oxidative stress in roots of *Medicago sativa* L *J Inorg Biochem* **101** 1-9

9

MECHANISM OF ARSENIC TOLERANCE AND HYPERACCUMULATION

MADHUMITA DASH AND C P MALIK

INTRODUCTION

Arsenic is a very common element that ranks 20th in abundance in the earth's crust, 14th in the seawater, and 12th in the human body (Woolson, 1975). Arsenic compounds are naturally present in the environment, and have been also used for many years in medicine, agriculture and forestry. Weathering of rocks converts arsenic sulphides to arsenic trioxide, which enters the arsenic cycle as dust or by dissolution in rain, rivers, or groundwater. So, groundwater contamination by arsenic is a serious threat to mankind all over the world. It can also enter food chain causing widespread distribution throughout the plant and animal kingdoms. However, fish, fruits, and vegetables primarily contain organic arsenic, less than 10% of the arsenic in these foods exists in the inorganic form, although the arsenic content of many foods (i.e. milk and dairy products, beef and pork, poultry, and cereals) is mainly inorganic, typically 65-75%. A few recent studies report 85-95% inorganic arsenic in rice and vegetables, which suggest more studies for standardization (Meharg, 2004; Norra *et al.*, 2005) The major anthropogenic sources of environmental pollution with arsenic are the burning of coal and industrial metal smelting, and more recently the semiconductor industry.

Humans are exposed to this toxic arsenic primarily from air, food, and water. Thousands and thousands of people are suffering from the toxic effects of arsenicals in many countries all over the world due to natural groundwater contamination as well as industrial effluent and drainage problems. Chronic effects of prolonged low level exposure have recently showed up. Arsenic, being a normal component of human body is transported by the blood to different organs in the body, mainly in the form of MMA after ingestion. It causes a variety of adverse health effects to humans after acute and chronic exposures such as dermal changes (pigmentation, hyperkeratoses, and ulceration), respiratory, pulmonary, cardiovascular, gastrointestinal, hematological, hepatic, renal, neurological, developmental, reproductive, immunologic, genotoxic, mutagenetic, and carcinogenic effects (WHO, 2001).

More than 80 per cent of the world's population depends on rice as a staple food but production is dropping in the rice paddies of Bangladesh parts of India and South and East Asia due to toxic levels of arsenic in the topsoil. Om Parkash of the University of Massachusetts Amherst leads a research team that uses genetic engineering to produce rice plants that block the uptake of arsenic which could increase production of this valuable crop and provide safer food supplies for millions

"By increasing the activity of certain genes we can create strains of rice that are highly resistant to arsenic and other toxic metals " says Parkash a professor of plant soil and insect sciences. "Rice plants modified in this way accumulate several-fold less arsenic in their above-ground tissues and produce six to seven times more biomass making the rice safer to eat and more productive." This could help alleviate the current world-wide rice shortage

Deep tube wells installed to provide drinking water in Bangladesh and other countries are producing water with naturally occurring levels of arsenic that greatly exceed safe limits in drinking water. Groundwater is then being used to irrigate rice paddies and this irrigation is causing a buildup of arsenic in topsoils that is toxic to the rice plants reducing the amount of rice that can be produced in a given area

According to Parkash arsenic builds up in all parts of the plant including the rice grains used for food creating health problems in hundreds of thousands of people including several forms of cancer. Arsenic is also present in the rice straw used as animal fodder causing arsenic to enter the food chain in dairy products and meat and affecting the health of animals

"Already on the Indian subcontinent particularly in Bangladesh and West Bengal there are more than 300 000 people who have developed cancer from arsenic poisoning by drinking contaminated water and eating contaminated food " says Parkash. "The World Health Organization has dubbed this one of the major environmental disasters in human history."

"Basically the companies will use our gene constructs in new or existing rice lines producing hybrid rice that will go through the cultivation and seed production stage" says Parkash "Then the new strains of rice will be commercialized and brought to market."

Inorganic *As* species are also phytotoxic and affects plants. But plants do vary in their sensitivity or resistance to arsenic (Meharg, 1994). Arsenate resistance has been identified in a range of plant species whereas some plant species such as *Pteris vittata* – a Chinese brake fern – have been identified as hyperaccumulator of *As* (Ma *et al.*, 2001).

Arsenic resistant and hyperaccumulating plant species offer viable opportunity for phytoremediation of *As* from the environment and the development of safe crops that can be grown in the presence of *As* contamination. Thus it is necessary to understand the mechanism of *As* tolerance and accumulation in plants in order to explore its further potential.

In a recent study growth performances and total proline and protein content of Bamboo (*Dendrocalamus strictus*) clonal cell cultures and plantlets; subjected to different concentration of NaCl and arsenic stressed *in vitro*, were analyzed. Bamboo cell lines were cultured in two types of media (MS_3 and MSD_3). One set (MSD_3) was grown in MS media supplemented with 3% (w/v) sucrose, 3 mg L^{-1} 2,4-D and 0.8% (w/v) agar with pH of 5.75. Another set (MS_3) was subjected to MS medium supplemented with 3% (w/v) sucrose, 3 mg L^{-1} 2,4-D, 1000 mg L^{-1} inositol, 40 mg L^{-1} Thiamine HCl and 0.8% (w/v) agar with pH of 5.75. Both the set of cell lines were grown in control and NaCl supplemented media. The aim was to assess the role of inositol and thiamine in NaCl tolerance. Similarly the plantlets were grown *in vitro* in MS media supplemented with 3% (w/v) sucrose and 0.8% (w/v) agar with pH of 5.75 in order to study NaCl tolerance both at tissue as well as whole plant level. NaCl stress leads to reduction in growth and total protein content and accumulation of proline. The cell line in MS_3 media with and without NaCl (control) showed higher proline content than MSD_3 media with and without NaCl. This suggests role of thiamine and inositol in proline formation. The total proteins of these cell cultures were analyzed by gradient SDS-PAGE to assess changes in protein profile out of salinity induced stress. Quantitative change in the protein profiles were observed in the samples treated with different concentration of salts with respect to control. Synthesis of specific salt induced proteins of 31.3 kD and 17.2 kD were observed in the NaCl treated samples.

Growth performance along with total protein content of control and arsenate treated Bamboo (*Dendrocalamus strictus*) plantlets and clonal gametophytes of Chinese brake fern (*Pteris vittata*) were analyzed. Both the plant system showed reduction in growth and total protein content with arsenic stress. *P. vittata* gametophytes demonstrated increased sporophyte formation under arsenic treatment. The one dimensional gradient

SDS PAGE analysis of their protein samples showed quantitative and qualitative changes. Synthesis of specific arsenic stress induced polypeptide bands were observed in treated samples. Three bands of 98.29 kD, 25.02 kD and 19.201 kD were found in treated *P. vittata* samples only. Similarly, specific protein bands of 55.22 kD, 50.01 kD and 25.22 kD were observed only in treated *D. strictus* samples only.

The total proteins of both plant systems were also analyzed by 16% 6M Tricine-Urea SDS PAGE for better resolution of low molecular weight polypeptides like phytochelatin. This analysis confirmed absence of any low molecular weight polypeptide suggesting limited role of phytochelatins in the above two plant systems.

Further study would assess the functional attributes of the specific polypeptides revealing their exact role in inducing salinity and arsenate tolerance in the plant systems.

Mechanism for Arsenic uptake and detoxification

Inorganic arsenic species are generally highly toxic to plants. Arsenate acts as a phosphate analogue and is transported across the plasma membrane via phosphate cotransport systems (Ullrich-Eberius *et al.*, 1989). Arsenic is present both as arsenite (AsIII) and arsenate (AsV) in the environment, with arsenate being more prevalent than arsenite in soils (Oremland and Stolz, 2003). The solubility of As(III) oxide in water is fairly low, but high in either acid or alkali. In water, arsenic is usually in the form of arsenate or arsenite. As(III) is systemically more poisonous than the As(V), and As(V) is reduced to the As(III) form without exerting any toxic effects. Organic arsenicals also exert their toxic effects *in vivo* in animals by first metabolizing to the trivalent arsenoxide form. Overall, it seems that in most organisms arsenic is more toxic in the trivalent form, as arsenite, than in the pentavalent form, as arsenate.

In the prokaryote *E. coli* there are two phosphate transporters, Pit and Pst (Rosenberg *et al.*, 1977). Both catalyze arsenate uptake, but the Pit system appears to be the predominant system for arsenate (Willsky and Malamy, 1980). Similarly, in the eukaryote *Saccharomyces cerevisiae* several phosphate transporters participate in arsenate uptake (Yompakdee *et al.*, 1996). *E. coli* and yeast have also been shown to uptake arsenate through an aquaglyceroporin (Sanders *et al.*, 1997; Wysocki *et al.*, 2001).

Plants face arsenical compounds mainly in the form of the anions arsenite and arsenate. Arsenate is an analogue of phosphate and thus interferes with essential cellular processes, for example replacing phosphate in ATP to form unstable ADP-As, and leads to the disruption of energy flows in cells (Meharg, 1994). Arsenite is also highly toxic to plants as it reacts with sulfhydryl groups (–SH) of enzymes and tissue proteins, leading to inhibition of cellular function and death (Ullrich-Eberius *et al.*, 1989). Mechanisms of arsenite uptake by plant roots are not clearly understood. The main route of AsV uptake in plants is through the phosphate transporters as a phosphate analogue (Asher and Reay, 1979), whereas AsIII is transported in the neutral $As(OH)_3$ form through aquaglyceroporins. Uptake into the ericoid mycorrhizal fungus *Hymenoscyphus ericae* is passive (Sharples *et al.*, 2000). It has recently been shown that arsenite uptake in rice is active, described by two additive hyperbolic functions over high affinity (low concentration) and low affinity (high concentration) ranges (Abedin *et al.*, 2002). Arsenite was as effectively transported across the plasmalemma as arsenate, although via a different uptake mechanism.

Once inside the cytoplasm arsenate competes with phosphate, for example replacing phosphate in ATP to form unstable ADP-As, and leads to the disruption of energy flows in cells (Meharg, 1994). However, as part of *As* detoxification arsenate is rapidly reduced to arsenite in plant tissue (Bertolero *et al.*, 1987), arsenate will not normally have high enough cytoplasmic concentrations to exert toxicity. Three independently evolved families of arsenate reductase enzymes have been recognized (Mukhopadhyay *et al.*, 2002). The first family includes arsenate reductase which is the product of the last gene of the *ars* operon of *E. coli* plasmid R773 (Chen *et al.*, 1986). A second family of arsenate reductases also widely distributed in bacteria is typified by the *arsC* gene product of *Staphylococcus aureus* plasmid pI258 (Ji and Silver, 1992). The third family of arsenate reductases has been characterized is is Acr2p from *S. cerevisiae* (Mukhopadhyay *et al.*, 1998). Recently, AR genes have been identified in plants, including *Arabidopsis* (*AtAsr/AtACR2*), *Holcus* (*HlAsr*) and *Pteris* (*PvACR2*) (Bleeker *et al.*, 2006; Dhankher *et al.*, 2006; Ellis *et al.*, 2006). Many ARs from plants show CDC25 tyrosine phosphatase activity and contain the conserved motif HCX_5R (Dhankher *et al.*, 2006).

This reduction of arsenate to arsenite, observed *in vitro*, is also brought about nonenzymatically by glutathione (Delnomdedieu *et al.*, 1994), which has high cellular concentrations in plant tissues (Alscher, 1989). Moreover both anions arsenate and arsenite have been reported to trigger the formation of phytochelatins (PCs) in plants (Grill *et al.*, 1987; Maitani *et al.*, 1996). PCs are heavy-metal-binding peptides derived from glutathione (GSH) and have the general structure $(\gamma\text{-Glu-Cys})_n\text{-Gly}$ ($n = 2–11$) (Grill *et al.*, 1985; Zenk, 1996). Their biosynthesis is due to the transpeptidation of γ-glutamylcysteinyl dipeptides from GSH by the action of a constitutively present PC synthase (Grill *et al.*, 1989; Chen *et al.*, 1997). Since the immobilized metals are less toxic than the free ions, PCs are considered to be part of the detoxifying mechanism of higher plants (Grill *et al.*, 1985; Zenk, 1996), algae (Gekeler *et al.*, 1988), and some fungi (Kondo *et al.*, 1985; Kneer *et al.*, 1992). Plants exposed to *As* substantially increase the synthesis of glutathione (GSH) and phytochelatins (PCs), the polymers of GSH (Schat *et al.*, 2002).

Leguminous plants synthesize homo-PCs (hPCs) in addition to PCs during *As* stress (Grill *et al.*, 2006). Augmented phytochelatin (PC) synthesis has been observed in tolerant species, such as *H. lanatus* (Hartley-Whittaket *et al.*, 2001), and many studies point to the essential role of PCs in both constitutive and adaptive tolerance to *As* (Schat *et al.*, 2002; Li *et al.*, 2004). Complexation of *As* with GSH and PCs has been demonstrated in various plants. Raab *et al.* (2004) reported the formation of AsIII–PC_3, GS–AsIII–PC_2, AsIII–GS_3, AsIII–$(PC_2)_2$ complexes and the newly identified monomethylarsonic–PC_2 [MA-AsIII–PC_2] complex in *H. annuus*. Similarly, Schmöger *et al.* (2000) reported rapid induction of the metal-binding PCs in cell suspension cultures of *Rauvolfia serpentina*, in seedlings of *Arabidopsis*, and in enzyme preparations of *Silene vulgaris* upon challenge to arsenicals. However, in the hyperaccumulator *Pteris vittata* most *As* appears to be in an unbound, inorganic form (Pickering *et al.*, 2006), suggesting that the role of PCs in *As* hyperaccumulation might be limited (Fig. 2).

The localization of *As*-PC complexes within plant tissue is still unknown. *As*-PC complexes are not stable at neutral or alkaline pHs, but are stable under acidic conditions present in the vacuole (Schmöger *et al.*, 2000; Sneller *et al.*, 2000). If *As*-PC complexes are also transported into the root vacuole, then under the acidic conditions present, they might remain stable preventing re-oxidation of *As* (III) and allowing accumulation of high concentrations of *As*-PC complexes in arsenate-resistant plants. The yeast *S. cerevisiae* and the protozoan parasite *Leishmania tarentolae* both form and sequester *As*-thiol complexes in intracellular compartments (Mukhopadhyay *et al.*, 1996; Ghosh *et al.*, 1999).

A final step in detoxification involves *As* sequestration in the vacuoles of root and shoot tissue. There is considerable evidence that at least part of the arsenic accumulated by plants is coordinated with glutathione and/or PCs, and that complexion is independent of the inorganic arsenic species originally present (Delnomdedieu *et al.*, 1994). *In vitro* nuclear magnetic resonance (NMR) studies have demonstrated that two molecules of GSH are required to reduce arsenate to arsenite, with GSH oxidized via the formation of a disulphide bond to give oxidized glutathione (GSSG) (Delnomdedieu *et al.*, 1994). This in turn is rapidly recycled to two GSH molecules (Cobbett, 2000). A ratio of 1: 1 of arsenate to GSH only resulted in the formation of GSSG, with no formation of complexes. This demonstrates that arsenate must be reduced to arsenite before formation of GSH complexes can occur, with three GSH molecules required to complex arsenite (Schmöger *et al.*, 2000). GSH can also reduce pentavalent dimethylarsinic acid to trivalent dimethylarsonic acid, with subsequent GSH complexes having a 1: 1 stoichiometry (Delnomdedieu *et al.*, 1994).

Fig. 1

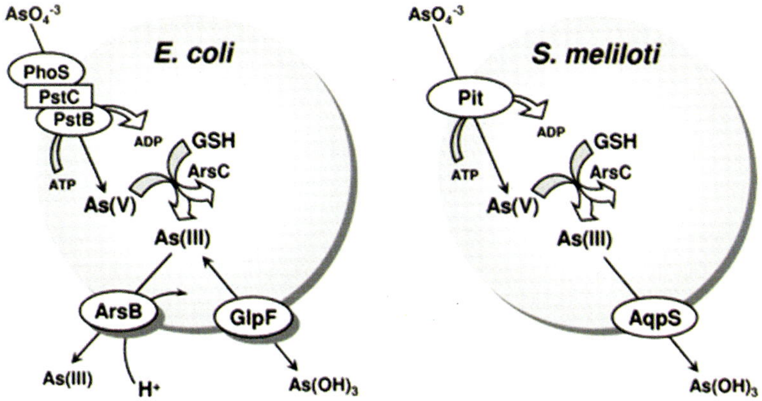

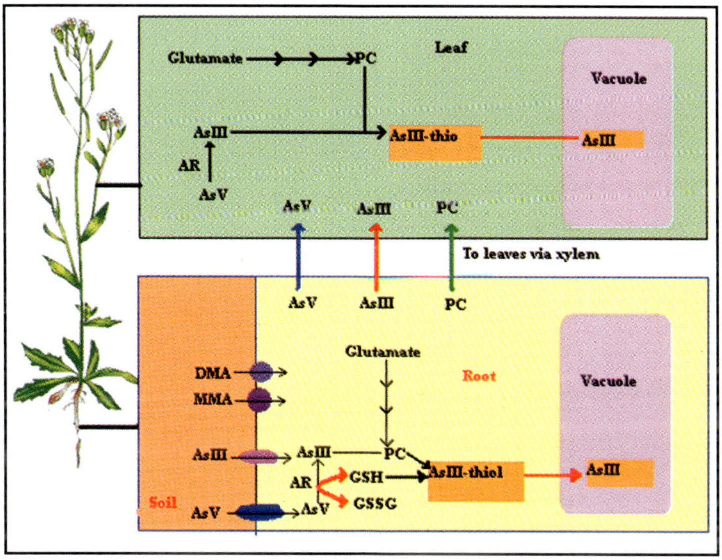

Fig. 2

Fig.1) Detoxification of arsenical in *E. coli* K-12 and *S. meliloti* 1021. In both *E. coli* and *S. meliloti* arsenate is brought into cells by the phosphate transporters. The first step of detoxification involves reduction of arsenate to arsenite by either *E. coli* or *S. meliloti* ArsC. Subsequent detoxification steps in *E. coli* involves removal of the trivalent form of the metalloid from the cytosol by active extrusion through the As(OH)3/H+ antiporter ArsB, while in *S. meliloti*, the AqpS channel facilitates downhill transport of As(III). Since arsenite can be taken up directly by cells, using either GlpF in E. coli or AqpS in *S. meliloti*, the detoxification mechanism functions when *S. meliloti* cells are exposed to arsenate. *(adopted from Yang et al., 2005)*

Fig. 2) Mechanism of arsenic uptake, translocation and detoxification in plants. Abbreviations: AsV= Arsenate; AsIII = Arsenite; PC = Phytochelatin; MMA = Monomethylarsonic acid; DMA= Dimethylarsinic acid; AR = Arsenate Reductase.

Arsenic Hyperaccumulation

A wide range of plant species have been identified on arsenic-contaminated land. Some of these plant species are reported to hyperaccumulate arsenic. *Pteris vittata,* the brake fern, was the first species shown to hyperaccumulate *As* from soil (Ma *et al.,* 2001). It could accumulate up to 22,630 mg *As* kg^{-1} in the shoot (frond) dry weight. Furthermore, the bioconcentration factor, defined as the ratio of shoot *As* concentration to soil *As* concentration, was greater than 10. The fern possesses three key features that are typical of metal/metalloid hyperaccumulator plants: an efficient root uptake, an efficient root to shoot translocation, and a much-enhanced tolerance to *As* inside plant cells. After the discovery of this first *As* hyperaccumulator, several other fern species, including *Pityrogramma calomelanos* (Francesconi *et al.,* 2002), *Pteris cretica, Pteris longifolia,* and *Pteris umbrosa* (Zhao *et al.,* 2002), have been added to the list of *As* hyperaccumulators. The hyperaccumulation trait of these ferns may be potentially exploitable in phytoremediation of *As* contaminated soils. Efficient uptake and translocation from root-to-shoot contribute greatly to hyperaccumulation of *As* in *P. vittata* (Ma *et al.,* 2001; Singh and Ma, 2006). Not only is far more *As* taken up but an exceedingly high proportion of *As* is translocated to shoot tissue. This fact emphasizes the importance of the transporters involved in *As* uptake, root-to-shoot translocation and vacuolar sequestration during hyperaccumulation. In general, the constitutive expression of genes that encode various transporters, and the biosynthesis of chelators, is higher in hyperaccumulator plants compared with non-accumulators (Singh and Ma, 2006).

Future Prospects

Many sites are contaminated with organic arsenic compounds from the now banned application of organic arsenic pesticides and through natural transformation of inorganic arsenic to organic arsenic in soils. The need to develop efficient strategies for cleaning *As*-polluted soil and water and also to restrict *As* contamination of food crops is clear.

1) Biomonitoring

The potential of metabolites, enzymatic processes and changes in plant performance as biomarkers in environmental assessment has been reviewed. Biomonitors may be used as an early warning system of specific or general stress at each biological level, from molecules to ecosystems. The sensitivity of a species and, thus, the efficiency of a biomonitor will depend on the degree of already present adaptation to environmental stress and on the homogeneity of the investigated population. Biomonitors for arsenic specific environmental stresses are scarce.

The use of *As* tolerant plants in biomonitoring constitutes an irreplaceable tool for investigation in ecological research. These plant biomonitors need to conform to certain required characteristics, not least being metal accumulators. Use of a suite of such biomonitors would allow recognition of the presence and relative magnitude of different metal sources. It is not valid to compare absolute accumulated arsenic concentrations in biomonitors interspecifically, although interspecific comparisons of rank orders do allow cross correlations of relative bioavailabilities of heavy metals to different biomonitors at the same sites. There is a need to identify widespread cosmopolitan biomonitors to allow intra-specific comparisons of arsenic bioavailabilities between geographical areas. Such cosmopolitan biomonitors may include both arsenic tolerant and hyperaccumulating species. A major caveat in the use of such cosmopolitan biomonitors remains the need for reliable, specific taxonomic identification.

2) Phytoremediation

Phytoremediation, a plant-based technology, however, holds great promise for the purification of contaminated soil and water (Meagher, 2000). The long-term goal is to develop and test vigorous, field adapted plant species that can prevent elemental pollutants from entering the food-chain by extracting them to above ground tissues, where they can be managed. Natural *As* hyperaccumulating plant species like *Pteris vittata* can be used for cleaning up *As* contaminated sites. Whereas, through genetic engineering the uptake and hyperaccumulation of other plants species can also be improved. The following strategies can be applied to increase arsenic accumulation for remediation:

a) Increasing As chelation: Chemical sinks like phytochelatins and glutathiones can bind toxic arsenic species and can increase the total arsenic concentration aboveground. Therefore, one of the strategies to improve heavy metal and/or metalloid remediation is to increase synthesis of phytochelatins through overexpression of phytochelatin synthase (PCS). But a recent study of Li *et al.* (2004) showed that overexpression of PCS on its own has limited effects and can easily lead to unwanted properties, such as toxicity. Whereas overexpression of *γ-ECS* in *A. thaliana* lead to 3- to 20-fold greater production of γ-EC, GSH and PCs in plants exposed to As; a markedly increased resistance to *As* and a slight increase in *As* accumulation (Li *et al.*, 2006). However, opting for multigenics approach with additional overexpression of components involved in GSH biosynthesis, such as γ-glutamylcysteine synthetase (*γ-ECS*) and glutathione synthetase (*GS*), might prove a more successful approach as this will lead to increased levels of GSH and the end-product, PCs.

b) Manipulating arsenate reduction: Most *As* resistant plants appear to have high levels of endogenous arsenate reductase activity, which reduces >95% of the arsenate they take up to arsenite (Dhankher *et al.*, 2002). Arsenite, which at neutral pH contains arsenic in the trivalent oxidation state [As(III)], and probably as the acid $HAsO_3^{2-}$, is highly reactive and readily forms As(III)-thiol complexes (Ramirez-Solis *et al.*, 2004). It appears that in most plants, arsenite is sequestered in roots, preventing it from moving up into stems, leaves, and reproductive organs. Thus, the enzymatic reduction of arsenate to arsenite in roots is a major barrier to engineering plants that will efficiently translocate arsenic to aboveground tissues (Dhankher *et al.*, 2002). Dhankher *et al.* (2006) demonstrated higher accumulation of *As* in the shoots by blocking the arsenate reduction activity by silencing *Arabidopsis thaliana* arsenate reductase 2 gene (*ACR2*). However, the transgenic lines were more sensitive than wild terrestrial plants when exposed to high concentrations of AsV, probably because of the negative effects of AsV on phosphate metabolism. By contrast, Bleeker *et al.* (2006) observed overexpression of *Arabidopsis asr AtASR* (equivalent to *AtACR2*) gene resulted in enhanced tolerance to mildly toxic levels of AsV but hypersensitivity to higher concentrations of AsV – probably owing to the fast conversion of AsV to the more toxic AsIII. Constitutive overexpression and suppression of AR appear, therefore, to result in plant hypersensitivity to high levels of AsV. Thus, to improve *As* tolerance and *As* accumulation without affecting phosphate metabolism, overexpression of AR might need to be limited to the roots and might need the simultaneous up regulation of the capacity for AsIII chelation, root-to-shoot transport and subsequent vacuolar sequestration.

c) Enhancing** As **uptake and root to shoot translocation: Efficient uptake and translocation from root-to-shoot is the key to increased *As* hyperaccumulation (Singh and Ma, 2006). The uptake and consequent root to shoot translocation requires expression of numerous genes. In general, the constitutive expression of genes that encode various transporters, and the biosynthesis of chelators, is higher in hyperaccumulator plants compared with non-accumulators (Kramer, 2005). Thus, the levels of expression of these genes may affect the level of *As* tolerance and hyperaccumulation.

d) Vacuolar sequestration of As: A concurrent suppression of vacuolar arsenic sequestration in roots and an enhanced vacuolar sequestration in shoots would ensure high levels of arsenic deposition in aboveground tissues. Tong *et al.* (2004) proposed that approaches using overexpression of vacuolar transporters might prove successful because, even with a modest increase in vacuolar transport capacity, substantially more substrate would be sequestered without the need for large amounts of chelating peptide.

3) Crop Development

Along with remediation of *As* contaminated sites there is need to develop crops that can be grown in *As* contaminated sites. For *As* tolerant crop development the major focus is to restrict *As* contamination of edible parts by preventing root to shoot translocation or complete cessation of *As* uptake in case of root crops. The main mechanisms to reduce crop arsenic contamination are:

a) Limiting uptake of arsenic : Phosphate being an analogue of arsenate competes with arsenate uptake. In some plant and bacterial species augmented phosphate uptake has lead to lower accumulation of arsenate (Lee *et al.*, 2003; Duker *et al.*, 2005) and hence increased *As* tolerance. An AsV tolerance gene, *AsTol*, has been identified on chromosome six in rice inbred lines of Bala × Azucena (Dasgupta *et al.*, 2004). *AsTol* was shown to be related to another QTL (quantitative trait locus), which is known to affect phosphorus uptake. This landmark research might help to identify orthologous genes in other varieties and thus achieve reduced uptake of AsV.

b) As extrusion: Some of the microbial genes responsible for extrusion of arsenite from cells and for their volatilization have been characterized in microorganism. Lee *et al.* (2003) observed *Arabidopsis* plants transformed with *E. coli ZntA* (encoding a PbII/CdII/ZnII transporting ATPase that pumps metals out of the cell) showed a reduction in heavy metal content and improved resistance to Pb and Cd. Moreover, the transgenic protoplasts showed faster release of preloaded Cd than wild type protoplasts. Similar approaches might prove successful in reducing *As* build up in edible plant parts such as grains. A high level of root AsIII extrusion capacity through ArsAB- and ACR3-like mechanisms in the plasma membrane, coupled to the action of gene products such as ArsM in shoot tissue – to promote *As* volatilization – might be desirable in this respect.

A variety of strategies adopted by *As* resistant plants can be manipulated to combat environmental arsenic stress. Attention should be focused to explore such molecular targets to engineer hyperaccumulating and fast growing resistant plant species for arsenic remediation along with non-accumulating crop species for non entry of arsenic into food chain.. Moreover, the possibility of extrusion of AsIII and *As* volatilization might be considered to reduce *As* load. Progress in these directions is unlikely to be swift because multiple targets might have to be combined. In any case *As* contents in rice may be minimized to acceptable limits by selecting pertinent varieties and through defined management of agronomic practices.

REFERENCES

Abedin MJ Feldmann J Meharg AA 2002 Uptake kinetics of arsenic species in rice (*Oryza sativa* L) plants *Plant Physiology* **128:** 1120-1128

Alscher RG 1989 Biosynthesis and antioxidant function of glutathione in plants *Physiologia Plantarum* **154:** 457–464

Asher CJ Reay PF 1979 Arsenic uptake by barley seedlings *Australian Journal of Plant Physiology* **6:** 459-466

Bertolero F Pozzi G Sabbioni E Saffiotti U 1987 Cellular uptake and metabolic reduction of pentavalent to trivalent arsenic as determinants of cytotoxicity and morphological transformation *Carcinogenesis* **8:** 803–808

Bleeker PM Hakvoort HWJ Bliek M Souer E Schat H 2006 Enhanced arsenate reduction by a CDC25-like tyrosin phosphatase explains increased phytochelatin accumulation in arsenate-tolerant *Holocus lanatus The Plant Journal* **45:** 979-929

Chen CM Misra TK Silver S Rosen BP 1986 Nucleotide sequence of the structural genes for an anion pump The plasmid-encoded arsenical resistance operon *Journal of Biological Chemistry* **261:** 15030-15038

Chen J Zhou J Goldsbrough PB 1997 Characterization of phytochelatin synthase from tomato *Physiologia Plantarum* **101:** 165-172

Cobbett CS 2000 Phytochelatin biosynthesis and function in heavy-metal detoxification *Current Opinion in Plant Biology* **3:** 211–216

Dasgupta T *et al* 2004 An arsenate tolerance gene on chromosome 6 of rice *New Phytologist* **163:** 45–49

Delnomdedieu M Basti MM Otvos JD Thomas DJ 1994 Reduction and binding of arsenate and dimethylarsinate by glutathione – a magnetic-resonance study *Chemico–Biological Interactions* **90:** 139–155

Dhankher O P Li Y Rosen BP Shi J Salt D Senecoff JF Sashti NA Meagher RB 2002 Engineering tolerance and hyperaccumulation of arsenic in plants by combining arsenate reductase and γ-glutamylcysteine synthetase expression *Nature Biotechnology* **20:** 1140-1145

Dhankher OP 2005 Arsenic metabolism in plants: an inside story *New Phytologist* **168:** 503–505

Dhankher OP Rosen BP McKinney EC Meagher RB 2006 Hyperaccumulation of arsenic in the shoots of *Arabidopsis* silenced for arsenate reductase (ACR2) *PNAS* **103:** 5413-5418

Ellis DR *et al* 2006 A novel Arsenate Reductase from the arsenic hyperaccumulating fern *Pteris vittata Plant Physiology* **141:** 1544-1554

Francesconi K Visoottiviseth P Sridokchan W Goessler W 2002 Arsenic species is an arsenic hyperaccumulating fern *Pityrogramma calomelanos*; a potential phytoremediator of arsenic-contaminated soils *The Science of the Total Environment* **281:** 27-35

Gekeler W Grill E Winnacker EL Zenk MH 1988 Algae sequester heavy metals via synthesis of phytochelatins complexes *Archives of Microbiology* **150:** 197-202

Ghosh M Shen J Rosen BP 1999 Pathways of *As* (III) detoxification in *Sachharomyces cerevisiae PNAS* **96:** 5001–5006

Grill E Winnacker E-L Zenk MH 1985 Phytochelatins: the principal heavy metal complexing peptides of higher plants *Science* **230:** 674–676

Grill E Winnacker EL Zenk MH 1987 Phytochelatins a Class of Heavy-Metal-Binding Peptides from Plants are Functionally Analogous to Metallothioneins *PNAS* **84:** 439-443

Grill E Loffler S Winnacker EL Zenk MH 1989 Phytochelatins the heavy-metal-binding peptides of plants are synthesized from glutathione by a specific γ-glutamylcysteine dipeptidyl transpeptidase (phytochelatins synthase) *PNAS* **86:** 6838-6842

Hartley-Whitaker J Ainsworth G Vooijs R Ten Bookum WM Schat H Meharg AA 2001 Phytochelatins are involved in differential arsenate tolerance in *Holcus lanatus* L *Plant Physiology* **126:** 299–306

Ji G Silver S 1992 Regulation and expression of the arsenic resistance operon from *Staphylococcus aureus* plasmid pI258 *Journal of Bacteriology* **174:** 3684–3694

Kneer R Kutchan TM Hochberger A Zenk MH 1992 *Saccharomyces cerevisiae* and *Neurospora crassa* contain heavy metal sequestering phytochelatin *Archives of Microbiology* **157:** 305–310

Kondo N Isobe M Imai K Goto T 1985 Synthesis of metallothionein-like peptides cadystin A and B occurring in a fission yeast and their isomers *Agricultural Biological Chemistry* **49:** 71–83

Krämer U 2005 Phytoremediation: novel approaches to cleaning up polluted soils *Current Opinion in Biotechnology* **16**: 133–141

Lee DA *et al* 2003 *ars1* an *Arabidopsis* mutant exhibiting increased tolerance to arsenate and increased phosphate uptake *Plant Journal* **35:** 637–646

Li Y *et al* 2004 Overexpression of phytochelatins synthase in *Arabidopsis* leads to enhanced arsenic tolerance and cadmium hypersensitivity *Plant and Cell Physiology* **45:** 1787-1797

Li Y *et al* 2006 The shoot specific expression of γ-Glutamylcysteine Synthetase directs the long-distance transport of thiol-peptides to roots conferring tolerance to mercury and arsenic *Plant Physiology* **141:** 288-298

Ma LQ Komar KM Tu C Zhang W Cai Y Kennelley ED 2001 A fern that hyperaccumulates arsenic *Nature* **409:** 579

Maitani T Kubota H Sato K Yamada T 1996 The composition of metals bound to class III metallothionein (phytochelatin and its desglycyl peptide) induced by various metals in root culture of *Rubia tinctorum Plant Physiology* **110:** 1145–1150

Meagher RB 2000 Phytoremediation of toxic elemental and organic pollutants *Current Opinion in Plant Biology* **3:** 153-162

Meharg AA 1994 Integrated tolerance mechanisms: constitutive and adaptive plant responses to elevated metal concentrations in the environment *Plant cell and environment* **17:** 989-993

Meharg AA 2004 Arsenic in rice – understanding a new disaster for South-East Asia *Trends Plant Science* **9:** 415–417

Mukhopadhyay R Dey S Xu N Gage D Lightbody J Ouellette M Rosen BP 1996 Trypanothione overproduction and resistance to antimonies and arsenicals in *Leishmania PNAS* **93:** 10383–10387

Mukhopadhyay R Rosen BP 1998 *Saccharomyces cerevisiae* ACR2 gene encodes an arsenate reductase *FEMS Microbiology Letters* **168:** 127-136

Mukhopadhyay R Rosen BP 2002 Arsenate Reductases in Prokaryotes and Eukaryotes *Environmental Health Perspectives* **110:** 745-748

Norra S Berner ZA Agarwala P Wagner F Chandrasekharam D **Stüben D 2005** Impact of irrigation with *As* rich groundwater on soil and crops: A geochemical case study in West Bengal Delta Plain India *Applied Geochemistry* **20:** 1890-1906

Oremland RS Stolz JF 2003 The Ecology of Arsenic *Science* **300:** 939-944

Pickering IJ *et al* 2006 Localizing the biochemical transformations of arsenate in a hyperaccumulating fern *Environmental Science Technology* **40:** 5010-5014

Raab A Feldmann J Meharg AA 2004 The nature of Arsenic-Phytochelatin complexes in *Holcus lanatus* and *Pteris cretica Plant Physiology* **134:** 1113-1122

Ramirez-Solis A Mukopadhyay R Rosen BP Stemmler TL 2004 Experimental and theoretical characterization of arsenite in water: Insights into the coordination environment of *As*-O *Inorganic Chemistry* **43:** 2954-2959

Rosenberg H Gerdes RG Chegwidden K 1977 Two systems for the uptake of phosphate in *Escherichia coli Journal of Bacteriology* **131**: 505-511

Schat H Llugany M Vooijs R Hartley-Whitaker J Bleeker PM 2002 The role of phytochelatins in constitutive and adaptive heavy metal tolerances in hyperaccumulator and non-hyperaccumulator metallophytes *Journal of Experimental Botany* **53:** 2381-2392

Schmöger MEV Oven M Grill E 2000 Detoxification of arsenic by phytochelatins in plants *Plant Physiology* **122:** 793–802

Sereno D Cavaleyra M Zemzoumi K Maquaire S Ouaissi A Lemesre JL 1998 Axenically grown amastigotes of *Leishmania infantum* used as an *in vitro* model to investigate the pentavalent antimony mode of action *Antimicrobial Agents Chemotherapy* **42:** 3097– 3102

Sharples JM Meharg AA Chambers SM Cairney JWG 2000 Mechanism of arsenate resistance in the ericoid mycorrhizal fungus *Hymenoscyphus ericae Plant Physiology* **124:** 1327–1334

Singh N Ma LQ 2006 Arsenic speciation and arsenic and phosphate distribution in arsenic hyperaccumulator *Pteris vittata* L and non-hyperaccumulator *Pteris ensiformis* L *Environment Pollution* **141**: 238–246

Sneller FEC van Heerwaarden LM Koevoets PLM Vooijs R Schat H Verkleij JAC 2000 Derivatization of phytochelatins from *Silene vulgaris* induced upon exposure to arsenate and cadmium: Comparison of derivatization with Ellman's reagent and monobromobimane *Journal of Agricultural and Food Chemistry* **48:** 4014–4019

Tong YP *et al* 2004 Vacuolar compartmentalization: a second-generation approach to engineering plants for phytoremediation *Trends Plant Science* **9**: 7–9

Tripathi RD Srivastava S Mishra Seema Singh N Tuli R Gupta DK Maathuis JM 2007 Arsenic hazards: strategies for tolerance and remediation by plants *TRENDS in Biotechnology* **25** :159-165

Ullrich-Eberius CI Sanz A Novacky AJ 1989 Evaluation of arsenate- and vanadate-associated changes of electrical membrane potential and phosphate transport in *Lemna gibba* G1 *Journal of Experimental Botany* **40:** 119–128

Willsky GR Malamy MH 1980 Characterization of two genetically separable inorganic phosphate transport systems in *Escherichia coli Journal of Bacteriology* **144**: 356-365

Woolson EA 1975 Bioaccumulation of arsenicals ACS Symp Ser 7: 97-107

Wysocki R *et al* 2001 The glycerol channel Fps1p mediates the uptake of arsenite nd antimonite in *Saccharomyces cerevisiae Molecular Microbiology* **40:** 1391-1401

Yang Hung-Chi Jiujun Cheng1 Finan Turlough M Rosen Barry P and BhattacharjeeY Hiranmo 2005 Novel Pathway for Arsenic Detoxification in the Legume *Symbiont Sinorhizobium meliloti Journal of Bacteriology* **187** 6991-6997

Yompakdee C Bun-ya M Shikata K Ogawa N Harashima S Oshima Y 1996 A putative new membrane protein Pho86p in the inorganic phosphate uptake system of *Saccharomyces cerevisiae Gene* **171:** 41-47

Zenk MH 1996 Heavy metal detoxification in higher plants- a review *Gene* **179:** 21-30

Zhao FJ Dunham SJ McGrath SP 2002 Arsenic hyperaccumulation by different fern species *New Phytologist* **156:** 27-31

10

COMPARTMENTATION AND METABOLITE FLUXES IN C_3-C_4 AND C_4 PLANTS IN RELATION TO METABOLIC ENGINEERING

SANJEEV K THIND

INTRODUCTION

A complex of different metabolite transport processes serve a common physiological function. CO_2-concentrating mechanism which mitigates the oxygenase activity of Rubisco, potentially improves the efficiency of carboxylation of water use and of nitrogen use in photosynthesis, with a number of important implications for plant performance and survival. The carboxylation and decarboxylation events which drive the CO_2-concentrating mechanisms of C_4 and CAM plants are similar, but they operate on different anatomical, physiological and biochemical principles. They are initiated by the fixation of HCO_3 via PEP carboxylase (PEPcase), leading to the synthesis of malate and aspartate, and the subsequent decarboxylation of these C_4 acids to generate CO_2 for Rubisco. In C_4 plants the PEPcase and Rubisco are spatially separated in outer mesophyll and inner bundle sheath cells of the Kranz complex. Simultaneous bidirectional fluxes of C_3 and C_4 metabolites in the symplasm connecting these cells are required to sustain C_4 photosynthesis, and the 4-C carboxyl carbon of malate and aspartate serves as a short-lived reservoir of CO_2 during transport. Small amounts of C_4 acids (a few m mol. per mg chlorophyll) are involved, and this transport pool turns over in about 10 s. In CAM plants, on the other hand, both PEPcase and Rubisco are present in all chloroplast-containing cells. The activities of these enzymes are thus regulated in time, rather than in space, with PEPcase active in the dark but inactive for much of the light period, and Rubisco active only in the light. In CAM plants the 4-C carboxyl of malate serves as a longer-lived, much larger temporary store of CO_2. Large amounts of malic acid (several hundreds of mmoles per mg chlorophyll) may accumulate in the large cell vacuoles of CAM plants and may have a turnover time of thousands of seconds. The tonoplast fluxes of malic acid, rather than symplastic transport, thus play the key role in the CO_2-concentrating mechanism of CAM, and also in the day-night regulation of enzyme activity.

In general, regulations or transport processes in C_4 or CAM plants are coordinated in space and time in several unique ways which permit biochemical, physiological and anatomical differentiation of subtypes of C_4 plants and CAM plants. In addition, it is now recognized that there are many intermediate forms between C_4 and C_3 metabolism, and between CAM and C_3 metabolism. Similarly, there are many genetically-determined variants of CAM, and the environmental control of this pathway leads to many manifestations intermediate between strict CAM and C_3 photosynthesis.

C_4 photosynthesis in tropical grasses such as sugarcane and maize depends on mesophyll cell chloroplasts which lack most of the enzymes of the Calvin cycle, and bundle sheath cells with poorly developed PSII activities (Glawischnig *et al.*, 2001). This means that photosynthetic CO_2 fixation in C_4 plants depends on inter- and intracellular regulation of metabolite transport within and between cells. The term 'cooperative photosynthesis' was introduced to describe the interdependencies of mesophyll and bundle sheath cells. The genome of the chloroplasts from the two cell types is probably the same but transcriptional regulation of gene expression by light seems to be responsible for the differentiation of photosynthetic properties (Poolman *et al.*, 2003). Differential expression of the gene for the enzymes peculiar to the C_4-cycle seems to involve light-modulation of transcript pools in the two cell types which parallel the development of etioplasts during greening, rather than the morphogenesis of the two cell types. Greening is associated with the loss of mRNA for Rubisco large subunit in mesophyll plastids and a light-dependent arrest of its transcription. In bundle sheath plastids, the expression of this gene, and of genes for the small subunit of Rubisco, is amplified by light.

Extensive surveys of enzyme distribution within and between cells in different C_4 plants led to the identification of three decarboxylase enzymes in the bundle sheath cells. Apart from the different patterns of intercellular movement of C_3 and C_4 compounds that these subtypes of C_4 photosynthesis imply, there are patterns of regulation in C_4 plants which differ from those in C_3 plants. One of the most striking examples is the way in which 3-PGA reduction is shared between the Calvin cycle in bundle sheath chloroplasts and the mesophyll chloroplasts in all subtypes of C_4 plants.

Critical to defining Photosynthesis in C_4 plants is understanding the intercellular and intracellular compartmentation of enzymes between mesophyll and bundle sheath cells in the leaf. This includes enzymes of the C_4 cycle (including three subtypes), the C_3 pathway and photorespiration. The current state of knowledge of this compartmentation is a consequence of the development and application of different techniques over the past three decades. Initial studies led to some alternative hypotheses on the mechanism of C_4 photosynthesis, and some controversy over the compartmentation of enzymes. Now, the ability to determine localization of photosynthetic enzymes conclusively, through *in situ* immunolocalization by confocal light microscopy and transmission electron microscopy, is providing further insight into the mechanism of C_4 photosynthesis and its evolution. There have been found interesting variations in the C_4 syndrome in leaves and cotyledons of species in the tribe Salsoleae of the family Chenopodiaceae, in relation to evolution and ecology (Edwards *et al.*, 2001).

C_4 photosynthesis has a number of distinct properties that enable the capture of CO_2 and its concentration in the vicinity of Rubisco, so as to reduce the oxygenase activity of Rubisco, and hence the rate of photorespiration. The Kranz leaf anatomy of C_4 photosynthesis and the division of the C_4- cycle between two cell types involves intercellular co-operation that requires modifications in regulation and transport to make C_4 photosynthesis work. Comparisons are made with the C_4- type photosynthesis found in single-celled C_4- type CO_2-concentrating mechanisms, found in the aquatic plant, *Hydrilla verticillata* and the single-celled C_4 system found in the terrestrial chenopod *Borszczowia aralocaspica* (Leegood, 2002).

Intercellular transport

In the majority of C_4 plants the photosynthetic cells are organized in two concentric cylinders. Thin-walled mesophyll cells with large intercellular spaces radiate from the thick-walled bundle sheath cells. In some C_4 plants, the cell walls between the bundle sheath and mesophyll contain suberin, but in all it seems likely that the thickened cell wall is itself a major barrier to the diffusion of solutes and gases. On the wall between mesophyll and bundle sheath cells there are extensive pit-fields with plasmodesmata which provide symplastic connections between cells. The plasmodesmata generally permit the rapid exchange of solutes between cells and slower exchange of gases in solution. The metabolite transport in C_4 photosynthesis could be sustained by diffusion, driven by gradients in the concentration of specific metabolites between the source and the sink cells. Thus, in the case of malate or aspartate, gradient with the concentration in mesophyll source cells 10-30 mM higher than in bundle sheath sink cells would be needed to sustain observed rates of photosynthesis.

It should be noted that effective surface area is only 3% of the mesophyll/bundle sheath interface. Plasmodesmata are frequent only in primary pit-fields at the areas of contact between mesophyll and bundle sheath cells. It has been found that the cross-sectional area of the sphincter in plasmodesmata, or of the desmotubule, occupies 1.5–3% of the pit-field of the surface between mesophyll and bundle sheath cells. Perhaps the greatest uncertainty, stems from the need to sustain bidirectional metabolite fluxes, when there is a net flux of sucrose from mesophyll to bundle sheath coincident with water flux in the other direction (Wiechert *et al.*, 1999).

The position of chloroplasts within the bundle sheath cells, have been used to predict the different decarboxylation types. Other features included are presence of cells which intervene between the metaxylem vessel elements, the chlorenchymatous bundle sheath cells and occurrence of an even or uneven outline to the bundle sheath cell walls, which influences the area of contact between the two cell types. At the subcellular level, other differences that are related to metabolic pathways arise. In $NADP^+$-malic enzyme ME plants, the bundle sheath chloroplasts contain few grana and show a deficiency of PSII proteins and function. In NAD^+ - ME and PEPCK plants which decarboxylate malate in the bundle sheath mitochondria, the mitochondrial frequency is between two and four times that in the bundle sheath of $NADP^+$ - malic enzyme type C_4 species.

The chloroplast position and the presence of other cells between the mesophyll and the bundle sheath will affect the distance over which metabolites must be transported. These cells also need to be considered in relation to translocation of sucrose from the mesophyll cells to the vascular tissue. The occurrence of a suberized lamella varies; it is absent in dicotyledonous species, and in grasses is present only in species with either an uneven bundle sheath outline or with centrifugally located chloroplasts.

Intercellular movement of C_3 and C_4 metabolites

Evidence of rapid metabolite movement may be inferred from the rapid transfer of ^{14}C from C_4 acids to 3-PGA and products (labelled in the bundle sheath) which occurs when $^{14}CO_2$ is supplied to leaves of C_4 plants. After a 2 s pulse of $^{14}CO_2$ (when the majority of the label is in C_4 acids), the cytoplasm of the mesophyll cells is clearly labelled, but a considerable amount of label is also found in the bundle sheath cells. The majority of the label is transferred to the bundle sheath during a 10 s chase. This indicates very rapid movement of C_4 acids, and also sizeable pools of these compounds within the bundle sheath cells.

The gradient of malate between the mesophyll and the bundle sheath is sufficient to generate a flux between the two cell types (7.2 mmol m^{-2} s^{-1}) which is much larger than the photosynthetic flux (0.5 mmol m^{-2} s^{-1}). There is predominance of pyruvate in the mesophyll compartment, so that the gradient of pyruvate appears to lie in the opposite direction to the expected flux, but this is a result of the intracellular transport of pyruvate within mesophyll cells. The amounts of 3-PGA and triose-P in leaves of maize are typically 10-20 times higher than in the leaves of C_3 plants. This reflects the opposing intercellular concentration gradients of these metabolites which are adequate to generate fluxes required by the assimilation rates.

From the mesophyll cells, C_4 species possess the enzymes for 3-PGA reduction in the mesophyll chloroplasts although most other enzymes of the Calvin cycle are absent.

The triose-P which is formed from 3-PGA in the mesophyll has two fates. The first is its conversion to sucrose or starch in the mesophyll cells, while the second is its return to the bundle sheath to regenerate RuBP. In $NADP^+$ - ME plants, NADPH generation from malic enzyme demands less PSII activity in the bundle sheath and hence O_2 evolution is reduced. This favours the carboxylation of RuBP over its oxygenation. Another role for 3-PGA reduction in the mesophyll could be coordination of Calvin cycle and C_4 cycle turnover. It may be a means of ensuring H^+ transport, between the two cell types. The reduction of 3-PGA to triose-P in the mesophyll consumes a proton, after which the triose-P is transported to the bundle sheath. The consumption of a proton in this reaction is necessary because a proton is released when CO_2 is hydrated and the resulting HCO_3^- is fixed by PEPcase.

In NAD^+ - ME plants amount of triose-P and 3-PGA were similar to those in maize, suggesting that extensive intercellular transport of these metabolites occur. In PEPCK plants, transport is likely to be rather more complex, because both malate and aspartate are transferred from the mesophyll to the bundle sheath, and PEP and alanine return to the mesophyll.

Intercellular movement of inorganic carbon

The symplastic pathway of metabolite movement also permits movement of inorganic carbon out of the bundle sheath. Direct measurements of the pool of HCO_3^- and CO_2 from radiotracer studies indicated that the total pool of inorganic carbon is about 0.6 mM in the bundle sheath of maize. Analysis of CO_2 concentrating mechanisms in the three different decarboxylation types indicated that the absence of carbonic anhydrase in bundle sheath cells is of critical importance. If the mesophyll inorganic carbon pool is assumed to be zero, then the flux of HCO_3^- via the plasmodesmata would be 0.03 mmol m^{-2} s^{-1}, which is less than 10% of the required flux for metabolites. These differences could depend on the amount of O_2 evolved in bundle sheath chloroplasts and/or differences in the leakiness of the bundle sheath compartment. The cells with higher inorganic carbon levels are more leaky. The more leaky systems are less efficient, in that futile cycling of the energetically expensive C_4 cycle leads to lower quantum yields.

C_3-C_4 intermediates

More than twenty species of plants exhibit photosynthetic characteristics that are intermediate between C_3 and C_4 plants in that they show reduced rates of photorespiration and CO_2 compensation points in the range 7-15 ml l^{-1}, compared with typical values in C_3 plants of 50 ml l^{-1} and in C_4 plants of less than 5 ml l^{-1}. There is little evidence for a functional C_4 acid cycle which donates CO_2 from C_4 acids to the Calvin cycle. The intermediate character with respect to CO_2 compensation point seems to be due to more efficient refixation of photorespiratory CO_2 in the bundle sheath cells. Like C_4 plants, the mesophyll cell mitochondria of 'intermediate' plants have low activities of glycine decarboxylase.

In the other type of 'intermediate' plants (such as *Flaveria anomala and Neurachne minor*), appreciable activities of the C_4 pathway enzymes PEPcase, pyruvate Pi dikinase and $NADP^+$ - ME are found. These plants show varying capacities to fix ^{14}C into C_4 acids during short-term exposure to $^{14}CO_2$, and to transfer this to products of the Calvin cycle, suggesting limited capacity for a C_4 pathway.

NADP-ME provides CO_2 to the bundle sheath chloroplasts, where it is fixed by ribulose-1, 5-bisphosphate carboxylase/oxygenase in different photosynthetic species. A 72 kD isoform, present to varying degrees in all species of *Flaveria* (C_3, C_3-C_4 intermediate, C_4-like, C_4), is predominant in leaves of C_3 sp. and is also present in stem and root tissue. It is mostly localized in the upper palisade mesophyll chloroplasts of C_3 photosynthetic tissue. Two other isoforms of the enzyme, with molecular masses of 62 and 64 kD, occur in leaves of certain intermediates having C_4 cycle activity. The 62 kD isoform, which is the predominant highly active form in the C_4 species, is localized in the bundle sheath chloroplasts. Among *Flaveria* sp. there is a 72 kD constitutive form, a 65 kD form that may have appeared during evolution of C_4 metabolism, and a 62 kD form that is necessary for the complete functioning of C_4 photosynthesis (Drincovich *et al.*, 1998).

By comparison, immediate export rates of ^{14}C-translocates during photosynthesis in *Panicum* "type 1" C_3- C_4 intermediates were greater than those in C_3 species. In contrast, immediate export was much lower in *Flaveria* "type II" C_3- C_4 intermediates than in their C_3 and C_4 relatives. In all species ^{14}C-reserves, primarily starch and sugars, sustained export during the chase periods, but the extent to which subsequent day-night export compensated for lower export during the feed was greater among Panicum than *Flaveria* species. The Panicum "type I" C_3- C_4 intermediates still behaved in an "intermediate" manner with respect to daily ^{14}C-export. In contrast, the *Flaveria* "type II" C_3-C_4 intermediates still had the lowest total ^{14}C-export by the next morning. Overall, in both genera daily export correlated better with immediate export rates than with photosynthesis itself and other leaf export parameters.

Immunological cross-reactivity of PEPC in leaf extracts of C_3, C_4 and C_3- C_4 intermediate of *Alternanthera* (along with a few other C_3- and C_4- plants) was studied using anti-PEPC antibodies. Anti-PEPC antiserum raised against *A. hypochondriacus* enzyme showed high cross-reactivity with PEPC in leaf extracts of *A. hypochondriacus* or *Amaranthus viridis* or *Alternanthera pungens* (all C_4 dicots), but limited cross-reactivity with that of *Zea mays*, Sorghum or *Pennisetum* (all C_4 monocots). Interestingly, PEPC in leaf extracts of *Alternanthera tenella, A. ficoides, Parthenium hysterophorus* (C_3-C_4

intermediates) exhibited stronger cross-reactivity (with anti-serum raised against PEPC from *Amaranthus hypochondriacus*) than that of *Pisum sativum, Commeline benghalensis, Altenanthera sessilis* (C_3 plants). Further studies on cross-reactivities of PEPC in leaf extracts of these plants with anti-PEPC antisera raised against PEPC from leaves of *Zea mays* or *Alternanthera pungens* confirmed two points–(i) PEPC of C_3-C_4 intermediate is distinct from C_3 species and intermediate between those of C_3-and C_4-species; and (ii) PEPC of C_4-dicots was closer to that of C_3-species or C_3-C_4 intermediates (dicots) than to that of C_4-monocots (Gayathri *et al.*, 2001).

The maximum catalytic activities of several photorespiratory and photosynthetic enzymes have been determined in leaf extracts of three C_3-C_4 intermediates (*Alternanthera ficoides, A. tenella and Parthenium hysterophorus)* and were compared to those of C_3 (*A. sessiles, Pisum sativum*) and C_4 (A. *Pungens, zea mays and Amaranthus hypochondriacus)* species. The activity levels of key photorespiratory enzymes, glycolate oxidase, catalase, NADH-hydroxypyruvate reductase and glycerate kinase were less (28 to 35% reduced) in intermediates than those of typical C_3 species. Similarly, the activities of photorespirator aminotransferases in the C_3-C_4 intermediates were also partially reduced (23 to 37% reduction.) The activities of phosphoenol pyruvate carboxylase (PEPC), pyruvate, orthophosphate dikinase and NAD-ME were higher (2 to 7 times) in leaf extracts of the intermediates than those of C_3 species. But the ratios of PEPC/Rubisco in the C_3-C_4 intermediates were more like C_3 than C_4 species (Devi and Raghavendra 1993). The partial reduction in enzyme activity of photorespiratory metabolism, could be an important factor for restriction of photorespiration in the C_3-C_4 intermediate species, in addition to enzyme compartmentation and/or operation of a 'C_4-like' cycle.

The intracellular transport of metabolites between the cytoplasm and organelles is a key component of C_4 photosynthesis in both mesophyll and bundle sheath cells. Leaf organelles in C_4 plants share the translocators of organelles in leaves of C_3 plants, but also contain translocators with unique or greatly altered kinetic properties. The observation that there is no overall gradient of pyruvate between the bundle sheath and mesophyll cells suggest that a gradient between the bundle sheath and the mesophyll might simply be obscured by accumulation of pyruvate within the mesophyll chloroplasts. Transport of pyruvate on a specific carrier occurs in chloroplasts of both C_3 and C_4 plants, but the translocator is much more active and is light-dependent in mesophyll chloroplasts of C_4 plants. Moreover, the light activation of pyruvate Pi dikinase seems to depend on maintenance of high pyruvate /PEP ratios, so it seems likely that pyruvate is maintained at high concentrations in the mesophyll chloroplasts.

The dicarboxylates, malate, oxaloacetate, 2-oxoglutarate, aspartate and glutamate are transported through the mesophyll chloroplast envelope in a carrier-mediated mode. These compounds undergo counter-exchange on the dicarboxylate translocator and the K_m for uptake of a particular dicarboxylic acid is similar to the K_i for the inhibition of uptake by other dicarboxylates.

The translocator in C_3 mesophyll chloroplasts can only organize certain organic phosphates with phosphate in the C_3–position. Thus, in addition to PEP, phosphoglycolate and 2-PGA are effective inhibitors of 3-PGA-dependent O_2 evolution in C_4 mesophyll chloroplasts but have no influence on 3-PGA-dependent O_2 evolution in barley chloroplast.

Two naturally occuring species of the genus *Alternanthera,* namely *A. ficoides* and A. *tenella,* which were identified as C_3-C_4 intermediates based on the leaf anatomy, photosynthetic CO_2 compensation point (G), O_2 response of G, light intensity response of G, and the activities of key enzymes of photosynthesis exhibited a less distinct Kranz-like leaf anatomy with substantial accumulation of starch both in mesophyll and bundle sheath cells. Photosynthetic CO_2 compensation points of these two intermediate species at 29°C were much lower than in C_3 plants and ranged from 18 to 22 µl L^{-1}. Although *A. ficoides* and *A. tenella* exhibited similar intermediacy in G, the apparent photorespirator component of O_2 inhabitation in *A. ficoides* is lower than in *A.tenella* (Rajendru *et. al.,* 1986). The relatively low apparent photorespiratory component in these two species is presumably the basis for the C_3- C_4 intermediate photosynthesis.

A two tiered, nested molecular phylogenetic study of panicoid grasses to explore character state transitions between the C_3 and C_4 adaptive syndromes was presented. A broad survey of 92 panicoid species was sampled for the grass-specific insert sequence in the chloroplast RNA polymerase locus (*rpo*C_2), an intensive phylogenetic investigation of one clade of seven species that included *Steinchisma hians,* which is notable for exhibiting intermediacy between the C_3 and C_4 photosynthetic types was conducted. A rigorous phylogenetic investigation of *S. hians* and 13 other species of *Panicoideae* included in the broad survey was performed with sequences of both *rpo*C_2 and the externally transcribed spacer region of the nuclear ribosomal repeat (Duvall *et al.,* (2003). These 14 species were selected to maximize representation among photosynthetic subtypes. Combined analysis resolved single origins of two photosynthetic subtypes. A reversion of C_4 to C_3 photosynthesis during the evolution of the lineage that includes *S. hians* was identified. These results indicated that repeated reversions from C_4 to C_3 have occurred.

C_4 Syndrome

Borszczowia aralocaspica (Chenopodiaceae) accomplishes C_4 photosynthesis in a unique, polarized single-cell system in leaves. Mature cotyledons have the same structure as leaves, with chlorenchyma cells having biochemical polarization of dimorphic chloroplasts and C_4 functions at opposite ends of the cell. Development of the single-celled C_4 syndrome in cotyledons was characterized. In mature seeds, all cell layers are

Fig. 1

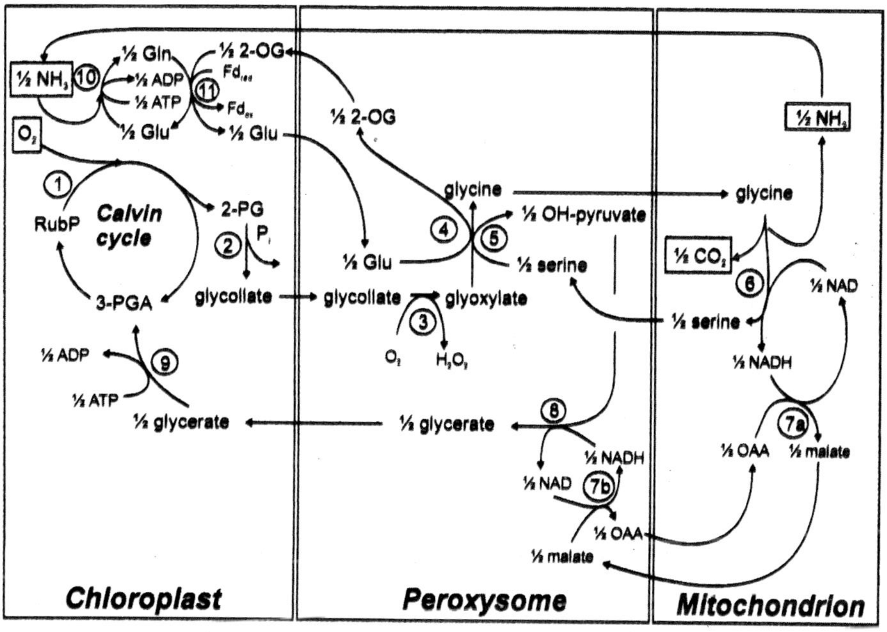

Fig. 2

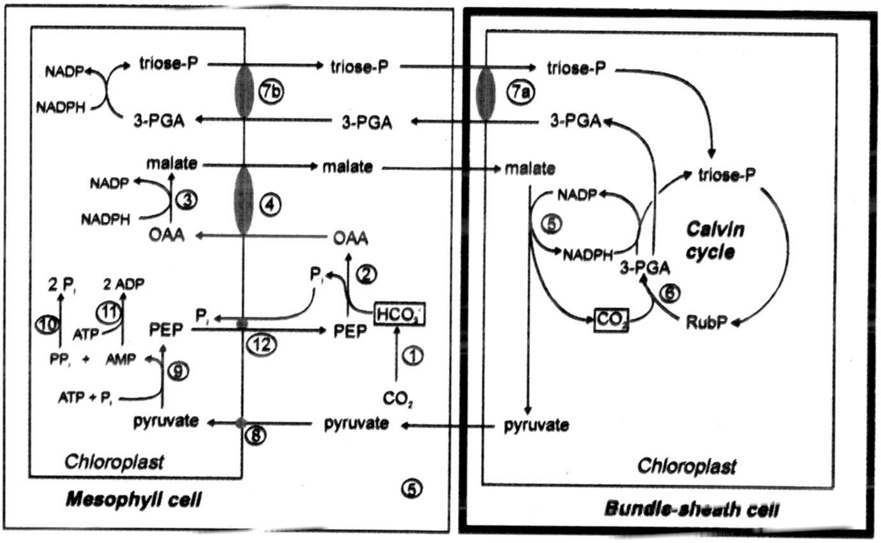

Fig. 1)

Fig. 2) CO_2 concentrating mechanism in a NADP malic enzyme C_4 plant, (maize, sugar cane or Sorgham). Various enzymes are : carbonic anhydrase (1), PEPC (2), NADP-MDH (3), malate/OAA transporter (4), NADP-ME (5), oxidative malate decarboxylase (6), TPT (7a), strorma TPT (7b), and reduced pyruvate transporter (8), PPDK (9), pyrophosphatase (10), adenylate kinase (11). PPT (12).

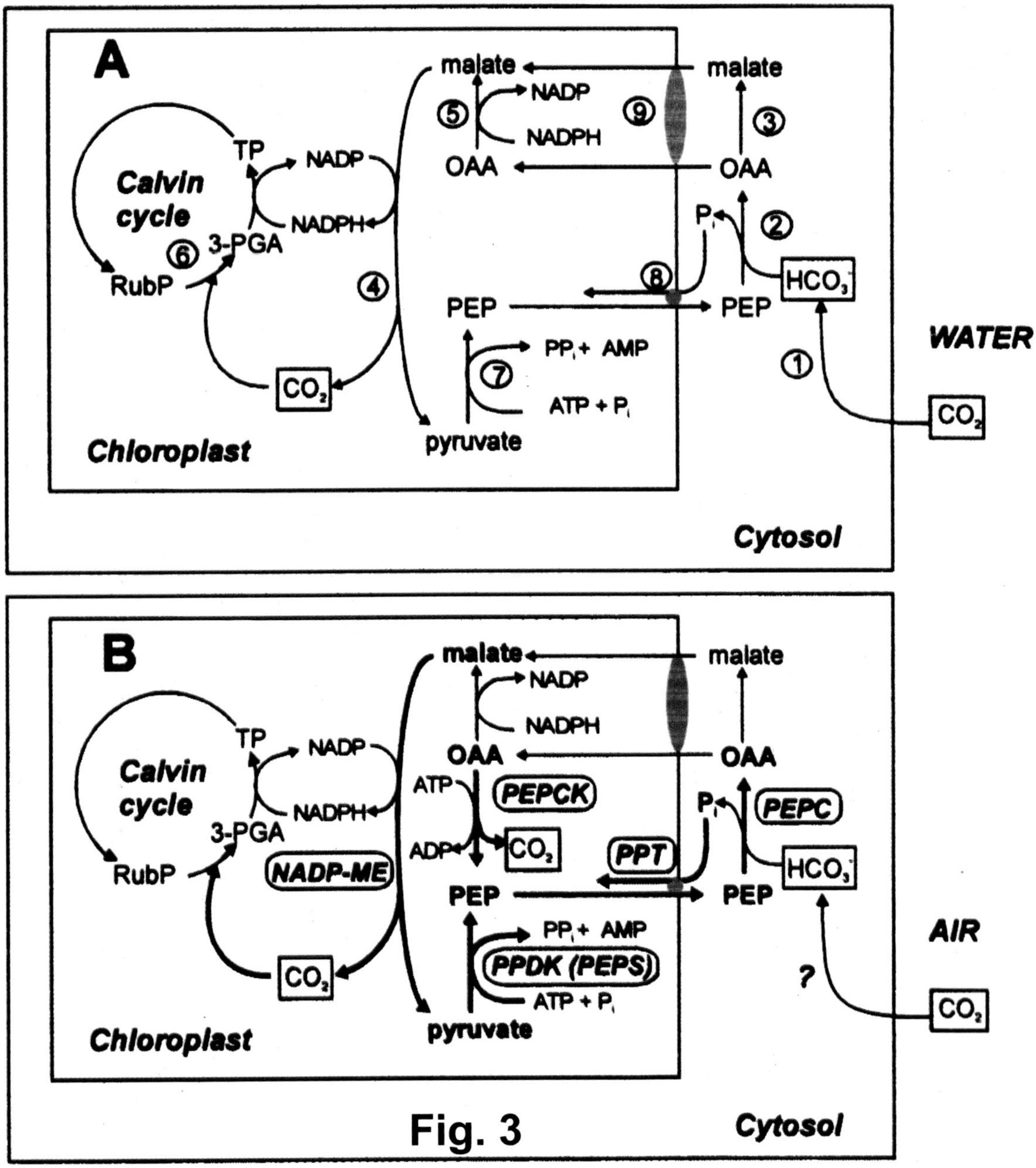

Fig. 3

Fig. 3) (A) Induced C_4 cycle in the submerged aquatic plant Hydrilla verticillata without 'kranz' anatomy. (1) CA, (2) PEPC, (3) NAD-MDH, (4) NADP-MDH, (5) NADP-ME, (6) Rubisco, (7) PPDK, (8) PPT. (B) Genetically engineered single cell CO_2 concentrating mechanism in transgenic tobacco plants. The transformants contain all combinations of C_4 cycle enzymes (PEPC, PEPCK, NADP-ME, PPDK or PEPS as well as the PPT).

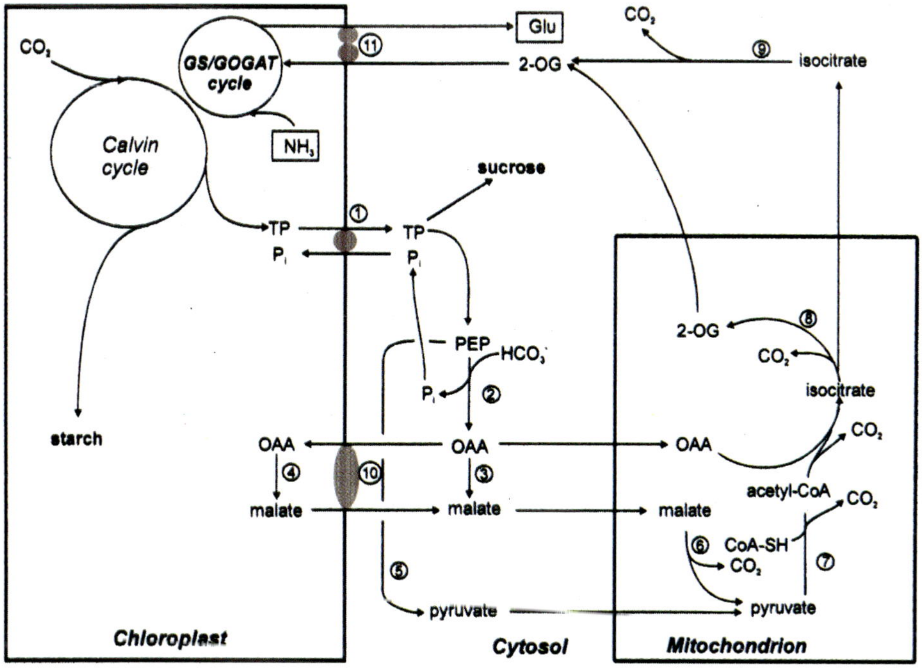

Fig. 4

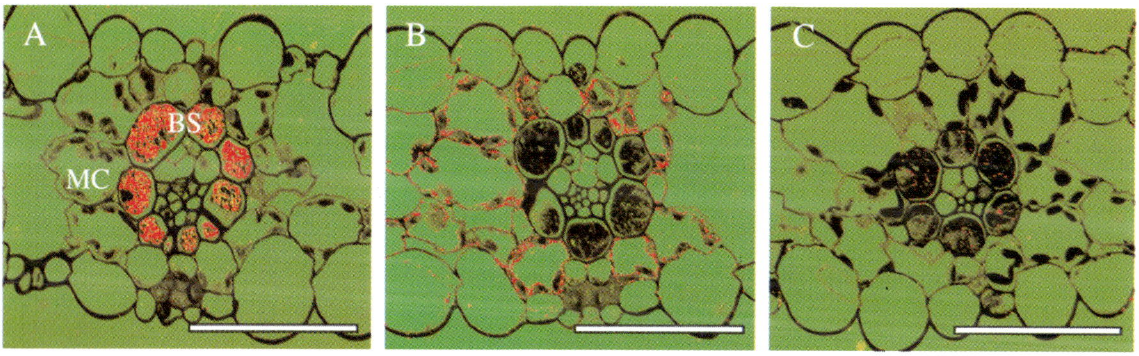

Fig. 6

Fig. 4)

Fig. 6) Reflected/transmitted confocal microscopy imaging of *in-situ* immunolocalization of photosynthetic enzymes in leaves of *Echinocholoa frumentacea*, a member of the NADP-ME type C_4 biochemical group. (A) Rubisco; (B) PEPC; (C) PEPC-CK. BS, bundle sheath; MC, mesophyll cells. Scale bars = 50 μm.

Fig. 5

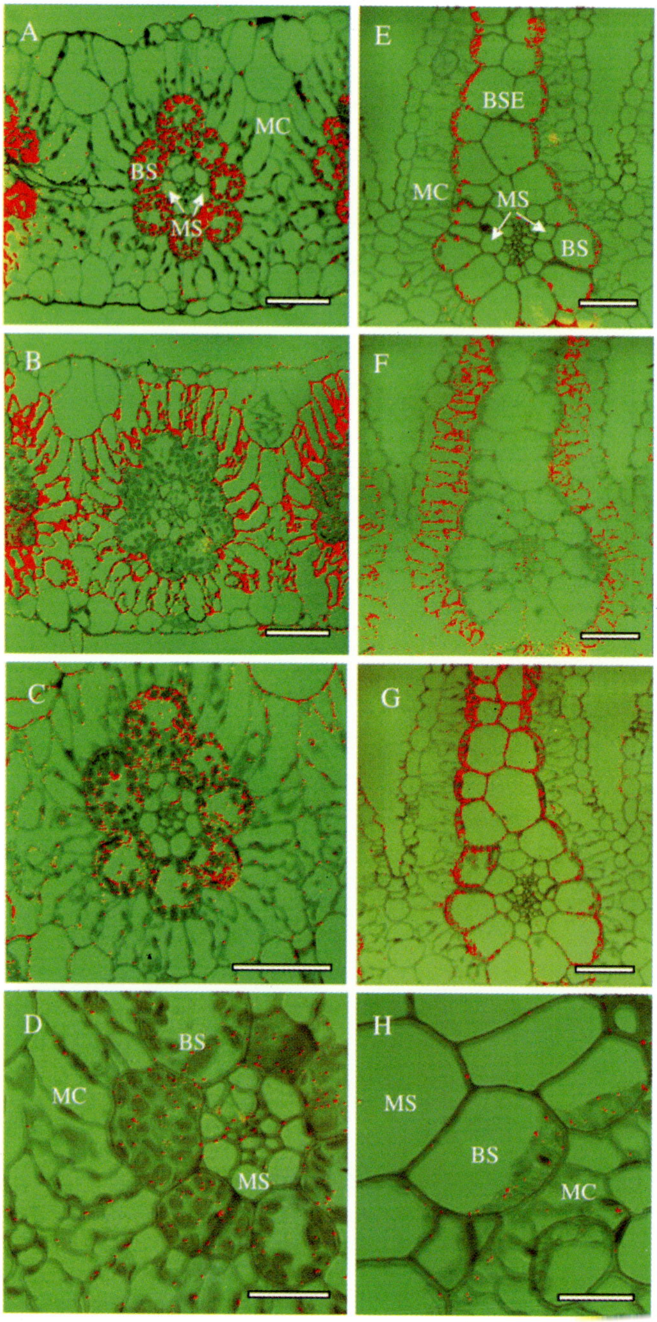

Fig. 5) Reflected/transmitted confocal microscopy imaging of *in-situ* immunolocalization of photosynthetic enzymes in leaves of grasses of the PEP-CK-type C_4 biochemical group. (A-D) *Urochloa texana*:(A) Rubisco; (B) PEPC; (C) PEP-CK; (D) NAD-ME; (E-H) *Spartina anglica*: (E) Rubisco; (F) PEPC; (G) PEP-CK; (H) NAD-ME. BS, Bundle Sheath; BSE, bundle sheath extension; MC, mesophyll cell; MS, mestome sheath. Scale bars: A-C and E-G = 50μ m; D and H = 20 μm.

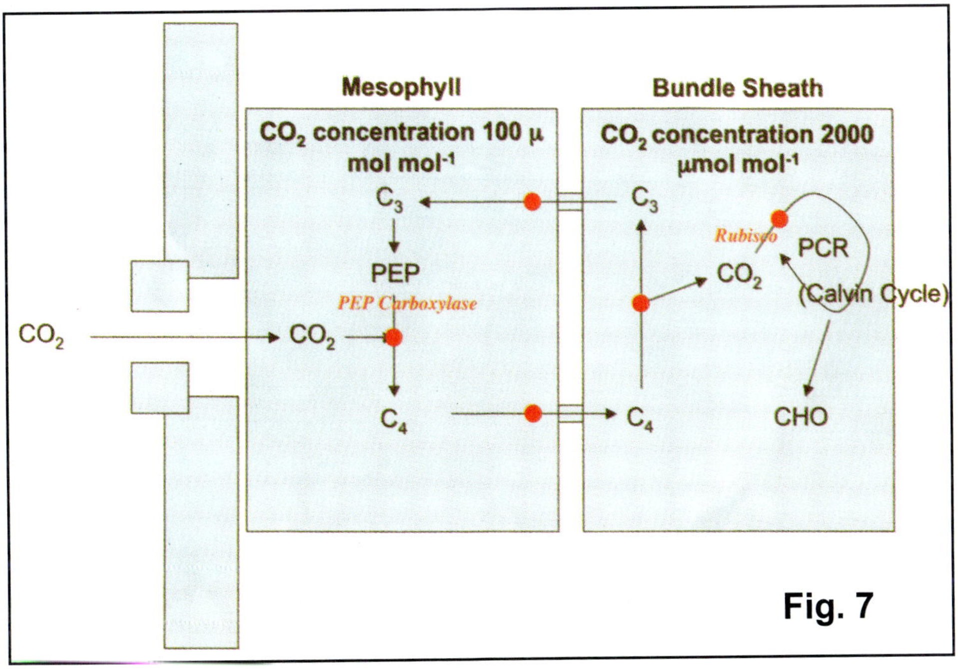

Fig. 7

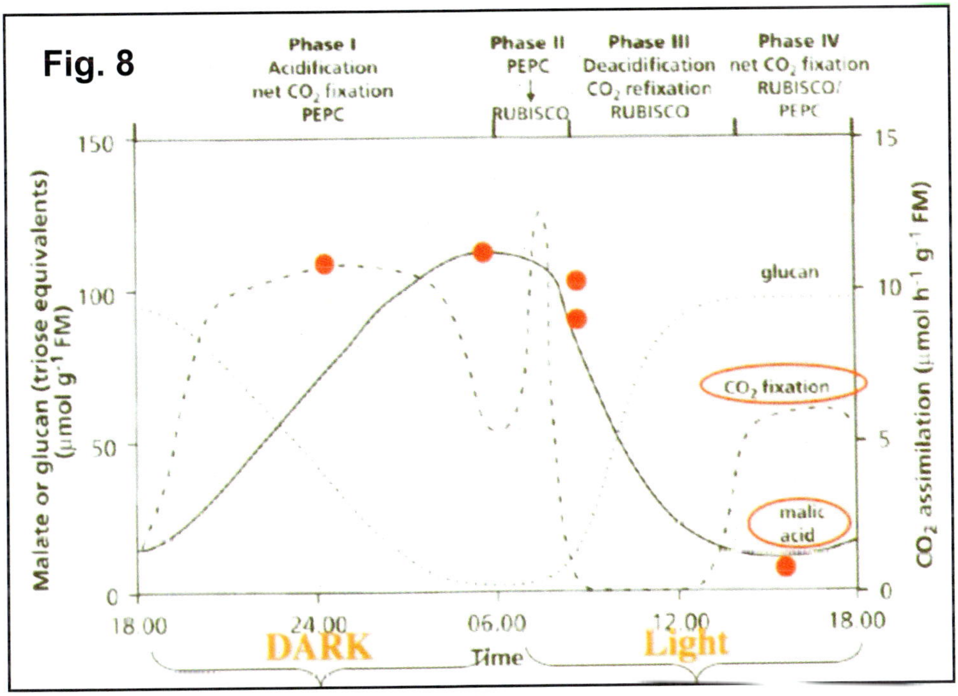

Fig. 7)

Fig. 8) CO_2 fixation in CAM plants, showing diurnal patterns for stomatal conductance, net CO_2 assimilation, malic acid concentration, and carbohydrate concentrations; PEP is PEP carboxylase.

already present in the cotyledons, which contain mostly lipids and little starch. The incipient chlorenchyma cells have a few plastids towards the centre of the cell. Eight days after germination and growth in the dark, small plastids are evenly distributed around the periphery of the expanding cells. Immunolocalization studies show slight labelling of Rubisco in plastids in seeds, includ-al and water storage, but not epidermal cells. After imbibition and 8 d of growth in the dark labelling for Rubisco progressively increased, being most prominent in chlorenchyma cells. There was no immunolabelling for the plastid C_4 enzyme pyruvate Pi dikinase under these conditions. Cotyledons developing in light show formation of chlorenchyma tissue, induction of the cytosolic enzyme phosphoenolpyruvate carboxylase and development of dimorphic chloroplasts at opposite ends of the cells. Proximal chloroplasts have well-developed grana, store starch and contain Rubisco; those located distally have reduced grana, lack starch and contain pyruvate Pi dikinase. Thus, the cotyledons developing in the dark have a single structural plastid type which expresses Rubisco, while light induces formation of dimorphic chloroplasts from the single plastid pool, synthesis of C_4 enzymes, and biochemical and structural polarization leading to the single-cell C_4 syndrome (Voznesenskay *et al.*, 2004).

Table 1 : Main differences between the three subtypes of C_4 species

Major decarboxylase in BSC	Decarboxylation occurs in	Major substrate moving from MC to BSC	Major substrate moving from BSC to MC	Photosystems
• NADP-malic enzyme	chloroplast	malate	pyruvate	I
NAD-malic enzyme	cytosol	aspartate	alanine	I and II
PEP carboxy-kinase	mitochondria	aspartate	PEP	I and II

* MC, mesophyll; BSC, vascular bundle sheath cells.

Development of characteristics of the C_4 syndrome was studied in cotyledons and leaves of NADP-ME type C_4 *Salsola richteri,* a desert shrub. This species had seeds in which the cotyledons contain chloroplasts, storage proteins, and lipid bodies but no starch. Following imbibition (day 0), tissue types are already apparent in the cotyledons, and the chloroplasts have extensive grana stacking, but Kranz type anatomy and C_4 photosynthesis have not developed. At day 0, there is high Rubisco and low PEPC content, distributed throughout all tissue. After 15 d of development in the dark, the cotyledons were in a C_3-like default condition with three distinct layers of chlorenchyma (hypodermis, mesophyll, and bundle sheath [BS]); all contained Rubisco in their chloroplasts. Light was required for development of the C_4 syndrome, including differentiation of chloroplasts in mesophyll and BS cell, development of thick BS cell walls, selective compartmentation of Rubisco in BS cells and PEPC in mesophyll cells, and development of intercellular air spaces around hypodermal and mesophyll cells. During cotyledon development in the light, Rubisco content decreases and PEPC content increases, and starch synthesis was associated with tissue-specific compartmentation of Rubisco. There was also a C_3 to C_4 transition during leaf development. After 3 d of leaf development in the light, cell division was still occurring, the tissue was not fully differentiated, and there was no cell-specific compartmentation of Rubisco or PEP. There was the beginning of tissue and organelle differentiation together with Rubisco and PEPC compartmentation in 5-d leaves. In mature leaves, Kranz anatomy was fully developed with selective compartmentation of PEPC in mesophyll.

Leaf mitochondria and C_4 photosynthesis

Since the mesophyll cell chloroplasts lack Rubisco, and because Rubisco in bundle sheath chloroplasts functions at high CO_2 concentrations, the flux of carbon through photorespiration is low in C_4 plants. Leaf mitochondria in these plants contain low activities of the glycine decarboxylase complex, compared with C_3 plants. However, in both the NAD^+ - ME and PEPCK sub-types of the C_4 pathway, bundle sheath cell mitochondria are the major sites of C_4 acid decarboxylation.

The carbon fluxes through the mitochondria in C_4 sub-types are equivalent to the rate of photosynthesis, i.e. several fold greater than those during photorespiration in C_3 plants and 10 to 20-fold greater than respiratory carbon fluxes in leaves. However, even these fluxes are in order of magnitude slower than those involved in thermogenesis in the *Arum spadix*. Both in the *Arum spadix* and in NAD^+ - ME–type C_4 plants it seems that these high carbon fluxes are freed of respiratory control by engagement of the alternative, cyanide-insensitive pathway of respiration. In PEPCK-type C_4 plants, on the other hand, a much more complicated interaction between malate decarboxylation in mitochondria and OAA decarboxylation in the cytoplasm is indicated. The ATP required for PEPCK in the cytoplasm may arise from the mitochondrial oxidation of malate and oxidative phosphorylation. These complex intracellular metabolite transport patterns are reminiscent of those which appear to operate during de-acidification in CAM plants.

Coordination of the Calvin and C_4 cycles

Coordination of the rate at which the Calvin and C_4 cycles fix CO_2 is necessary if photosynthesis is to proceed efficiently under different environmental conditions. Coordination could occur in a variety of ways, all of which are linked to metabolite fluxes between the bundle sheath and the mesophyll cells:

1. In $NADP^+$ - ME species such as maize, the C_4 cycle is obligatory coupled to the Calvin cycle, because NADPH generated by $NADP^+$-ME is reoxidized by the reduction of 3-PGA in bundle sheath chloroplasts.

2. Electron transport in the mesophyll chloroplasts not only powers conversion of pyruvate to malate in the C_4 cycle, but also drives 3-PGA reduction in all C_4-types.
3. Interconversion of 3-PGA and PEP, catalyzed by phosphoglycerate mutase and enolase in the mesophyll cytoplasm, provides metabolic communication between the C_4 cycle and the Calvin cycle.
4. Products of 3-PGA exported from the Calvin cycle, such as triose-P, and ultimately hexose-P, act as positive effectors of PEPcase, and relieve inhibition by malate.

The regulation of PEPcase in the mesophyll

Coordination of the C_4 cycle with the Calvin cycle during photosynthesis is achieved by metabolite modulation of PEPcase in the mesophyll cytoplasm. This enzyme, and the enzyme which generates its substrate, pyruvate Pi dikinase, are known to be modulated by light. Evidence of coordination between the C_3 and C_4 cycles is seen in the relationship between the assimilation rate and photosynthetic intermediates in *Amaranthus edulis* and maize in response to the intercellular concentration of CO_2. As the rate of photosynthesis increases with increasing CO_2, the RuBP pool falls, but the amounts of PEP and metabolites of the C_4 cycle increases. This behaviour suggests that a feedback loop from the Calvin cycle is operative *in vivo*. The simplest explanation is that the amount of PEP declines when the assimilation rate falls because the amount of 3-PGA exported to the mesophyll also declines. Activation of PEPcase by triose-P and hexose-P is important in determining the response to the supply of 3-PGA, the rate of triose-P utilization by sucrose synthesis in the mesophyll cytosol, and to triose-P consumption in the Calvin cycle. In leaves of maize and *Amaranthus edulis,* the amount of triose-P is always closely related to the assimilation rate whether the flux is changed by alterations in irradiance, CO_2 or temperature. Sucrose synthesis is regulated by fructose-2, 6-bisphosphate (Fru-2, 6-P_2) which ensures that the formation of sucrose is highly responsive to the supply of triose-P because the synthesis of hexose-P is triggered by an appropriate threshold concentration of triose-P. Increasing the concentration of hexose-P then increases the activity of PEPcase. On the other hand, the tendency for diversion of 3-PGA to malate by excessive rates of PEP carboxylation will be ameliorated both by accumulation of the inhibitors of PEPcase, malate and aspartate, and by decrease in the levels of the activators, triose-P and hexose-P.

The contents of aspartate, alanine, glycerate-3-phosphate and triose phosphates were present in sufficient quantities to support transport by diffusion, whereas pyruvate and oxaloacetate were not likely to contribute appreciably to the flux of carbon between the mesophyll and bundle sheath cells of *Amaranthus edulis* L. The amounts of ribulose-

1,5-bisphosphate were high at low intercellular partial pressures of CO_2, and fall rapidly as the CO_2-assimilation rate increased with increasing intercellular partial pressures of CO_2, indicating that bundle sheath CO_2 concentrations fall at low intercellular partial pressures of CO_2. In contrast, the amount of phosphoenolpyruvate and of C_4-cycle intermediates declined at low intercellular partial pressures of CO_2 to the co-ordination of carbon assimilation between the Calvin and C_4 cycles (Leegood and Caeimmerer, 1988).

The activities of starch synthase, of the branching enzyme and of ADP-glucose pyrophosphorylase are higher in the bundle sheath of maize than in the mesophyll. On the other hand, the enzymes of starch degradation, starch phosphorylase and amylase, are evenly distributed and may be slightly higher in the mesophyll. The regulation of starch synthesis has also been modified in C_4 plants. The metabolite gradients which develop during photosynthesis lead to lower 3-PGA/Pi ratios in the mesophyll than in the bundle sheath which, together with the relatively low activities of enzymes of starch synthesis, would appear to limit synthesis of starch in the mesophyll relative to the bundle sheath.

Intracellular metabolite transport in CAM plants

The distinctive features of succulent plants that have CAM metabolism are the nocturnal fixation of CO_2 via PEPcase, accumulation of malic acid in the vacuole and subsequent diurnal de-acidification in which CO_2 released from malic acid is fixed via Rubisco during photosynthesis. This day/night pattern occurs within all chloroplast-containing cells of CAM plants, and the expression of its many components is very sensitive to biological and environmental factors. The essentials of intracellular transport of malic acid and the analogies in metabolic regulation between C_4 and CAM plants are thus best discussed by dissecting the patterns of metabolism into four phases. Phase I embraces dark acidification, involving the stoichiometric conversion of carbohydrates (glucans or soluble sugars) to PEP, carboxylation via PEPcase reduction to malic acid and acid accumulation in the vacuole. Phase II : Upon illumination, CO_2 fixation often increases transiently, involving both PEPcase and Rubisco, before malic acid efflux, de-acidification and stomatal closure. Phase III is the stoichiometric antithesis of phase I, in which de-acidification can be accomplished in 4 to 8 hours, depending on light and temperature. Phase III : The CO_2 which is released by decarboxylation of malic acid accumulates to high concentrations and is refixed by Rubisco. To sustain CAM it is essential that the C_3 residue from malic acid decarboxylation should be conserved during phase III. Phase IV: stomata open again and CO_2 fixation by Rubisco predominates, some CO_2 fixation via PEPcase occurs.

CAM plants like C_4 plants can be divided into major subtypes on the basis of their decarboxylation pathways during phase III. However, unlike C_4 plants, the malic enzyme type (ME-CAM plants) have a cytoplasmic $NADP^+$-ME which, with mitochondrial NAD^+ - ME, participate in decarboxylation. Gluconeogenic recovery of the C_3 residue from decarboxylation requires pyruvate Pi dikinase in the chloroplast. CAM plants seem not to be further differentiated into $NADP^+$ - ME or NAD^+- ME types. PEPCK-type CAM plants have very low malic enzyme activities and no pyruvate Pi dikinase, but high activities of cytoplasmic PEPCK. The implications for metabolic transport between organelles in ME-type CAM plants are easily discerned, but not in PEPCK-type CAM plants.

Simple considerations of volume, pH, and osmotic pressure imply that the massive amounts of malic acid that may be accumulated in CAM plants during phase I must be transported to the cell vacuole. The active transport of malic acid to the vacuole must occur and many studies have demonstrated both an ATP-dependent, and PPi-dependent. H^+ transport in tonoplast membranes and vesicles of CAM plants. The rapid accumulation of such vast quantities of malic acid is unique to the vacuole of CAM plants and the proton-ATPase transport system shares many similarities with proton pumps in other systems.

The energy requirements for malic acid transport to the vacuole dominate the nocturnal respiratory activity of CAM plants. Malic acid synthesis is energetically self-contained, and depending on the carbohydrate utilized, may even lead to net ATP production which can be used for transport. Thus, in ME-CAM plants such as *Kalanchoe*, glycolysis of glucans is initiated by hexose-P production via phosphorylase, and the overall equation for glycolysis, carboxylation and reduction to malate in these plants is:

$$Glucan + P \rightarrow Hexose\text{-}P$$

$$Hexose\text{-}P + ADP + 2CO_2 \rightarrow 2\ malate + ATP$$

In PEPCK-type CAM plants, including the bromeliad *Aechmea*, soluble hexoses are the principal carbohydrate sources for glycolysis. In these, the additional ATP required to synthesize Hexose-P means that the overall equation for glycolysis, carboxylation and malate reduction is simple.

Assuming a similar stoichiometry of the proton ATPase in both *Kalanchoe* and *Aechmea*, one can predict that the ATP demands on respiration for malic acid transport will be about twice as large in the latter as in the former. It was confirmed that molar ratios of O_2 uptake to malic acid was about 6 in *Kalanchoe* and 12 in *Aechmea*. In pineapple, which can use both glucan and soluble sugar in glycolysis, the ratio was intermediate.

In C_3 plants, the metabolite Fru-2,6-P_2 has an important role in the regulation of carbon partitioning during photosynthesis. To investigate the impact of Fru-2, 6-P_2 on carbon metabolism during CAM, transgenic plants of *Kalanchoe daigremontiana* were generated containing a modified rate liver 6-phosphofructo-2-kinase gene. In the plants analyzed, the activity of 6-phospho-fructo-2-kinase ranged from 175% to 198% of that observed in wild-type plants, resulting in Fru-2,6-P_2 concentrations that were 228% to 350% of wild-type plants after 2 h of illumination. A range of metabolic measurements were made on these transgenic plants to investigate the possible roles of Fru-2,6-P_2 during sucrose, starch, and malic acid metabolism across the deacidification period of CAM. The results suggest that Fru-2,6-P_2 plays a major role in regulating partitioning between sucrose and starch synthesis during photosynthesis. However, alterations in Fru-2,6-P_2 levels had little effect on malate mobilization during CAM fluxes (Truesdale *et al.*, 1999).

The electrochemical potential gradient across the tonoplast can be calculated from measurements of cell sap pH and the transmembrane electrical potential at the tonoplast, assuming that malic acid transport at the tonoplast serves to maintain the cytosol at pH 7.5, when malate is fully dissociated. In several CAM plants, at the highest malic acid concentrations, the calculated electrochemical potential gradient is equivalent to a Gibbs free energy of 20-30 kJ mol^{-1}. This corresponds to almost half the free energy of ATP hydrolysis, suggesting that the proton transport system involved in malic acid accumulation in CAM plants is a $2H^+$/ATPase.

In contrast to the active transport of protons and malate into the vacuole, the return passage of free malic acid during phase III is thought to be a passive process. During nocturnal malate accumulation in the vacuole, the dissociation equilibrium in the vacuole shifts towards the undissociated acid.

$$\text{malate}^{2-} \; + H^+ \quad \underset{-H^+}{} \quad \text{H malate}^{1-} \; + H^+ \quad \underset{-H^+}{} \quad H_2\text{ malate}$$

$$pK_a\ 4.25 \qquad\qquad pK_a\ 3.18$$

At high malic acid contents, efflux across the tonoplast is likely to occur via a passive lipid solution mechanism driven simply by the high permeability of lipid membranes to the undissociated acid.

The circadian rhythms of CO_2 fixation via PEPcase in CAM plants in continuous light are consistent with an active malate influx, passive efflux model. The notion that a critical turgor might precipitate a change in membrane properties and facilitate malic acid efflux is contradicted by sustained diel malate influx and efflux in wilted leaves. Thus, it seems likely that the malic acid storage capacity of the vacuole is set by the equilibrium between active influx and passive efflux. A corollary of this is that cytoplasmic malic acid levels also increase, and the rate of malic acid synthesis is impaired due to malate inhibition of PEPcase. Under natural conditions, this situation is reached at the end of 10 to 14 hour dark period. The efflux is presumably amplified following the activation of decarboxylase enzymes during illumination in phase II. Net efflux is sustained throughout phase III by malic acid decarboxylation in the cytoplasm or mitochondria, and the sustained malate inhibition of malate synthesis via PEPcase.

Organelle and metabolite transport processes in CAM plants

CAM plants share the same organelle translocator systems as C_3 and C_4 plants, but they tend to be deployed differently. Thus, the energetic self-sufficiency of malic acid synthesis in those CAM plants which metabolize glucans requires the chloroplast Pi translocator and the dicarboxylic acid shuttle work in phase I and the chloroplast Pi translocators have a high affinity for PEP. This translocator works in the opposite direction in phase III. The gluconeogenic recovery of pyruvate via chloroplastic pyruvate Pi dikinase during phase III presumably depends on an electrogenic pyruvate translocator which is analogous to that in mesophyll chloroplasts of C_4 plants. Subsequently, in phase IV, Calvin cycle CO_2 fixation predominate and chloroplasts which share the properties of mesophyll chloroplasts from C_4 plants then behave as those of a C_3 plant.

Rubisco, PEPcase and several decarboxylase enzymes are present at high activities in all chloroplast-containing cells of CAM plants. The regulation of these activities by metabolites is essential if CAM is to proceed. Thus, for malic acid to accumulate in phase I, PEPcase must be active, and Rubisco and the decarboxylase systems must be inactive. If de-acidification and the Calvin cycle are to proceed in phase III, decarboxylases and Rubisco must be active and PEPcase inactive. At the same time, glycolysis and gluconeogenesis must be regulated to ensure substrate supply and restoration for the cyclic carbon fluxes in CAM.

The amount of PEPcase does not vary throughout the day-night cycle. However, as in C_3 and C_4 plants, malic acid is a potent inhibitor of PEPcase. The PEPcase of CAM plants has a higher affinity for PEP, and a higher sensitivity to malic acid, than the enzyme from C_4 plants. Hence, malic acid synthesis continues in phase I because this inhibitory end product of PEPcase is efficiently transported from cytoplasm to vacuole, and then declines as flux equilibrium is approached. In the light, at least during phase III, futile cycles of malic acid decarboxylation and resynthesis are prevented because PEPcase is inhibited by high malic acid concentration in the cytoplasm during de-acidification (Chollet *et al.*, 1996).

This simple hypothesis has been augmented by evidence that PEPcase is sensitized to malate inhibition in the light, thus greatly amplifying the feedback control. When PEPcase is extracted from CAM plants in the dark, it shows a high affinity for PEP, is susceptible to glucose-6-P stimulation, and has a low sensitivity to malic acid. If extracted from the same tissues during the light period, the enzyme has a low affinity for its substrate, low sensitivity to effectors and is much more sensitive to malic acid. The malic acid insensitive form of the enzyme isolated in the dark is phosphorylated, and this covalent modification of the protein is responsible for changes in kinetic properties. An analogous phosphorylation of PEPcase from C_4 plants in the light has also been reported (Jeanneau *et al.*, 2002).

The circadian clock plays a central role in controlling many of the metabolic, transport and physiological components of CAM. The level of control exerted by the clock can range from transcriptional through to post-translational regulation, depending on the genes, proteins, and even plant species under consideration. A further layer of control is provided by metabolites, including organic acids and carbohydrates, which show substantial reciprocal fluctuations in content over the diel cycle (Borland and Taybi, 2004). The resistance to diffusive flux of metabolites between mesophyll and bundle sheath cells of C_4 plants viz. *Panicum miliaceum, Urochloa panicoides, Atriplex spongiosa,* and *Zea mays* was examined (Weiner *et al.*, 1988). Diffusive flux of metabolites into isolated bundle sheath cells was monitored by following their metabolic transformation and rapid fluxes occurred via functional plasmodesmata. Diffusion of metabolites into these cells was unaffected by a wide variety of compounds including respiratory inhibitors, monovalent and divalent cations, and plant hormones, but it was interrupted by treatments inducing cell plasmolysis.

It is often observed that the rate of CO_2 fixation in the dark declines as the malic acid content of the vacuole reaches a maximum level, towards the end of phase I. This could reflect feedback inhibition due to malate and/or the onset of decarboxylation. Fumarase randomization of the (4-C) carboxyl implies that much of the total pool moves in and out of the mitochondria before entering the vacuole. There is evidence that mitochondrial NAD-ME is inactive in the dark in some CAM plants, but not in others. In these circumstances, inhibition of mitochondrial NAD-ME could prevent futile cycling.

Both Rubisco and the decarboxylase enzymes are active in the light. After malic acid is consumed in photosynthesis, the relief of PEPcase inhibition permits complex parallel CO_2 fixation via both carboxylases in phase IV. However, because net CO_2 fixation in phase IV shows the O_2 sensitivity normally associated with Rubisco, it seems that PEPcase activity is quantitatively unimportant.

Enzyme compartmentation and organelle metabolite translocators play key roles in the regulation of carbohydrate metabolism in CAM plants. These roles can be identified in the glucan-utilizing ME-CAM plants, but are quite uncertain in PEPCK-CAM plants which convert soluble carbohydrates to malic acid and vice-versa. In ME-CAM plants decarboxylation of malic acid produces pyruvate and CO_2. Gluconeogenic recovery of pyruvate can only take place in the chloroplast after conversion of pyruvate to PEP. Chloroplasts of CAM plants, like mesophyll chloroplasts of C_4 plants, evidently contain a specific pyruvate translocator which could ensure the stoichiometric transport of pyruvate into the chloroplast during de-acidification. However, gluconeogenic recovery of PEP involves further transport of PEP from, and E-PGA into, the chloroplasts. CO_2 released from malic acid is fixed via the Calvin cycle and sucrose synthesis proceeds in the cytoplasm. This presumably involves metabolite and Pi exchanges across the chloroplast envelope analogous to those in C_3 plants and similar regulatory interactions (Fischer *et al.*, 1997).

Metabolic flux analysis

Isotopic tracers are used to both trace metabolic pathways and quantify fluxes through these pathways. The use of different labelling methods recently led to profound changes in the views of plant metabolism. Examples are taken from primary metabolism, with sugar interconversions, carbon partitioning between glycolysis and the pentose pathway, or metabolite inputs into the tricarboxylic acid (TCA) cycle, as well as from secondary metabolism with the relative contribution of the plastidial and cytosolic pathways to the biosynthesis of terpenoids. Short time labelling is adequate to study limited areas of the metabolic network. Long-term labelling, when designed to obtain metabolic and isotopic steady-state, allows to calculate various fluxes in large areas of central metabolism. After longer labelling periods, large amounts of label accumulate in structural or storage compounds: their detailed study through the retrobiosynthetic method gives access to the biosynthetic pathways of otherwise undetectable precursors (Dieuaide-Noubhani *et al.*, 2007).

An experimental approach combining IH-NMR and ^{13}C-NMR spectroscopy was proposed to investigate metabolite flux in cells under physiological conditions and presented a mathematical model used for determining the absolute and/or relative values of five fluxes involving pyruvate, ethanol, acetyl-CoA and glutamate via the Krebs cycle in glucose-grown repressed *Saccharomyces cerevisiae* cells fed with (1-^{13}C) glucose and/or unlabelled ethanol. The glucose consumption and the production of various compounds such as ethanol, glycerol, trehalose etc. were studied qualitatively and/or quantitatively as a function of time. The 13C fractional enrichment of (2-13C) ethanol was determined by observing the proton resonance of the methyl group. Addition of 25mM unlabelled ethanol shows no significant effect on the glucose consumption or the production of any metabolites. However unlabelled ethanol exerts a strong influence on the enrichment of glutamate C_4, but only induces an insignificant change on glutamate C_2 and C_3. Apart from the fact that ethanol is a potential precursor of acetyl-CoA as expected, these results indicate that (a) the probability for citrate and 2-oxoglutarate to make one turn or more in the Krebs cycle is negligible and (b) the scrambling between C_4 and C_3 via the glyoxylate shunt is virtually absent. The flux of ethanol formation from pyruvate acetyl-CoA formation, respectively, from pyruvate via pyruvate dehydrogenase. Without addition of unlabelled ethanol, the ratio of the integrated resonance of glutamate $(C_2+C_3)/C_4$ reflecting the activity of pyruvate carboxylase relative to that of citrate synthase, is about 1.1. By comparing the absolute values of the different fluxes, it was found that 88% of the glucose was used to synthesize ethanol but the observed concentration of ethanol in the supernatant represents only 58% of the glucose consumption (Tran-Dinh *et al.*, 1991).

The evaluation of metabolic flux is instrumental in understanding carbon partitioning in plant metabolism. Since fluxes provide a quantitative depiction of carbon flow through competing metabolic pathways (Ratcliffe and Shachar-Hill, 2001), they are an important physiological characteristic akin to levels of transcripts, proteins and metabolites (Saur, 2004). Flux measurements and comparisons of fluxes between phenotypes can provide insights toward selection of appropriate metabolic engineering targets (Stephanopoulos, 2002; Glawischnig *et al.*, 2002) and toward the construction of predictive models of plant metabolism (Ratcliffe and Shachar-Hill, 2001), the necessity for which has been emphasized recently (Girke *et al*, 2003; Katagiri, 2003; Raikhel and Coruzzi, 2003).

Although the importance of flux measurement in plants has often been stressed (Shachar-Hill, 2002), it has received rather limited attention in plant science as compared to profiling technologies for transcript, protein, and metabolite levels (Kruger and von Schaewen, 2003). This is principally due to the fact that fluxes have to be quantified by back-calculating them from their effect on the distribution of a labelled substrate, and such calculation requires a detailed mathematical model if it is to be accurate. Mathematical models relating labelling data to fluxes are often nontrivial, particularly in the case of compartmented metabolism inherent in plants. Consequently, flux measurement technology in plants remains underdeveloped (Ratcliffe and Shachar-Hill, 2001; Sweetlove *et al.*, 2003).

Most of the reports of labelling studies in plants have focused on the qualitative goal of inferring which pathways are in operation (identification of metabolic network topology) but not on the mathematically involved endeavor of evaluating how much carbon is processed by those pathways (quantification of flux). Wheeler *et al.*, (1998) delineated the pathway of ascorbic acid synthesis in higher plants from ^{14}C labelling data, and Krook *et al.*, (1998) showed that two separate oxidative pentose phosphate pathways (oxPPP) operate in the cytosol and the plastid, using ^{13}C enrichment data from *Daucus carota* cell suspensions. More recently, a suite of articles by Eisenreich and co-workers (Bacher *et al.*, 1999; Glawischnig *et al.*, 2002) reported the abundances of isotopomers of several isolated sink metabolites in maize (*Zea mays*) kernels. Although these articles demonstrated advances in label measurement technology, the inferences from them were either qualitative or semiquantitative.

Two recent pioneering research efforts have concentrated on quantification of fluxes in plants. In the first, Raymond and co-workers calculated fluxes through glycolysis, oxPPP, tricarboxylic acid (TCA) cycle, and anaplerotic reactions using ^{13}C atom enrichment data of metabolites isolated from tomato (*Lycopersicon esculentum*) suspension cells (Rontein *et al.*, 2002) and maize root tips (Dieuaide-Noubhani *et al.*, 1995). However, ^{13}C isotopomer abundance data provide richer metabolic information than ^{13}C atom enrichments. Isotopomers are isomers of a metabolite that differ in the labelling state (^{13}C or ^{12}C) of their individual carbon atoms. Isotopomer measurements provide information on carbon-carbon connectivities in metabolites, in systems supplied with a mixture of U- ^{13}C-labelled and U- ^{12}C-labelled substrates (Szyperski, 1998). Therefore, isotopomer abundance data facilitate statistically superior estimation of metabolic fluxes and reaction reversibilities than atom enrichment data (Wiechert *et al.*, 1999). The second important development in flux quantification in plants was by Ohlrogge and co-workers, who used mass isotopomer data of metabolites separated from developing *Brassica napus* embryos to calculate fluxes (Schwender *et al.*, 2003), although only the glycolysis and oxPPP were considered.

^{13}C-labelling experiments were performed on developing soybean embryos and obtained exhaustive labelling data from sink metabolites by 2-D NMR (Sriram *et al.*, 2004). It was possible to quantify carbon partitioning through several metabolic processes, including glycolysis, oxPPP, gluconeogenesis, anaplerotic pathways, TCA cycle, and glyoxylate and GABA shunts. Fluxes in different compartments were distinguished based on labelling data of sink metabolites known to be synthesized in separate compartments.

Fluxes were calculated and statistical analysis was performed. A high flux was found through the oxidative pentose phosphate pathway (19.99 ± 4.39 μmol d^{-1} cotyledon^{-1}, or 104.2 carbon mol ± 23.0 carbon mol per 100 carbon mol of Sucrose uptake). Separate transketolase and transaldolase fluxes could be distinguished in the plastid and the cytosol, and those in the plastid were found to be at least 6-fold higher. The backflux from triose to hexose phosphate was also found to be substantial in the plastid (21.72 ± 5.00 μmol d^{-1} cotyledon^{-1}, or 113.2 carbon mol ±26.0 carbon mol per 100 carbon mol of Sucrose uptake). Forward and backward directions of anaplerotic fluxes could be distinguished.

The Metabolic modelling

The information used to construct a model is ultimately based on experimental observation, and that the behaviour of the resulting model is determined by those observations. If the model behaviour is at variance with expectation, there are two possible causes: either a mistake has been made in the definition of the model, or the expectations are incorrect. It is under these conditions that insight is generated. Either of these cases may be put to good use, if, for example, it is possible with elementary modes analysis to show that a given set of reactions cannot catalyse a net reaction shown to be present experimentally, then the presence of at least one extra reaction in the system has been demonstrated. Identifying missing reaction(s) would be rather less easy, but could be approachable by further modelling studies. Both the theory and technology that underpin metabolic modelling continue to develop. Current challenges include the ability to make the connection between modelling and bioinformatics, and to build models containing all expressed enzymes. This is starting to become feasible for structural modelling, although the possibility of dealing with kinetic models of such magnitude appears to be some way off.

To date, progress has been made predominantly within the field of metabolic engineering with the technique known as metabolic flux analysis. Lack of space precludes a detailed description, but, briefly, manipulation of equation means that if certain input or output fluxes can be measured, it is possible to determine the rates of at least some of the internal reactions.

The most important challenge is to re-integrate the work of modellers with that of experimentalists. The full potential, not only of modelling but also the spectacular experimental techniques of the post-genomic era, are unlikely to be realized if the two groups are unable to find a common language.

REFERENCES

Bacher A Rieder C Eichinger D Arigoni D Fucher G Eisenreich W 1999 Elucidation of novel biosynthetic pathways and metabolite flux patterns by retrobiosynthetic NMR analysis FEMs *Microbiol Rev* **22** 567-598

Borland AM and Taybi T 2004 Synchronization of metabolic processor in plants with crassulacean acid metabolism *J Exptl Bot* **55**(400) 1255-1265

Chollet R Vidal J O'Learg MH 1996 Phosphoenol pyruvate a ubiquitous highly regulated enzyme in plants *Annu Rev Plant Physiol plant Mol Bio* **47** 273-298

Devi MT Raghavendra AS 1993 Partial reduction in activities of photorespiratory enzymes in C_3-C_4 intermediates of *Alternanathera* and *Parthenium* Oxford University Peru

Dieuaide–Noubhanbi M Alonso AP Rolin D Eisenreich W Raymond P 2007 Metabolic flux analysis Recent advances in carbon metabolism in plants *Experientia supplementum* **97** 217-243

Dieuaide-Noubhani M Raffard G Canioni P Pradet A Raymond P 1995 Quantification of compartmented metabolic fluxes in maize root tips using isotope distribution from ^{13}C–or ^{14}C labelled glucose *J Biol Chem* **22** 13147-13159

Drincovich MF Casati P Andreo CS Chessin SJ Frnacesch VR Edwards GE Ku Maurrice SB 1998 Evolution of C_4 photosynthesis in *Flaveria Species Pl Physiol* **117** 733-744

Duvalli MR Saar DE Grayburn WC Holbrook GP 2003 Complex transitions between C_3 and C_4 photosynthetic during the evolution of paniceal A photogenetic case study emphasizing the position of *Steinchisma hians* (Poaceae) a C_3-C_4 intermediate- *International J Sci* **164** 949-958

Edwards GE Franceschi VR Maurice S Elena KB Voznesenskay V Pyankoo VJ Anreo CS 2001 Compartmentation of photosynthesis in cells and tissues of C_4 plants *J Exptl Bot* **52**(356) 577-590

Fischer K Kammerer B Gutensohu M Arbinger B Weber A Hausler RE Flugge UI 1997 A new class of plastidic translocators a putative link between primary and secondary metabolism by the phosphoenol pyruvate/ phosphate antiporter *Plant cell* **9** 453-462

Gayathri J Parvathi K Chinthapalli B Westhoff P Raghavendra AS 2001 Immunological characteristics of PEP carboxylase from leaves of C_3- C_4- and C_3-C_4 intermediate species of *Altenathera*–comparison with selected C_3- and C_4- plants *Ind J Exp Biol* **39**(7) 643-649

Girke T Ozkan M Carter D Raikhel NN 2003 Towards a modelling infrastructure for studying plant cells *Plant Physiol* **132** 410-414

Glawischinig E Gierl A Tomas A Bacher A Eisenreich W 2001 Retrobiosynthetic nuclear magnetic resonance analysis of amino acid biosynthesis and intermediary metabolism metabolic flux in developing maize kernels *Plant Physiol* **125** 1178-1186

Glawischnig E Gierl A Towar A Bacher A Eisenreich 2002 Starch biosynthesis and intermediary metabolism in maize kernals quantitative analysis of metabolite flux by nuclear magnetic resource *Plant Physiol* **130** 1717-1727

Jeanneau M Vidal J Gausset- Dupout A Lebouteiller B Hodges M Gerentes D Perez P 2002 Manipulating PEPcase levels in plants *J Exp Bot* **53** 1837-1845

Katagiri F 2003 Attacking complex problems with the power of systems biology *Plant Physiol* **132** 417-419

Krook J Vrengenhil D Dijkema C Vasdev P as LHW 1998 Sucrose and starch metabolism in carrot (*Daucus carota)* cell suspensions analyzed by ^{13}C labelling indication for a cytosol and a plastid-localized oxidative pentose phosphate pathway *J Exp Bot* **49** 1917-1924

Krugar N J Von Schaewen A 2003 The oxidative pentose phosphate pathway Structure and organization *Curr opin Plant Biol* **6** 236-46

Leegood RC Caeimmerer SV 1988 The relationship between contents of photosynthetic metabolite and the rate of photosynthetic carbon assimilation in leaves of *Amaranthus edulis* L *Planta* **174**(2) 253-262

Leegood RC 2002 C_4 photosynthesis principles of CO_2 concentration and prospects for its introduction into C_3 plants *J Expl Bot* **53**(369) 591-590

Poolman MG Assmus HE Fell DA 2003 Application of metabolic modelling to plant metabolism *J Exptl Bot* 55(400) 1177-1186

Raikhel NV Coruzzi GM 2003 Plant systems Biology *Plant Physiol* **132** 403

Rajendru G Prasad JSR Dama Das VS 1986 C_3-C_4 intermediate species in *Alternanthera (Amaran thaceal) Pl Physiol* **80**(2) 409-414

Ratcliffe RG Sachar I Hill Y 2001 Probing plant metabolism with NMR *Annu Rev Plant Physical Plant Mol Biol* **52** 499-526

Routein D Dieuaide Noubhani M Dufurce EJ Raymond P Ralin D 2002 The metabolic architecture of plant cells stability of central metabolism and flexibility of anabolic pathways during the growth cycle of tomato cells *J Biol Chem* **277** 43948-43960

Sachar-Hill Y 2002 Nuclear magnetic resource and plant metabolic engineering *Metabol Engg* **4** 90-97

Saur U 2004 High thoughtput phenomics experimental methods for mapping fluxomes *Curr Opin Biotechnol* **15** 58-63

Schwender J Ohlrogge JB Sachar Hill Y 2003 A flux model of glycolysis and the oxidative pentose phosphate pathway in developing *Brassica napus* embryos *J Bio chem* **278** 29442-29453

Sriram G Fulton DB Iyear VV Peterson JM Zhou R Westgate ME Spalding MH Shanks JV 2004 Quantification of compartmented metabolic fluxes in developing soybean embryos by employing biosynthetically directed fractional ^{13}C labeling two dimensional (^{13}C ^{1}H) nuclear magnetic resonance and comprehensive istopomer balancing *Plant Physiol* **136** 3043-3057

Stephanopoulos G 2002 Metabolic engineering perspective of a chemical engineer *AICHEJ* **48** 920-926

Sweetlove LJ Last RL Fernie AD 2003 Predictive metabolic engineering a goal for systems biology *Plant Physiol* **132** 433-448

Szypoerski T 1998 ^{13}C NMR MS and metabolic flux balancing in biotechnology research *Q Rev Biophys* **31** 41-106

Tran-Dinh S Herve M and Wietzerbin J 1991 Determination of flux through different metabolite pathways in *Saccharomyces cerevisiae* by ^{1}H-NMR and ^{13}C-NMR spectroscopy *Eur J Biochem* **201** 715-721

Truesdale MR Toldi O Scoot P 1999 The effect of elevated concentrations of fructose 2 6- bisphosphate on carbon metabolism during deacidification on the crassulacean acid metabolism plant *Kalanchoe daigremoutiana Pl physiol* **121** 957-964

Voznesenskay EV Franceschi VR Edwards GE 2004 Light-dependent development of single cell C_4 photosynthesis in cotyledons of *Borszczowia aralocaspica* (Chenopodiaceae) during transformation from a storage to a photosynthetic organ *Ann Bot* **93** 177-187

Wheeler GL Jones MA Smirnoff N 1998 The biosynthetic pathway of vitamin C in higher plants *Nature* **393** 365-369

Wiechert W Mollney M Isermann N Murzel M de Graff A A 1999 Biodirectional reaction steps in metabolic networks III Explicit solution and analysis of isotopomer labelling systems *Biotechnical Bioengg* **66** 69-85

11

PHYSIOLOGICAL PROCESSES, NITROGEN FIXATION, GROWTH AND PRODUCTIVITY OF LEGUMES IN RESPONSE TO ABIOTIC STRESSES

PUSHP SHARMA AND VIRENDER SARDANA

INTRODUCTION

Legumes are important component of conservation agriculture. Some of the unique characteristics of legumes are deep rooting, nitrogen fixation capacity, leaf shedding and ability to mobilize insoluble soil nutrients. Shift towards intensive cereal-cereal cropping systems in order to feed increasing population has met its basic objective of self-sufficiency in food grain supply. On the other hand, it has resulted in the deterioration of soil health, leading to the unsustainably of the system. By improving physical, chemical and biological environment in soil, legumes can arrest the declining trend in productivity. Inclusion of legumes in intensive cereal based cropping systems in itself can act as a component of integrated nutrients supply. Therefore, legumes have become a viable alternative to improve soil health, conserve natural resources and sustain agricultural productivity. Currently no amount of chemical fertilizers can fully compensate for crop rotation benefits that included a pulse and analysis of these individual factors generally does not explain the entire yield responses associated with this system. Legumes help in solubilizing insoluble P in soil and N derived from the legumes, either grown in sequence or in intercropping systems benefit the following or associated crop. Grain legumes namely soybean, cowpea, pigeonpea and groundnut as preceding crops had beneficial residual effect on the yield of following cereal crop (Subba Rao *et al.*, 2007). So, legumes should essentially form a stable and remunerative proposition in cropping systems with potentials of higher productivity and saving on the cost of chemical fertilizer.

Plants are often faced with numerous environmental factors, which hamper their growth and performance. In general, crops are limited to about 25% of their potential yield by the impacts of environmental stresses (Boyer, 1982). Environmental stress can be defined as a set of conditions that cause an aberrant change in physiological processes, which eventually leads to injury of system. These stresses are broadly divided into two major group i.e. abiotic (physico-chemical) and biotic (pathogens, herbivores etc.). Generally biotic stresses endanger an active diffusive response and may cause modifications in the plant developmental plan. Abiotic stresses faced by legumes such as changes in rhizospheric atmosphere, drought, salinity, waterlogging, temperature extremities etc., make plants to respond with some sort of defence mechanisms but chronic exposure typically modifies their developmental patterns (Abbas *et al.*, 1999). Physiological responses of pulse crops for survival in stressful environment are based on their ability to express pre-existing defense programme and/or adaptation during which plants adjust to the stress in an environmental oriented manner. In this review, we have discussed the legumes physiology, growth, productivity and N_2 fixation in relation to abiotic stresses. Table 1 outlines the relative significance of abiotic factors according to different zones in India. Drought is the most serious factor at all the places. High temperature is important cause of low yield in region from 0 to 25° latitude, whereas extrememly low temperature at 25 to 45° latitude affects the crop performance seriously. High moisture content is occasionally important at 20 to 30° latitude. Soil toxicity can be serious factor in any region.

Table 1 : Relative importance of Abiotic Stresses (Deshmukh *et al.*, 2006)

Abiotic Stress	Zone (Latitude)				
	I(0-20°)	II(20-25°)	II(25-30°)	IV(30-45°)	
Drought	1	1	1	-	1
High temperature	2	2	4	-	2
Low temperature	-	-	3	1	-
High moisture	-	4	5	2	-
Soil toxicities	3	3	2	-	-

Where 1 = Extremely important, 5 = Occasionally important

Leguminous plants belonging to family Fabaceae (Leguminosae) are of outmost importance by virtue of their ability to scavenge atmospheric nitrogen through symbiotic association with certain soil microorganisms belonging to genera *Rhizobium*, *Bradyrhizobium* etc. Legumes possess certain specialised structures called root nodules and the symbiotic interaction between plants and microorganisms leads to the synthesis of ammonia, which is ultimately assimilated to amino acids and proteins. The physiological processes that influence legume rhizobium symbiosis depend on number of factors, which are discussed vividly in this review.

1. SALINITY STRESS

Salinity invariably affects the germination and growth of plants through a check on the nitrogen-fixing machinery, protein synthesis, carbohydrate metabolism, enzyme synthesis etc., which in turn have a tremendous negative effect on the harvest index of the crops. It is important to note how these commonly encountered stresses, involving water relationship of the plant with that of the environment would affect the system at the species or at the varietal level, in order to screen relatively salt tolerant lines. In this context the efficacy of different physiological processes under adverse conditions must be appraised and analysed in details.

1.1 Effect on Germination, Growth Rates and Productivity

Salt injury influences several physiological processes and productivity, which has gained immense importance in the last couple of years. It is the toxicity of ions penetrating the cells as well as augmentation in the osmotic potential of the medium, which subsequently hinder the penetration of water into the roots (Munns *et al.*, 1983). It was believed that the accumulation of salts within plant tissues caused a disturbance in the plant growth and development thus, altering the degree of hydration of cytoplasm during the period of osmotic adaptation (Sands and Clarke 1977). The extent of injury varied with plant species and within same species, it varied with the type of predominant ions, their concentration, the physiological stage of growth at which plant is exposed to salinity (Kumar, 1973).

1.1.1 Osmotic effects: The mechanism of salinity damages in plants, for almost half a century, has been attributed to "physiological drought". However, many plants adapt themselves osmotically on being exposed to salinity, allowing water flow into the plant. Concomitantly resistance of leaves to water loss increased with increasing salinity. These adjustments that tended to overcome the increased resistance to water uptake in the roots resulted partially either from active or passive salt accumulation in the cells or breakdown of polysaccharides or the release of K^+ ions from bound sites within the cells. However, such processes are energy intensive and expenditure during osmotic adjustment to salinity was one of the main factors responsible for reduced plant growth (Greenway 1973). Earlier studies laid greater emphasis on the changed osmotic potential while in recent years, attempts have been made to understand the direct effects of salinity on metabolic processes.

It is difficult to distinguish between the osmotic and specific ion effects of salinity on various metabolic processes. The most common method employed to distinguish between secondary osmotic and primary salt injury is to compare the effects of isotonic solutions of salts with those of organic substances. Polyethyleneglycol (PEG), a high molecular weight substance (4000 daltans) is the most effective and widely used osmoticum with no evidence of toxicity. Strogonov and Lapina (1964) found that some plants were relatively insensitive to the osmotic effects while others to the primary salt effects. In some cases, osmotic effects of PEG were more deleterious on growth (Kuhad and Garg, 1983; 1984) while in others, ionic effects of salt solutions prove to be more detrimental (Sharma and Garg 1983; Kuhad *et al.*, 1987). Osmotic environment reduced the rate of sugar and amino acids uptake from the attached seed coats of soybean (Walswinkel and Ammerlaan 1988). It was concluded that the low osmotic potential of the seed apoplast was plausibly one of the most important factors controlling the rate of assimilate transport into the developing seeds. Barnet *et al.* (1996) reported reduction in dry matter accumulation, number and weight of nodules of *Medicago sativa* with application of NaCl. These detrimental effects seemed to be mainly due to the toxicity of the salts rather than by the osmotic stress as induced by PEG treatments. The imposition of water stress (10 per cent w/v PEG 6000) on the non-structural carbohydrates in soybean (*Glycine max*) plants inoculated with two different nitrogen fixing strains of *Bradyrhizobium japonicum* caused reduction in transpiration, nodule dry weight, nitrogen fixation as well as the activities of sucrose synthase and alkaline invertase (sucrose consuming enzymes). The decrease was relatively less in the nodules colonized by strain USDA 110 as compared to those with 61A (Muller *et al.*, 1996) *Arachis hypogaea* cultivar JL 24 showed greater tolerance compared to TMV 2 under various levels (0.6 to 0.1 MPa) of PEG stress (Purushotham *et al.*, 1998) and correlated the tolerance character with higher values of growth, turgor potential, amino acids and proline accumulation.

1.1.2 Nutritional effects/ionic effects : Salinisation of soil may be due to excess of cations (Ca^{2+}, K^+, Mg^{2+} and Na^+) anions (Cl^-, SO_4^{2-}, NO_3^-, CO_3^-) or both. However, their proportion may vary from soil to soil. The excess of some ions in the soil would prevent the uptake or availability of other ions present in relatively smaller amount. The NaCl induced salinization, locked the uptake of Ca, Mg and K in plants. Hasson-Porath *et al.* (1972) showed that K^+ content was slightly higher in roots of pea plants which were exposed to NaCl as compared to those supplied with sodium sulphate. Sashidhar *et al.* (1981) reported a varietal variation regarding better performance in groundnut following $CaCl_2$ treatment which was linked to higher endogenous K^+ concentration under stress. Sheoran and Garg (1983) observed more growth reduction in *Cicer arietinum* under sodium sulphate salinity than chloride salinity which might be due only to toxic ion accumulation in case of chloride, whereas under sulphate salinity, both the ion accumulation and reduced turgor seemed to be responsible. Lal and Bhardwaj (1984) observed decreased contents of total nitrogen, potassium and magnesium and increased contents of phosphorus, calcium, sodium and chloride upon salinity treatment in field pea and reported that salinity tended to affect the absorption, mineral composition, accumulation and balance of different major nutrients in the organs of pea.

El-Sayed *et al.* (1992) reported that cells on *Glycine max* adapted at 680 mM NaCl accumulated Na^+ and Cl^-as principal solute for extensive osmotic adjustment. Salinity levels of 7.25 dSm^{-1} and 9.00 dSm^{-1} led to an increase in the contents of Na^+ with a concomitant decrease in K^+ causing seedling death after 25 days of growth in *Cajanus cajan* L. (Gill and Sharma, 1993). Saline conditions resulted in shrivelling of *Phaseolus radiatus* L. Wilczek seeds leading to a significant reduction in the level of Na, K and Cl than healthy seeds. Although Na and Cl content increased with salinity in both healthy and shrivelled seeds, increase in K was evident only in the former (Dhingra and Sharma 1993). Gulati and Jaswal (1994) observed increase in sodium concentration in the roots of tolerant cultivars of *Phaseolus radiatus* as compared to those of salt sensitive cultivars and ascribed this to Na accumulation at the shoot or cellular level. Dhingra *et al.* (1994) reported the accumulation of Na and Cl in chickpea seeds without any effect on K content under saline conditions. *Vicia faba* L. plants treated with salinity levels of 1.5 upto 13.7 dsm^{-1} showed decreased water and osmotic potential of leaves and an increased influx of Na^+ and Cl^-ions into the stem and leaves with a concomitant reduction in K^+ ions thereby causing salt injury to the plants (Sharma, 1995).

Olmas and Hellin (1996) reported increase in the intracellular levels of Na^+ and Cl^- in cell cultures of *Pisum sativum* adapted to 85.5 mM NaCl with an increase in the concentration of sugars (solute for osmotic adjustment). Cordovilla *et al.* (1996) reported that an addition of 8 mM KNO_3 to the medium increased the plant tolerance in *Vicia faba*. 0.5 mM of $CaCl_2$ ameliorated the negative effect of 200 mM NaCl in callus cultures of *Arachis hypogaea* (Gupta and Singhal, 1996). Nodules of *Lupinus albus* exposed to 150 mol m^{-3} NaCl for 6 days showed an exclusion of Na^+ ions from roots and leaves with only a few Cl^-ions penetrating the region, which might be possible reason for the relative salt tolerance of the plant (Fernandez-Pascual *et al.*, 1996). Increase in salinity up to 7.9 dsm^{-1} led to an increase in the concentrations of Na^+ and Cl^-in chickpea genotypes (Sharma 1997). Dua (1998) studied the effects of 80 mM saline solution on 20 day old salt tolerant CSG 88101 and salt sensitive CSG 8890 genotypes of chickpea and reported deleterious effects on growth and root/shoot water content with a marked increase in Na^+ and Cl^-ions in the sensitive genotypes while the shoot water content remained almost constant indicating an appropriate index of salinity tolerance. Tolerant cultivar attained higher K^+/Na^+ ratio as compared to the sensitive one and such difference in the cation distribution could be used for identifying resistant cultivars.

Pulses or grain legumes have been classified as moderately sensitive to sensitive in response to salt stress (Table 2).

Table 2 : Salt tolerance of pulses (Francois and Maas, 1999)

Crop	Tolerance based on	Thre- ECE (dsm^{-1})	Slope shold	Rating (% per ds m^{-1})
Common bean (*Phaseolus vulagaris* L.)	Seed or pod yield	1.0	19	Sensitive
Chickpea (*Cicer arietinum* L.)	Seed yield	-	-	Moderately sensitive
Cowpea (*Phaseolus unguiculata* L. Walp.)	Seed yield	4.9	12	Moderately tolerant
Mungbean (*Phaseolus radiatus* L. Wilczek)	Seed yield	1.8	20.7	Sensitive
Peas (*Pisum sativum* L.)	Seed fresh weight	3.4	10.6	Moderately sensitive

ECE : Electrical conductivity of the saturated soil extract

Chickpea is comparatively moderately sensitive to saline conditions and genetic variation in response to salinity has been reported by many workers. The mechanism to cope saline conditions involves genoty pe's inherent ability to exclude salt or accumulate it in plant system. Tolerant genotype have an ability to exclude sodium. Therefore, screening for tolerance to salinity is largely done by growing plants at graded level of salinized soil. The response of chickpea to salinity varies with the moisture availability in the soil. Reduction in chickpea yield under saline conditions occurs both as a result of osmotic and specific ion effects. The significant interaction of variety, salinity and moisture level was observed for yield (Bhardwaj, 1962). Since, interaction occurs between salt tolerance and nitrogen source, selection under both symbiotic and nitrogen fed conditions has been thought desirable.

Salinity reduced N content of nodules, while Na^+/K^+ ratio and Cl^-content increased significantly (Kumar *et al.*, 2003). Urdbean genotypes IP 98/120 and IP 98/126 grown under salinity conditions (0-12 mS cm^{-1}) showed accumulation of osmotic metabolites like glycine betaine, choline and proline (Rathore and Vershney, 2003). Salinity regulates the biosynthesis of plant stress hormones ABA and causes accumulation of reactive oxygen species (Hernandez *et al.*, 2001). An osmosensory histidine kinase cascadi probably regulates osmotic homeostasis and ROS scavenging. Osmotic stress and absicisic acid mediated regulation of LEA (late embryogensis abundant) type proteins also play important roles in plant salt tolerance (Viswanath *et al.*, 2005).

1.1.3 Germination : It is generally believed that salinity delays germination and reduces final germination percentage. Prisco and O'Leary (1970) reported that below-8 bars only osmotic effects operated, while above that both osmotic and toxic effects of ions caused in the inhibition of germination of *Phaseolus vulgaris*. However, the contribution of each factor in affecting germination might depend upon the salinity level, ionic species and plant species or even varieties. Sharma *et al.* (1971) reported that germination in some varieties of green gram was more sensitive to NaCl, $CaCl_2$ and Na_2SO_4 while others to Na_2HCO_3. Kumar (1973), Sheoran (1975) and Sheoran and Garg (1983) reported that Na_2SO_4 resulted in reduced germination whereas NaCl, KCl and K_2SO_4 delayed germination in cowpea, mung and gram. However, different salts failed to influence germination of *Medicago* (Kuhad *et al.*, 1987) and *Trifolium* (Gonzalez-Murna *et al.*, 1985). Garg *et al.* (1988) reported progressive degeneration in the germinability and early seedling growth of mung seeds with increasing salt concentration of the medium. Nautiyal *et al.* (1989) reported significant suppression of germination and seedling growth of four cultivars of *Arachis hypogaea* L. by salinity level 20.5 EC. Garg (1992) observed significant reduction in the germination percentage and early seedling growth in four legume genera namely pigeonpea, soybean, mungbean and chickpea proving to be most salt sensitive and soybean relatively the most salt tolerant. In soybean with increasing NaCl concentrations of 20, 40 and 60 mM, a corresponding inhibitory effect on germination were reported by Murthukumarasamy and Panneerselvam (1995). A significant reduction in the germination percentage, shoot and root lengths as well as fresh mass was observed in two cultivars of *Vigna radiata* L. Wilczek namely Sujata and K 851 when the same were grown in 0.5, 1, 2 and 3 per cent NaCl salinity. NaCl (100 mM) was sufficient to cause a reduction of about 65.4 per cent in the seedling length of *Vigna unguiculata* (Filho Eneas *et al.*, 1995). Significant reduction in germination has been reported in chickpea seedlings. Reduction in germination of mungbean due to concentration of salts of sodium and calcium was reported by Sharma *et al.* (1971) and Kheradnam and Ghordshy (1973). Chloride – dominant salinity was found to be more toxic to chickpea then sulphate salinity (Manchanda *et al.*, 1981).

Irrigation water containing electrical conductivity of 4 mmhos cm^{-1} caused injury to Mungbean (Reddy, 1982). Mungbean seeds could tolerate 6 mmhos cm^{-1} salinity as compared to 3 mmhos cm^{-1} by urdbean while concentration of 8.3 mmhos cm^{-1} reduced plant height, leaf area and plant dry matter production of mungbean with 5 levels of salinity. Reduction in germination and seedling development was observed in 22 mungbean and 22 cultivars of urdbean as salinity increases (Maliwal and Paliwal 1982).

1.1.4 Growth and Development : Salinity affects all the stages of development in a majority of crops while in a few crops, the sensitivity varies with the growth stage. Saline conditions, in general, cause stunting, restrict lateral shoot development, reduce the size of leaves and fruits and decrease the number, size, fresh and dry weights of different plant organs ultimately reducing yield to varying degrees. In addition to the degree of salinity, the type of salt responsible for salinition has an important role in the growth and development of plants. Hassan Porath *et al.* (1972) reported that at-4 bars Na_2SO_4 was not deleterious whereas NaCl was detrimental to pea seedlings. Garg (1976) observed that during germination of pea seeds, bicarbonates were more depressive while carbonates were more detrimental during the subsequent growth.

Salinity caused constriction of pea roots above the apex (Solomon *et al.*, 1986) whereas in chickpea salt primarily inhibited growth through the accumulation of Na^+ and Cl^-in the shoots (Lauter and Munns 1987).

Sharma and Kumar (1990) reported that Pusa 209 of chickpea was affected more in terms of visible injury, growth and yield than BG 312, which proved to be more salt tolerant. Romero and Maranon (1994) reported reduced growth during early and middle stages and root-shoot allometric coefficients in the generative phase of *Melilotus segetalis* grown in saline environment. Saradhi *et al.* (1996) observed a significant reduction in the growth of 5 day old seedlings of *Vigna radiata* under salt stress. NaCl, KCl, Na_2SO_4 or $CaCl_2$ at 130 OSM reduced plant dry matter production of *Medicago sastiva* by 50 per cent after 55 days of growth (Banet *et al.*, 1996).

Muthukumarasamy and Panneerselvam (1997) concluded that NaCl concentrations of 100 and 150 mM were sufficient to reduce growth of green gram, while 50 mM did not have any adverse effect. Sangwan *et al.* (1997) observed that in two genotypes of *Cicer arietinum*, increasing salinity levels upto 60 mol^{-1} led to a continuous declension in relative growth of the calli with the sensitive genotype showing more reduction in relative growth rate as compared to the tolerant.

1.2 Effect on Physiological and Biochemical processes

1.2.1 Carbohydrate metabolism: Metabolism of carbohydrates in plants is affected by an increase in the level and type of salinity. Inhibition in amylase activity and the decreased sugar content in horsegram (*Dolichos biflorus*) has been reported by Sudhakar *et al.* (1987). Salinity checked the absorption and incorporation of externally supplied ^{14}C-glucose into the macromolecules of pea root tips and NaCl proved to be more depressive than Na_2SO_4 (Hasson-Porath and Poljakoff-Mayber, 1973). An analysis of relative amounts of different sugars, in the roots and aerial parts of crops showed that control plants contained mainly glucose with small amounts of sucrose and only traces of fructose, while plants grown in NaCl contained no detectable amounts of fructose and glucose but a markedly higher sucrose content Strogenov (1973). The total sugar content was much lower under saline conditions.

Cells of *Glycine max* L. adapted to NaCl salinity showed an increase in the average concentration of soluble sugars and total free amino acids as a function of the level of adaptation towards salinity stress (El-Sayed *et al.*, 1992). Saline conditions resulted in shrivelling of *Vigna radiata* L. Wilczek seeds leading to a significant decrease in total soluble sugars as well as reducing sugars (Dhingra and Sharma, 1993). A similar decrease in the soluble sugars and starch has also been reported by Dhingra *et al.* (1994) in the seeds of 3 promising genotypes of chickpea viz. ICCV 88102, H-82-2 and C-235. Singh and Singh (1995) reported a decrease in the reducing, non-reducing sugars and total saccharides in *Pisum sativum* with increasing soil exchangeable sodium percentage. Siddiqui and Krishnamoorthy (1995) observed that salinity decreased the amylase activity and starch content with concomitant accumulation of soluble carbohydrates in the cotyledons of *Vigna unguiculata.* Sheokand *et al.* (1995) reported an increase in the total soluble carbohydrates in 45-days old chickpea plants subjected to 50 and 100 mM NaCl treatments. Salinity levels of 20, 40 and 60 mM inhibited carbohydrate metabolism in the seedlings of soybean (Muthukumarasamy and Panneerselvam 1995). Mishra *et al.* (1996) also reported an accumulation of total soluble saccharides in *Vigna radiata* L. seedlings under 0.5, 1, 2 and 3 per cent NaCl. Relatively tolerant calli of *Pisum sativum* accumulated more reducing sugars indicating osmotic adaptation via reducing sugars (Olmos *et al.*, 1996). NaCl concentration of 150 mol m^{-3} decreased nodular starch content with an increasing sucrose content suggesting osmotic regulation in *Lupinus albus* L. Accumulation of total as well as reducing sugars was reported by Garg *et al.* (1996).

1.2.2 Nitrogen metabolism : Although nitrogen metabolism is affected by various levels/types of salinity and the plant species/cultivars, there is lack of understanding regarding effect of salinity on N-metabolism.

Nodulation and N-fixation in relation to salt stress: Salinity depresses nodulation and nitrogen fixation in almost all the N fixing systems. The deleterious effects of salinity may be associated with the survival of *rhizobium,* the infection process, the nodule initiation, its development and nitrogen fixation *per se.* Salinity induced through NaCl decreased nodule size and number in *Vigna sinensis* and many other cultivated beans (Hendry, 1918), caused poor nodulation in pasture legumes (Cope, 1958) and complete failure of nodule development as reported by Kling (1954). Significant reduction in nodulation and nitrogen fixation and nodule dry matter in alfa-alfa and soybean (Bernstein and Ogata, 1966), nodulation in pea (Kumar and Garg, 1981) has been well documented. Salinity and/or alkalinity reduced the production of leghaemoglobin with a concomitant enhancement in the greening of nodules, reduced leaf area, total nitrogen and dry matter content of pea plants (Kumar and Garg, 1981). Wilson (1970) reported an inhibition in the initiation of new nodules with a simultaneous reduction in nitrogen fixation by the existing nodules in *Glycine wightii*. However, the nodules already developed were surprisingly resistant and retained their pigmentation and nitrogen fixing capacity. NaCl applied to detached nodules or whole root system of soybean inhibited N_2 fixation and acetylene reduction by small nodules was stopped by 100 mel^1 NaCl (Sprent, 1972). High salt concentrations (rooting medium) markedly inhibited nodulation and nitrogen fixation (nitrogenase activity) in *Pisum sativum* and *Vicia faba* (Yousel and Sprent, 1986). Total nodule number, weight and nitrogen content per plant decreased due to salt treatment which interfered with the nodule initiation but not with their further development in cowpea and mungbeans (Balasubramanian and Sinha, 1976). The decline in nitrogen fixing efficiency of nodules was more in mungbean than that of cowpea. In clusterbean and mothbean salt stress decreased nodule number, dry weight and nitrogenase activity (Garg *et al.*, 1984). Siddiqui *et al.* (1985) observed poor development of leghaemoglobin in the nodules with concomitant enhanced senescence while desalinization reversed these adverse effects to varying levels. Zahran and Sprent (1986) reported that polyethylene glycol (PEG) significantly reduced root hair diameter and proportion of root hair containing infection threads of *Vicia faba* plants as compared to NaCl. After 3 weeks of treatment, both NaCl and PEG inhibited nodule number by about 50 per cent and nodule weight by more than 60 per cent. Gill (1987) observed that poor yield in susceptible cowpea was due to the alkalinity around root zone which caused its adverse effects on the stem and leaves decreased the chlorophyll, protein, RNA and DNA contents in these active parts. Similarly, Rai (1987) observed a significant decline in the number and weight of nodules, reduced nitrogenase activity, nodulin content, leghaemoglobin content and proline concentration in five genotypes of lentil (*Lens culinaris*) with increasing salinity.

Kumar and Sharma (1989) have isolated and characterized callus cultures of *Vigna radiata*, which were resistant to high concentrations of NaCl (300 mol m^{-1}) and had lower K^+ content as compared to the normal media.

Sheokand *et al.* (1995) reported that 45 days old chickpea plants when subjected to NaCl (50 and 200 mM) treatments for 7 and 14 days accelerated nodule greening and decreased the leghaemoglobin (Lb) content by 41 per cent and 27 per cent of the control and decreased the acetylene reduction activity (ARA) by 63 and 70 per cent after 7 DAT. However, 50 mM NaCl treated plants showed recovery in leghaemoglobin, total soluble carbohydrates and ARA, 14 days after treatment suggested the probable role in osmoregulation. Comba *et al.* (1997) worked on two soybean cultivars, one NaCl tolerant (377) and the other relatively NaCl sensitive (411) and reported a 50 per cent decrease in leghaemoglobin (Lb) as well as ARA in the latter with no significant change in tolerant cultivar.

Root nodules of mungbean irrigated with saline water recorded significant reduction in osmotic potential, relative water content, increase in proline content and a sharp decline in leghaemoglobin content with increasing salinity (Kumar *et al.*, 2003).

Salinity reduces nodule formation in *Vigna* group. However, once the nodules are formed, salinity does not affect its functioning if plants maintain reasonable photosynthetic activity. Remarkable increase in ethylene evolution was noticed from nodulated roots of mungbean with the increasing level and duration of salinity (Nandwal *et al.*, 2000).

Proteins and nucleic acids: Salinity, by and large, depressed protein synthesis, accelerated its degradation and distributed the soluble amino acids protein ratio in plants. Both uptake and incorporation of labeled amino acids into proteins were reduced by the presence of salts in the incubation mixture in pea roots. Sulphate salinity was found to be more deleterious than chloride. Breakdown of polysomes accompanied by an increased level of RNAse during stress has been demonstrated (Marin and Vieira De-Silva, 1972).

Moderate salinity in some cases may increase total protein or non-protein nitrogen but innumerable reports revealed decrease in protein and nucleic acetic acid content in roots, leaves and embryo axis of mungbean, chickpea, pea and cowpea under NaCl and Na_2SO_4 salinity (Kumar, 1973; Sharma, 1974; Sheoran, 1975 and Garg and Garg, 1980). Exposure to NaCl (50 mM) reduced the incorporation of ^{14}C-proline and completely inhibited ^{3}H-hydroxyproline into the proteins in pea roots. Further salinisation of medium (with NaCl and $Na_2SO\text{-}_4$) significantly increased yield of the total peptide fraction from pea plants. Isolation of peptides, specific for salinisation from the peptide fraction, revealed stimulated synthesis of growth inhibiting peptides on a salinised background and inhibition of synthesis of growth stimulating ones (Gabre, *et al.*, 1978). The NaCl and Na_2SO_4 influenced the incorporation of ^{14}C-Leucine into protein of whole leaves of horse bean (*Vicia faba*) in which protein synthesis was stimulated upto sodium level of 4.0 per cent above which, a phase of inhibition was set in (Leonova and Shevyako, 1975). Salinity increased incorporation of ^{14}C-uracil into RNA in the excised embryo axis of mungbean with sulphate proved to be more effective than chloride (Gabre *et al.*, 1978). Bar-Nun and Poljakoff-Mayber (1979) observed increased leucine content in protein of a salt tolerant variety of *Pisum sativum* and suggested a correlation between the leucine content in the protein and salinity tolerance of a plant. Salinity disturbed the accumulation pattern of nitrogenous fraction in pea leading to decrease in total as well as protein nitrogen (Lal and Bhardwaj, 1987). Differential response of NaCl and Na_2SO_4 salinity was noticed in *Cajanus cajan* for total nitrogen and protein content where these contents were higher under NaCl but reduced under Na_2SO_4, indicating the specific ion effects on nitrogen metabolism.

Garg and Garg (1982) reported significant reduction in the contents of RNA, DNA and proteins with increasing concentrations of salts in *Phaseolus aureus*. Saline conditions resulted in shriveling of *Vigna radiata* seeds leading to significant decrease in proteins, free amino acids as well as proline (Dhingra and Sharma, 1993). Singh and Singh (1995) reported a decrease in the protein content in three cultivars of *Pisum sativum* with increasing soil exchangeable sodium percentage. El-Sayed *et al.* (1992) reported that the relative salt tolerance of *Glycine max* cultivars was due to the accumulation of soluble proteins, amino acids and proline. Saline conditions led to decrease in the protein content with a concomitant increase of protease activity in the root, shoot and leaf of chickpea seedlings. However, a reversal of the trends was observed in the cotyledons. Soybean callus cultures treated with 150 mM of NaCl showed significant increase in the protein content (El-Enamy, 1997).

Free amino acids and proline : The response of salinity on total free amino acids has been found to be variable with contents of some amino acids like proline increased, whereas that of glutamate decreased (Saxena, 1971). The increase in proline content was reported for the first time by Kamble and Macpherson (1954), amounted to as high as one per cent of the leaf dry matter in many species (Singh *et al.*, 1972). Biosynthesis of sulphur containing amino acids was inhibited under both NaCl and Na_2SO_4 salinisation.

However, NaCl salinisation increased ^{35}S-methionine incorporation into cysteine and sulfoxide of S-methyl cysteine whereas, Na_2SO_4 inhibited the same (Leonova and Shevyakova, 1975). Similarly, Pluennke and John (1972) observed significant increase in asparagines, arginine and methionine contents of plants grown under high Na concentration. Totawat and Saxena (1971) found a decrease in lysine, arginine, histidine, proline and cysteine content in *Phaseolus aureus* with increasing salinity of irrigation water. Stewart and Lee (1974) opined that proline might function as a source of solute for intracellular osmotic adjustment under saline conditions. Chu *et al.* (1976) found that plants grown on salt solutions accumulated less proline than on iso-osmotic solution of PEG indicating role of proline as a storage compound for reduced carbon and nitrogen during stress. Sheoran and Garg (1978) also concluded that accumulation of proline in chickpea and mungbean was determined by both osmotic as well as specific ion effects. Bar-Nun and Poljakaff-Mayber (1979) observed reduction in basic amino acids with a concomitant increase in aspartic acid, glutamic acid, threonine, serine, leucine and isoleucine among different varieties of *Pisum sativum* with increasing salinity. The most tolerant cultivar showed significant accumulation of free proline indicating a direct correlation with salinity tolerance. The pre-treatment of 10-day old *Pisum sativum* with proline (10^{-6}M or 10^{-5}M) for 24 hours before salinisation and subsequently with 50 mM NaCl for 2-days resulted in an increase in endogenous free proline content which alleviated the inhibitory effect of NaCl resulting in decreased Na^+ and Cl^-accumulation in the shoot (Fedina *et al.*, 1993). A direct correlation between proline accumulation and salt tolerance was reported by Sashidhar *et al.* (1981) in groundnut and Moftah and Michel (1987) in soybean. The increase in the level of salt concentration (50, 100, 150 and 200 meql^{-1}) in the 20-days old three salt tolerant lines of *Vicia faba* L. fed to accumulation of proline in the roots and stems making it an index for plant tolerance towards aggressions of salinity (Bekhodja *et al.*, 1993). Belkhadja (1996) further observed that the amount of proline differs in different plant organs depending upon the development stage, species of plants and salt treatment and increases proportionally with the increasing salt concentration in the some tolerant varieties of *Vicia faba*. Saline conditions resulted in shriveling of *Vigna radiata* seeds leading to a significant reduction in proteins, free amino acids as well as proline compared to healthy seeds grown under non-saline conditions (Dingra and Sharma, 1993). Differential response of two cultivars of *Glycine max* to salinity led to an increase in the protease activity, amino acids and proline accumulation, which probably served as an osmoticum to protect the cell organalles and enzymes under saline conditions (Durga Prasad *et al.*, 1996). Abd El-Samad *et al.* (1997) also attributed the salt tolerance in two cultivars of soybean to the accumulation of amino acids and proline with cultivar Clark showing tolerance to NaCl salinity upto-1800 Kpa. The negative effect of NaCl could be ameliorated by $CaCl_2$ (0.5 mM) in *Arachis hypogaea* indicating that a specific Ca^{2+}/Na^+ ratio was required for the cell growth (Gupta and Singhal 1996). This ratio also signified an increase in the proline content indicating a direct link between proline content and improved growth. At flowering stage, increase in the salt concentration upto 10 dSm^{-1} decreased chlorophyll content with a concomitant increase in RNA and free proline in the leaves of clusterbean, mungbean and mothbean. However, clusterbean was relatively more tolerant than the other two. Garg *et al.* (1996) and Mishra *et al.* (1996) also observed 4-5 times higher proline accumulation in the roots of mungbean seedlings when grown under 0.5, 1, 2 and 3 per cent NaCl. In another study, 5-day old *Vigna radiata* seedlings exposed to NaCl (200 mM) showed 2 fold higher level of proline (Saradhi *et al.*, 1996), whereas *Medicago sativa* and *Glycine max* showed 5 fold proline and 2 fold trigonelline increase (Tramontano *et al.*, 1997). An increase in proline, glycine and betaine over control along with a marked reduction in growth and chlorophyll content was observed in the root, stem and leaf of green gram at 100 and 150 mM NaCl treatments (Muthukumarasamy and Panneerselvam 1997). Urdbean cultivars IP 98/120 and IP 98/126 grown under the salinity conditions (0-12 $mScm^{-1}$) showed accumulation of osmotic metabolites like glycine betaine, choline and proline. Higher content of glycine betaine was observed in the plants as compared to choline, which might`be due to quicker metabolization of accumulated choline (Rathore and Varshney 2003). Garg (1991) and Muthukumarasamy *et al.* (1997) reported that saline conditions enhanced the proline content over control in all parts of chickpea seedlings.

Enzymes **:** Effects of salinity on plant metabolism can be visualized at the sub-cellular level since the same is known to reduce the synthesis of RNA, DNA and proteins and increase their degradation by affecting the hydrolytic enzyme such as nucleases and proteases. Salinity responses of various enzymes depend on the intensity, duration, types of ions, stage of plant growth and assay conditions. The activity of some enzymes decreased and that of others increased while a few remain immune in the presence salt. Nitrate reductase (NR), nitrogenase and phenylalanine ammonia lyase (PAL) are probably the enzymes most dramatically affected by the salt stress. Plaut (1974) reported a significant decrease in the level of nitrate reductase both *in vivo* and *in vitro* by the presence of sodium chloride in nutrient solution. Similarly, Klyshev (1974) recorded significantly more suppression of nitrate reductase activity by NaCl than Na_2SO_4 at the same osmotic concentration in pea (*Pisum sativum*). Total as well as specific activity of glutamate dehydrogenase (GDH) and glutamine synthetase were inhibited by salts in pea (Rakova *et al.*, 1978) probably due to reduction in the contents of glutamic acid. Hassan-Porath and Paljakoff-Mayber (1979) established that salts reduced the ratio of NAD and NADH under NaCl and Na_2SO_4 salinisation in root tips of pea.

A significant suppression in the activity of NR and GDH with increasing salinisation has also been in *Pisum sativum* (Lal and Bhardwaj, 1987). Greenway and Osmond (1972) showed that enzymes malate dehydrogenase, aspartate transaminase, glucose 6-phosphate dehydrogenase and isocitrate dehydrogenase extracted from salt sensitive (*Phaseolus vulgaris*) and salt tolerant (*Atriplex spongiosa*) species grown in saline cultures showed no significant changes in specific activity with an increase in salt concentration. Similar response for several other enzymes from leaves, roots and stem of pea was reported by Weimberg (1970) under NaCl, KCl, Na_2SO_4 and K_2SO_4 salinisation. Sheoran and Garg (1978) reported inhibition of RNase and DNase activities in the cotyledons and roots but an increased activities in the embryo axis and leaves without any effect on the protease activity in any of the organs during the early seedling growth in mungbean under saline conditions. Variable response of different plant parts to peroxidase activity was also reported by these workers. Activities of phosphorylase and protease were enhanced while that of proline dehydrogenase and amylase were inhibited in cotyledons of horsegram (*Dolichos biflorus*) under NaCl or Na_2SO_4 salinity (Sudhakar *et al.*, 1987). However, Rouxel *et al.* (1987) observed no change in the subcellular distribution of RNase activity under NaCl salinity.

Studies on the effect of increased NaCl concentrations on seed germination of two *Medicago* species revealed an appearance of glyoxysomal enzyme activity leading to the disappearance of storage lipids (Ben-Miled-Daoud and Cherif, 1992). Garg *et al.* (1988) reported a significant increase in peroxidase activity in the nodules of mungbean with increasing levels of salinity in the rooting medium. However, salinity levels of 12 dSm^{-1} enhanced the IAA-oxidase activity in the leaves of soybean with a marked reduction in the peroxidase activity (Zaidi and Singh 1995). Huang and Chen (1995) observed enhanced activities of glutathione reductase (GR) and its related enzymes like ascorbate reductase in salt acclimated plants of soybean. They further observed that GR kept a high ratio of GSH/GSSG in the cytosol, which probably enabled the plants to tolerate the salt stress environment. Filho Eneas *et al.* (1995) observed reduced activities of α- and β galactosidases under 100 mM NaCl treatment in the cotyledons of *Vigna unguiculata*, suggesting that salinity probably delayed enzyme solubilization Cordovilla *et al.* (1996) reported decline in the activities of (GS, NADH-GOGAT and NADH-GDH) in nodules of *Vicia faba* by 8 mM KNO_3 and this concentration of KNO_3 increased the salinity tolerance even at 100 mM NaCl treatment. Nitrate reductase (NR) and nitrite reductase (NiR) activities substantially declined in the leaves and roots of two soybean genotypes grown under a range (0-12 dSm^{-1}) of NaCl and Na_2CO_3 salt (Khan 1996). Olmos and Hellin (1996) while working revealed the higher presence of enzymes related to sugar metabolism (glucokinase, fructokinase and acid invertase) in two lines of *Pisum sativum* in response to NaCl. Sodium chloride stress lowered the DH activity in mungbean at 2-3 per cent NaCl. Garg *et al.* (1997) reported a significant reduction in the levels of nitrate reductase and glutamine synthetase under salinity upto 12 dSm^{-1} in clusterbean. Continuous decline in α-amylase activity was reported in two genotypes of *Cicer arietinum* subjected to 60 mel^{-1} salinity (Sangwan *et al.*, 1997). Bishnoi *et al.* (1997) revealed a significant decrease in catalase, ARA and peroxidase activity in soybean callus cultures treated with 150 mM concentration of NaCl whereas total sulfhydryl group (-SH) and GR concentrations increased. The GSH/GSSG ratio increased significantly as a result of GR activity, which seems to be related to the adaptive response of soybean to NaCl (El-Enamy, 1997). Comba *et al.* (1997) reported that NaCl concentration of 150 mM decreased acetylene reduction activity (ARA) by 50 per cent in sensitive cultivar, without having any effect on tolerant cultivar of soybean. Furthermore, decrease of 25 per cent in GSH, 50 per cent in glutathione reductases, 30 per cent in ascorbate peroxidase, 20 per cent in catalase and 25 per cent in superoxide dismutase activities in the sensitive cultivar and an increase of 15, 20, 30 and 18 per cent respectively in the tolerant ones. Rajan *et al.* (1988a) observed that NaCl salinity decreased accumulation of sugar content and activities of α and β-amylase, whereas soaking of NaCl stressed seeds in triadimefan (15 mg l^{-1}) alleviated the adverse effects of salt stress by increasing the sugar content and α and β-amylase activity. These results thereby indicated a correlation between N_2-fixation and oxidative stress induced by salt treatment.

Photosynthesis: Chlorophyll content of leaves decreased under salt stress conditions (Sharma, 1974; Garg and Garg, 1982). *Cajanus cajan* plants when treated with salinity levels of 7.25 and 9 dSm^{-1} showed a marked declension in the total chlorophyll content (Gill and Sharma, 1993). Application of 100 mM NaCl salinity in *Vigna radiata* resulted in decline in chlorophyll content to the tune of 50 percent, 57 per cent increase in chlorophyllase activity and 51 per cent decrease in Hill reaction activity as compared to control (Chakravarti and Mukherji, 1994). Significant decrease in the chlorophyll 'a' and 'b' content was reported by Singh *et al.* (1994) in *Vigna radiata*. Salinity levels upto 20 dSm^{-1} inhibited total chlorophyll content, chlorophyll a/b ratio, net assimilation rate (NAR) and relative growth rate (RGR) at 25 and 50 days after sowing in soybean (Zaidi and Singh, 1995) and green gram (Muthukumarasamy and Panneerselvam, 1997). On the contrary, an increase in chlorophyll content under salt stress was reported by Nieman (1962) in pea leaves. Weimberg (1975) reported that saline conditions lowered the rate of chlorophyll formation in etiolated pea seedlings exposed to light. The commonly observed effect of salinity as induced by various salts in many of the crop plants has been decrease in photosynthesis per unit leaf area. Shone and Gale (1983) observed that NaCl salinity diminished both gross and net photosynthesis in *Medicago sativa*. Salt stress, in general, reduced the rate of photosynthesis in both C_3 (*Phaseolus vulgaris* and *Xanthium strumarium*) as well as C_4 (*Zea mays* and *Atriplex halimus*) plant species which could be partially overcome by supplementing an extra dose of CO_2 (Schwarz and Gale 1984). Six week old chickpea genotype when treated with four levels of salinity showed a reduction in the rate of photosynthesis as well as CO_2 assimilation (Sharma 1997). Ball and Anderson (1986) while probing into the inner facets of photosynthesis as moderated by salinity in mangrove and pea observed a linear decline in photosystem II (index O_2 evaluation) with increasing NaCl concentration with both the species suffering in more or less equal magnitude.

Lapina *et al.* (1976) observed that iso-osmotic (2-3 atm) concentrations of sodium chloride and sodium sulphate reduced the cyclic and non-cyclic photophosphorylation in *Phaseolus vulgaris*. The 10-day old seedlings of *Pisum sativum* when treated with NaCl (50 mM) showed an increase in CO_2 compensation, concentration, photorespiration with an inhibition of about 60 per cent in the net photosynthetic rate (P_N) and $^{14}CO_2$ fixation in the protoplasts (Fedina *et al.*, 1993).

Respiration : The phenomenon of accentuated respiration by roots in the presence of high concentration of mineral salts usually referred as salt respiration has been established since 18th century. High concentration of sodium chloride has often, but not always, been reported to increase respiration of roots and other tissues in beans (Neiman, 1962). Respiration was stimulated by about 33 per cent in the roots of pea seedlings grown in 77 mM NaCl (Livne and Levin, 1967). Decrease in respiration rate has also been reported (Sarin and Rao 1958).

Stimulation or repression of respiration was governed by the salt concentration with lower salt concentration generally stimulating, but higher level of salinity decreasing the respiration. Like most of the other salinity responses, there appeared to be large differences among various species depending upon the salt level, plant organ, the stage of plant growth, etc.

The route of respiration was changed from glycolytic to pentose phosphate shunt in pea root tips under saline conditions (Goris 1969). Chloride salinity depressed CO_2 evolution from C_6 and not from C_1 whereas sulphate salinity depressed both. Salinity depressed activities of most of the glycolytic enzymes by about 10 folds. Greenway *et al.* (1977) studied glycolysis *in vitro* using extracts from germinating pea seeds and showed that KCl and NaCl reduced steady state rates of glycolysis by weakening the control of glycolysis by enzyme 6-phospho-fructokinase. Subsequently, high NaCl and KCl concentrations had two main effects :

- Transient reduction in the rate of glycolysis, which disappeared when substrate of certain enzymes increased above the initial low level due to accumulation.
- Permanent inhibition, which persisted despite high substrate concentration.

Salinity levels of 100 mol m^{-3} NaCl significantly increased the O_2 diffusion resistance in the nodules of *Lupinus albus* over a 6-day period (Fernandez-Pascual *et al.*, 1996). The 5-day old *Vigna radiata* seedlings exposed to 200 mM NaCl stress significantly increased the level of NADH, H^+ with marked decrease in the electron transport activities than control (Saradhi *et al.*, 1996).

1.3 Plant Growth Regulating Substances (PGRs) in relation to Growth and Metabolism

It is believed that in addition to genetic manipulations physiological tolerance to salinity can also be improved through an indepth analysis of legume-rhizobium physiology as well as their metabolism. In this context, the application of growth regulators to manoeuvre the efficacy of different plant physiological processes under adverse conditions should be appraised in detail.

One of the most common effects of salinity is stunting of growth, often without any other sign of damage such as chlorosis of leaf burn. Kahane and Poljakoff Mayber (1968) reported reduction in amino acid incorporation into protein in the cells of pea roots, which were exposed to sodium chloride or sodium sulphate. It was further observed that the effects of NaCl but not of Na_2SO_4 could be partially counteracted by exogenously applied kinetin. Sarin and Uprety (1965) reported an improvement in flower induction and consequently the yield by exogenous application of phosphon-D, GA_3 and B-nine under saline conditions. A check on the deleterious effects of salinity by growth regulators may be due to protection provided by them against protein breakdown. Prisco (1970) reported that benzyladenine and gibberellic acid sprays increased fresh and dry weight of shoots of intact salinised plants, but did not affect control plants. Kinetin was, however, found to be ineffective. NaCl increased the resistance of roots to water flow and that of leaves to water vapour diffusion. Kinetin, benzyladenine and GA reduced leaf resistance to water vapour diffusion. Shokraii and Fahimi (1969) found that foliar spray on plants growing in saline media, with 5-25 mg l^{-1} of GA, IAA, kinetin or cycocel intensified the suppression of growth. Foliar sprays of ascorbic acid reversed the effects of salinity on nodulation, nitrogen fixation and growth of pea plants (Kumar, 1973).

Plant growth regulators, especially kinetin and abscisic acid (ABA), appear to play an important role in plant water relations through their effect on stomata. Kinetin promoted and ABA reduced opening of stomata. Other hormones may also be involved in the response of plants to salinity. Indole acetic acid (IAA) was probably not directly involved as it did not affect stomatal aperture (Johansen, 1954), although it might affect plant water balance by modifying the number and size of leaves. GA_3 application (100 ppm) counteracted the negative effects of salinisation on growth, photosynthesis and translocation in bean plants (Strack *et al.*, 1975). Bozuck (1981) observed that kinetin applied simultaneously with sodium chloride to the seeds reduced the stress response significantly and caused breakage of osmotically induced dormancy. He further reported that in salt stressed bean plants, zeatin and GA_3 partially re-established the monovalent : divalent ion rate, disturbed under salt stress.

Chakrabarti and Mukherji (1994) reported that NaCl (100 mM) application reduced the Hill activity by 50 per cent and chlorophyll content by 51 per cent in *Vigna radiata*, which was completely reversed by pre-treatment of 10^{-6} M kinetin. Singh *et al.* (1994) observed that kinetin @ 50 and 100 ppm was able to partially reverse the inhibitory effect of salinity on chlorophyll content and resulted in lesser proline accumulation in *Vigna radiata*. Dhingra *et al.* (1995) reported that both NAA (20 ppm) and BAP (1 ppm) improved seed proteins and starch content in chickpea under saline conditions. Agarwal and Gupta (1995) while working on callus cultures of cowpea reported that kinetin and IAA (30 mg l^{-1}) suppressed the salt induced injuries by lowering the osmotic potential and maintaining the cellular turgidity with an increase in the level of proline, soluble amino acids, sugars, insoluble sugars and proteins thereby maintaining the growth processes under salt stress. Zaidi and Singh (1995) reported that growth regulators (GA_3 and IAA) alleviated the detrimental effects of NaCl salinity on leaf growth, biochemical compositions, relative growth rate (RGR) and net assimilation rate (NAR) in *Glycine max*. Further, salt stressed soybean plants when treated with growth regulators viz. GA_3 and IAA, showed a reduction in IAA-oxidase activity with a simultaneous increase in the peroxidase activity in the leaves and roots with GA_3 proving to be relatively more effective. Cytokinin (10^{-5} M) treatments administered to *Cicer arietinum* and *Glycine max* neutralized the harmful effects of salinity (upto 6 dsm^{-1}) and significantly improved the nitrogen fixing capacity, leghemoglobin, nitrogen content, total sugars, total free amino acids etc. in soybean while proving to be relatively less effective in ameliorating stress in case of chickpea indicating relatively more salt tolerance of soybean then chickpea (Garg *et al.*, 1995). Zaidi and Singh (1996) observed progressive enhancement in the Na^+/K^+

and Na^+/Ca^{2+} ratios along with reduced root and leaf growth in soybean. Further presoaking of soybean seeds with 100 and 200 ppm solutions of GA_3 and IAA significantly mitigated the adverse effects of salinity. Plant growth retardants (triazoles and CCC) when applied as soil drench improved not only growth and photosynthetic activity but also grain yield of mungbean under salinity. It was due to increase in chlorophyll, protein content and activity of catalase enzymes in leaves (Saha and Gupta, 1998). Salt stress caused by NaCl lowered the dry weight, chlorophyll and protein content besides inhibiting the activities of proteases and proline oxidase in urdbean. However, amino acids particularly proline and glycine betaine increased. It was further observed that paclobutrazol treatment ameliorated the deleterious effects of salt stress by increasing amino acid, protease, proline oxidase activities, protein and chlorophyll content and hence produced high dry matter in urdbean (Paneerselvam *et al.*, 2000).

It may be concluded that changes in the levels of hormones also appear to be involved in the response of plants to salinity, especially by way of their effect on plant water balance. The levels of endogenous hormones, known to be promotory to stomatal opening, are lowered while that of ABA are raised under saline conditions and this phenomenon has gained lot of weightage in last few years. However, little is known about the mechanism of salt injury in the plant and every piece of research has attempted to add a new dimension/explanation to the existing state of knowledge.

2. TEMPERATURE EXTREMITIES

2.1 High temperature

2.1.1 Photosynthesis : Plant productivity is directly related to the rate of photosynthesis which is one of the most heat sensitive processes. For a specific plant there exists an optimal temperature at which the net rate of carbon dioxide fixation is maximal. The optimum temperature for photosynthesis is generally higher than the growth temperature. Further, the optimum is also dependent on the environmental temperature. When plants are exposed to temperature above the normal physiological range photosynthesis gets inactivated. High temperature affects photosynthesis by changing the structural organization of the thylakoids and thus, altering the excitation, energy distribution, changing the activity of calvin cycle and other metabolic processes such as photorespiration and photosynthesis. However, plants have evolved a number of adaptive mechanisms allowing photosynthetic apparatus to cope with changes in temperature. These mechanisms may involve changes in membrane fluidity and lipid composition, increased thermal energy dissipation, decreased absorption, synthesis of heat shock proteins (HSPs). High temperature can inhibit photosynthesis well before other symptoms of heat-stress are detected. The temperature sensitivity of photosynthesis is dependent upon plant species, developmental history of leaves, their time of exposure to and intensity of new temperature regime. A long period of exposure to moderately high temperature may be as injurious as a brief exposure to an extreme temperature (Georgieva and Brugnoli, 2003). Young, developing leaves appeared to be more resistant to temperature changes since they are plastic in their photosynthetic response to sudden changes in temperature whereas mature, fully expanded leaves exhibit a reduced capacity to adjust (Karim *et al.*, 1999).

2.1.2 Chloroplast structure : Heat stress induces significant modifications in the composition of the chloroplast membrane lipids and proteins together with the structural changes of the thylakoid membranes.

Thylakoid membranes mainly contain monogalactosyl diacylglycerol (MGDG), digalactosyl diacylglycerol (DGDG), sulfoquinovosyl diaylglycerol and phosphatidyl glycerol, both of which are composed of a large proportion of highly unsaturated fatty acids. Heat stress was shown to decrease MGDG/DGDG ratio and increase the incorporation of DGDG with saturated fatty acids into the thylakoid membrane. Reduced lipid content and a moderate increase in saturation for total lipids have been observed after exposure to high temperature. The level of free fatty acids particularly unsaturated (linoleic and hexadecatrienoic acids) increased by 2 to 3 folds after incubation of isolated membranes to high temperature (Santarius, 1980). This indicates the activation of endogenous lipases, thereby increasing the content of free acids with a resultant increase in polar interactions at the membrane surface, thus, reducing the thermostability of membrane structure. Increasing temperatures cause an increase in the fluidity of membrane lipids followed by the formation of non-bilayer lipid structure. Presumably one of the main consequences of these lipid changes is a destabilization of lipid-protein interactions perturbating the organization and function of photosystems. It was suggested that the strength of the hydrophilic interactions linking the light harvesting antenna with photosystem II (PSII) complexes decreased with increasing temperature, while that of the hydrophobic interactions increased, so the pigment-protein complexes tend to associate more with the lipids than each other, resulting in their dissociation. High temperature increases the membrane permeability which decreases the protein gradient across the thylakoid membrane, thus, inhibiting the energy dependent non-photochemical quenching under high temperature stress.

2.1.3 Photosynthetic activity : It is universally accepted that PS II electron transport activity is very sensitive to high temperature and the primary site of damage is within the PS II complex. The most likely target of heat is the oxygen-evolving complex. PS I is rather stable at high temperature and can compensate partially for reduced PS II activity. The higher PS I thermostability may also be associated with an increased capacity for cyclic electron flow around PS I, which may be an adaptive process, producing ATP under conditions when PS II activity is severely inhibited. This ATP synthesis could be important for the survival of plants and necessary for repair of stress damaged processes.

2.1.4 CO_2 assimilation : Inhibition of photosynthetic CO_2 assimilation by high temperature is well documented in many plant species (Georgieva *et al.*, 2003; Pastenes and Horton, 1996). Temperature limits CO_2 availability either because of physiological responses of leaves resulting in increase in gas diffusion resistance or because of the relative solubility of CO_2 and O_2 competing for the same acceptor at the level of the enzyme Rubisco. However, inhibition of carbon assimilation at high temperature occurs under both photorespiratory and non-photorespiratory conditions. Thus, the temperature induced inhibition of CO_2 fixation cannot be explained solely by increased photorespiration via enhanced oxygenase activity of Rubisco at high temperature.

At moderately high temperature, the internal concentration of CO_2 was maintained nearly constant and photosynthesis was not influenced. Stomatal conductance is also dependent on temperature although the relationship between temperature and CO_2 conductance has not always been positively correlated, mainly because of the thermal effect on stomatal conductance, which is also dependent on the vapour pressure difference between leaf and air particularly in C_3 plants.

2.1.5 Acclimation : It is known fact that increasing the growth temperature of plants leads to upward shift of the optimal temperature of photosynthesis in numerous species and consequently renders the photosynthetic apparatus more tolerant to heat-stress. This phenomenon is termed as "acclimation". Acclimation can also be referred to as acclimatization or hardening (Lewitt, 1980). The ability of plants to withstand high temperatures is due to a variety of mechanisms acting at different levels of plant organization. These include increased evapotranspiration to maintain lower leaf temperature, modification of photosynthetic system to avoid photoinhibition, free radical damage and synthesis of HSPs in plant cells (Vierling, 1991). Acclimation to a new growth temperature is not sudden but requires a certain time period. Under stress, plants undergo initially a destabilization followed by normalization and then stability is enhanced when limits of tolerance are not exceeded and adaptive capacity persists. The acclimation capacity differs widely among species and is also dependent on previous environmental conditions.

High temperature affects plant growth, development and productivity particularly during the active pod filling stage (Marshall, 1982). Relatively little is known about the genetic diversity for tolerance to this stress. Saxena *et al.* (1988) reported that cool season food legumes are extremely sensitive to high temperatures, especially at full bloom stage. A few days of exposure to high temperatures (30-35°C) causes heavy yield loss. Summerfield *et al.* (1984) observed that the greater the exposure during the reproductive period to hot days (30-35°C), the lower is the yield of chickpea.

Alexandrov (1964) reviewed several methods for assessing heat resistance in plants. Among the methods evaluated were vital staining, depression of protoplasmic streaming, leakage of electrolytes, changes in protoplasmic viscosity, luminescence of chloroplasts, infiltrated flurochromes, retardation or cessation of respiration and photosynthesis. Sullivan and Ross (1979) suggested that electrolyte leakage from leaf discs subjected to heat-shock can serve as a simple, rapid and reliable technique for measuring heat tolerance. Expression of heat shock protein can also be used as screening technique. However, field screening method for heat tolerance is still not available.

The length of vegetative phase, initiation of flowering and appearance of leaves on main stem of warm season pulses such as mungbean and urdbean follow a linear function of accumulated temperature that is modified by day length and relative humidity. Increasing the mean temperature hastens flowering (Aggarwal and Poehlman, 1977). Dry mungbean seeds may tolerate upto 70°C for one hour without affecting seedling growth, but seedling growth is retarded if exposure of seed exceeds 70°C (Vora and Patel, 1975). Exposure of seeds to 40-50°C for one to three weeks causes cracks in the mungbean seeds thereby reducing the number of hard seeds and improving germination percentage (Manohar *et al.*, 1969).

Pigeonpea has both warmer base temperature and a greater thermal sum requirement for emergence relative to other tropical grain legumes (Table 3). Germination of several pigeonpea genotypes was high over a relatively broad temperature range (19-43°C), while germination was rapid and even over temperature range of 26-43°C (de Jabrum *et al.*, 1981). The optimal range for rate of hypocotyl elongation was narrower (29-36°C). No germination occurred at 7.1°C or 46.5°C.

Table 3: Relative sensitivity of pulse crops to temperature during emergence. Estimated time (days) to emergence at mean temperatures of 18°C and 25°C for pigeonpea and other tropical grain legumes based on experimental estimates of base temperatures (T_b) and thermal time to emergence (day degrees > T_b)

Crop	Days to emergence 18°C	Days to emergence 25°C	Basal temperature (T_b °C)	Thermal time (day degrees > T_b)
Pigeonpea (*Cajanus cajan*)	11.2	4.8	12.8	58.2
Cowpea (*Vigna unguiculata*)	6.1	3.1	11.0	43.0
Green/black gram (*Vigna radiata/mungo*)	6.9	3.5	10.8	49.6
Navy bean (*Phaseolus vulgaris*)	7.0	3.6	10.6	52.1
Soybean (*Glycine max*)	8.7	3.6	9.9	70.5
Groundnut (*Arachis hypogaea*)	16.2	6.5	13.3	76.3

Source : Angus et al., 1981

High root zone temperature has been shown to strongly affect bacterial infection and N_2 fixation in several legume species including soybean (Munewar and Wallum, 1982), peanut (Kishinevsky *et al.*, 1992), cowpea (Rainbird *et al.*, 1983) and beans (Panwar *et al.*, 1988). Elevated temperatures may delay nodule initiation and development and interfere with nodule structure and functioning in temperate legumes, whereas in tropical legumes, N_2 fixation efficiency is mainly affected. The effect of high temperature depends on plant species, cultivar and strain of *Rhizobium*. Hernandez-Armenta *et al.* (1989) found that transferring nodulated bean plants from 26 to 35°C daily temperature markedly inhibited N_2 fixation. Soybean plants appear somewhat more tolerant, nitrogen fixation only being severely inhibited by daytime temperatures greater than 41°C (La Favre and Eaglesham, 1987). Root temperature above 30°C reduced nitrogenase activity in peanut (Nambiar and Dart, 1983). Peanut-Bradyrhizobium symbiosis was completely inhibited by a root temperature of 40°C (Krishinevsky *et al.*, 1992). The effect was both due to failure of nodulation and inability of the nodule to function at this temperature even if they were formed. It is well known that respiration increases with higher temperatures resulting in less carbon available for symbiosis.

2.2 Low temperature

The production of cool season grain legumes is constrained by low temperature. In most agricultural species, suboptimal temperatures can be divided into chilling range and freezing range. Freezing range temperature for chickpea as below –1.5°C, which is the typical freezing point of plant tissue (Graham and Patterson, 1982) and chilling range temperature for chickpea is between –1.5°C and 15°C. Temperatures upto 15°C have been demonstrated to cause flower and pod abortion in parts of the Indian subcontinent and Australia (Srinivasan *et al.*, 1988; Clarke, 2001).

2.2.1 Freezing stress : The need for greater tolerance to freezing range temperatures has arisen where distribution has expanded to higher altitudes or the growing season has been altered to include colder conditions.

Mechanisms of freezing injury : Freezing sensitive plants are damaged/killed by temperatures below –1.5°C. Damage commonly occurs due to ice formation within the intercellular spaces. Intracellular ice formation is lethal (Guy, 1990). The rigid ice lattice structure extends with decreasing temperature and may penetrate cellular walls and membranes to an extent that is irreparable by cell processes (Andrews, 1996). Most of the crop species do not tolerate freezing e.g. chickpea. The elongation regions are often the first to be affected by freezing which is manifested in terms of sigmoidal curves around elongation point, commonly referred as 'hockey stick' symptoms. Depending on the minimum temperature and the duration of the frost, plants may be partially damaged or killed, resulting in lower yield or crop failure.

The freezing tolerance of a plant varies greatly between different tissues, i.e. upper and lower leaves of the plant canopy, stem, meristems or roots. Tolerance to freezing is often associated with mechanisms at cellular level, including increased membrane fluidity and osmotic adjustment as well as supercooling without ice nucleation.

Acclimation : For many plant species tolerance to freezing range temperature is not static, but can change seasonally or when the temperature and other environmental conditions are varied. Seasonal dynamics of freezing tolerance is directly related to process of cold acclimation in response to low temperatures, resulting in lowering of the temperature at which a plant is damaged by freezing (Prasad, 2001).

Many plant species can, however, tolerate extracellular ice formation which is relatively non-harmful and can act as a nucleation site of water drawn out of the cell, promoting cell desiccation. Although ice driven desiccation can continue to the point of irrevocable cellular damage, it is in a stable desiccated state that most cold-stressed tissues overwinter at subzero temperatures (Andrews, 1996). Cellular and metabolic changes that occur during cold acclimation include increased levels of sugars, soluble proteins, prolines and organic acids as well as appearance of new isoforms of proteins and altered lipid membrane composition (Huges and Dunn, 1996). The ability to acclimate is essential to freezing tolerance of the cool season grain legumes (Wery *et al.*, 1994). This ability seems to decrease with plant age and is reliant on a slow rate of plant growth in early winter sown crops in colder temperatures. Winter sown chickpea are exposed to decreasing photoperiods and temperatures that fall gradually as the season progresses from autumn to early winter. Therefore, seedlings of winter planted chickpea have a possibility of acquiring some degree of tolerance to moderate subzero temperatures, as has been shown for faba beans by Herzog (1978).

Winter Hardiness : Winter hardiness of chickpea represents tolerance to a complex of winter rigours. Winter survival of plants often requires tolerance to factors other than cold, for example, frost-heaving, waterlogging, frost drought and diseases (Murray *et al.*, 1988). The primary causes of winterkill are heaving caused by the formation of ice in the soil, either as a solid or as capillary needles or columns, which push the plants upwards, breaking and exposing the roots smothering, physiological drought and freezing of the plant tissue (Grafius, 1981). Snow cover itself can be positive environmental factor because the thermal and radiative properties of snow result in a distinct thermal profile, where its surface temperature is at near equilibrium with air temperature (below zero), while temperature gradually increases with snow depth, often to near zero at the soil surface. Thus, snow can act as an effective insulator for underlying plants (Blum, 1988). Cultural practices such as planting date, fertilizer regime, disease and insect control, tillage practice that influence snow capture and drainage, combine to influence the overwintering ability of grain legumes (Murray *et al.*, 1988).

Effect on phenology : The effect of freezing range temperatures which can vary depending upon phenological stage of plant (Summerfield and Roberts, 1985) are briefly described below.

Pre-emergence stage : The germination of dicotyledonous seeds, such as chickpea, starts the growth process of a quiescent or dormant embryo and is evidenced by the growth of embryonic axis (de Rueda *et al.*, 1994). Good germination and seedling emergence are important pre-requisites for a successful crop which are greatly influenced by temperature (soil and air). In chickpea, 0°C has been proposed to be the base temperature for germination (Calcagno and Gallo, 1993). Thus, in freezing soils, chickpea will not germinate.

Seedling to early vegetative stage : Plants at the autotrophic stage are more vulnerable to temperature stress than those at the heterotrophic stage prior to germination because their endosperm resources have been exhausted. In chickpea, tolerance to freezing range temperatures decreases as the plant progresses from seedling stage (most tolerant) to flowering (least tolerant) (Singh *et al.*, 1995). The main effects of freezing temperatures on the developing seedling are related to membrane injury which causes reduced respiration and photosynthesis, loss of turgor resulting in wilting and temperature induced drought stress.

For early winter sown chickpea, the early vegetative stage encounters gradually decreasing temperatures and photoperiods. As the winter progresses, night temperature falls below zero and remains low for a long period. Although some lines of chickpea can withstand temperatures of-12°C in the post-emergence vegetative stage, the minimum temperature at which chickpea generally seems to survive is-8°C (Wery, 1990). However, it is difficult to determine an exact threshold temperature for field grown chickpea because snow cover can protect plants, while wind can increase injury. Yield superiority of winter sown chickpea as compared to spring crop may be ascribed to the longer vegetative growth periods leading to a larger vegetative structure. This larger vegetative structure intercepts photosynthetically active radiation (PAR) more effectively in spring and supports a proportionally larger reproductive sink with adequate partitioning of dry matter (Singh *et al.*, 1997). The reproductive phase of winter sown chickpea is also longer than spring sown chickpea contributing to higher seed yield. Better utilization of PAR increases total biomass of winter sown chickpea, while retaining a similar harvest index to that of spring sown chickpea (Siddique and Sedgley, 1986; Singh *et al.*, 1997).

Late vegetative stage : Unlike the seedling to early vegetative stage, which exhibits a relatively high degree of freezing tolerance, there is a high sensitivity to freezing damage at the more advanced vegetative stage (Singh *et al.*, 1995) as observed in field pea and faba bean (Scarascia-Mugnozza and Margi, 1971; Murray *et al.*, 1988). Increased sensitivity to freezing range temperatures was attributed to the inability of plants to set osmoregulation mechanisms in motion during the active growth phase (Wery, 1990).

In many crops, the microclimate has been shown to significantly influence the degree of damage from low temperatures. In chickpea, stem elongation during vegetative stage means that the shoots are in colder air strata than that nearer to the soil surface. It has been suggested that propensity to branch may be a pre-requisite for tolerance to freezing range temperatures in chickpea (Murray *et al.*, 1988) and that rosette rather than upright growth habit may be a means of escaping the worst of the freezing temperatures (Lawes *et al.*, 1983). The extent of frost damage is affected by microclimate; great variability occurring within paddocks even on the same plant, can be amplified by climatic conditions such as clear sky, dry atmosphere, windless conditions, soil type, soil moisture, position in the landscape, crop density and crop nutrition. In chickpea, whether the duration of the vegetative phase of growth is short or long, photoperiod is negatively related to the mean diurnal temperature irrespective of the genotype (Roberts *et al.*, 1985). Freezing temperatures at this phenological developmental stage can cause considerable damage.

Reproductive stage : The onset and duration of flowering are functions of genotype, photoperiod and temperature (Roberts *et al.*, 1985). During reproductive stage, chickpea is far more likely to encounter temperatures within the chilling range as opposed to temperatures within the freezing range. At this stage, plants do not tolerate freezing temperatures, such as those encountered via a late frost event. Undoubtedly, this is the case for extended periods of freezing range, but, due to the indeterminate nature, chickpea may be able to flower and set pods following an isolated frost event provided soil moisture conditions are conducive during subsequent periods.

The time and duration of flowering affects tolerance and the ability of plant to compensate after the frost has occurred. Frosts that occur towards the end of reproductive period following pod set are more damaging, resulting in pod abortion.

2.2.2 Chilling stress : Temperature within the chilling range for chickpea (-1.5°C to 15°C) is common throughout the growing season across almost all chickpea growing areas. The upper limits of the chilling range are quite acceptable and even optimum for early growth in some genotypes, but the reproductive processes can become susceptible to damage from temperature of 15°C and lower (Khanna-Chopra and Sinha, 1987; Clarke, 2001). Exposure to prolonged periods of low temperature (chilling range) can cause poor germination, slow growth, flower shedding and pod abortion and in severe cases cell necrosis and plant death. The shift in planting time means the chickpea crop is exposed to far colder temperatures during the early growing stages than when planted in summer. Increased tolerance to chilling range temperatures at flowering has been identified as a highly desirable trait in Australia and parts of the Indian subcontinent due to a smaller window of opportunity for spring/summer sowing (Srinivasan *et al.*, 1999; Siddique *et al.*, 1999). In these growing regions, flower shed and pod abortion due to chilling range temperatures are major cause of poor yield.

Mechanisms of chilling injury : Chilling sensitivity is a characteristic of plants of tropical or sub-tropical climates (McWilliam, 1983). Varietal differences in susceptibility to chilling injury have been reported for a number of species (Lyons *et al.*, 1979). Plants which are able to survive and grow at temperatures between –1.5°C and 15°C are chilling tolerant. However, no sharp distinction exists between chilling intolerant and tolerant plants because the phenological stage of the plant at the time of exposure can have such a large effect on the sensitivity of plants. In the absence of detailed information and little research regarding the mechanisms of chilling injury in chickpea, the effect of chilling at the cellular level on the function of sensitive plants in general are summarized in Table 4.

At cellular level, the processes by which chilling injures plant cells are likely to have commonalities between species.

Table 4 : Effects of Prolonged Chilling Range Temperatures at the Cellular level in Susceptible Plant Species

Cause	Effect	Reference (s)
Enzymes and Oxidative Stress		
Enzymes and Oxidative Stress	Impairment of photosynthesis and respiration	Wolk and Herner, 1982; Van Heerden and Kruger, 2000
Antioxidant enzymes suppressed	Excessive level of free radicals-Reactive Oxygen Species (ROS)	Raison, 1991; Pastori *et al.*, 2000; Prasad, 2001
Oxidative stress-ROS	Lipid peroxidation by ROS and DNA damage	Shewfelt & Erickson, 1991; Prasad, 1996; Prasad, 2001
	Protein oxidation by ROS. Proteases inactivated, toxic levels of oxidized proteins since no degradation	Pacifici and Davies, 1990; Stadtman and Oliver, 1991; Stadtman, 1993
Altered Membrane Function		
Membrane transition from liquid crystalline to solid gel phase as affected by lipid composition	Membrane permeability changes	Lynch, 1990; Nuotio *et al.*, 2001; Shewfelt, 1992; Riken *et al.*, 1993; Murray *et al.*, 1988; Chen *et al.*, 1983
As affected by lipid peroxidation	Membrane shrinkage and vacuole disruption	Blum, 1988; Ishikawa, 1996
As affected by water content of tissue	Leakage of electrolytes from cell	Wery *et al.*, 1994
	Loss of cell turgour	Levitt, 1980; Wang, 1982; Markhart, 1986; Collins *et al.*, 1995
	Impairment of cytoplasmic streaming	Woods *et al.*, 1984
	Impairment of photosynthesis; particularly in strong light (photoinhibition)	Li *et al.*, 1990; Wery *et al.*, 1993; Van Heerden and Kruger, 2000; Nuotio *et al.*, 2001; Smillie *et al.*, 1988
	Alteration/inhibition of protein synthesis and function	McWilliam, 1983; Cooper and Ort, 1988
	Vacuolisation, disruption of cytoplasmic reticulatum, and organelle disorder in general	Christiansen and St John, 1984; Blum, 1988
Reduced ATP availability		
	Reduced protein synthesis	Prasad *et al.*, 1994;
	Inhibition of protein transport (e.g. from cytosol into chloroplast)	Leheny and Theg, 1994;
	Reduced respiration in mitochondria	Prasad *et al.*, 1994;
	Respiration by alternative pathway-heat evolution	Moynihan *et al.*, 1995

Pre-Emergence Stage : Chickpea seeds sown early in the season in temperate areas are frequently exposed to chilling (3° to 8°C) or even freezing temperatures during germination, resulting in its reduced stand and low seedling vigour (Chen *et al.*, 1983). A number of interacting factors mediate seed response to low germination temperatures which include the temperature to which seed is exposed, the duration of exposure, the germination period, seed moisture content prior to the start of imbibition (water influx) and the genotype. Generally, the longer, a germinating seed of a sensitive species is exposed to a chilling temperature, the greater it will sustain (Wolk and Herner, 1982). Rate of germination of chickpea seed is inversely related to daily mean temperature (Roberts *et al.*, 1980) and a positive correlation between the number of seedlings emerged and the average soil and air temperature (Ellis *et al.*, 1986).

Prior to water imbibition, the seed remains unaffected by even the very low temperatures. It is during the process of water imbibition into the seed that sub-optimal temperatures become a potentially damaging. Chickpea, along with many other chilling-sensitive species, is very sensitive to 'imbibitional chilling injury' resulting from chilling range temperatures during water influx (Tully *et al.*, 1981). The greatest sensitivity to cold corresponds to the first 30 minutes of imbibition (Chen *et al.*, 1983). When the dry seeds first begin to imbibe water, a variety of intracellular solutes such as amino acids, sugars, organic acids, gibberellic acid, phenolics and phosphates leak-out (Simon, 1979), suggesting a general leakage of cellular contents. Solute leakage damages the cells on the cotyledonary surface in tissue death (Powell, 1989). Consequence of chilling-induced solute leakage is the establishment of an excellent medium around the seed for the growth of soil pathogens. Thus, under field conditions chilled seed is often subjected to extensive infestation by soil organisms, leading to a reduction in seedling survival.

Desi types of chickpea generally suffer less damage from low temperature at germination than Kabuli types. The rapidity of imbibition is controlled principally by the thickness of testa. Kabuli types generally have a thinner testa than desi types, resulting in more rapid imbibition of water and consequently greater levels of imbibitional damage. Phenolic content of the seed is another factor affecting germination success at cold temperatures, which presumably confers fungistatic properties.

Post-emergence of seedling : Sensitivity of freezing and chilling range temperatures increases as the plant progresses from germination to flowering. The visual symptoms of chilling injury at the seedling stage include the inhibition of seedling growth, accumulation of anthocyanin pigments, water-logged appearance with browning of mesocotyle, browning and desiccation of coleoptiles and undeveloped leaves (Prasad, 2001). Non-visible symptoms include alterations in gene expression, membrane properties, proteins, lipids, carbohydrate composition and solute leakage. The main effects of chilling range temperatures on the developing seedling are related to membrane injury, reduced respiration, photosynthesis and loss of turgor resulting in wilting and cold induced water stress. Damage caused by rapid inbibition of water into cold affected seeds carries over to the developing seedling, resulting in reduced vigour.

RESPIRATION

The death of large numbers of cells in the cotyledons following imbibitions damage and a change in membrane associated enzymes of the respiratory system leads to reduced respiration.

PHOTOSYNTHESIS

Damage to Photosystem II and reduced stability of chloroplast membranes leads to a reduction in photosynthesis in chilling affected plants.

MEMBRANES

Primary damage of chilling injury is to the membranes. Damage leads to increased membrane permeability due to breakdown of the structural integrity and stability of the cellular membranes.

FOOD RESERVES

Decreased rate of food transport from cotyledons to growing axis resulting from death of cells due to rapid imbibition at germination

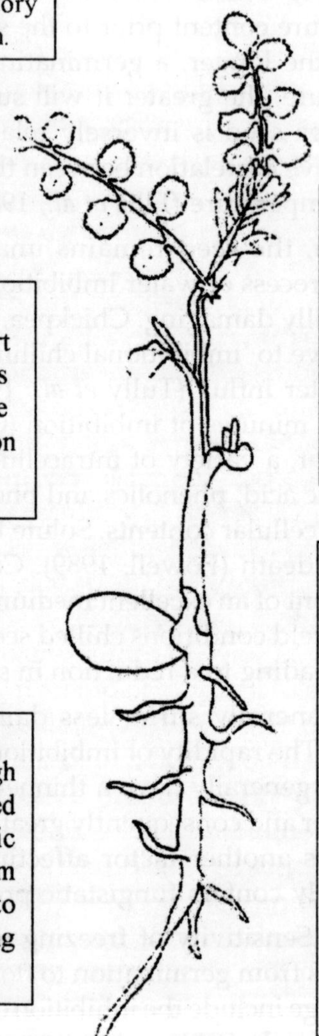

WILTING

Changes in the membrane permeability allows H_2O and soluble material to leak into intercellular spaces and evaporate. Additional effects such as slow closure of stomatal openings on transfer to chilling temperatures and reduced root hydraulic conductivity combine to result in wilting. If exposure to chilling temperatures is prolonged tissue necrosis and plant death can occur.

ROOTS AND SHOOTS

Decreased water uptake through roots and shoots due to decreased metabolic activity and hydraulic conductivity resulting from membrane damage. Leads to reduced water potential/wilting and chilling-induced drought.

Fig. 1 : Effect of chilling injury on chickpea at the seedling to vegetative phenological stage [illustration of chickpea seedling adapted from Cubero, 1987)]

Exposure to chilling range temperatures during early growth of established seedlings can exert macroscopic formative effects on leaf shape, size, plant height, root development and floral initiation (Christiansen and St. John, 1984). Growth of kidney bean (*Phaseolus vulgaris* L.) was reduced by 10°C chilling at the 3-leaf stage and intermittent chilling during the lifecycle was shown to reduce yield by as much as 34 per cent (Wierenga and Hagan, 1966). Other effects of chilling on seedling growth include water loss due to increased membrane permeability, slow closure of stomatal openings, reduced root hydraulic conductivity and water uptake through both roots and shoots (Fig.1, Wolk and Herner, 1982; McWilliam *et al.*, 1982). In sensitive species chilling reduces the conductivity of the plasmalemma and tonoplast of the guard cells, rendering the stomata less responsive to changes in leaf water potential. The combination of reduced water uptake and slow closure of stomata under conditions of continued evaporative demand in the light causes a reduction in water potential leading to wilting and ultimately to severe tissue dehydration. In controlled environment, treatments can be applied to ameliorate wilting in cold affected plants. These include affecting stomatal closure by spraying leaves with ABA or holding plants in dark prior to chilling to reduce transpiration and retain a more favourable plant water balance. Maintaining a water-saturated atmosphere around the shoots of chilling-sensitive seedlings at low temperature is another means of restricting water loss through transpiration to prevent wilting. The reduction in the growth of seedlings from damaged seeds is associated with a slower rate of food reserve transfer from the cotyledons to the growing axis. Mobilization of food reserves is reduced in tissues in which dead cells occur, because enzyme activity is reduced (Powell and Matthews, 1978).

Acclimation of seedlings : Seedlings may be conditioned/hardened by exposure to temperature slightly above the chilling range. Plants conditioned in this way are more resistant to subsequent chilling than seedlings transferred directly to chilling temperatures. In readily hardened, chilling sensitive plants, the degree of fatty acid unsaturation associated with the phospholipid fraction increased during low temperature conditioning. Hardening does not cause increase in total leaf fatty acids or in the degree of unsaturation of glycolipids. The use of low temperature to harden plants has also been shown to prevent the loss of ATP, which normally occurs during chilling (Wilson, 1978). Acclimated seedlings were found to have an increased antioxidant defense system that scavenged reacting oxygen species (ROS) during acclimation followed by chilling stress. It appears likely that ROS-induced lipid peroxidation is, at least in part, responsible for increased chilling sensitivity in seedlings and perhaps, all the chilling sensitive crop plants (Prasad, 2001).

Mid to late vegetative stage : Chilling range temperatures at the mid to late vegetative stage retard growth rate and reduce plant vigour. Siddique *et al.* (1983) described a close linear correlation between air temperature and the appearance of expanded leaves in chickpea. Khanna-Chopra and Sinha (1987) reported the optimum temperature for achieving maximum leaf growth as 10°C to 25°C. These effects are due to the same mechanisms that affect post-emergent seedling growth, that is, reduced respiration and photosynthesis and in severe cases a loss of turgor and subsequent water stress. Wilson and Crawford (1974) demonstrated the degree of fatty acid unsaturation and the weight of phospholipids decreased with age of leaves of plants grown at 25°C which may be related to an increase in sensitivity of older leaves to chilling injury.

Reproductive stage : Air temperature and photoperiod have a major influence on the timing of reproductive events in legumes especially chickpea, with the rate of progress to flowering being a linear function of mean temperature (Summerfield *et al.*, 1980). Roberts *et al.* (1985) demonstrated that longer photoperiods at any temperature results in faster accumulation of the thermal sun required for flowering. The earliest flowering genotypes are least responsive to photoperiod, and there is no apparent correlation between relative sensitivity to temperature and photoperiod. This fact, together with the lack of interaction between these environmental parameters, suggests that the response to both temperature and photoperiod both affect time of flowering, even though, they are under separate genetic control.

Non-optimum temperatures constrain reproductive development in higher plants at various stages, including pollen development, transfer of viable pollen to the stigma, pollen germination and tube growth, ovule fertilization and seed development. Recent research by Sharma *et al.*, (2005) indicated that pollen germination and vigour is affected by chilling range temperatures. Sensitive and tolerant cultivars can be ascertained under chilling range temperatures from the relative germination of pollen and growth of pollen tubes. Reproductive phase in long duration pigeonpea coincided with low temperature of winter season in North Eastern Plain Zone, as a result, the formation of floral buds and anthesis were adversely affected if the mean temperature falls below 15°C (Choudhary, 2007). Additional demonstrated and hypothesized effects of chilling range temperatures on chickpea reproduction are summarized in Table 5.

Srinivasan *et al.* (1999) opined that sensivity of chickpea to cold temperatures might be ascribed to its evolution as a spring/summer crop in West Asia, where flowering and podding occur in progressively increasing temperatures. Consequently, there has been no selection of chickpea for cold tolerance at reproductive stages. Factors other than cold like water stress (both inadequate or excess), cloudy weather, which can reduce incident solar radiation and is often associated with rainfall and high humidity can also result in abortion of reproductive structures. A number of authors have reported that reduced light intensity on its own can increase abortion of reproductive structures in chickpea due to premature bursting of the anther sacs or the adverse effect of the weather on the pollinated stigma and pollen (Chandrasekharan and Parthsarathy 1963; Khanna-Chopta and Sinha, 1987).

Table 5 : Effect of Chilling Range Temperatures on Chickpea Reproduction

Effect	Description	Author(s)
Flower shedding/ floral abortion Pod shedding/drop	Sudden low temperatures (0° to 10°C) during flowering induce flower shedding, which causes partitioning of assimilates to vegetative growth, resulting in lowered Harvest Index. Major cause of low pod and seed set in subtropical South Asia and Australia.	Sinha, 1977; Varma and Kumari, 1978; Saxena, 1980; Savithri *et al.*, 1980; Siddique and Sedgley, 1986; Saxena *et al.*, 1988; Saxena and Johansen, 1990; Srinivasan *et al.*, 1998
Interrupted pollen tube growth	Temperatures upto 25°C shown to interrupt pollen tube growth. Failure of fertilization results from poor germination and slow growth of pollen tubes in susceptible genotypes at low temperatures.	Savithri *et al.*, 1990; Calcagno and Gallo, 1993; Clarke, 2001
Lowered pollen viability	Pollen in tolerant genotypes more viable (90%) compared to susceptible genotypes (60%). Two stages of pollen sensitivity at 5 and 9 days before anthesis have been identified.	Srinivasan *et al.*, 1999; Clarke, 2001; Sharma *et al.*, 2005
Reduced ovule size	Ovules in flowers opening on cool days were 9-45% smaller than warm day ovules-more pronounced in chilling susceptible than tolerant genotypes.	Srinivasan *et al.*, 1999
Reduced pistil size	Heterostyly-the distance between the anther and stigma at the time of flower opening is greater in sensitive than in tolerant genotypes.	Srinivasan *et al.*, 1999
Reduced stigmatic esterase activity	Reduced esterase activity was identified in susceptible genotypes suggesting the stigmas were less receptive to pollen tube growth.	
Delayed anther dehiscence	Anther dehiscence is delayed by chilling temperatures, reducing fertilization events.	
Reduced pollen germinability	Possibly due to smaller amount of storage material in pollen of sensitive genotypes.	Srinivasan *et al.*, 1999
Reduced pollen turgor	Turgidity is an absolute requirement for germination. Pollen cells with leaking membrances cannot become turgid and germinate.	Srinivasan *et al.*, 1999

Low temperatures delay root hair infection, decrease nodulation and nitrogenase activity. Low temperature was also shown to adversely affect nodule development in *Trifolium* spp. and *Vicia faba* without altering the sequence of anatomical changes (Fyson and Sprent, 1982). Effective nodule development and nitrogenase activity were completely inhibited in alfalfa at 8°C (Cralle and Heichel, 1982). Sub-optimal root zone temperature decreased N_2 fixation by directly decreasing the activity of nitrogenase enzyme complex and suppressing and/or delaying root infection and nodulation (Walsh and Layzell, 1986). The greater degree of sensitivity to low root zone temperature below 17.5°C was due to events related to or occurring before infection thread penetration to the base of root hair (Zhang and Smith, 1994). Lie (1974) and Fyson and Sprent (1982) concluded that low temperatures delayed bacteroid tissue formation. The increased individual nodule mass and reduced specific nodule activity at low temperatures have been attributed to inhibition of nodule cell differentiation and bacteroid tissue development (Pankhurst and Layzell, 1984). In soybean *Bradyrhizobium japonicum* symbiosis, suboptimal root zone temperature slowed down nodule development by disruption of signal exchange between the host plant and *Bradyrhizobium* (Zhang and Smith, 1994). Lower temperature resulted in marked reduction in nodule efficiency and ARA per nodule dry weight in soybean. More of the N fixed at lower root temperature was accumulated in the root, presumably due to decreased N translocation (Lynch and Smith, 1993). Sensitivity of cultivar/varieties to low temperature is also evident in the results of Rennie and Kemp (1981) who showed that the bean variety Kentwood failed to nodulate at 10°C and took 32 days to initial nodualtion at 12°C, while cultivar Aurora nodulated in 23 and 21 days at 10 and 12°C, respectively. Strain differences in nodulating ability at low temperatures have been shown between arctic and temperature zone rhizobia (Schulman *et al.*, 1988).

3. DROUGHT/WATER STRESS

Drought refers to moisture deficit for plants which results when the amount of water available in soil is not sufficient to meet the demands of potential evapo-transpiration. It would be imperative to briefly outline different mechanisms of drought such as drought tolerance, drought escape, drought avoidance, drought resistance etc.

A substantial period of time without significant rainfall can be defined as a drought by meteorological criteria. However, shortage of rainfall does not become a drought in crop production systems until the water shortage begins to limit the plant growth and development. There are two broad options available for improving yields under limited water availability. First, agronomic practices such as conserving soil moisture optimizing sowing time to escape terminal drought, high temperature and to avoid detrimental effect of low temperature, use of supplementary irrigation, optimizing plant density, intercropping involving crops of different growth habits and root systems. Secondly, there are genetic options such as development of short duration cultivars to escape terminal drought, incorporation of drought tolerance in high yielding genotypes. Generally plants can adapt in a limited way to drought by the mechanism of escape or tolerance.

To distinguish different plant responses to drought/water stress, researchers sometimes divide drought adaptation into several categories such as drought escape (shortening lifecycle), drought avoidance (growing deeper roots, depositing leaf wax and closing stomata) and drought tolerance (production of osmolytes, antioxidants and other stress relieving agents).

Pulses are relatively tolerant to drought and hence are adapted to wide range of soil moisture regimes compared to cereals. Usually *Vigna* species adapt to moisture deficit by escape mechanism, which involves phenological synchromy or developmental plasticity allowing the crop to complete its lifecycle or at least its reproductive phase before a seasonal period of low water availability sets in. Tolerance mechanism enables a plant to either postpone or reduce dehydration at lower water content (Table 6).

Table 6 : Useful traits acquired by crops for yield improvement and ease of screening under drought conditions (Deshmukh *et al.*, 2006)

Characteristic	Useful	Ease of screening
Early vigour	Variable	Easy
Root density and depth	High	Very difficult
Phenology	Very high	Easy
Developmental plasticity	High	Easy
Stomatal control	High	Difficult
ABA accumulation	Questionable	Difficult
Osmotic adjustment	High	Difficult
Membrane stability	Low	Easy
Lethal water potential	Medium	Difficult
Proline accumulation	Questionable	Easy
Harvest Index(assimilate redistribution)	High	Easy

Pigeonpea is well adapted to drought-prone environments. In India where nearly 90 per cent of world production of pigeonpea occurs, it is grown mainly as a rainy season crop, where substantial period of its growth, particularly reproductive phase occurs in the dry season on stored soil water. The crop as such is exposed to periods of transient drought during much of its vegetative ontogeny, followed by a terminal drought during most, if not all, of its reproductive ontogeny. Although reported to be drought tolerant, experimental evidences suggested that these crops might also avoid drought by shortening the growth duration (Hughes and Keatinge, 1983; Chauhan *et al.*, 1987).

Efficient monitoring of environmental soil and climatic factors influencing crop growth and plant factors that contribute towards higher yield and survival is important part of drought management that enables us to characterize drought prone areas as well as to assess the effect of drought on plant attributes.

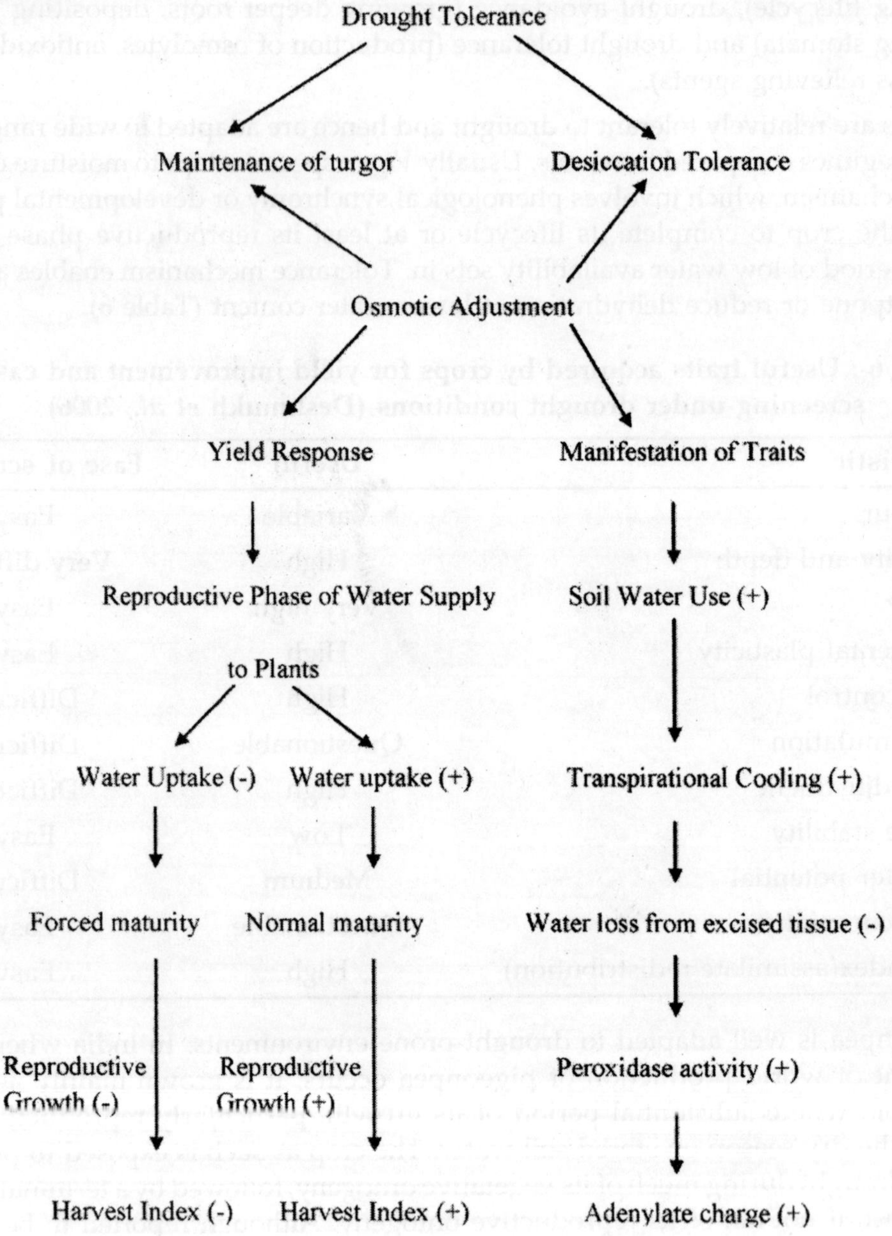

Fig. 2 : Possible role of osmotic adjustment in drought tolerance and expression of traits and seed yield in chickpea under terminal drought conditions. The (+) or (-) symbol indicates increase or decrease respectively [Illustration adapted from Singh et al. (1990) with suggestive modification]

3.1 Mechanism of drought tolerance

Two physiological approaches have been identified to improve breeding efficiency for drought tolerance (Turner, 1999). The first "Drought Resistance Framework" identifies specific physiological and biochemical traits that lead to improved yields in drought prone environments through drought escape, dehydration postponement and dehydration tolerance. The second framework "Yield Component Framework" considers yield variation in terms of plant characteristics that affect water use, water use efficiency and harvest index (Passioura, 1977).

3.1.1 Drought escape (early phenology) : Matching phenology to water supply is one of the important factors determining yields under water limited environment. Drought escape involves identification of plant with early flowering and pod initiation, short growth periods (rapid flowering, podding and maturity) to efficiently utilize available moisture (Siddique *et al.*, 1997). Genetic variation for early vigour, faster canopy development, days to flowering, podding, seed filling and maturity has been reported in chickpea by Kumar *et al.* (1999). In environments where terminal drought is likely to occur, selection for early flowering may not always increase yields in areas prone to frost damage at flowering or poor fertilization of flowers associated with cool temperatures (Srinivasan *et al.*, 1998). As flowering followed by pod setting decelerates growth of foliage because of competition between reproduction organs and vegetative growth early flowering may not always compensate. Thus, genotype having short vegetative phase often produces low yield, which can be compensated by increasing plant population per unit area. This will further increase competition where initial conserved soil moisture remains relatively high.

In chickpea, the seed yield was generally higher at medium degree of osmotic adjustment due to better partitioning of dry matter from root-shoot to seed (Singh *et al.*, 1990). The higher degree of osmoregulation showed positive correlation with water use from sub-soil (90-180 cm depth) during vegetative stage, but negative association with water use during reproductive phase of growth. This resulted in the reduction of maturity duration and HI in chickpea, probably due to drying of sub-soil and degeneration of roots during late reproductive phase of growth in chickpea. High temperature during grain filling might have caused forced maturity of chickpea. Thus, in long duration chickpea only medium degree of osmotic adjustment will be beneficial for proper rationing of the available soil water throughout the crop life cycle to attain HI index and seed yield under dryland/rainfed conditions (Fig.2).

3.1.2 Dehydration postponement : Another mechanism to combat drought is either by reducing water loss primarily through stomatal control or by increasing water use through deep and vigorous root system, or by accumulation of solutes to maintain turgor as water shortage develops i.e. osmotic adjustment. Open stomata result in transpirational cooling of leaf. Hence canopy of any genotype having higher stomatal conductance should show lower temperature than those having close stomata or low stomatal conductance. Osmotic adjustment has been shown to maintain stomatal conductance and photosynthesis at low leaf water potential, delay leaf senescence, reduce flower abortion and improve root growth and water extraction from soil (Ludlow, 1987).

High root biomass besides providing accessibility to deeper soil moisture maintains high leaf turgor and leaf water potential, lower canopy temperature thereby enhancing the vegetative biomass. Root size, morphology, depth, length, density, hydraulic conductance are basic to meet the transpirational demands of the shoot (Passioura, 1982).

Deep penetration of roots, adequate root density and adequate longitudinal conductance in main roots contribute to maximizing extraction of soil moisture.

The water uptake pattern across the depth of rooting zone is not uniform. In general, nearly 40 per cent of the total water uptake occurs over the first one-fourth of the root zone, 30 per cent over the second, 20 per cent over the third and 10 per cent over the last one-fourth of the total rooting depth. Stress induced plants often maintain higher root length densities than required in the surface layer to facilitate rapid uptake of recent rain water before it evaporates, to provide reserve capacity in case of disease and pest damage, to extract relatively immobile nutrients like phosphorus and to compete with weeds or other neighbouring plants for both water and nutrients.

3.1.3 Dehydration tolerance : Ability of cells to continue metabolism at low leaf water status is termed as dehydration tolerance. It is related to the degree of osmotic adjustment (Hsiao *et al.*, 1984). Drought tolerance mechanism involves stability of membrane and the membrane disorder often causes leakage of solutes from the cell (Blum, 1988).

3.1.4 Proline accumulation and osmotic adjustment : Accumulation of proline in water stressed plant tissue is another mechanism, that protects and stabilizes protein structures as cell dehydrates and serves as an organic nitrogen source and/or energy source for utilization during recovery from the stress. Four metabolic mechanisms to explain the increase of proline have been hypothesized :

1. decreased rate of proline oxidation by mitochondrial proline oxidase
2. lower rate of proline assimilation into proline (less likely)
3. slowed rate of phloem translocation of proline from the leaves
4. enhanced and preferential synthesis from glutamate (not protein breakdown). It is considered to be 'compatible solute' that accumulates in cytosol under moisture stress conditions. Proline accumulation was more in shoots than in roots of mungbean genotypes, indicating the role of light in proline accumulation during stress. A possible mechanism outlining the involvement of photoinduction/photoactivation of key enzymes involved in synthesis of proline has been suggested (Arora *et al.*, 1995).

3.2 Effects of drought

3.2.1 Physiological processes : Water stress invariably decreases several vital processes of the plant and at the same time, modifies a number of morpho-physiological characters in a manner that plant can thrive well under drought. The effects of drought on seed yield depend on the aggregate response of morpho-physiological processes such as photosynthesis, leaf area expansion, leaf senescence, partitioning of assimilates and biomass production. It also depends on the timing of the stress within the growth period. Selection for drought tolerance is, therefore, complicated by many processes and their interactions with the environment. Reduction of photosynthesis is considered to be primary yield limiting factor under water stress. Jefferies and MacKerran (1989) showed that dry matter accumulation is most affected by drought largely as a consequence of poor canopy expansion. Sekhon *et al.*, (2007) observed that water stress during reproductive phase in chickpea (initiation of flowering to seed formation) severely reduced the seeds per pod and 100 seed weight. Drought stress does not affect all plant functions similarly; cell elongation and N-fixation are more susceptible than photosynthesis and transpiration (Table 7).

Table 7 : Effect of drought stress on the processes involved in seed yield and nitrogen balance of grain legumes (Wery *et al.*, 1993).

Parameters showing increasing drought stress	Physiological activity affected	Crop parameter affected	Yield component affected
Full turgor potential	Cell elongation, nodule activity, nitrogen fixation	Organ size, nodule-number and size	Total biomass, Nitrogen
Beginning of stomata closure	Photosynthesis	Carbon balance	-
Increase in canopy temperature	Translocation	Grain filling, death of plant organs (leaves, flowers and pods)	Grain number, yield, weight vigour

3.2.2 Photosynthesis and metabolism : Water stress usually affects stomatal conductance, photosynthetic activity and the consumption of assimilates in expanding leaves and decreases the amount of photosynthate exported from leaves. Drought during early pod growth reduces assimilation capacity of mungbean leaves both during stress and during the subsequent pod growth stage (Venkateswarlu and Subramanian, 1993). Reduced leaf water potential, transpiration and increased leaf temperature and stomatal conductance were observed in urdbean under drought stress which resulted in decreased photosynthesis. Maximum reduction in these biophysical parameters were observed when the mungbean plants experienced stress at flowering stage, while, stress at vegetative and pod development stages had lesser effect (Chetti and Sirohi, 1998).

3.2.3 Sensitivity to water stress : The response of cultivated crops to drought varies with species. The extent of reduction in biomass, photosynthesis and other physiological traits due to water deficit may differ within and among different species and thus, variation largely depends upon the water retention capacity of the tissue which involves water and osmotic potential, bulk modules of elasticity, relative water content (RWC) and to a greater extent, the root capacity to extract water from soil.

In dry land conditions, where water is the main limitation, productivity in terms of water use, rather than PAR assumes greater significance. Productivity under such conditions can be expressed in terms total water used, its efficiency of use in producing biomass (WUE) and HI. In pigeonpea, the main emphasis should be to maximize use of available water by the crop, which in turn will maximize the potential biomass and minimize drought constraints to partitioning and improve HI. It remains to be demonstrated whether any significant gains can be made through increasing the physiological efficiency of water use (Lawn, 1988).

The overall strategy, which is already in practice where pigeonpea matures on residual soil water, is to grow genotypes whose duration matches the available water supply to escape the effects of drought (Lawn and Williams, 1987). A related escape mechanism for variable water supply from year to year, phenological plasticity, whereby the plant can adjust its duration, particularly of the reproductive phase to water supply.

Several mechanisms that contribute to the short term survival of pigeonpea during periods of transient drought, with greater or lesser effects on productivity during and after the drought, include low epidermal conductance, stomatal closure, leaf rolling, paraheliotropic leaf movements, less rapid rates of leaf area development, leaf abscission/ shorter leaf area duration, all contributing to reduced energy interception and/or slower rates of water loss and prolong the period of plant survival under limited water supply, but with successively greater impacts on productivity once stress is over. The deep rooting ability of pigeonpea is an asset enhancing its ability to make most effective use of stored soil water. Pigeonpea also has low epidermal conductance relative to other tropical legumes (Sinclair and Ludlow, 1986). Leaf rolling is very strongly expressed (Muchow, 1985a). However, there is little documented evidence of extent of genotypic variation for any of these traits in pigeonpea and their potential remains unexploited in any systematic way in improvement programmes. Two mechanisms, which may be related and which contribute to strong tolerance of tissue water deficits in pigeonpea are high level of osmotic adjustment and low critical or lethal RWC of leaves (Flower and Ludlow, 1986; Sinclair and Ludlow, 1986). Genotypic variation has been demonstrated in the levels of osmotic adjustment (Flower and Ludlow, 1987), although the range among the genotypes tested was relatively small, and the potential for further improvement remains uncertain. Various mechanisms which postpone leaf and/or plant mortality are of great potential value in the context of transient drought and are also important in a terminal stress situation, because any prolonged survival could conceivably enable greater partitioning of assimilate to seed before the crop dries off and thus, minimize the effects of drought on HI.

Physiological mechanisms/traits which might conceivably be manipulated to enhance productivity of pigeonpea in rainfed/dry land environments are listed in Table 8.

3.2.4 Root shoot signals : At the onset of stress conditions, rapid changes in the endogenous level of hormones have been commonly observed. Hormones act as chemical signals in perception to response phenomenon under moisture stress conditions. As soon as soil perceives drying, it sends signal through root to shoot in the form of an enhanced ABA level and stomatal closure. Plants sense drying of soil around the root and communicate this information to the shoot by a reduction in flux of water to the shoot. Positive and negative signals have been observed in plants under stress conditions. Positive signals are those chemicals (e.g. ABA), the concentration of which increases in the roots and xylem sap inducing the stomatal closure even before there was an appreciable decrease in leaf water potential. Negative signals are those chemicals (cytokinin and potassium), the level of which comes down during moisture stress in root and xylem sap inducing stomatal closure.

Table 8 : Physiological mechanisms or traits potentially contributing to drought escape or drought survival in pigeonpea

Strategy	Mechanism/trait	Reference
Escape	Matching phenology to water supply (phenological adjustment, plasticity)	Lawn 1981, 1988
Survival Reduced water use	Stomatal closure Slower leaf development, Shorter leaf area duration Paraheliotropic movement, leaf rolling Low epidermal conductance	Sinclair and Ludlow, 1986; Sinclair and Ludlow, 1986; Muchow 1985a Muchow, 1985a; Sinclair and Ludlow, 1986;
Improved water uptake	Deep rooting	Sheldrake and Narayanan, 1979
Dehydration tolerance	Osmotic adjustment, Low critical relative water content	Flower and Ludlow, 1987; Sinclair and Ludlow, 1986

3.2.5 Growth and development : Growth and development are complex processes, which involve various morphological and biochemical changes during stress conditions. Though the growth and yield of mungbean are reduced by moisture stress at any stage of plant development, the moisture stress during vegetative stage is most detrimental (Sadasivam *et al.*, 1988). Ali and Alam (1973) observed in pot experiments that drought stress caused reduction in vegetative growth, initiation and retention of floral buds and finally seed yield in mungbean which developed more rapidly from flowering to maturity (drought escape).

Some of the earliest responses to stress appear to be mediated by biophysical events rather than changes in chemical reactions caused by dehydration. As the water content of plant decreases, the cell shrinks and the cell wall relaxes resulting in lowering of water potential. This leads to slower cell expansion and reduced leaf expansion thereby reducing leaf area of the plants. Moisture stress at vegetative stage caused reduction in plant height, root growth, leaf area, number of pods and dry matter accumulation in mungbean and urdbean (Rao *et al.*, 1996). Similar observations have been reported by Sharma *et al.*, (2004, 2007). Haqqani and Pandey (1994) suggested that decrease in LAI and increase in specific leaf weight (SLW), leaf water potential and root length density are some of the drought avoidance mechanisms in mungbean, which help in the survival of the crop during the reproductive phase in the event of moisture depletion.

Pigeonpea has remarkable resilience to water deficits prior to flowering (drought tolerance). Water deficits subsequent to flowering can induce the abscission of flower buds and small pods, or, if encounted towards maturity, may promote premature leaf loss and reduce seed size. In addition to constraints on the economy of the crop. The N accumulation in pigeonpea can be reduced under drought conditions (Chapman and Muchow, 1985) because symbiotic N_2 fixation is sensitive to water deficits (De. Vries, 1986). Thus, where pigeonpea is grown in the dry season, the tendency for longer duration genotypes to have smaller HI and nitrogen harvest index (NHI) would be accentuated to the extent that water deficit increases as the dry season progresses.

Sink growth and development : Traits such as total biomass, pod number, branch number, stem dry matter and total foliage dry matter which have significant correlation with seed yield are reduced under water stress. However, 100-seed weight and harvest index are seldom affected which implies that poor yield under drought condition could be due to source limitation rather than sink resulting in significant decrease in number of effective pods rather than seed size. High pod filling rate and sucrose synthase activity in developing grains under moderate stress, suggests strong sink activity, however, little is known about the mechanisms involved in high sink activity and efficient HI under water stress.

Yield determinants : *Harvest index, transpiration and water use efficiency* : A useful conceptual framework for analyzing yield (Y) in grain legumes is provided by the relationship Y = TDM x HI where TDM is total above ground dry matter and HI is harvest index. Based on this framework, various other crop models have been proposed, some of these are Y= C x Dr x P (Duncan *et al.*, 1978), Y= RI x RUE x HI (Monteith, 1977),Y = T x TE x HI (Passioura, 1977) and Y = T x TE x HI (Fischer, 1981), where seed yield (Y) is considered to be function of crop growth rate (C), duration of reproductive growth (Dr), proportion of C partitioned to yield (P), the amount of radiation intercepted by Crop (RI), efficiency of conversion of radiation into dry matter production (RUE), partitioning of dry matter into the reproductive component (HI), the water passing through the crop in transpiration (T) and the efficiency with which this transpiration water is utilized to produce dry matter i.e. transpiration efficiency (TE). Based on these physiological studies, genetic variation in WUE has been worked out for improving yield under drought condition.

3.3 Effect of drought on relative nitrogen fixing efficiency and nodule respiration

Adverse effects of moisture deficit on N_2-fixation in legumes have been reviewed in the past by Swaraj and Bishnoi (1994). The extent of inhibition in N_2 - fixation depended on the intensity and duration of moisture stress, rapidity of stress development, plant species and stage of growth.

Moisture deficit in nodules occurs when the root system is unable to supply sufficient water both to export materials from the nodules to root xylem sap and also to compensate for losses from nodule surface in the drying soil. At vegetative and flowering stages, nodule specific activity declined by 30-40 per cent with 5 per cent decrease in nodule moisture over the control and 95 per cent when water loss exceeded 10 per cent (Venkateswarly *et al.*, 1989). A gradual drought by withholding water/nutrient supply resulted in lowering of water potential of leaves and nodules and a consistent significant reduction in N_2-fixation. However, the amount of nitrogenase components 1, 2 remains virtually unaffected by drought.

Under natural conditions water stress usually occurs gradually and progressively as moisture is exhausted from the soil. Under such conditions in the field, root growth of groundnut (*Arachis hypogaea* L.) and soybean (*Glycine max* L. Merr.) was stimulated enabling maintenance of pre-stress stomatal conductance, leaf water potential and N_2-fixing ability until water depletion in the subsoil occurred (Swaraj *et al.*, 1996). Under intermittent drought, clover nodules confined to drying soil profile may become dessicated and show significant decrease in N_2-fixation (Schubert *et al.*, 1995). Similar observations have been confirmed by Swaraj *et al.* (1996) in Soybean raised in sand culture. Thus, it appeared that nodules replenish their water supply by surface absorption from the immediately surrounding soil/rooting medium.

A decline in N_2—ase activity under water stress was accompanied by a concomitant decrease in nodule respiration. However, the reduction in N_2-ase activity was relatively faster than that of nodule respiration. As a result, when energetics of N_2-fixation was calculated in terms of moles of CO_2 evolved per mole of C_2H_4 evolved, the energy requirement for equivalent amount of N_2-fixed increased with increase in the intensity of stress. Efficiency of N_2-fixation, (mg C utilized per mg N-fixed), by nodulated roots and nodules of chickpea was always more when plants were grown continuously under limited water supply.

3.3.1 Factors limiting N_2-fixation : Most of the adverse effects of drought on N_2-fixation in soybean and clusterbean could be explained on the basis of reduced C-fixation or reduced photosynthate supply. Though, the rate of N_2-fixation is often closely linked to the rate of photosynthesis and supply of current photosynthates to the nodules but recent evidence indicates that the two processes operate independently (Sangakkara, 1994). More soluble sugars were accumulated within cowpea nodules dried independently of the host plant as well as in soybean nodules experiencing whole plant water deficit. Inhibition of nodule activity under long term moisture stress was not caused solely by the decreased photosynthetic CO_2 exchange rates but also due to changes in photosynthate pool size.

Many workers considered that the direct effects of moisture stress on nodules brought down the rate of N_2-fixation and attributed it to decreased permeability of the nodules to O_2. Increased O_2 limitation in drought stressed nodules can be ascribed to increased resistance to O_2 diffusion resulting from collapse of lenticels and outer cortical layers or to decrease in leghaemoglobin concentration in the nodules. Lower rates of water movement out of the nodule during drought stress may restrict the export of products of N_2 – fixation, thus, inhibiting nitrogenase activity via feedback mechanism. Inhibition of nitrogenase activity by drought stress could be explained in terms of product inhibition of the enzyme as an efficient feedback regulation mechanism triggered by decreased N demand of plant (Schubert *et al.*, 1995). Legume species that transported high concentration of ureides (allantoin and allantoic acid) are more drought sensitive compared to species with low or no ureides (amide transporters) (Sinclair and Serraj, 1995). The feedback in those legumes that export ureides from the nodules seems to be especially aggravated by the accumulation of ureides in the plant. However, specific feedback signal compound or mode of action is yet to be understood. The process of the legume-rhizobium symbiosis is extremely sensitive to drought stress (Sharma and Sharma 2007). Consequently, legume productivity can be greatly depressed both by intermittent drought, which could occur at any time during the growing season when rainfall is inadequate and by terminal drought which occurs when stored soil moisture is depleted resulting in crop senescence. Oti-Boateng and Salisbury (1993) suggested a feedback control of nodule activity mediated through the pool of soluble N in the plant. Parsons *et al.* (1993) developed this concept further by suggesting that nodule formation, nitrogenase activity and nodule permeability to oxygen might be controlled by the concentration of reduced N compounds entering the nodule through the phloem. It is not obvious which function of the moisture stressed plant is actually affecting the nodule (Streeter, 1993). The effect could be due to (i) reduced photosynthesis, causing reduced availability of carbohydrates; (ii) less water for the transport of N-products away from the nodule; (iii) direct effects on nodule permeability; (iv) the alteration of nodule metabolic activity.

There is a direct correlation between the decreased nodule water potential and the decline of nitrogenase activity (Pankhurst and Sprent, 1975). Durand *et al.* (1987) suggested that water stress decreased nitrogenase-linked respiration. With soybean, Purcell and Sinclair (1995) showed changes in nodule permeability as a consequence of reduced nodule respiration and such changes have been directly related to the export of nitrogen components (Walsh, 1995). Depending on nodule morphology i.e. determinate and indeterminate structure, legume species may vary in their response to soil drying and recovery from stress when water deficits are removed. Indeterminate nodules of pea plants showed a marked inhibition in their ability to reduce acetylene even at a moderate water stress due to an increase in resistance of oxygen diffusion at the barrier of the nodule cortex. For determinate nodules (soybean) Diaz del Castillo and Layzell (1995) found that internal O_2 concentration remained quite high in mildly stressed nodules and most of the reduction in enzymatic activity could not be restarted by raising the external O_2 concentration. This implies that the nodules of water stressed plants no longer had the metabolic capacity to support N_2-fixation even in the presence of sufficient oxygen. Purcell and Sinclair (1995) found that nitrogenase activity was affected prior to nodule permeability. When PEG treatment was used to impose a reverse and rapid water stress, thus, suggesting that factors other than gas permeability could be involved in the initial decline in the nitrogenase activity. However, Serraj and Sinclair (1996) suggested the initial effect was due to oxygen limitation with a secondary metabolic limitation occurring after 24 hours.

3.3.1.1 ***Effects on Leghaemoglobin :*** Decrease in leghaemoglobin (Lb) content under stress has been reported in a number of legumes. Moderate conditions of stress which increased water saturation deficit upto 8 per cent did not seriously affect Lb content of nodules. Under severe stress conditions, increase in water saturation deficits upto 34.6 per cent considerably decreased Lb content of the nodules. However, rehydration of nodules from stress led to further denaturation of Lb in soybean. Venkateswarly *et al.* (1989) observed a sharp increase in Lb content of groundnut nodules under water stress, but no definite trend could be found in cowpea. Ahlawat (1990) observed that when nodules were exposed to prolonged periods of water stress, acceleration in their rate of senescence was accompanied by conversion of red functional form of Lb to green non-functional form. Swaraj *et al.* (1996) reported that not only content but protein composition of Lb was also affected by limited moisture availability.

3.3.1.2 ***Nodule metabolism***

Ammonia assimilation : Decrease in leaf water potential and nodule moisture content in mungbean was accompanied by decreased activity of asparagine synthetase, glutamine synthetase (GS), xanthine dehydrogenase and uricase and increased the activity of glutamate dehydrogenase whereas activity of allantoinase and allantoicase remained unaffected (Irigoyen *et al.*, 1992). Enzymes of ammonia assimilation were reported to be more sensitive to drought as compared to enzymes of ureide metabolism. In pigeonpea nodules also, ammonia assimilatory enzymes were adversely affected by drought (Kaur *et al.*, 1995). Koundal and Chopra (1989) reported that both GS and NADH-GOGAT showed a decline in activity under water stress in chickpea nodules, but the latter was relatively more sensitive. Contrarily, Gonzalez *et al.*, (1995) observed that gradual drought that withheld water/nutrient supply did not result in a significant change in GS and aspartate amino transferase activity in soybean nodules.

Carbon metabolism : Enzymes of carbohydrate metabolism in pigeonpea nodules were unaffected differentially under decreasing moisture level. The most sensitive enzymes were PEP-carboxylase (PEPc) and invertase while the activities of other enzymes were affected to a lesser extent. Irigoyen *et al.* (1992) observed that at moderate stress (ψnod-2.0 Mpa), showed a reduction *in vitro* PEPc, but severe stress (nod lower than –2.0 MPs) stimulated it. Gonzalez *et al.* (1995) did not observe any significant decrease in PEPc in soybean nodules under gradual drought.

In alfalfa nodules, the content and activity of sucrose synthase (SS) declined dramatically within a few days of withholding of water. Coincident with this decline, there was significant increase in the sucrose, total free amino acids and ureides in the nodule. These reports suggested that SS may play a key role in the regulation of nodule carbon metabolism and influence N_2-fixation under drought stress conditions. Malate dehydrogenase (MDH) activity rose in slightly stressed soybean nodules (ψnod-1.3 Mpa) whereas greater moisture deficits resulted in a sharp decline in MDH activity (Irigoyen *et al.*, 1992). Nodule malate content followed the pattern of MDH. The plant fraction of the soybean nodules constitute alcohol dehydrogenase (ADH) which was stimulated at slight and moderate drought level accompanied by accumulation of ethanol. Bacteroid β-OH-hydroxybutyrate dehydrogenase (β-OH-BDH) activity was stimulated in slightly and moderately stressed soybean nodules. Increased activity of this enzyme has been related to an excess of oxidizable carbon sources. Koul *et al.* (1991) observed high activity of pyrroline-5-carboxylate reductase and proline dehydrogenase, the enzymes associated with proline synthesis in soybean nodules, which might help the nodules to accumulate proline for osmoregulation.

Anti-oxidants : Ascorbic acid and glutathione (reduced) content of nodules decreased sharply at all the growth stages of plants exposed to stress. Phenol content of leaves as well as nodules increased 1.5-2.0 fold in the plants subjected to permanent wilting. Gogorcena *et al.* (1995) correlated drought caused inhibition of N_2-fixation in pea nodules with antioxidant defense mechanism against activated oxygen and recorded drop in nodule water potential from –0.29 Mpa (control) to approximately –2.03 Mpa (stressed plants), decreased activity of catalase (25%), ascorbate peroxidase (18%), dehydroascorbate reductase (15%), glutathione reductase (31%) and superoxidase dismutase (30%). Water stressed nodules contained only approximately 40 per cent ascorbate and reduced glutathione and 60 per cent oxidized glutathione in comparison with the controls. NAD^+ and NADH decreased by 43 percent, $NADP^+$ by 31 per cent and NADPH by 17 percent. It was suggested that the decline in antioxidant capacity of nodules may result from a restricted supply of NADP(P)H *in vivo* for the ascorbate-glutathione pathway and Fe-catalyzed Fenten reactions of ascorbate and glutathione with activated oxygen leading to formation of highly damaging hydroxyl free radicals, as 2-fold increase in catalytic Fe, was observed in stressed nodules. Oxidative damage to the nodules under drought reflected itself in augmented levels of lipid peroxides (2.4 fold) and oxidatively modified proteins (1.4 fold).

It is, thus, clear that legumes cultivated under dryland rainfed conditions possess following characteristics of N_2-fixing system.

i) Nodulation is well spread over the root system during growth period of the plant. Few but large nodules are produced.

ii) Nodules are of indeterminate type having prolonged meristematic activity.

iii) The synthesis of different components of leghaemoglobin is so balanced that the protein exhibits maximum efficiency in O_2 – transport.

iv) The synthesis of proteins that trigger nodule senescence is delayed.

4. WATERLOGGING

It is well established that most of the agricultural crops are not adapted to wetland conditions. Flooding reduces the shoot and root growth, dry matter accumulation and ultimately the crop yield. When a soil is temporarily flooded, a number of physical and chemical changes may profoundly influence the soil properties leading to reduced growth. In addition, incompletely oxidized metabolic products may accumulate to toxic levels decreasing the availability of essential nutrients. Reduced absorption and closure of stomata may lead to a lower rate of photosynthesis. Subsequent changes include decreased permeability of roots, reduced or increased mineral uptake, alteration in growth hormone balance, leaf epinasty, chlorosis and abscission.

Legumes are generally intolerant of waterlogged or poorly aerated soils. Detrimental effects of waterlogging tend to be greater in plants depending totally on N_2-fixation for their nitrogen requirement than those having access to combined nitrogen. Soil pores under waterlogged or flooded conditions are filled with water and diffusion of O_2 is blocked. Under these conditions, roots are unable to obtain sufficient oxygen from the soil for their aerobic respiration.

Urdbean and mungbean are generally unsuited in wet tropics where annual precipitation exceeds 1,000 mm. These plants are readily damaged by heavy rains and windstorms. Prolonged rainy periods during ripening may result in moulding of the seeds or even sprouting of the seeds in the pods as they do not possess seed dormancy. Mungbean does not tolerate waterlogging and is inferior to urdbean but a physiological explanation for this difference is still wanting/ellusive (Poehlman 1991). During long periods of waterlogging, nodules and outer root tissues slough off, causing a temporary nitrogen deficiency. Verma and Rao (1975) reported that population of soil rhizobia, number and weight of nodules markedly increased with the increase in soil moisture content from 25 to 50 per cent on dry weight basis. Further increase in moisture content reduced the population, being lowest at 100 per cent soil moisture. Trung *et al.* (1985) reported that waterlogging for 3-days recorded reduced mungbean yield by 82 per cent at flowering, 60 per cent at pod formation, 40 per cent at pod development stage and 17 per cent at grain maturity. Waterlogging resulted in disappearance of leghaemoglobin in root nodules and cessation of nitrogen fixation activities, reaching almost zero during the 3-days of watrlogging. Islam *et al.* (1994) observed that leaf area, CGR, NAR, leaf canopy structure, dry matter production and yield in mungbean were more susceptible to waterlogging than to drought. Minchin and Summerfield (1976) reported enhanced phellogen activity producing large lenticels in cowpea and soybean nodules under waterlogged conditions, which is an adaptation to reduced O_2 availability. Cowpea was much less tolerant to waterlogging than soybean and may lose large proportion of its nodule population. However, subsequent drainage resulted in fast resumption of nodule growth. Plant growth was depressed as a result of competition for available carbohydrates. A 4-day flooding treatment of soybean plants irreversibly altered the ultrastructure of central infected regions of nodules located at the middle and lower parts of the root system. The bacteroids were broken into fragments followed by their lysis within the peribacteroid space. The infected cells of nodules located on upper part of the root near the water surface showed relatively less damage probably due to diffusion of some oxygen to these tissues. *Pisum sativum* nodules developed under waterlogged conditions showed depressed production of vascular transfer cells accompanying decreased N_2-fixation activity (Minchin and Pate, 1975).

Pigeonpea is adapted to drier environments. In India, exposure of pigeonpea to waterlogging is most likely and most detrimental during seedling growth in cracking clay soils, when monsoon rains lead to short-term flooding. Waterlogging damages the root system through anoxia and possibly, the effects of ethylene, and produces symptoms of water deficit (Chauhan, 1987). Under waterlogged conditions, the reduced nitrogenase activity in soybean nodules was attributed to enhanced ethanol production (Sprent and Zahran 1986). Plants are also extremely vulnerable to attack by the fungus *Phytophthora drechsleri* f. sp. *Cajani* that causes blight, and to N deficiency through the inhibition of N_2-fixation (Thompson *et al.*, 1981). Genotypic differences in response to waterlogging have been reported (Chauhan, 1987; Dubey and Asthana, 1987). Average survival after 6-days of waterlogging was 96 per cent in a tolerant genotype compared to 28 per cent in a susceptible genotype of pigeonpea. Combined tolerance to waterlogging and salinity has also been observed in pigeonpea (Chauhan, 1987).

4.1 N_2-fixation

Soil flooding is known to repress formation and N_2-fixing activity in root nodules of legumes. Andreeva *et al.* (1985) observed strong inhibition of N_2-fixation within 4-days of flooding of soybean plants. Waterlogging inhibited the nitrogen fixing activities of nodule (Sung, 1993). This may be due to decreased synthesis of nitrogenase by the bacteriods or by curtailed carbohydrate metabolism of host tissue in the nodule. However, nitrogenase activity recovered completely from these adverse effects within 4-5 days after draining of the water. Waterlogging (7 and 14 days) decreased the nitrogenase activity by 37.6 per cent and 67.2 per cent in peanut (Bishnoi and Krishnamoorthy, 1990) and 32 per cent and 50.3 per cent in chickpea (Bishnoi and Krishnamoorthy, 1991). Experiments conducted by Minchin and Pate (1975) on pea grown in sand culture suggested that sand water contents from 2 per cent (sand dry weight) to permanent wilting at 1.2 per cent and field capacity at 1.5 per cent upto fully waterlogged conditions at 17 per cent led to drastic reduction in N_2-fixation. But within range of 2-15 per cent water content had little effect on N_2-fixation. Nodulated plants raised in water culture fixed N_2 maximally when aerated with air (20% O_2), Decreasing O_2 concentration in the aerating gas stream resulted in a considerable decline in N_2-fixation. Verma and Rao

(1975) reported that excess soil moisture (75 and 100% of soil dry weight) was more detrimental to nodulation, N_2-fixation and yield of mungbean than limited soil moisture (25% of soil dry weight), N deficiency through the inhibition of N_2-fixation has been reported in pigeonpea by Thompson *et al.* (1981).

Waterlogging induced soil aerobiosis is harmful to nodule formation and functions in several legume species. Bisseling *et al.* (1980) found that production of the iron protein of the nitrogenase complex was repressed under waterlogging conditions. Reduced specific nodule activities may also be due in part to concomitant decline in content of leghaemoglobin, which facilitates the O_2 diffusion from the nodule surface to the bacteriod. Buttery (1987) reported that soybean plants grown on nitrate were less sensitive to a 30-day flooding stress than plants relying on N_2-fixation. The positive effect of nitrate on flooding tolerance may be due to a lesser O_2 requirement for nitrate uptake and assimilation compared with N_2-fixation which may mimic N deficiency and subsequent decrease in biomass accumulation (Bacanamwo and Purcell 1999). These results indicate that improving N availability under flooding or selecting for genotypes that maintain high rates of N assimilation may be effective strategies for improving tolerance to flooding stress.

4.2 Nodule respiration and aeration

Inhibited development and reduced nitrogenase activity of root nodules under flooded or anoxic soil conditions were primarily thought to be caused by reduced respiration. Respiration and N_2-fixation studies on detached nodules exposed to different oxygen tension showed that the efficiency of carbohydrate consumption in fixation was the highest at near atmospheric levels of O_2 supply. The effect of aeration N_2-fixing nodules is complicated because of considerable variation in the ability of plants to transport O_2 from shoots to roots. Only those legume species/cultivars will support N_2-fixation under waterlogged conditions which can develop aerenchyma in the lower part of the stem, roots and nodules to face reduced O_2 – supply on flooding. Waterlogging and flooding frequently lead to development within plant tissues of anoxic zones that co-exist with zones receiving sufficient O_2 for oxidative phosphorylation. In waterlogged and flooded soils, both tissue zones would be exposed to increased CO_2 and ethylene concentrations (Greenway and Gibbs, 2003). Fig.3 speculates on responses to the energy crisis inherent to anoxia. The scheme proposes mechanisms of survival as well as events leading to cell death. Most anoxia tolerant and intolerant species share a decrease in pH_{cyt} from 7.5 to between 6.8 and 7.2, which stimulates ethanol fermentation, hence anerobic ATP synthesis, and synthesis of enzymes involved in anerobic catabolism while decreasing synthesis of a large number of proteins. A key acclimative mechanism in anoxia tolerant tissues would be reduction in energy requirements for maintenance, which would allow down regulation of ethanolic fermentation, thus, prolonging carbohydrate reserves (Fig.4). However, such regulation of energy flow to processes essential to survival is largely unexplored.

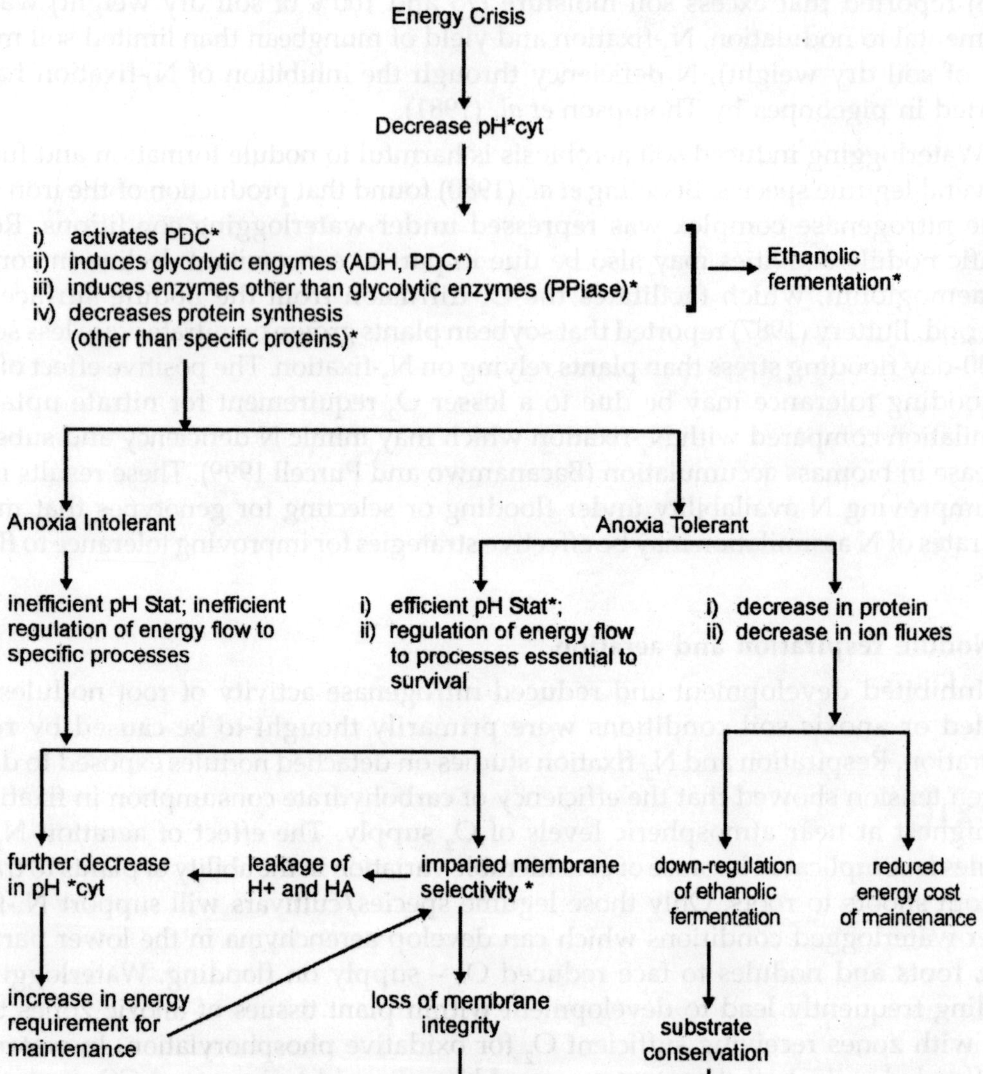

Fig. 3 : A scheme for events that follows the onset of anoxia, and therefore an energy crisis in plant tissues, indicating possible mechanisms of tolerance, and causes of intolerance to anoxia [Greenway and Gibbs, 2003]

Asterisks indicate processes for which substantial evidence exists. Other processes indicated in the diagram are supported by limited evidence and in some cases are speculative only
PDC : Pyruvate decarbonylase, ADC : alcohol dehydrogenase, PPiase : Pyrophosphatase

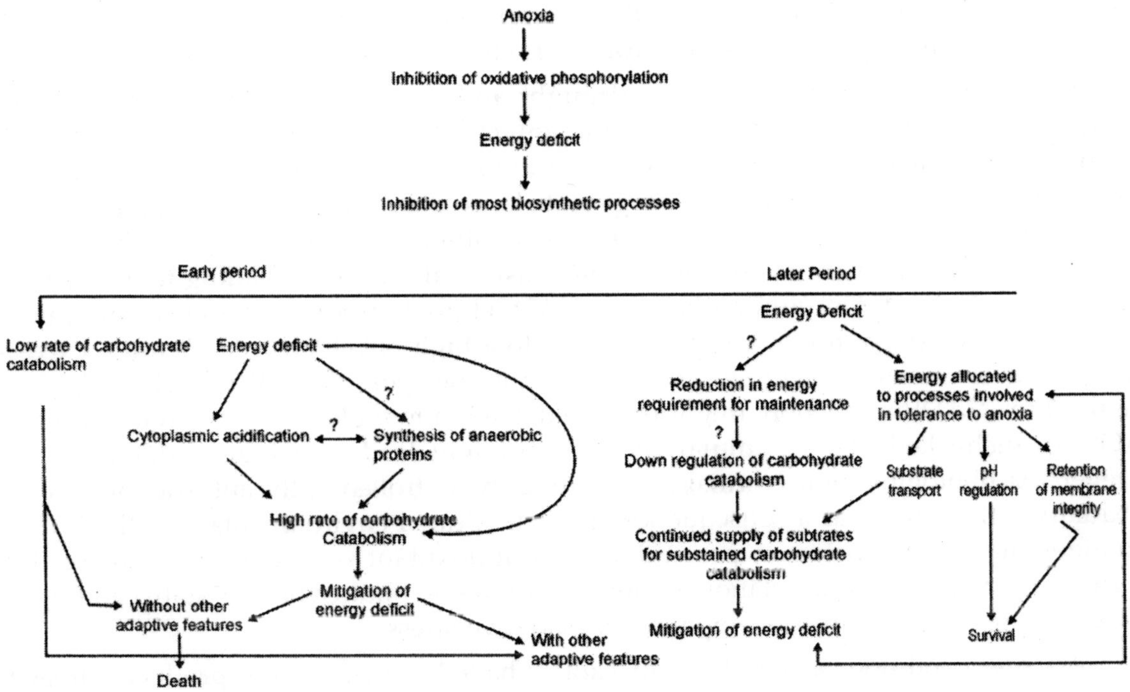

Fig. 4 : Response to short and long term anoxia. [Greenway and Gibbs 2003]

5. OXIDATIVE STRESS

It is widely accepted that many adverse growth conditions cause oxidative damage to plant cell. At molecular level, oxygen radicals could selectively attack proteins, nucleic acids or lipids (Valentine *et al.*, 1998). In response, protecting mechanisms at the whole plant level are switched on, including scavenging of oxygen-centered radicals, maintenance of ion uptake and water balance, reactions altering carbon and nitrogen allocation. In most situations, the response is moderate and an imbalance occurs between production of toxic oxygen derivatives and the antioxidant defense systems of the plant (Foyer *et al.*, 1994).

The photosynthetic electron transport chain is somewhat intriguing. It is the ultimate source of energy and reducing power that supports metabolism, but it is also the major source of the active oxygen species (AOS) responsible for most of the cellular injury, especially when plants are exposed to adverse conditions. This self-contradictory behaviour is caused by the nature of the components of the photosynthetic electron transport chain. Light absorbed through the photosystems driven to generate high energy electrons should be readily used in a productive manner i.e. for $NADP^+$ photoreduction. If key intermediate components are kept in a high-energy state for enough time, they could interact with molecular oxygen and produce AOS such as singlet oxygen ($^1O_2$)

and superoxide radical (O^-_2). Moreover, these components could also be severely damaged; rendering the photosynthetic electron transport chain inoperative (Palatnik *et al.*, 2003).

Many adverse situations such as drought and extreme temperatures have been claimed to produce at least part of their deleterious effects through increased accumulation of AOS. In fact, whenever plants are exposed to light energy in excess to that required for photosynthesis, the result is an energy imbalance that might lead to oxidative stress and photoinhibition. Current models of photosynthetic operation suggest that stress conditions diminish plant assimilatory metabolism (calvin cycle), leading to a decrease in NADPH demand. Uncoupling between NADPH photoproduction and consumption may cause a shortage in the $NADP^+$ pool due to a high $NADPH/NADP^+$ ratio. Excess energy and electrons in the photosynthetic electron transport chain, that could no longer be used for $NADP^+$ reduction, might then be delivered to molecular oxygen to generate AOS. If unchecked, the accumulation of AOS might lead to oxidative damage and irreversible photoinhibition. Normally AOS are easily controlled by the antioxidant system. Oxidative stress arises as a consequence of a sub-optimal environmental condition that changes the balance between prooxidant and antioxidant reactions in favour of the former. Therefore, the interrelation among environment, photosynthesis status and light intensity are crucial to the development of oxidative stress.

Adverse conditions as tabulated in Table 9 have been claimed to produce damage through AOS accumulation. Increased production of toxic oxygen derivatives may be considered a universal feature of stress conditions and plant productivity is often limited by oxidative stress.

Table 9 : Factors Provoking Oxidative Stress in Plants

Factors	Reference(s)
High light	Gillet *et al.*, 1998
Redox-cycling herbicides : Paraquat, diquat	Halliwell, 1991
Drought	Navari-Izzo *et al.*, 1996 Iturbe-Ormaetxe *et al.*, 1998
Air pollutants (SO_2, NO, NO_2, Ozone)	Hippeli and Elstnes, 1996
Hyperoxia	Scandalios, 1992 Simontacchi *et al.*, 1995
Mineral deficiency	Polle *et al.*, 1992; Zu and Scandalios, 1992
Pathogen infection	Grant and Loake, 2000
Aluminium	Richards *et al.*, 1998
Heavy metals	Dal *et al.*, 2000
Low temperature	Pastori *et al.*, 2000
High temperature	Filek *et al.*, 1997; Dat *et al.*, 1997
Senescence	Miller *et al.*, 1999; Weaver *et al.*, 1998

5.1 Glutathione and oxidative stresses

The tripeptide glutathione (GSH) is the major non-protein thiol in most plants. GSH is a versatile antioxidant and is involved in the enzymatic detoxification of activated oxygen species (AOS) and participate in the ascorbate GSH cycle for peroxide removal in the chloroplasts of plants.

Legumes are an interesting plant material to study thiol metabolism for different reasons. Firstly, root nodules exhibit an active ascorbate-GSH cycle, which require a continuous supply of GSH to protect nitrogen fixation against toxic oxygen species. Secondly, the leaves, roots and seeds of some legumes contain a thiol tripetide homolog, homoglutathione (hGSH : γGlu-Cys-β Ala) in place of or in addition to GSH. The synthesis of hGSH is thought to involve γECS (-glutamylcysteinyl synthetase, γECS:EC 6.3.2.2) and a specific hGSH synthetase. Thirdly, GSH is believed to be involved in plant morphogenesis, cell division, control of redox status and signaling of stress and pathogen attack with some modifications. All these processes are important in nodule formation and functioning and thus, GSH is likely to be critical molecule of nodules. Nodules thus, are a main site of GSH and hGSH synthesis within the plant and provide indirect evidence that thiol compounds play a crucial role in the process of nitrogen fixation. High concentrations of thiol tripeptides are present in nodules as well, but the information available is based on an enzymatic assay that does not distinguish between GSH and hGSH (Griffith 1980), Matamoros *et al.* (1999) found that glutathione (GSH) was the major non-protein thiol in all indeterminate nodules examined, as well as in determinate nodules of eight legumes. Evans *et al.* (1999) reported hGSH is more abundant than GSH in soybean nodules and concluded that natural senescence in soybean nodule is an oxidative stress process. As the thiol content decreased, catalytic iron, thiol oxidation and oxidative damage increased.

6. ELEVATED ATMOSPHERIC CO_2

Elevated CO_2 is likely to affect carbon cycling and crop productivity by stimulating photosynthesis. Some authors have postulated that legumes, which can fix atmospheric N_2- would have an advantage over other plants. Legumes could respond to a potential C:N imbalance caused by elevated CO_2 by introducing N through symbiotic N_2 fixation. One obvious hypothesis in predicting the response of symbiotic N_2 fixation to elevated atmospheric CO_2 would be that as the rate of photosynthesis increased, more photosynthates would be delivered to nodules and thus, nitrogenase actively increases (Hardy and Havelka, 1976). The long term increase of total nitrogenase activity per unit nodule dry weight was not greater than that in the control (Ryle *et al.*, 1992). The CO_2 enhanced plant growth may result in an increased N-sink strength, as a result of nitrogenase activity would be upregulated to meet the increased N demand.

Low rhizosphere CO_2 concentration below 100 ppm results in significant decline in nitrogenase activity of legume nodules. However, increasing the partial pressure of CO_2 in the rhizosphere above the ambient did not stimulate the nitrogenase activity of soybean. Therefore, direct affects of altered rhizosphere partial CO_2 on nitrogenase activity are unlikely. Nitrogenase activity is rather oxygen than carbon limited (Hunt and Layzell, 1993). Increase in CO_2 concentration has shown not only to increase photosynthesis, plant biomass, stomatal resistance, leaf water potential and transpiration rate but also to decrease respiratory CO_2 in mungbean (Rao and Sengupta, 1997). On the other hand, the higher atmospheric temperature due to green house effect can adversely affect the physiological processes and productivity of pulses. To understand the role of rising atmospheric CO_2 under high atmospheric temperatures, an interactive study of CO_2 and temperature is required to be taken up in future. The mechanisms may possibly the CO_2 induced enhancement in the growth and dry matter production in mungbean plants. (a) Besides effecting photosynthesis, the reduced CO_2 efflux rate would increase their ability to act as a sink for carbon; (b) The increase in stomatal resistance and reduction in transpiration with a consequent beneficial effect on leaf water potential would help in leaf expansion and photosynthesis; (c) The greater allocation of carbon to roots in elevated CO_2 would improve its potential for exploring the soil volume for water nutrients. All these factors increase the productivity of pulses in general.

Sharma and Sengupta (1997) observed that physiological age not only affected the response of net photosynthesis to elevated CO_2 concentration in leaves at different positions on the main stem but also showed variations even when they were of the same physiological age. Allocation of extra carbon towards starch synthesis or current export demands depends upon the amount of extra carbon fixed. The carbon is diverted to starch synthesis only when sucrose synthesis is saturated. Increased starch content in source was also observed by Ghildiyal *et al.* (1998) in response to increased CO_2 in mungbean. Differential responses were observed under short term and long term elevated CO_2 while increased temperature caused detrimental effects, thus, modifying the positie effects of CO_2 enrichment. Short term effect of elevated CO_2 results in increased carboxylation which favours higher accumulation of sugars and more dry matter production. But, the long term effect of CO_2 often ends in increased temperature which increased the oxygenase activity of Rubisco relative to carboxyation by decreasing the affinity of Rubisco for CO_2 and aqueous solubility of dissolved CO_2 to that of O_2 (Deshmukh *et al.*, 2007). Increased temperature also affects the reproductive biological processes like tapetal degeneration, pollen germination, pollen tube growth, pollen viability, micro and mega sporogenesis. There are several reports indicating significant imbalanaces in growth promoters and inhibitors due to increased temperature leading to flower and fruit drop, causing significant reduction in productivity of pulses. Therefore, comprehensive studies on growth regulators under enriched CO_2 and elevated temperatures in inverse pulses will provide indepth understanding of some crucial aspects leading to enhanced productivity under changing environmental conditions.

7. UV-B RADIATIONS

The increase in concentration of chloro-fluoro carbon (CFC_S), methane, nitrous oxide and other gases in the atmosphere due to man-made activities is responsible for the depletion of stratospheric ozone layer, which is a sole attenuator of solar ultraviolet B (UB-B, 180-320 nm). The C_3 plants appear to be more sensitive to UV-B radiation than C_4 plants. Among C_3, species belonging to Leguminosae family are categorized as most sensitive to UV-B radiation. *Vigna* species have been reported to be affected by current level of ambient UV-B radiation. Pal *et al.* (1998) reported that mungbean plants exposed to UV-B radiation showed significant reduction in plant height, leaf area, total biomass, photosynthetic rate, chlorophyll, soluble protein and carbohydrate content. However, UV-B radiation increased the content of flavonoids, anthocyanin, total free amino acid and PAL activity. Seedlings of *Vigna sinensis* when grown under additional UV-B radiation at 20-30°C temperature showed gradual decrease in RUBPc activity, soluble protein content and $^{14}CO_2$ fixation. These parameters showed parallel increase with increasing period of UV-B treatment at 10°C.

The polypeptide pattern of crude leaf extract showed a progressive decline in the level of 55 and 13 KDa polypeptides with the time of UV-B treatment at 20 and 30°C (Kulandaevelu and Nedunchezhian, 1993). Among the *Vigna* species, maximum reduction for net photosynthesis was observed with increased UV-B radiation equivalent to 15 per cent ozone depletion in urdbean whereas the greater reduction in nitrogenase activity was observed in mungbean (Singh, 1997).

8. ACID RAINS AND SOIL ACIDITY

Acid rain has been termed as a cumulative and secondary form of air pollutant. The adverse effects of acidic deposition on aquatic, wetland and forest ecosystems have been well documented (Evans, 1988). Though, it was considered a problem of heavily industrialized regions in Europe and America, signs of its occurrence in Asian countries have been noticed recently (Rhodhe *et al.*, 1992). India is considered as an acid rain hot spot and several metropolitan cities have been experiencing acidic precipitation (Mohan and Kumar, 1998). So far, studies on the impact of acidic rains on crop plants have been confined to temperate regions only (Evans, 1988). The photosynthetic pigments, soluble proteins, reducing and total sugars and starch content were higher in mungbean cultivars with simulated sulphuric acid rain (SAR) of 5.5 pH and slowly decreased with increasing levels of acidity (Kumaravelu and Ramanujani, 2004).

Soil acidity affects all aspects of nodulation and nitrogen fixation viz., survival and multiplication of the rhizobia in the soil, infection, nodulation and nitrogen fixation. The failure of legumes to nodulate under acid soil condition is common especially in soils with pH less than 5.0. The inability of some rhizobia to persist under such conditions is one of the causes of nodulation failure (Graham *et al.*, 1982; Lowendorf and Alexander 1983). In chickpea root nodule bacteria survived quite well and enhanced lateral nodulation was observed under acidity (soil pH 4.2) but the inability of the host plant to tolerate acidity might limit productivity (Scott and Howieson, 2000). Species of *Rhizobium* were known to differ in their tolerance to soil acidity, the slow growing *Bradyrhizobium* strains being generally more acid tolerant than fast growing especially *R. meliloti* (Munns and Keyser, 1981). Rossum *et al.* (1994) observed that the soil acidity enhanced nodulation in groundnut particularly with indigenous *Bradyrhizobium* population. However, the enhancement varied in groundnut cultivars with acid tolerant and sensitive cultivars differing in their content and concentration of root exudates (Raman, 1992).

9. AGROCHEMICALS

Pesticides have become an essential input in agriculture. Pesticides effect symbiotic processes. Fungicides normally at their recommended dose do not effect symbiosis. Rhizobial strains show differential tolerance to fungicides (Gibson, 1971). Carboxin + Captan at the recommended level did not affect plant growth, nodulation and N_2-fixation but carboxin alone or mixed with Captan at 10 times dose was detrimental (Mallik and Tesfai, 1985). Application of fungicide carbendazim to cowpea caused no significant effect on number of nodules but increased the fresh weight of nodules and total N content (Singh *et al.*, 1986). However, it decreased the number of *Bradyrhizobium japonicum* surviving on seeds, nodulation and yield of soybean. But, when associated with oxine copper, carbondazim had no detrimental effect on *B. japonicum* population or soybean growth (Catroux and Arnand, 1991). Thiram had no effect on chickpea nodulation (Welty *et al.*, 1988). *Rhizobium* strains displayed different sensitivities to thiram when associated with carbondazim. Abdel-Gaffar (1987) who concluded that terbutryn @ 2.5 l/ha significantly increased dry weight of nodules while trifluralin @ 4.76 l/ha applied 21-days after sowing, stimulated plant growth and total nitrogen fixed. Herbicides when used at the recommended rates, generally have little effect on symbiosis.

Herbicides may inhibit growth, deform stems and roots and may kill the rhizobia present in the soil, thus, reducing nodulation and nitrogen fixation (Bollich *et al.*, 1984). However, the stimulatory or inhibitory effects of herbicides on N_2-fixation are dependent on herbicide concentration, site of action and mechanism. Herbicides affecting root development or chlorophyll synthesis may adversely affect the symbiosis. Oxyfluorfen and oxadiazon at recommended rates of application adversely affected the nodulation and N_2 fixation of lentil (Sandhu *et al.*, 1991). Similarly pendimethalin (0.75 to 1.25 kg ha^{-1}) and metribuzin (0.25 and 0.75 kg ha^{-1}) adversely affected the nodulation and nitrogenase activity of chickpea (Pahwa and Prakash, 1992).

Normally insecticides at recommended doses do not show any adverse effect as reported by Pareek and Gaur (1970) in *Vigna radiata*. Acephate, diazinon and texaphene at recommended and 5 times more dose reduced N_2 fixation and total N content but not growth and nodulation (Mallik and Tesfai, 1985). Dimethoate @ 1% reduced nodule number by 63 per cent in black gram and 64 per cent in Pea. Leghaemoglobin concentration, nitrogenase activity and size of the bacterial region were also reduced by the higher rates of dimethoate (Soam and Agrawal, 1989). At lower concentration, the insecticide phorate enhanced the activity of nitrogenase but inhibited it at higher concentration (Sehon *et al.*, 1985; Singh and Mathur, 1989 and Nayak *et al.*, 1989). The higher concentration of phorate caused a drastic fall in the activity of nitrogenase, indicating toxic effect on the bacteriod and/or permeability of nodule. The increase at lower concentration of phorate increased concentration of gibberellins (Singh *et al.*, 1982), which stimulated nitrogenase activity (Sehon *et al.*, 1985). Stimulation of nodulation and N_2 fixation in *Vicia faba* by some herbicides were reported by Abdel-Gaffar (1987).

CONCLUSION

Under stress conditions, reductions in biomass production to the extent of 30 to 60 per cent depending on intensity and duration of stress have been recorded. Traits like root and seed growth rate, membrane injury index and relative water content could be used in identifying suitable parents for breeding high yielding genotypes under stress environments. Regulation of processes limiting sink development, triggering the productivity by making critical use of nutrients, growth promoters and retardants which influence plant architecture and growth habits need manipulation for breaking the yield plateau. Areas where further understanding is particularly needed include research to define quantitatively and predictively the environmental control of reproductive ontogeny, so that cultivars can be better matched to their environments; research to critically evaluate environmental and physiological constraints to partitioning of biomass and N to seed, so that HI and NHI can be improved, research to explore the opportunities for exploiting physiological traits which assist survival during stress and/or basis of tolerance mechanisms against various abiotic stresses. Symbiotic N_2 fixation by legume crops depends on interaction between plants and bacteria, which is affected by both above and below ground environment. Therefore, study is required to ascertain the reason for the failure of nodulation and/or N_2-fixation under certain environmental conditions, in order to minimize the detrimental effects on these processes. Further screening of germplasm under various unfavourable abiotic conditions will be useful for better nitrogen fixation and yield of pulse crops. The field of molecular biology is developing very fast and, therefore, molecular markers closely linked with important sink development processes need to be identified. To achieve the goal of higher and stable productivity, the development of ideotypes and management technology should be location and season specific with major emphasis on the improvement of WUE and soil health.

REFERENCES

Abdel-Gaffar S (1987) Factors affecting nitrogen fixation and yield of fababean under Egyptian field conditions A Review *Arid Soil Res Rehabit* **1** : 65-75

Abd-El-Samod HM Shaddad MAK (1997) Salt tolerance of soybean cultivars *Biol Plant* (Prague) **39** : 263-269

Agarwal RK Gupta SC (1995) Plant growth substances as osmoregulants under salt stresses in callus cultures of cowpea *Ind J Plant Physiol* **38** : 325-327

Aggarwal VD Poehlman JM (1977) Effect of photoperiod and temperature on flowering in mungbean (*Vigna radiata* (L) Wilczek) *Euphytica* **26** : 207-219

Ahlawat S (1990) MSc Thesis Haryana Agricultural University Hissar

Alexandrov V Y (1964) Cyto-physiological and cyto-ecological investigations of heat resistance of plant cell towards the action of high and low temperatures *Quarterly Rev Biol* **39** : 35-37

Ali A Alam K (1973) Effect of soil moisture stress on some growth characters of mung *Pak J Sci Res* **25** : 255-61

Andrews CJ (1996) How do plants survive ice? *Ann Bot* **78** : 529-536

Angus J F Cunningham R B Moncur M W Mackenzie D H (1981) Phasic development in field crops I Thermal response in the seedling phase *Field Crops Res* **3** : 365-378

Arora S Daradhi PP Sandeep A (1995) Light induced proline enhancement of *Vigna radiata* exposed to environmental stress *Aust J Plant Physiol* **22** : 11-16

Bacanamwo M Purcell LC (1999) Soybean drymatter and N accumulation responses to flooding stress N sources and hypoxia *J Exp Bot* **50** : 689-696

Balasubramanian V Sinha SK (1976) Effect of salt stress on growth nodulation and nitrogen fixation in Cowpea and Mungbeans *Physiol Plant* **36** : 197-200

Ball MC Anderson JM (1986) Sensitivity of photosystem II to NaCl in relation to salinity tolerance comparative studies with thylakoids of salt tolerant mangrove *Avicennia marina* and salt sensitive pea (*Pisum sativum*) *Aust J Plant Physiol* **13** : 689-698

Banet G Smadar W Hanna B Bruria Ben D Yehezkal F Yoram K (1996) Toxic and osmotic effects of salinity on growth and nodulation of *Medicago sativa* Symbiosis **21** : 209-222

Barnet G Smadar W Hanna B Bruria BD Yehezkal Friedman Yoram K (1996) Toxic and osmotic effects of salinity on growth and nodulation of *Medicago sativa* Symbiosis **21** : 209-222

Bar-Nun N Poljakoff-Mayber A (1979) Intervarietal differences in the amino acid composition of pea roots as related to their response to salinity *Annals Bot* **44** : 309-314

Belkhodja M (1996) Salt concentration effect on the proline amount of old organs of 3 faba bean lines (*Vicia faba* L) during the development of plant *Acta Bot Gallica* **143** : 21-28

Belkhodja Moulay Meriem AS (1993) Effect of salinity on the proline amount in the young organs of the 3 Faba bean lines (*Vicia faba* L) *Acta Bot Gallica* **140** : 473-478

Ben-Miled Daoud D Cheriff (1992) Effect du NaCl sur "utilization des lipids et les activities enzymatiques glyoxysomales au cours de la germination de deux especes de Medicago" *Can J Bot* **70** : 876-883

Bernstein L Ogata G (1966) Effect of salinity on nodulation nitrogen fixation and growth of soybeans and alfalfa *Agron J* **58** : 201-203

Bhardwaj SN (1962) Physiological studies on salt tolerance in crop plants XVIII Influence of iso-osmotic concentrations of NaCl and sucrose on early seedling growth of wheat *Agra Uni J Res Sci* **11** : 69-74

Bishnoi N R Singh H Swaraj K (1997) Influence of sodium chloride on nitrogen fixation and enzyme associated with scavenging hydrogen peroxide in cluster bean root nodules *Ind J Expt Biol* **35** : 193-196

Bishnoi NR Krishnamoorthy HN (1990) *Pl Physiol Biochem* **28** : 663-666

Bishnoi NR Krishnamoorthy HN (1991) *Ind J Exp Bot* **29** : 291-293

Bisseling T Van Stavern W VanKammen A (1980) The effect of waterlogging on the synthesis of nitrogen components in bacteriods of *Rhizobium leguminosarum* in root nodules of *Pisum sativum Biochem Biophy Res Comm* **93** : 687-693

Biswas A Patra AP (2007) Study on the effect of phosphorus vesicular arbuscular mycorrhizae (VAM) and phosphate solubilizing bacteria (PSB) on the performance of summer greengram [*Vigna radiata* (L) Wilczek] In : National Symposium on Legumes for ecological Sustainability : Emerging Challenges and Opportunities Nov 3-5 2007 Ind Inst Pulses Research Kanpur Abstr : Natural Resource Management 76 p 72

Blum A (1988) Plant Breeding for stress environments CRC Press Boca Ratons Florida USA

Bollich PK Duningan EP Hager TR Kitchen LM (1984) Effects of herbicides on nodulation nitrogen fixation and seed yields of soybeans in Louisiana *Louisiana Exp Station Bull* **762** : 3-15

Bozuck S (1981) *Annals Bot* **48** : 81-84

Buceiarelli B Hanan J Palmquist D Vance CP (2006) A standardized method for analysis of *Medicago truncatula* phenotype development *Plant Physiol* **142** : 207-219

Buttery BR (1987) Some effects of waterlogging and supply of combined nitrogen on soybean growth *Can J Plant Sci* **67** : 69-77

Calcagno F Gallo G (1993) Physiological and morphological basis of abiotic stress resistance in Chickpea In KB Singh MC Saxena eds Breeding for stress tolerance in cool-season food legumes West Sussex UK John Wiley & Sons pp 293-309

Catroux G Arnand F (1991) Compatibility of a soybean peat incoulant with some seed applied fungicides and microgranular insecticides *Toxicol Environ Chem* **30** : 229-239

Chakrabarti N Mukherji S (1994) Effect of Indole-3-acetic acid gibberellic acid and kinetin as pretreatment or foliar spray on chlorophyll content chlorphyllase enzyme activity photosynthetic non-cyclic electron transport and CO_2 uptake in C_3 glycophyte *Vigna radiata* (L) *Wilczek under salt stress* **21** : 59-64

Chandrasekharan SN Parthasarathy N (1963) Botany of useful plants – Indian gram *Madras Agric J* **16** : 5-11

Chapman AL Muchon RC (1985) Nitrogen accumulated and partitioned at maturity by grain legumes grown under different water regimes in a semi-arid tropical environment *Field Crop Res* **11** : 69-79

Chauhan YS Venkataratnam N Sheldrake AR (1987) Factors affecting growth and yield of short-duration pigeonpea and its potential for multiple harvests *J Agri Sci Cambridge* **109** : 519-529

Chen THH Yamamoto S D K Gusta L V Slinkard A E (1983) Imbibitional chilling during chickpea germination *J Am Soc Horti Sci* **108** : 944-948

Chetti MB Sirohi GS (1998) Effect of low leaf water potential at deifferent stages on biophysical parameters in mungbean In National Symposium on Future Goals of Physiological Research for the Improvement of Plant Resources Annamalai University Annamalainagar India Dec 18-20

Choudhary A K (2007) Selection criteria for low temperature tolerance in long duration Pigeonpea In : National Symposium on Legumes for ecological Sustainability : Emerging Challenges and

Opportunities Nov 3-5 2007 Ind Inst Pulses Research Kanpur Abstr : Genetic Enhancement - 100 p 266

Christiansen MN St John JB (1984) Chemical modification of plant response to temperature extremes In : Bioregulators : chemistry and uses Amer Chem Soc Washington DC USA Washington USA USDA pp 235-243

Chu TM Aspinall D Paleg LG (1976) Stress metabolism VIII Specific ion effects on proline accumulation in barley *Aust J Plant Physiol* **3** : 503-511

Clarke HJ (2001) Improving tolerance to low temperature in chickpea In : 4th European Conference on Grain Legumes towards the sustainable production of healthy food feed and novel products Cracow Poland July 8-12 2001 European Association for Grain Legume Research Paris pp 34-35

Comba Maria E Maria P Benavides E L (1997) Relationship between nitrogen fixation and oxidation stress induction in nodules of salt treated soybean plants *Phyton* **60** : 115-126

Cooper P Ort DR (1988) Changes in protein synthesis induced in tomato by chilling *Plant Physiol* **88** : 454-461

Cope F (1958) Catchment salting in Victoria Publ Victoria Soil Cons Auth NoTSI (Quoted from Wilson 1970)

Cordovilla MP Ligero F Lluch C (1996) Growth and nitrogen assimilation in nodules in response to nitrate levels in *Vicia faba* under salt stress *J Exp Bot* **47** : 203-210

Cralle HJ Heichel GH (1982) Temperature and chilling sensitivity of nodule nitrogenase activity of unhardened alfalfa *Crop Sci* **21** : 300-304

Cubero JI (1987) Morphology of chickpea In The chickpea MC Saxena KB Singh eds Oxon UK CAB International pp 40

Dat J Vandenabeele S Vranova E Van Montagu M Inze D Van Breusegem F (2000) Dual action of the active oxygen species during plant cell responses *Cell Mol Life Sci* **57** : 779-795

Dat JF Lopez-Delgado H Foyer CH Scott IM (1998) Parallel changes in H_2O_2 and catalase during thermotolerance induced by salicylic acid or heat acclimation in mustard seedlings *Plant Physiol* **116** : 1351-1357

De Jabrun PLM Byth DE Wallis ES (1981) Imbibition by and effects of temperature on germination of mature seed of Pigeonpea In : Proc Int Workshop on Pigeonpea **2** : 15-19 ICRISAT Center India Patancheru AP India : ICRISAT pp 181-187

De Rueda PM Matilla AJ Sanchez-Calle IM Buene M Gallardo M (1994) Thermoinhibition alters the polyamine levels in cotyledons and embryonic axes during germination of Stratified chickpea seeds *Plant Sci* **101** : 143-150

De Vries JD (1986) Water relations and nitrogen fixation of three grain legumes during water stress M Sc thesis University of Florida USA pp 107

Deshmukh PS Kumar P Ksuhwaha SR Singh TP (2006) Physiology and Productivity In Ali M Kumar S eds Advances in Mungbean and Urdbean IIPR pp 230-256

Deshmukh PS Uprety DC Diwedi N Arun Kumar R Kiran B Talwar S Kushwaha SR Singh TP (2007) Impact of climate change on productivity of pulses In : National Symposium on Legumes for ecological Sustainability : Emerging Challenges and Opportunities Nov 3-5 2007 Ind Inst Pulses Research Kanpur Abstr : Plenary Lecture p 266

Deshmukh RB (2005) Advances in major pulse crops research – success stories Souvenir 4th International Food Legumes Research Conference Oct 18-25 held at New Delhi India pp 25-33

Dhingra HR Chhabra S Kajal N Varghese TM (1995) Salinity and growth regulators induced changes in seed quality of chickpea *Ind J Plant Physiol* **38** : 322-324

Dhingra HR Sharma PK (1993) Biochemical and mineral composition of young healthy and shriveled mungbean (*Vigna radiata* L Wilczek) seeds in response to salinity *Ind J Plant Physiol* **36** : 115-117

Dhingra HR Varghese TM Kajal N (1994) Salinity induced change in carbohydrate metabolism and inorganic constituents in the developing chickpea seedes *Ind J Plant Physiol* **37** : 21-24

Diaz del Castillo L Layzell DB (1995) Drought stress permeability to O_2 diffusion and the respiratory kinetics of soybean root nodules *Plant Physiol* **107** : 1187-1194

Dua RP (1998) Salinity tolerance in chickpea (*Cicer arietinum* L) *Ind J Plant Physiol* **3** : 102-106

Dubey SD Asthana AN (1987) Selection of plant type for resistance to waterlogging in Pigeonpea In Wallis E S Byth D E Eds Food Legume Improvement for Asian Farming Systems : Proceedings of an International Workshop held in Khon Kaen Thailand 1-5 Sept 1986 ACIAR Proceedings No18 ACIAR Canberra pp 311

Duncan WG McCloud DE McGraw RL Boote KL (1978) Physiological aspects of Peanut yield improvement *Crop Sci* **18** : 1015-1020

Durand JL Sheehy JE Minchin FR (1987) Nitrogenase activity photosynthesis and nodule water potential in soybean plants experiencing water deprivation *J Exp Bot* **38** : 311-321

Durga Prasad KMR Muthukumarasamy M Panneerselvam R (1996) Changes in protein metabolism induced by NaCl salinity in soybean seedlings *Ind J Plant Physiol* **1** : 98-101

El-Enamy A E (1997) Glutathione metabolism in soybean callus-cultures as affected by salinity *Biol Plant* (Progue) **39** : 35-39

Ellis RH Covell S Roberts EH Summerfield RJ (1986) The influence of temperature on seed germination rate in grain legumes XI Intraspecific variation in chickpea (*Cicer arietinum* L) at constant temperatures *J Exp Bot* **37** : 1503-1515

El-Sayed Hameda Ralph C Kirkwood (1992) Solute accumulation on soybean (*Glycine max* L) cells adopted to NaCl salinity *Phytron* (Horn) **31** : 233-249

Evans LS (1988) Effect of acidic deposition on vegetation; State of Science In D N Rao KJ Ahmad Md Yunus S M Singh eds Perspective in Environmental Botany New Delhi India Today and Tomorrow's Printers and Publishers **2** : 73-119

Evans PJ Gallesi D Mathieu C (1999) Oxidative stress occurs during soybean nodule senescence *Planta* **208** : 73-79

Fedina IS Tsonev T Guleva E I (1993) The effect of pre-treatment with proline on the responses of *Pisum sativum* to salt stress *Photosynthetica* **29** : 521-527

Fernandez-Paşcual M De Lorenzo C De Felipe MR Rajalakshmi S Gordon AJ Thomson BJ Minchin FR (1996) Possible reaons for relative salt stress tolerance in nodules of white Lupin CV Multolupa *J Expt Bot* **47** : 1709-1716

Filek M Baczek R Niewiadomska E Pilipowicz M Koscienlniak J (1997) Effect of high temperature treatment of *Vicia faba* roots on the oxidative stress enzymes in leaves *Acta Biochem Pol* **44** : 315-321

Filho Eneas Joaquim Osmundo Brilhante De Oliveira Neto Jose Tarquinio Prisco Eneas GF Cynthia MN (1995) Effects of salinity *in vivo* and *in vitro* on Cotyledonary galactosidases from *Vigna ungiculata* (L) Walp during seed germination and seedling establishment *Revista Brasileria de Fisiologia Vegetal* **7** : 132-142

Fischer RA (1981) Optimizing the use of water and nitrogen through breeding of crops *Plant & Soil* **58** : 249-278

Flower DJ Ludlow MM (1987) Variation among accessions of Pigeonpea (*Cajanus cajan*) in osmotic adjustment and dehydration tolerance of leaves *Field Crops Res* **17** : 229-243

Foyer CH Descourvieres P Kunert KJ (1994) Protection against oxygen radicals : an important defense mechanism studied in transgenic plants *Plant Cell Environ* **17** : 507-523

Francois LE Maas EV (1999) Crop response and management of salt affected soils In : M Pessarali Hand Book of Plant and Crop Stress New York Marcel Dekker Inc pp 169-201

Fyson A Sprent JI (1982) The development of primary root nodules on *Vicia faba* grown at two root temperatures *Ann Bot* **50** : 681-692

Gabre AI Raafat A Shalaby A (1978) RNA and protein in tomato leaves after spraying CCC under saline conditions *Garten bauwissens-chaft* **43** : 182-184 (Quoted from Crop Physiol Abst **5** : 52)

Garg BK Venkateswarlu B Vyas SP Lahiri AN (1984) Nodulation and nitrogenase activity in cluster bean and mothbean under salt stress *Ind J Expt Biol* **22** : 511-512

Garg BK (1976) Effect of saline alkaline conditions on plant growth and metabolism Ph D Thesis HAU Hissar

Garg BK Garg OP (1982) Growth and some metabolic changes in maize leaves as affected by saline-alkaline conditions due to sodium carbonate and sodium bicarbonate *Ind J Plant Physiol* **25** : 220-225

Garg BK Venkateswarlu B Vyas SP Lahiri AN (1984) Nodulation and Nitrogenase activity in cluster bean and Mothbean under salt stress *Ind J Exp Biol* **22** : 511-512

Garg N Dua IS (1996) Effect of saline conditions on the nitrogen fixing potential of Pigeonpea (*Cajanus cajan* L) Mill sp and its dependence on the availability of cytokinin In T S Sharma S S Saini M L Trivedi M Sharma eds *Curr Res Plant Sci* pp 131-136

Garg N Dua IS (1998) Nitrogen fixation in Leguminous crops under saline conditions and the manoevrability of their response through plant growth regulators In : A Hemamtaranjan ed Advances in Plant Physiology Scientific Publishers India pp 1-15

Garg N Dua IS Sharma SK Garg OP (1988) Nodulation and nitrogen fixing capacity in relation to variable salinity levels in *Vigna radiata* L (mung) *P U Res Bull* (Sci) **39** : 187-196

Georgieva K Brugnoli E (2003) Influence of high temperature on the photosynthetic apparatus In Hemantaranjan A ed Plant Physiol Mol Biol New Millennium Scientific Publsiehrs India Jodhpur pp 57-74

Ghildiyal MC Poonam S Khan FA (1998) Photosynthetic accumulation to elevated CO_2 in relation to saccarides content in wheat and mungbean leaves *Ind J Exp Biol* **36** : 217-200

Gibson AH (1971) Factors in the physical and biological environment affecting nodulation and nitrogen fixation by legumes *Plant and Soil* Special Vol (1971) : 139-152

Gill KS (1987) Effect of soil alkalinity on growth yield and some biochemical parameters at grain filling stage in Cowpea and *Sesbania cannabina Ind J Plant Physiol* **30** : 38-43

Gill KS Sharma PC (1993) Mechanism of salt injury at seedling and vegetative growth stage in *Cajanus cajan* (L) Mill sp *Plant Physiol Biochem* **20** : 49-52

Gillet B Beyly A Peltier G Rey P (1998) Molecular characterization of CDSP 34 a chloroplastic protein induced by water deficit in *Solanum tuberasom* L plants and regulation of CDSP 34 expression by ABA and high illumination *Plant J* **16** : 257-262

Gogorcena Y Iturbe-Ormaetxe I Escuredo PR Becana M (1995) *Plant Physiol* **108** : 753-59

Gonzalez EM Gordon AJ James CL Arres-Igorc (1995) The role of sucrose synthase in response of soybean nodules to drought *J Exp Bot* **46** : 1515-1528

Gonzalez-Murna Sanchez-diaz CM Aparicio-Tejo Munoz-Rueda P Reid JSG (1985) The effect of NaCl and water stress on germination and b-galactosidase activity in germinated seeds of *Medicago sativa Trifolium repense* and *Brachycalycinum J Plant Physiol* **119** : 317-326

Goris I Ya (1969) Respiratory metabolism of seeds germinating in salt solution *Sel khoz Biologia* **4** : 246-251

Grafius JE (1981) Breeding for winter hardiness In CR Olien MN Smith eds Analysis and Improvement of Plant Cold Hardiness Boca Raton Florida CRC Press pp 161-174

Graham D Patterson BD (1982) Response of plants to low non-freezing temperatures : proteins metabolism and acclimation *Ann Rev Plant Physiol* **33** : 347-372

Graham PH Viteri SE Mackie F (1982) Variation in acid soil tolerance among strains of *Rhizobium phaseoli Field Crops Res* **5** : 121-128

Grant JJ Loake GJ (2000) Role of reactive oxygen intermediates and cognate redox signaling in disease resistance *Plant Physiol* **124** : 21-29

Greenway H (1973) Salinity plant growth and metabolism Field Crop Abst **27** : 614-665

Greenway H Gibbs J (2003) Mechanism of anoxia tolerance in plants II Energy requirement for maintenance and energy distribution to essential processes *Functional Plant Biology* **30** : 999-1036

Greenway H Osmond CB (1972) Salt responses of enzymes from species differing in salt tolerance *Plant Physiol* **49** : 256-259

Greenway H Setter T Hackling K (1977) Some general characteristics of glycolysis *in vitro* and its inhibition by high concentrations of chloride salts *J Expt Bot* **28** : 534-544

Griffith OW (1980) Determination of glutathione and glutathione disulfide using glutathione reductase and 2-vinylpyridine *Anal Biochem* **106** : 207-12

Gulati A Jaiwal PK (1994) Cellular and whole plant responses of *Vigna radiata* to NaCl stress *Biol Plant* (Progue) **36** : 301-307

Gupta A Singhal GS (1996) Effect of sodium and calcium chlorides ABA and proline on callus cultures of *Arachis hypogaea* L *Biol Plant* (Prague) **38** : 525-529

Guy CL (1990) Molecular mechanisms of cold acclimation In : F Katterman ed Environmental injury to plants London Academic Press Inc pp 35-61

Halliwell B (1991) Oxygen radicals : their formation in plant tissues and their role in herbicide damage In NR Baker MP Percival eds Herbicides Elsevier Amsterdam pp 87-129

Hanson AD Nelson CE Everson EH (1977) Evaluation of free proline as an index of drought resistance using two contrasting barley cultivars *Crop Sci* **17** : 720-726

Haqqani AM Pandey RK (1994) Response of mungbean to water stress and irrigation at various growth stages and plant densities I Plant and growth parameters *Tropical Agri* **77** : 281-288

Hardy RWF Havelka UD (1986) Photosynthate as major factor limiting nitrogen fixation by field grown legumes with emphasis on soybean In PS Nutman ed Symbiotic N_2 fixation Cambridge Univ Press Cambridge UK pp 421-436

Hasson-Porath E Kahane I Poljakoff-Mayber A (1972) The effect of chloride and sulphate types of salinity on growth and osmotic adaptation of Pea seedlings *Plant Soil* **36** : 449-459

Hasson-Porath E Poljakoff-Mayber A (1973) The effect of salinity on glucose absorption and incorporation by pea roots *Plant Cell Pysiol* **14** : 361-363

Hendry GW (1918) *Am Soc Agron* **10** : 246-249 (Quoted from Wilson 1970)

Hernanadez JA Ferrev MA Jimenez A Barcelo AR Sevilla F (2001) Antioxidant systems and O_2/H_2O_2 production in the apoplast of pea leaves : its relation with salt induced necrotic lesions in minor veins *Plant Physiol* **127** : 817-831

Hernandez-Armenta R Wien HC Eaglesham ARJ (1989) Maximum temperature for nitrogen fixation in commonbean *Crop Sci* **29** : 1260-1265

Herzog H (1978) Growth and cold tolerance of *Vicia faba* under various test conditions 2 CO_2 assimilation and its change after a freezing stress *Z Acker Pflanzenbau* **147** : 111-120

Hill JO Simpson RJ Moore AD Chapman DF (2006) Morphology and response of roots to pasture species to phosphorus and nitrogen nutrition *Plant Soil* **286** : 7-19

Hippeli S Elstner EF (1996) Mechanisms of oxygen activation during plant stress Biochemical effects of air pollutants *J Plant Physiol* **148** : 249-257

Hsiao TC Toole JC Yambao EB Turner NC (1984) Influence of osmotic adjustment on leaf rolling and tissue death in rice (*Oryza sativa* L) *Plant Physiol* **95** : 338-341

Huang Chi-Ying Yiu-Mei Chen (1995) Role of glutathione reductase and related enzymes in salt-tolerant mechanism of soybean plants grown under salt stress condition *Taiwania* **40** : 21-34

Huges MA Dunn MA (1996) The molecular biology of plant acclimation to low temperature *J Exp Bot* **47** : 291-305

Hughes G Keatinge JDH (1983) Solar radiation interception dry matter production and yield in pigeonpea (*Cajanus cajan* (L) Millspaugh) *Field crops Res* **6** : 171-178

Hunt S Layzell DB (1993) Gas exchange of legume nodules and the regulation of nitrogenase activity *Ann Rev Plant Physiol Plant Mol Biol* **44** : 483-511

Irigoyen JJ Emerich DW Sanchez-Diaz M (1992) Phosphenol pyruvate carboxylase malate and alcohol dehydrogenase activities in alfalfa nodules under stress *Physiol Plant* **84** : 61-66

Ishikawa HA (1996) Ultrastructural features of chilling injury : injured cells and the early events during chilling of suspension-cultured mungbean cells *Am J Bot* **83** : 825-835

Islam MT Agala W Kubota F Hamid A (1994) Evaluation of characteristics related to photosynthetic rate and leaf area production in mungbean genotypes *Japanese J Crop Sci* **63** : 64-67

Iturbe-Ormaetxe I Escuredo PR Arrese-Igor C Becana M (1998) Oxidative damage in Pea plants exposed to water deficit or paraquat *Plant Physiol* **116** : 173-181

Javier FP Estela MV Nestor C (2003) Oxidative stress and damage in chloroplasts from dawn to dusk In A Hemantaranjan ed Plant Physiol Mol Biol New Millennium 2003 Scientific publishers (India) Jodhpur pp 75-88

Jefferies RA MacKerren DKL (1989) Radiation interception and growth of irrigated and droughted potato (*Solanum tuberosum* L) *Field Crop Res* **22** : 101-112

Johansen S (1954) Effect of Indole-acetic acid on stomata and photosynthesis Physiol Plant **7** : 531-537

Kahane I Poljakoff-Mauber A (1968) Effect of substrate salinity on the ability of protein synthesis in Pea roots *Plant Physiol* **43** : 1115-1119

Kajjidoni ST Nagaral I Patil P Shoba I (2007) Genetic variability for phosphorus uptake under P-stress condition in blackgram *In* : National Symposium on Legumes for ecological Sustainability : Emerging Challenges and Opportunities Nov 3-5 2007 Indian Institute of Pulses Research Kanpur Abstr : Natural Resource Management 89 p 78

Kamble AR Macpherson HT (1954) Liberation of amino acids of perennial rye grass during wilting *Biochem J* **58** : 46-49

Karim M Frachebound Y Stamp P (1999) Photosynthetic activity of developing leaves of *Zea mays* is less affected by heat stress than that of developed leaves *Physiol Plant* **105** : 685-693

Khanna-Chopra R Sinha SK (1987) Chickpea : physiological aspects of growth and yield In : M C Saxena K B Singh eds The Chickpea Wallingford UK CAB International pp 163-189

Kheradnam M Ghorashy SR (1973) Salt tolerance of chickpea varieties during germination *Agron J* **65** : 329

Kishinevsky BD Sen D Weaver RW (1992) Effect of high root temperature on *Bradyrhizobium*-peanut symbiosis *Plant Sci* **143** : 275-282

Kling FG (1954) The physiology of plants on saline soils (Quoted from Stragonov 1964)

Klyshev LK (1974) Regulation of plant metabolism in saline medium *Crop Physiol Abstra* **2** : 92

Kohl DM Kennelly EJ Suhubert KR Shearer G (1991) *J Exp Bot* **42** : 831-839

Koundal KR Chopta RK (1989) *Biochem Physiol* **185** : 69-74

Kuhad MS Garg OP (1983) Separation of osmotic and ionic effects on plant growth water relation and biochemical analysis in Pearl millet (*Pennisetum Americana* L) *Curr Agric* **7** : 181-186

Kuhad MS Garg OP (1984) Osmotic and specific ionic effect on germination and nitrogen metabolism in wheat (*Triticum aestivum* L) *Curr Agric* **4** : 63-67

Kuhad MS Sheoran IS Santosh Kumari (1987) Alleviation and separation of osmotic and ionic effects during germination and early seedling growth in pearl millet by presoaking the seeds with growth regulators *Ind J Plant Physiol* **30** : 139-143

Kulandevelu G Nedunchezhian N (1993) Synergistic effects of UV-B radiation and growth temperature on $RUBP_C$ and CO_2 fixation in *Vigna sinensis* L *Photosynthetica* **29** : 377-383

Kumar N Singh S Nandwal AS Sheoran IS Sharma MK (2003) Nodule functioning in two genotypes of *Vigna radiata* L under saline irrigation In : National Symposium on improving crop productivity in an ecofriendly environment : Physiological and Molecular Approaches GB Pant University of Agriculture and Technology Pantnagar India Oct 15-17

Kumar S (1973) Effect of saline-alkaline conditions and applied ascorbic acid on nodulation nitrogen fixation and growth of some leguminous crops PhD Thesis HAU Hissar

Kumar S Garg OP (1981) Effect of shift to saline-alkaline conditions on nodulation nitrogen fixation and growth in Pea (*Pisum sativum* L) *Ind J Plant Physiol* **24** : 212-217

Kumar S Van Rheenen HA Singh O (1999) Genetic analysis of different components of crop duration in Chickpea *J Genet & Breed* **53** : 189-200

Kumar V Sharma D R (1989) Isolation and characterization of sodium chloride resistant callus cultures of *Vigna radiata* L Wilzcek Van radiata *J Expt Bot* **40** : 143-147

Kumaravelu G Ramanujam MP (2004) Growth and yield response of green gram cultivars to stimulated acid rain stress *Ind J Plant Physiol* **9** : 79-81

La-Favre AK Eaglesham ARJ (1987) Effects of high soil temperatures and starter nitrogen on the growth and nodulation of soybean *Crop Sci* **27** : 742-745

Lal R K Bhardwaj S N (1987) Studies on nitrogen metabolism of salinity stressed field Pea (*Pisum sativum*) Linn Var Arvensis *Ind J Plant Physiol* **30** : 165-169

Lal RK Bhardwaj SN (1984) Effect of salinity on mineral composition and ionic balance in field Pea (*Pisum sativum*) Linn var *Arvensis Ind J Plant Physiol* **27** : 351-354

Lambers HY Shane WB Cramer MD Pearse SJ Veneblaas EJ (2007) Root structure and functioning for efficient acquisition of phosphorus : matching morphological and physiological traits *Ann Bot* (London) : 693-713

Lapina LP Bikmukhametova SA Murashov IN (1976) Effect of sodium sulphate and NaCl on activity of photosynthetic phosphorylation in plants of different salt tolerance *Sov Plant Physiol* **23** : 236-241

Lawes DA Bond DA Poulson M H (1983) *In* The Faba Bean PD Hebblethwaite ed Cambridge UK Cambridge University Press pp 23-26

Lawn RJ (1981) The potential contribution of physiological research to pigeonpea improvement In Proceedings of the International Workshop on Pigeonpea Vol **1** 15-19 Dec 1980 ICRISAT Center India Patancheru A P India : ICRISAT pp 151-164

Lawn RJ (1988) Interpretive summary of Part 4; Breeding for improved plant performance in drought-prone environments In F R Bidinger C Johansen eds Drought Research Priorities for the Dryland Tropics Patancheru A P India : ICRISAT pp 213-219

Lawn RJ Williams JH (1987) Limits imposed by climatic factors In E S Wallis D E Byth eds Food Legume Improvement for Asian Farming Systems : Proceedings of an International Workshop held in Khon Kaen Thailand 1-5 September 1986 ACIAR Proceedings No18 ACIAR Canberra pp 83-98

Leheny EA Theg SM (1994) Apparent inhibition of chloroplast protein import by cold temperatures is due to energetic considerations not membrane fluidly *Plant Cell* **6** : 427-437

Leonova TG Shevyakova NI (1975) Effects of NaCl and sodium sulphate on incorporation of 14-labelled C-leucine into protein of horse bean leaves *Sov Plant Physiol* **22** : 116-119

Levitt J (1980) Response of plants to environmental stresses : chilling freezing and high temperature stresses I Academic Press New York pp 497

Li G Knowles PF Murphy DJ Marsh D (1990) Lipid protein interactions in thylakoid membrances of chilling resistant and sensitive plants studied by spin label electron skin resonance spectroscopy *J Biol Chem* **265** : 16867-16872

Lie TA (1974) Environmental effects on nodulation and symbiotic N_2 fixation In : A Quisel ed The Biology of Nitrogen Fixation North Holland Publishing Co Inc Amsterdam ISBN 0-7204-7135-4 pp 555-582

Livne A Levin N (1967) Plant Physiol **42** : 407-414

Lowerdorf HS Alexander M (1983) The identification of *Rhizobium phaseoli* strains that are tolerant or sensitive to soil acidity *App Environ Microbial* **45** : 737-742

Ludlow MM (1987) Contribution of osmotic adjustment to the maintenance of photosynthesis during water stress In J Biggins ed Progress in Photosynthesis Research Martinus Nijhoff Dordrecht the Netherlands

Lynch D V (1990) Chilling injury in plants : the relevance of membrane lipids In : F Katterman ed Environmental Injury to Plants Academic Press London pp 17-34

Lyons JM Raison JK Stepkonus PL (1979) The plant membrane in response to low temperature : an overview In JM Lyons D Graham JK Raison eds Low temperature stress in crop plants : The role of the membrane Academic Press New York pp 1-24

Maliwal GS Paliwal KV (1982) Salt tolerance of some mungbean (*Vigna radiata*) urdbean (*Vigna mungo*) and guar (*Cyamposis tetragonoloba*) varieties at germination and early growth stage *Legume Res* **5** : 23-30

Mallik MAB Tesfai K (1985) Pesticidal effect on soybean-rhizobia symbiosis *Plant and Soil* **85** : 33-41

Manchanda HR Sharma SK Dixit ML Samandar S (1981) Possibility of growing gram (*Cicer arietinum*) under chloride and sulphate dominant saline conditions Paper presented at the All India Seminar on Water Resources – Its Development and Management Chandigarh India

Manohar MS Mishra DK Mathur MK (1969) Effect of temperature and moisture stress on germination of seeds *Lab dev J Sci & Tech* **7B** : 39-43

Marin B Vieira De Silva J (1972) Influence de la carence hydrique sur la repartition cellulaire de acide rebonucleique foliaire chez le cotonnier *Physiol Plant* **27** : 150-155

Markhart A H III (1986) Chilling injury : a review of possible causes *Euphytica* **110** : 1-6

Matamoros MA Moran JF Ormaetxe II Rubio MC Becana M (1999) Glutathione and homoglutathione synthesis in legume root nodules *Plant Physiol* **121** : 879-888

McWilliam JR (1983) Physiological basis for chilling stress and the consequences for crop production In : CD Raper Jr PJ Kramer eds Crop Reactions to Water and Temperature stresses in Humid Temperature Climates Westview Press Boulder Colorado pp 113-132

Miller JD Arteca RN Pell EJ (1999) Senescence associated gene expression during ozone induced leaf senescence in Arabidopsis *Plant Physiol* **120** : 1015-1024

Minchin FR Pate JS (1975) Effect of water aeration and salt regime on nitrogen fixation in a nodulated legume-definition of an optimum root environment *J Exp Bot* **26** : 60-69

Minchin FR Summerfield RJ (1976) *Plant Soil* **45** : 113-127

Misra AN Murmu B Singh P Misra M (1996) Growth ands accumulation of Proline in Mungbean seedlings as affected by NaCl *Biol Plant* (Progue) **38** : 531-536

Moftah AE Burlyn E Michel (1987) The effect of sodium chloride on solute potential and proline accumulation in soybean leaves *Plant Physiol* **83** : 238-240

Mohan M Kumar S (1998) Review of acid rain potential in India : future threats and remedial measures *Curr Sci* **75** : 579-593

Monteith JL (1977) Climate and the efficiency of crop production in Britain Philos Trans Royal Soc London Ser B **281** : 277-294

Moynihan MR Ordentlich A Raskin I (1995) Chilling induced heat evolution in plants *Plant Physiol* **108** : 995-999

Muchow RC (1985a) Canopy development in grain legumes grown under different soil water regimes in a semi-arid tropical environment *Field Crops Res* **11** : 99-109

Muller J Boller T Wiemken A (1996) Pools of non-structural carbohydrates in soybean root nodules during water stress *Physiol Plant* **98** : 723-730

Munevar F Wallum AG (1982) Response of soybean plants to high root temperature as affected by plant cultivar and *Rhizobium* strain *Agron J* **74** : 138-142

Munns DN Keyser HH (1981) Tolerance of rhizobia to acidity aluminium and phosphate *Soil Sci Soc Am J* **34** : 519-523

Murray GA Eser D Gusta LV Etene G (1988) Winter hardiness in pea lentil faba bean and chickpea In : RL Summerfield ed World crops : Cool-Season Food Legumes Academic Dordrecht pp 831-845

Muthukumarasamy M Panneerselvam R (1995) Effect of NaCl salinity on carbohydrate metabolism in germinating soybean seeds *J Physiol Res* **8** : 195-199

Muthukumarasamy M Panneerselvam R (1997) Growth and biochemical changes of green gram under NaCl stress *Geobios* **24** : 107-111

Nambiar PTC Dart PT (1983) Factors influencing nitrogenase activity of root nodules of groundnut *Peanut Sci* **10** : 26-29

Nandwal AS Godara M Kamboj DV Kandu BS Kumar B Sharma P (2000) Nodule functioning of two phenotypically different genotypes of mungbean as influenced by salinity In : National Seminar on Plant Physiology – an interface of Agri-horticulture and Industry Rajasthan Agri Uni Udaipur India Dec 30 1999-Jan 1 2000

Nautiyal PC Ravindra V Joshi YC (1989) Germination and early seedling growth of some groundnut (*Arachis hypogaea* L) cultivars under salt stress *Ind J Plant Physiol* **32** : 251-253

Navari-Izzo F Quartacci MF Segherri CM (1996) Superoxide generation in relation to dehydration and rehydration *Biochem Soc Trans* **24** : 447-451

Nayak DN Pasalu IC Rao VR (1980) Influence of natural and synthetic insecticides on nitrogen fixation in rice rhizosphere *Curr Sci* **49** : 118-119

Nieman RH (1962) Some effects of sodium chloride on growth photosynthesis and respiration of twelve crop plants *Bot Gaz* **123** : 279-285

Nuotio S Heino P Tapio Palva E (2001) Signal transduction under low temperature stress In : A S Prasad ed Crop Responses and Adaptations to Temperature Stress The Haworth Press Inc New York pp 151-155

Olmos Enrique Eladio H (1996) Mechanisms of salt tolerance in cell line of *Pisum sativum* : Biochemical and physiological aspects *Plant Sci* (Shannon) **120** : 37-45

Oti-Boateng C Salisbury JH (1993) The effect of exogenous amino acid as acetylene reduction activity of *Vicia faba* cv *Forid Ann Rev Bot* **7** : 71-74

Pacific RE Davies KJA (1990) Protein degradation as an index of oxidative stress *Methods Enzymol* **186** : 485-502

Pahwa SK Prakash J (1992) Effect of some herbicides on the growth nodulation and nitrogen fixation in chickpea *Ind J Plant Physiol* **35** : 207-212

Pal M Sengupta UK Srivastava AC Jain V Meena RC (1998) Changes in growth and photosynthesis of mungbean induced by UV-B radiation *Ind J Plant Physiol* **4** : 79-84

Paliwal KV Maliwal GL (1980) Growth and nutrient uptake relationship of some crops in relive substrate *Annals Arid Zone* **19** : 225-258

Paneerselvam R Gopi R Sujata BM (2000) Alleviation of salt stress by paclobutrazol in black gram seedlings In : National Seminar on Plant Physiology – an interface of Agri-horticulture and Industry Rajasthan Agricultural University Udaipur India Dec 30 1999 to Jan 1 2000

Pankhurst CE Sprent JI (1975) Effect of water stress on the respiratory and nitrogen fixing activity of soybean nodules *J Exp Bot* **26** : 287-304

Panwar JDS Chandra R Sirohi GS (1988) Effect of temperature on growth nodulation and nitrogen fixation in summer mung (*Vigna radiata* (L) Wilczek) *Ann Plant Physiol* **2** : 121-129

Pareek N Chandra R (2007) Enhancing nitrogen use efficiency through biofertilizers and foliar spray of urea in lentil In : National Symposium on Legumes for ecological Sustainability : Emerging Challenges and Opportunities Nov 3-5 2007 Ind Inst Pulses Research Kanpur Abstr : Natural Resource Management 64 p 67

Pareek R P Gaur A C (1970) Effect of dichlorodiphenyl-trichloro-ethane (DDT) on symbiosis of *Rhizobium* sp with *Phaseolus aureus* (Green gram) *Plant and Soil* **33** : 297-304

Parsons R Stanforth A Rouen JA Sprent JL (1993) Nodule growth and activity may be regulated by a feedback mechanism involving phloem nitrogen *Plant Cell Environ* **16** : 125-136

Passioura JB (1977) Grain yield harvest index and water use of wheat *J Aust Inst Agric Sci* **43** : 117-120

Pastori G Foyer CH Mullineanux P (2000) Low temperature induced changes in the distribution of H_2O_2 and antioxidants between bundle sheath and mesophyll cells of maize leaves *J Exp Bot* **51** : 107-113

Plant Z (1974) Nitrate reductase activity of wheat seedlings during exposure to and recovery from water stress and salinity *Physiol Plant* **30** : 212-217

Pluennke RH John HE (1972) The influence of low substrate Na levels up on free amino acid content of cotton leaves *Plant Physiol* **49** : 502-505

Poehlman JM (1991) Climatic requirement In The Mungbean New Delhi India Oxford and IBH pp 24-50

Polle A Chakrabarti K Chakrabarti S Siefert F Schramel P Rennenberg H (1992) Antioxidants and manganese deficiency in needles of Norway Spruce (*Picea abies* L) trees *Plant Physiol* **99** : 1084-1089

Powell AA (1989) The importance of genetically determined seed coat characteristics to seed quality in grain legumes *Ann Bot* **63** : 169-175

Powell AA Matthews S (1978) The damaging effect of water on dry Pea embryos during inhibition *J Exp Bot* **29** : 1215-1229

Prasad TK (1996) Mechanisms of chilling induced oxidative stress injury and tolerance : changes in antioxidant system oxidation in proteins and lipids and protease activities *Plant J* **10** : 1017-1036

Prasad TK (2001) Mechanisms of chilling injury and tolerance In AS Basra ed Crop Responses and Adaptations to temperature stress Hawthorne Press New York pp 1-52

Prasad TK Anderson MD Martin BA Stewart CR (1994) Evidence for chilling induced oxidative stress in maize seedlings and a regulatory role for hydrogen peroxide *Plant Cell* **6** : 65-74

Prisco JT O'Leary JW (1970) Osmotic and toxic effect of salinity on germination of *Phaseolus vulgaris* L seeds *Turriaeba* **20** : 177-184

Purcell LC Sinclair TR (1995) Nodule gas exchange and water potential response to rapid imposition of water deficit *Plant Cell Environ* **18** : 179-187

Purushotham MG Patel V Raddey PC Prasad TG Vajranabhaiah SN (1988) Development of *in vitro* PEG stress tolerant cells lines in two groundnut (*Arachis hypogaea* L) genotypes *Ind J Plant Physiol* **3** : 49-51

Rai R (1987) Chemotaxis of salt tolerant and sensitive *Rhizobium* strains to root exudates of lentil (*Lens culinaris* L) genotypes and symbiotic N-fixation proline content and grain yield in saline calcareous soil *J Agric Sci* **108** : 37-45

Rainbird RM Atkins CA Pate JS (1983) Effect of temperature on nitrogenase functioning in cowpea nodules *Plant Physiol* **73** : 392-394

Rajan S N Karikalan L Paneerselvam R (1998a) Effect of triadimefon on carbohydrate metabolism in the NaCl stressed green gram In : National Symposium on Future Goals of Physiological Research for the Improvement of Plant Resources Annmali Uni Annamalainagar India Dec 18-20 pp 50-51

Rakova NM Klyshev LK (1973) Effect of sodium sulphate and NaCl on activity of enzymes of primary ammonium nitrogen assimilation in plant roots *Sov Plant Physiol* **25** : 26-29

Raman R (1992) Effect of soil acidity factors on aspects of symbiotic N_2 fixation of *Lens culinaris* in acid soils *J Gen Applied Microbiol* **38** : 391-406

Rao PR Sengupta UK (1997) An indigenous technique for conducting carbon dioxide enrichment studies using open top chambers *J Res ANGRAU* **25** : 27-31

Rao VK Rao AP Sataiah V (1996) Effect of soil moisture regimes on phosphorus uptake by mungbean (*Vigna radiata* (L) Wilczek) *Legume Res* **9** : 108-110

Rathore S Vershney KA (2003) Accumulation and translocation of *Glycine betaine* and choline in salinized black gram (*Vigna mungo* L Hepper) plants In National symposium on improving crop productivity in an ecofriendly environment : Physiological and Molecular Appraches G B Pant University of Agriculture and Technology Pantnagar India Oct 15-17

Reddy CV (1982) Note on the effect of saline water irrigation on green gram *Curr Agr* **14** : 45-51

Richards KD Schoff EJ Sharma YK Davis KR Gardner RC (1998) Aluminium induces oxidative stress genes in *Arabidopsis thaliana Plant Physiol* **116** : 409-418

Riken A Dillwith JW Bergman DK (1993) Correlation between the circadian rhythm of resistance to extreme temperatures and changes in fatty acid composition in cotton seedlings *Plant Physiol* **101** : 31-36

Roberts EH Hadley P Summerfield RJ (1985) Effects of temperature and photoperiod on flowering in chickpeas (*Cicer arietinum* L) *Ann Bol* **55** : 881-892

Roberts EH Summerfield RJ Minchin FR Hadley P (1980) Phenology of chickpea (*Cicer arietinum*) in contrasting aerial environments *Exp Agric* **16** : 343-360

Rodhe H Gallaway J Dianwu Z (1992) Acidification in South-east Asia-Prospects for the coming decades *Ambion* **21** : 148-150

Romero JM Maranon T (1994) Long term responses of *Melilotus segetalis* salinity : I Growth and Partitioning *Plant Cell and Envi* **17** : 1243-1248

Rossum DV Muyotcha A DeHoop BM Van Verseveld HW Stouthamer AH Boogered FC (1994) Soil acidity in relation to groundnut – *Bradyrhizobium* symbiosis *Plant Soil* **163** : 165-175

Rouxel MF Billard P Bucaud J (1987) Effect of NaCl salinity *in vivo* and *in vitro* on RNase activity in the halophyte (*Suaeda maritime* var Macrocarpa) *Physiol Plant* **69** : 330-336

Ryle GJA Powell CE Davidson IA (1992) Growth of white clover dependent on N_2 fixation in elevated CO_2 and temperature *Ann Bot* **70** : 221-228

Sadasivam R Nataranjan N Babu RC Muralidharan V Rangaswami SRS (1988) Response of mungbean cultivars to soil moisture stress at different growth phases In S Shanmugasundaram B T Mclean eds Mungbean proceedings of the Second International Symposium Shanhua Taiwan : Asian Vegetable Research and Development Centre

Saha K Gupta K (1998) Effect of triazoles and CCC on growth yield and metabolism of mungbean plants under salinity stress *Ind J Plant Physiol* **3** : 107-111

Sandhu PS Dhingra KK Bhandri SC Gupta RP (1991) Effect of hand-hoeing and application of herbicides on nodulation nodule activity and grain yield of *Lens culinaris* (Medic) *Plant and Soil* **135** : 293-296

Sands R Clarke ARP (1977) Response of radiata pine to salt stress I Water relations osmotic adjustment and salt uptake *Aust J Plant Physiol* **4** : 637-646

Santarius KA (1980) Membrane lipids in heat injury of spinach chloroplasts *Physiol Plant* **49** : 1-6

Sarin MN Rao IM (1958) Influence of sodium sulphate on seedling respiration in wheat and gram *Ind J Plant Physiol* **1** : 30-38

Sarin MN Uprety DC (1965) Interactive effects of Phosphon-D nine gibberellic acid and salinity on *Vigna ungiculata* (L) Wall Proc Seminar (Botany) PU Chandigarh

Sashidhar VR Mekhri AA Krishna Sastry KS (1981) Potassium content and proline accumulation following seed treatment with calcium chloride in groundnut varieties *Ind J Plant Physiol* **24** : 89-92

Savithri KS Ganapathy PS Sinha SK (1980) Sensitivity to low temperature in pollen germination and fruit set in *Cicer arietinum* L *J Exp Bot* **31** : 475-481

Saxena MC Saxena NP Mohammed AK (1988) High temperature stress *In* RJ Summerfield eds World Crops : cool-season food legumes Kluwer Academic Publisher Dordrecht The Netherlands

Saxena NP (1980) Pod setting in relation to temperature in Hisar Int Chickpea Newsl **2** : 11

Saxena NP Johansen C (1990) Realised yield potential in chickpea and physiological consideration for further genetic improvement In : SK Sinha PV Sane SC Bhagrava PK Agarwal eds Proc Int Congress Plant Physiol Vol 1 15-20 Feb 1988 New Delhi India Indian Soc Plant Physiol and Biochem

Saxena NP Johansen C Sethi SC Talwar HS Krishnamurthy L (1988) Improving harvest index in Chickpea through incorporation of cold tolerance *Int Chickpea Newsl* **19** : 17-19

Scandalios JG (1992) Regulation of the antioxidant defense genes Cat and SoD of maize In J G Scandalios ed Molecular biology of free radical scavenging systems Cold Spring Harbor Laboratory Press Cold Spring Harbor pp 117-152

Scarascia-Mugnozza GT Marzi V (1979) Retrospective and prospective views for the *Vicia faba* crop in Southern Italy Commission of the European Communiities pp 7-22

Schubert S Serraj R Plies-Balzer E Mengel K (1995) Effect of drought stress on growth sugar concentrations and amino acid accumulation in N_2 fixing alfalfa *J Plant Physiol* **146** : 541-546

Schulman HM Lewis MC Tipping CM Bordeleau LM (1988) Nitrogen fixation by three species of leguminoceae in Canadian High Arctic Tundra *Plant Cell Environ* **11** : 721-728

Schwarz M Gale J (1984) Growth response to salinity at high levels of carbon dioxide *J Expt Bot* **35** : 193-196

Scott J Howieson J (2000) Pulse research and industry development in western Australia Chickpea Nodulation Survey In : Crop Updates pp 32

Sehon BS Kumar S Singh R (1985) Nitrogen metabolism and plant nutrition Proceedings of National Seminar on Plant Molecular Biology pp 15

Sekhon HS Sharma P Singh G (2004) Effect of irrigation on growth and yield on Kharif mungbean (*Vigna radiata* (L) Wilczek) genotypes *J Pl Sci Res* **20** : 102-105

Serraj R Sinclair TR (1996) Inhibition of nitrogenase action and nodule oxygen permeability by water deficit *J Exp Bot* **47** : 1067-1073

Sharma DC Puntamaker SS Mehta KC Seth SP (1971) Note on the effect of different common salt of sodium and calcium on the germination of green gram (*Phaseolus aureus* Roxb) varieties *Ind J Agric Sci* **41** : 636-638

Sharma P (1974) The effect of pre-sowing seed treatments with salts on the physiology of salt tolerance of some genotypes of moong (*Phaseolus aureus* Roxb) MSc Thesis HAU Hissar

Sharma P Sekhon HS Sandhu JS (2005) Cold toleralnce studies in chickpea (*Cicer arietinum* L) In proceedings of 4th International Food Legumes Research Conference (IFLRC-IV) Oct 18-22 IARI New Delhi (In Press)

Sharma P Sharma Poonam (2007) Biological nitrogen fixation : legume-Rhizobium Physiology Phytohormones mediated response to nodulation and nitrogen fixation In CP Malik ed Advances of Phytomorphology (In Press)

Sharma P Sekhon HS Singh G Sharma Poonam (2007) Effect of date of sowing on flower drop dry matter accumulation and growth parameters in Kharif mungbean *Envr and Ecol* **25** (4) : 838-842

Sharma SK (1995) Effect of salinity on growth performance and internal distribution of Na^+ K^+ and Cl^- in *Vicia faba* L *Ind J Plant Physiol* **38** : 69-72

Sharma SK (1997) Plant growth photosynthesis and ion uptake in chickpea as influenced by salinity *Ind J Plant Physiol* **2** : 171-173

Sharma SK Garg OP (1983) Comparative study of osmotic and salt stress effects on nitrate assimilation in wheat *Curr Agric* **7** : 36-40

Sharma SK Kumar S (1990) Effect of salinization on growth and distribution of Na^+ and Cl^-in two genotypes of chickpea *Ind J Plant Physiol* **33** : 269-274

Sheldrake AR Narayanan A (1979) Growth development and nutrient uptake in Pigeonpea (*Cajanus cajan*) *J Agri Sci Cambridge* **92** : 513-526

Sheokand Sunita Seema Dhand Kumari S (1995) Studies on nodule functioning and hydrogen peroxide scavenging enzymes under salt stress in chickpea nodules *Plant Physiol Biochem* (Montrouge) **33** : 561-566

Sheoran I S (1975) Effect of salinity on some important aspects of plant metabolism PhD Thesis HAU Hissar

Sheoran IS Garg OP (1978) Effect of salinity on the activities of RNase DNase and protease during germination and early seedling growth of mungbean *Physiol Plant* **44** : 171-174

Sheoran IS Garg OP (1983) Effect of different types of salinities on gram (*Cicer arietinum* L) during germination I Seedling growth and water relations *Ind J Plant Physiol* **26** : 363-369

Shewfelt RL (1992) Response of plant membrances to chilling and freezing In YY Leshem ed Plant Membranes The Netherlands Kluwer pp 192-219

Shewfelt RL Erickson MC (1991) Role of lipid peroxidation in the mechanism of membrane associated disorders in edible plant tissue *Trends Food Sci Technol* **2** : 152-154

Shokraii EH Fahimi H (1969) Interaction of phytohormones with salinity on the growth of beans Abstr 11th Int Bot Congr Seattle 1969 pp198

Shone MGT Gale J (1983) Effect of sodium chloride stress and nitrogen source on respiration growth and photosynthesis in Lucerne (*Medicago sativa* L) *J Exp Bot* **34** : 1117-1125

Siddique KHM Loss SP Reagen KL Jettner RL (1999) Adaptation and seed yield of cool season grain legumes in Mediterranean environment of Southwestern Australia *Aust J Agric Res* **50** : 375-387

Siddique KHM Loss SP Thomson BD (1997) Cool season grain legumes in dryland Mediterranean environments of western Australia : Significance of early flowering In N P Saxena C Johansen eds Management of Drought in Grain Legumes Aleppo Syria pp 189-99

Siddique KHM Sedgley RH (1986) Chickpea *Cicer arietinum* L a potential grain legume for southwestern Australia seasonal growth and yield *Aust J Agric Res* **37** : 245-261

Siddiqui S Krishnamoorthy HN (1995) Effect of B9 on the carbohydrate metabolism of Cowpea (*Vigna unguiculata* L) seedlings under saline conditions *Ind J Plant Physiol* **38** : 164-167

Siddiqui S Kumar S Sharma HR (1985) Studies on the effects of salinisation of nodulation and nitrogen fixation in Pea (*Pisum sativum* L) *Ind J Plant Physiol* **28** : 369-375

Simantacchi M Caro A Puntarulo S (1995) Oxygen dependent increase of antioxidants in soybean embryonic ones *Int J Biochem Cell Biol* **27** : 1221-1229

Simon EW (1979) Seed germination at low temperature In : JM Lyons D Graham JK Raison eds Low Temperature Stress in Crop Plants Academic Press New York

Sinclair TR Ludlow MM (1986) Influence of soil water supply on the plant water balance of four tropical grain legumes *Aust J Plant Physiol* **13** : 329-341

Sinclair TR Serraj R (1995) Dinitrogen fixation sensitivity to drought among grain legume species *Nature* **378** : 344-349

Singh A (1997) Increased UV-B radiation reduces N-fixation in tropical leguminous crops *Envi & Pollution* **95** : 289-291

Singh GR Sharma SR Anil Kumar TB (1986) Effects of biloxazol Carbandazim and triadimefon on nodulation and nitrogen fixation in Cowpea *Zentralbi Mikrobil* **141** : 35-37

Singh KB Malhotra RS Saxena MC (1990) Sources for tolerance to cold in *Cicer* species *Crop Sci* **30** : 1136-1138

Singh KB Malhotra RS Saxena MC (1995) Additional sources of tolerance to cold in cultivated and wild *Cicer* species *Crop Sci* **35** : 1491-1497

Singh KB Malhotra RS Saxena MC Bejiga G (1997) Superiority of winter sowing over traditional spring sowing of chickpea in the Mediterranean region *Agron J* **89** : 112-118

Singh M Singh S (1995) Sodicity induced changes in saccharides free proline and protein content of different Pea genotypes *Ind J Plant Physiol* **38** : 109-113

Singh NP Das SP (2007) Effect of nitrogen and phosphorus on field pea (*Pisum sativum*) under Tripura condition In : National Symposium on Legumes for ecological Sustainability : Emerging Challenges and Opportunities Nov 3-5 2007 Indian Institute of Pulses Research Kanpur Abstr : Natural Resource Management 62 p 66

Singh TN Aspinall D Paleg LG (1972) Proline accumulation and varietal adaptation to drought in barley a potential metabolic measure of drought resistance *Nature New Biol* **236** : 188-190

Singh VK Mathur M Mathur SN (1982) Phytotoxicity of insecticide phorate on germination of *Vigna mungo Agril Biol Chem* **46** : 1681-1682

Singh VK Mathur SN (1989) Studies with phorate an organophosphate inseciticide on growth nodulation and some enzymes of nitrogen metabolism in blackgram *Ind J Plant Physiol* **32** : 314-319

Sinha SK (1977) Food Legumes : Distribution Adaptability and Biology of yield Rome FAO

Smillie R M Hetherington S E He J Not F R (1988) Photoinhibition at chilling temperatures *Aust J Plant Physiol* **15** : 207-222

Soam SK Agrawal PK (1989) Effect of dimethoate on nodulation and nitrogenase activity of nodules in pea and blackgram *Ind J Agril Sci* **59** : 824-825

Solomon M Gedalovich E Mayer AM Poljakoff-Mayber AP (1986) Changes induced by salinity to the anatomy and morphology of excised Pea (*Pisum sativum*) cv Alaska roots in culture *Annals Bot* **57** : 811-818

Sprent JI Zahran HH (1988) Infection development and functioning of nodules under drought and salinity In DP Beck LA Materon eds Nitrogen fixation by legumes in Mediterranean Agriculture Martinus Nijhoff Dordrecht ISBN 90-247-2405-8 pp 145-151

Sprent JI (1972) The effect of water stress on nitrogen fixing root nodules IV Effects on whole plants of *Vicia faba* and *Glycine max New Phytol* **71** : 603-611

Srinivasan A Johansen C Saxena NP (1998) Cold tolerance during early reproductive growth of chickpea (*Cicer arietinum* L) : characterization of stress and genetic variation in pod set *Field Crops Res* **57** : 179-191

Srinivasan A Saxena N P Johansen C (1999) Cold tolerance during early reproductive growth of Chickpea (*Cicer arietinum* L) : genetic variation in gamete development and function *Field Crops Res* **60** : 209-222

Stadtman ER (1993) Oxidation of free amino acids and amino acid residues in proteins by radiolysis and by metal-catalyzed reactions *Ann Rev Biochem* **62** : 797-821

Stadtman ER Oliver CN (1991) Metal catalyzed oxidation of proteins Physiological consequences *J Biol Chem* **266** : 2005-2008

Stewart CR Lee JA (1974) The rate of proline accumulation in halophytes *Planta* **120** : 279-289

Strack Z Karwowska R Kraszewska E (1975) *Acta Soc Bot Pol* **44** : 567-588

Streeter JG (1993) Translocation a key factor limiting the efficiency of N_2 fixation in legume nodules *Physiol Plant* **87** : 616-623

Strogonov BP (1973) Structure and function of plant cells in saline habitats New trends in the study of salt tolerance PPST Jerusalem London

Strogonov BP Lapina LP (1964) A possible method for the separate study of toxic and osmotic action of salts on plants *Fiziol Rast* **11** : 674-680

Subba Rao A Srinivasarao Ch Ganeshamurthy AN Masood A (2007) Improving soil health and nutrient use efficiency through legumes In National Symposium on Legumes for Ecological Sustainability : Emerging Challenges and Opportunities Nov 3-5 2007 Indian Institute of Pulses Research Kanpur pp 33

Sudhakar C Reddy PS Veeranjaneyulu K (1987) Biochemical changes during early seedling growth of horsegram (*Dolichos biflorus* L) subjected to salt stress *Ind J Expt Biol* **25** : 479-482

Sullivan CY Ross WM (1979) Selecting for drought and heat resistance in grain sorghum In : H Mussel R Staples eds Stress Physiology in Crop Plants John Willey New York USA

Summerfield RJ Hadley P Roberts EH Minchin F R Rawsthrouw S (1984) Sensitivity of chickpea (*Cicer arietinum*) to hot temperatures during the reproductive period *Expt Agri* **20** : 77-93

Summerfield RJ Minchin FR Roberts EH Hadley P (1980) The effects of photoperiod and air temperature on growth and yield of Chickpea In J M Green Y L Nene J B Smithson eds Proceedings of the International Workshop Chickpea Improvement India ICRISAT pp 212

Summerfield RJ Roberts EH (1985) *Cicer arietinum* In A H Halevy ed Handbook of Flowering CRC Press Boca Raton Florida pp 92-99

Sung FJM (1993) Waterlogging effect on nodule nitrogenase and leaf NR activities in soybean *Field Crops Res* **35** : 183-189

Swaraj K Bishnoi NR (1994) Adverse effects of water deficit on nitrogen-fixation in legumes In Dryland Resources and Technology LLI Somani MN Bansal RN Sharma eds Scientific Publishers Jodhpur Vol 8 pp 21-33

Swaraj K Bishnoi NR Pascual MF Singh P (1996) Water stress and symbiotic nitrogen fixation In Khurana AL Behl RK Dogra RC eds Natural resource management CCC HAU Hisar and MMB New Delhi pp 107-120

Thompson JA Kumar Rao JVDK Dart PJ (1981) Measurement of inoculation response in Pigeonpea In : Proceedings of the International Workshop on Pigeonpeas Vol 1 15-19 Dec 1980 ICRISAT Center India Patancheru AP India ICRISAT pp 249-253

Totawat KL Saxena SN (1971) Effect of variations in salinity and alkalinity of applied irrigation waters on the amino acid make up of *Phaseolus aureus Plant Soil* **34** : 43-48

Tramontano William A Diana J (1997) Trigonelline accumulation in salt stressed legumes and the role of other osmoregulators as cell cycle control agents *Phytochemistry* **44** : 1037-1040

Trung BC Yoshida S Kobayashi Y (1985) Influence of excess soil moisture on the nitrogen nutrition and grain productivity of mungbean *Japanese J Crop Sci* **54** : 79-83

Tully RE Musgrave ME Leopold AC (1981) The seed coat as a control of imbibitional chilling injury *Crop Sci* **21** : 312-317

Turner NC (1999) Drought resistance : a comparion of two frame works In N P Saxena C Johansen Y S Chauhan R C N Rao eds Management of Agricultural Drought : Agronomic and Genetic options Oxford and IBH Publishing Co New Delhi India pp 545-57

Valentine JS Wertz DL Lyons TJ Liou LL Goto JJ Gralla EB (1998) The dark side of dioxygen biochemistry *Curr op chem Biol* **2** : 253-262

Van Heerden PDR Kruger GHJ (2000) Photosynthetic limitation in soybean during cold stress *South Afr J Sci* **96** : 201-206

Vance CP Uhde-Stone C Allan DL (2003) Phosphorus acquisition and use : critical adaptation by plants for securing a non-renewable resource *New Phytol* **157** : 423-447

Varma SK Kumari P (1978) Nature extent periodicity and intensity of flower and pod shedding in gram (*Cicer arietinum* L) *Legume Res* **1** : 108-114

Venkateswarlu B Maheswari M Saharan N (1989) *Plant Soil* **114** : 69-74

Venkateswarlu S Subramanian VB (1993) Photosynthesis during reproductive phase of rainfed green gram *Ind J Plant Physiol* **36** : 174-177

Verling E (1991) The roles of heat shock proteins in plants Ann Rev Plant Physiol Plant *Mol Biol* **42** : 579-620

Verma AK Rao NSS (1975) Effect of different levels of soil moisture on growth yield and some physiological aspects of mungbean *Indian J Agri Sci* **45** : 11-16

Viswanath C Jegendrf A Zhu JK (2005) Understanding and improving salt tolerance in plants *Crop Sci* **45** : 437-448

Vora AB Patel MK (1975) Effect of heat treatment and GA feeding on seedling growth of mungbean *Curr Sci* **44** : 749-750

Walsh KB Layzell DB (1986) Carbon and nitrogen assimilation and partitioning in soybeans exposed to low root temperature *Plant Physiol* **80** : 249-255

Walsh RB (1995) Physiology of legume nodule and its response to stress *Soil Biol Biochem* **27** : 637-655

Walswinkel PS Ammerlann A (1988) Effect of osmotic environment of the seed coat on sucrose and amino acid transport in developing seeds of *Lunaria annua Acer pseudoplatanus* and *Glycine max Physiol Plant* **74** : 262-269

Wang CY (1982) Physiological and biochemical responses of plants to chilling stress *Hort Sci* **17** : 173-186

Weaver LM Gan S Quirino B Amasino RM (1998) A comparison of the expression patterns of several senescence associated genes in response to stress and hormone treatment *Plant Mol Biol* **37** : 455-469

Weimberg R (1970) Enzyme levels in Pea seedlings grown on highly salinised media *Plant Physiol* **46** : 446-470

Weimberg R (1975) Effect of growth in highly salinised media on the enzymes of the photosynthetic apparatus in pea seedlings *Plant Physiol* **56** : 8-12

Welty LE Prestbuye LS Hall JA Mathre DE Dilterline RL (1988) Effect of fungicide seed treatment and *Rhizobium* inoculation on chickpea production *App Agric Res* **3** : 17-20

Wery J (1990) Adaptation to frost and drought stress in chickpea and implication in plant breeding In : MC Saxena JI Cubero J Wery eds Present status and future prospect of chickpea crop production and improvement in the Mediterranean Countries Spain CIHEAM pp 77-85

Wery J Knights SN Malhotra RS Cousin R (1994) Screening techniques and sources of tolerance to extremes of moisture and air temperature in cool season food legumes *Euphytica* **73** : 73-83

Wery J Turc O Lecoeur J (1993) Mechanisms of resistance to cold heat and drought in cool season legumes with special reference to chickpea and Pea In : K B Singh M C Saxena eds Breeding for stress tolerance in cool-season food legumes John Wiley & Sons Chichester

Wierenga P J Hagan R M (1966) Effects of cold irrigation water on soil temperature and crop growth *Calif Agric* **20** : 14

Wilson J M Crawford R M M (1974) Leaf fatty acid content in relation to hardening and chilling injury *J Exp Bot* **25** : 121-131

Wilson J R (1970) Responses to salinity in Glycine VI some effects of a range of short term salt stresses on the growth nodulation and nitrogen fixation of *Glycine wightii Aust J Agric Res* **21** : 571-582

Wilson JM (1978) Changes in the fatty acids and water status of *Phaseolus vulgaris* leaves at temperatures inducing chill hardiness Opportunities for chemical Plant Growth Regulation British Crop Protection Council Croydon UK Monograph **21** : 69-74

Wolk WD Herner RC (1982) Chilling injury of germinating seeds and seedlings *Hort Sci* **17** : 169-173

Woods CM Reid MS Patterson BD (1984) Response to chilling stress in plant cells I Changes in cyclosis and cytoplasmic structure *Protoplasma* **121** : 8-16

Yousel AN Sprent JI (1983) Effect of NaCl on growth nitrogen incorporation and chemical composition of inoculated and ammonium nitrate fertilized *Vicia faba* L plants *J Expl Bot* **34** : 941-950

Zahran HH Sprent JI (1986) Effect of sodium chloride and polyethylene glycol on root hair infection and nodulation of *Vicia faba* L plants by *Rhizobium leguminosarum Planta* **167** : 303-309

Zaidi PH Singh BB (1995) Modulation of adverse effects of salinity by growth regulators in soybean I photosynthetic area pigment efficiency and plant growth *Plant Physiol Biochem* **22** : 136-142

Zhang F Smith DL (1994) Effects of low root zone temperatures on the early stages of symbiosis establishment between soybean and *Bradyrhizobium japonicum J Exp Bot* **45** : 1467-1473

Zhu D Scandalios JG (1992) Expression of the maize MnSod (Sod3) gene in Mn SoD deficiency yeast rescues the mutant yeast under oxidative stress *Genetics* **131** : 803-809

12

BIOLOGICAL NITROGEN FIXATION : LEGUME-RHIZOBIUM PHYSIOLOGY, PHYTOHORMONES MEDIATED RESPONSE TO NODULATION AND NITROGEN FIXATION

PUSHP SHARMA AND POONAM SHARMA

INTRODUCTION

Biological nitrogen fixation (BNF), converts atmospheric dinitrogen (N_2) into plant usable form through soil microbes possessing nitrogenase (N_2ase) enzyme. BNF helps to maintain soil nitrogen reserves and can substitute for fertilizer to attain higher crop yields. It is an excellent, economically and environmentally sound source to substitute fertilizer.

1. Leguminous plants

Leguminous plants belongs to the family *Fabaceae (Leguminoseae)* with three major botanical subfamilies the Ceasalpinioideae, the Mimosoideae and the Papilionoideae. There are nearly 750 genera and 18,000-19,000 species of leguminous plants of which 500 genera and approximately 10,000 species belong to the sub family Papilionoideae. Only 16 percent of *leguminoseae* have so far been examined for nodulation of which 95 percent of Mimosoideae, 26 percent of Ceasalpinioideae and 90 percent of Papilionoideae possess root nodules. Inclusion of legumes in cereal based cropping system (Rice- wheat) has two major advantages as they directly introduce atmospheric (N_2) into the agricultural system and also control certain crop diseases. In addition, N_2 fixed through BNF is less likely to be lost than inorganic fertilizer N_2. Nitrogen fixation rates vary enormously, upto 600 kg N_2 fixed/ha/yr have been reported in forage legumes. In general grain legumes particularly pulses fix less N_2 than do forage legumes because they are grown for a shorter period and have fewer roots on which nodules can form. On the global basis this symbiotic association between legume and *Rhizobium* may reduce upto 70 million tons of atmospheric nitrogen to ammonia per annum which amounts to about 40 percent of all the biological fixed nitrogen per year (Burns & Hardy, 1975).

Table 1. Estimate of N_2 fixed by some legumes

Crop	*N_2 fixed (kg ha^{-1} $crop^{-1}$)*	*Crop*	*N_2 fixed (kg ha^{-1} $crop^{-1}$)*
Alalfa	100-300	Vetch	80-138
Blackgram	119-140	Pigeonpea	4-200
Clovers	50-300	Ricebean	32-97
Sweet clover	223-267	Soybean	49-450
Red cover	76-169	Fababean	176
Chickpea	23-97	Peas	46-99
Cluster bean	37-196	Kidneybean	3-91
Common bean	3-57	Fenugreek	44
Cowpea	9-125	*Sesbania rostrata*	70-458
Groundnut	27-206	Sesbania cannabina	119-188
Lentil	35-147	Crotalaria	190-195
Greengram	50-66		

(Sekhon *et al.*, 2002)

The atmosphere contains 78 percent N_2, which is the primary source of nutrients for plants and other biological systems, whether fixed biologically or non-biologically. According to an estimate, above every hectare of land at sea level there is 78000 t of inert N_2 gas (Wani *et al.*, 1995). Chemical fertilizers have substantially increased food production in the recent past ushering into green revolution. Since the 1970s, however the rising cost of fertilizers (about 1.3 t oil or equivalent amount of energy is required to fix 1 t of NH_3) (Ladha & Reddy, 1995) and concern for environmental pollution has awakened the agriculturists. Legumes are endowed with a noble trait for BNF (Table 1) in association with *Rhizobium*, which is vital for sustaining high productivity and soil fertility. Estimates of total annual terrestrial BNF inputs range from 139-170 million tones N_2, with 25-30 percent contribution from legume rhizobia symbiosis in aerable land. The greatest assest 3-458 kg N_2/ha that legumes carry, is the atmospheric nitrogen fixation capacity, which makes them more useful in the present world energy crisis. Indeed, BNF, which is the microbial conversion of atmospheric dinitrogen (N_2) gas into plant usable ammonia, plays a key role in crop production as it helps in soil fertility by using N_2 which is in abundance in the atmosphere (Parsons, 1997). Historically, BNF system has been an essential component of many farming systems since long. Fred *et al.* (1932) reported agricultural use of legumes dating back more than 4000 years. BNF can be exploited in farming system by using them as green manures or by including legumes in crop rotation.

The quantities of nitrogen given in Table 1 are only approximate and can not be taken as accurate, because the methods of estimation adopted by different workers may not be of the same value. Various legumes vary in their ability to fix nitrogen as it is dependent upon physical, environmental, nutritional and biological factors. In general, while estimating BNF, plant roots and fallen leaf material are not taken into account which results in under estimation of the quantity of nitrogen fixed. It is essential that BNF in roots and fallen plant material should also be considered while, estimating the amount of nitrogen fixed by legumes. Net nitrogen balances can be calculated as given below (Sekhon *et al.*, 2000).

Net N_2 Balance = Total plant N uptake – (N derived from BNF + N derived from fertilizer + N added to soil through plant roots and fallen plant)

The contribution of different legumes is estimated to be about 25.2 percent to the total 139 million tones of biologically fixed N_2 annually in various terrestrial ecosystems (Burns & Hardy, 1975). Symbiotic partnerships between pulse crops and *Rhizobium* under best compatibility and crop management supplies approximately 80-90 percent of total N_2 required by the crop. However, the N_2 contribution depends on the effectiveness of *Rhizobium* strains, its compatibility, biomass accumulation of host, soil and environmental factors. Recent estimates of N_2 fixation by pulse crops indicated fixation of 458 kg N_2 ha^{-1} under different agroclimatic conditions (Singh & Sekhon, 2005). The pulse crops utilize majority of the fixed N_2 for their growth, and also leave some residual amount of N in soil for the succeeding crops.

2. Discovery of symbiotic nitrogen fixation

Historically, BNF system has been an essential component of many farming systems since long. Fred *et al.* (1932) reported Agricultural use of legumes dates back to more than 4000 years. In 1679, Malpighi provided the first diagrams and descriptions of root nodules of plants but considered them as insect galls. In 1866, Woronin saw fungal hyphae in root nodules which looked like threads and are now known as 'infection thread'. Subsequent observations led to the belief that root nodules were caused by fungi. Hellrigel & Wilfarth (1888) first established clearly it were microbes in the root nodules that allowed legumes to obtain nitrogen from the air while other plants could not (Quispel, 1988; Schilling, 1988). In same year, Beijerinck, (1888) succeeded in isolating and culturing the bacteria from root nodules. Frank, (1879) examined nodules and like other, thought the symbiont was a fungus. He named it as *Schinzia leguminosarum*. When Beijerinck, (1888) isolated the legume nodule bacterium he called it *Bacillus radicicola*, but the following year Frank (1889) named it *Rhizobium leguminosarum*. The development of root nodules from the root hair to the origin of nodule protrusions has been described long back.

The fact that legume plants enriched the soil was known to Romans. The Greeks and Romans were also aware of the value of leguminous plants as green manure. The Greek philosopher, Theophrastus (370-285 B.C.) in his 'Enquiry into plants' spoke of leguminous plants as 'reinvigorating' the soil. The Roman farmers knew the value of leguminous crops as green manure. Varrode in (37 B.C.) said, Legumes should be planted in light soils, not so much for their own crops as for the good they do to subsequent crops". Fuchsius in 1542 was the first to observe the nodules on the roots of leguminous plants. Delechamps in his Historia Generalis Plantarum published in the year 1586-87 gave the name nodules to the malformations on the roots.

In 1838, J.B. Boussingault conducted some experiments and found evidence that nitrogen enters legume plants by some unknown process, but not the non-legumes. This was questioned by many workers like J. Liebig and others. In 1858, P.G. Lachmann studied the nodules in the roots of legumes and observed the bacteria. In 1866, M Woronin further studied the nodule bacteria and found them to be rod shaped and motile. In 1879, Frank showed that the root nodules are formed due to infection by the organism which he named *Rhizobium*. In 1884, W.O. At water conducted some studies with peas and came to the conclusion that the leguminous plants could fix free nitrogen of the air. In the same year, Hellriegel and Wilfarth brought out, the connections between the capacity of legume plants to increase nitrogen content and the bacteria which are present in the nodules of these legumes, thus associating nitrogen fixation with legume nodules. However, the nature of the organism was understood only when M.W. Beijerinck in 1888, isolated the bacterium into pure culture and studied its physiology. The mechanism of root infection by the bacterium was studied by A. Prazmowski in 1889. The differences in the amounts of nitrogen fixed in different plants and strain variations and cross-inoculation groups in *Rhizobium* was also observed by Raugaswami & Bagyaraj (2002)

As early as in 1879, A.B. Frank described differences in the morphology of the bacterium occurring in different hosts. In 1887, A. Tschirch also reported on similar differences and Beijerinck in 1888, differentiated X and Y shaped forms, rod and pear shaped forms from three sets of host plants. Later, workers substantiated these observations on the morphological differences in the bacterium which led to the identification of seven different species of *Rhizobium* (Table 2).

Table 2. Cross inoculation groups of *Rhizobium*

Cross inoculation group	*Rhizobium species*	*Host it can nodulate*
Alfalfa group	*R. meliloti*	*Medicago, Melilotus, Trigonella*
Clover group	*R. trifolii*	*Trifolium*
Pea group	*R. leguminosarum*	*Pisum, Vicia, Lathyrus, Lens*
Bean group	*R. phaseoli*	*Phaseolus*
Lupine group	*R. lupini*	*Lupinus, Ornithopus*
Soybean group	*R. japonicum*	*Glycine, Crotolaria, Arachis*
Cowpea group	*Rhizobium spp.*	*Cajanus, Dolichos, Pueraria,* Cicer

The symbiosis between the non legume *Parasponia* and *Rhizobium* is an extreme example of boundary jumping by rhizobia.

3. Taxonomy or classification of *Rhizobium*

Fred *et al.,* (1932) reviewed earlier history in their classic and comprehensive book 'Legume Nodule Bacteria and Leguminous Plants'. Beijerinick (1888) was the first to isolate and cultivate micro-organisms from the nodules and named it *Bacillus radicicola* which is now placed in Bergey's Manual of Determinative Bacteriology under the genus *Rhizobium.* The genus *Rhizobium* includes such diverse families as *Azotobacter*iaceae, *Mycobacteriaceae,* and *Pseudomonadaceae.* Speciation of *Rhizobium* based on the Linnaean concept has proved difficult and therefore, the cross inoculation grouping based on the classical studies of Fred *et al.,* (1932) is being generally followed. The principle of cross inoculation grouping is based on the ability of an isolate of *Rhizobium* to form nodules in a limited number of legume species related to one another. All rhizobia that could form nodules on roots of certain legume types have been collectively taken as species. This system of classification has provided a workable basis for the agricultural practice of legume inoculation. Under this scheme, seven species are generally recognized (Table 2). The groups listed in the table are generally confined to their own set of host plants, but under certain conditions, they may cross over to different groups. The system of cross inoculation grouping of rhizobia is not perfect since bacteria have been found to cross infect or interchange between groups. However, until a better system of classification is perfected, it appears that we have to be contented with the cross inoculation grouping as a convenient and workable method of classifying root nodule bacteria into species.

International committee on systematic bacteriology included the Genus *Rhizobium* in ninth edition of Bergey's Manual of Determinative Bacteriology. Family Rhizobiaceae has included now five genera (Young & Haukka, 1996).

Table 3. Classification of *Rhizobium*

1	*Rhizobium* (fast growing)	*Rhizobium leguminosorum* *biovar trifolli, biovar phaseoli, biovar viceae*
2	*Sinorhizobium*	*Sinorhizobium meliloti* *Sinorhizobium fredii, Sinorhizobium saheli* *Sinorhizobium teranga*
3	*Mesorhizobium*	*(Rhizobium) loti* *(Rhizobium) huakuii* *(Rhizobium) ciceri* *(Rhizobium) tianshanense* *(Rhizobium) mediterraneum*
4	*Bradyrhizobium* (slow growing)	*Bradyrhizobium japonicum* *Bradyrhizobium elkani* *Bradyrhizobium liaoningense*
5	*Azorhizobium*	*Azorhizobium caulinodans*
6	(Another genus)	*(Rhizobium) galegae*

4. Nodule formation and development

Nodulation is an essential event of the legume *Rhizobium* symbiosis. It is a very complex process and depends on the genetic apparatus of both the partners. Nodule formation is controlled by plant genes. Nodule shape and size are controlled by the host plant and varies from spherical, oblate, cylindrical to elongate, single and multiple branched. In most of legumes nodules are formed on root, whereas in legume like *Sesbania rostrata, Aschenomyne, Neptunia* etc. nodules are formed on root as well as on stem. All the members of *Fabaceae* do not form nodules while *Parasponia*, a non-leguminous plant also nodulate with *Bradyrhizobium* species.

In 1887, Mashall ward reported bright spots on the tip or sides or root hair. In 1890, A.B. Frank stated that the roots of leguminosae secreted a substance which attracted the bacterium. Infection occurs through amyloid spots at tips of very young root hair.

4.1 Process of nodulation

It involved the following steps

i. Root colonization

ii. Root hair curling and host specificity

iii. Root infection

iv. Nodule development

Physiological Events for Infection thread (Subba Rao, 2000)

Normal root hair

Exudation of organic substances by roots

↓

Colonization of *Rhizobium* in the rhizosphere

↓

Tryptophan to indole aceticacid

↓

Root hair curling and deformation

↓

Involvement of lectins in rhizobial recognition

↓

Interaction of nod D gene product with plant root secreted flavonoids

↓

Incorporation of *Rhizobium* into cell wall

↓

Invagination of root hair cell to form an incipient infection thread

↓

Thread containing rod shaped bacteria extending into root hair cell guided by the nucleus of the root hair

↓

Entry of infection thread into root cortex and its branching

4.1.1 Root colonization

Rhizobia colonize the roots of different legumes and non-legumes. Colonization of roots is affected by survivability of rhizobia in soil, chemotaxis and attachment. Rhizobial colonization is stimulated by biotin which is a component of root exudates.

4.1.2 Root hair curling and infection mechanism

After attachment of rhizobia root hair curling is an important factor. Rhizosphere organism convert tryptophan in the root exudates to Indole Acetic Acid (IAA), which brings about the non-specific curling of root hairs. The capsular polysaccharides (CPS) of homologous strains of rhizobia play a role in inducing a site of infection on root hair known as "Shepherd's Crook", which led to the formation of thread like structure known as 'infection thread'. Pectin methyl esterase produced by rhizobia and IAA caused the loosening of the cell wall at shepherd's crook. Due to the production of polygalacturonase by rhizobia, invagination of root hair takes place, resulting in the formation of infection thread through which rhizobia enter the legume host. Host nucleus of root hair cell guides the path of infection thread till it reaches the tetraploid cell (site of developing nodule) of root cortex to stimulate for intense meristematic activity. Infection thread bursts open to liberate the bacteria in the host cytoplasm. Each bacterium so liberated gets surrounded by a membrane, pinched off from the infection thread. The development of nodule is initiated when infection thread reaches a tetraploid cell in the cortex. This release process is essentially endocytosis. Rhizobia multiply very rapidly accompanied by enlargement of membranous envelope surrounding them (peribacteriod membrane). Motile rhizobia now enlarge, lose their power of division and are now termed as bacteroids. These visible changes are accompanied by biochemical changes. : nitrogenase (N_2ase) is synthesized, the bacterial enzymes of ammonia assimilation are repressed and N_2 fixation begins. As soon as the rhizobia are released in the plant cytoplasm, the translucent capsule glycogen or poly B-hydroxy butyrate granules disappear rapidly and the rhizobial cells change into bacteroids, size and morphology of which is governed by both host plant and microsymbiont. The bacteroids of temperate legumes undergo remarkable change during maturation of infected cells. Bacteroids of *Vigna* spp. and *Phaseolus* spp. do not undergo much morphological changes. Apart from morphological change the bacteroids undergo significant physiological changes also.

4.1.2.1 Host specificity

Specific recognition between *Rhizobium* strain and its homologous host is due to plant lectins (proteins). This recognition phenomenon is due to plant lectins which specifically bind to carbohydrate receptors on the rhizobial cell. This plant *Rhizobium* recognition theory has generated considerable evidence on specificity in *Rhizobium*-legume symbiosis. The complementary protein carbohydrate molecules Cross-Reactive Antigen (CRA) interact resulting in the recognition of the two symbionts with each other at site of infection. These phases of attachment of rhizobia to root hairs appears to be controlled by nod genes.

4.1.2.2 Role of nod genes and nod factors

Rhizobium genetics revealed several gene-controlled steps in entire symbiotic process. In the initial stage host specificity becomes evident. This is controlled by the nodulation (nod genes). Rhizobia are attracted to the legume rhizoplane by small molecular weight product of roots (root exudates) called flavonoids/isoflavonoids. These compounds interact with nod D protein to activate nod D gene which regulates the expression of other nod genes.

There are three type of nod/nolgenes :

(i) Common nod genes (nod ABC) including regulatory gene (nod D) : Common nod genes are found in all rhizobial species and are essential for synthesis of nod factors (NFs). Inactivation of common nod genes results in complete loss of the ability to elicit infection and nodulation in all the rhizobia studied. Common nod genes are involved in early stage of every legume-rhizobia symbiosis. The regulatory nod D genes is present closely to the common nod genes. The nod D protein interacts with plant signal molecules like flavonoids and isoflavonoids and activate the nod D gene which in turn allows the transcription of other nod genes.

(ii) Host specific nod genes (hsn) (nod EFGHQ & genes of M. *meliloti*, nod EFLMN of R. *leguminosarum* and nod Z of *B. japonicum*) : Host specific nod genes or hsn genes are rhizobial genes required to nodulate specific host plant. The nol genes are specific in cowpea and mungbean nodulating rhizobia.

(iii) Genotypic specific nod genes (gsn) Genotypic specific nod genes or gsn genes are the rhizobial genes required to nodulate specific plant genotypes with in the given legume specific e.g. *R. leguminosarum* bv. Viceae strain TOM is reported to possess nod X and can nodulate cultivar Afghanistan pea but other strains of the *R. leguminosarum* b.v. viciae fail to nodulate this variety. There are about 44 nod genes responsible for the process of nodulation. Current evidence indicates that nod/nol genes are involved in the synthesis of Nod factors (NFs), the signal molecules compatible to host to induces root hair curling and cortical cell division. NFs play the major role in host specificity and they have other important roles : (i) they prepare the way for the invading bacteria by stimulating the development of pre-infection thread and nodule primordial (ii) they stimulate flavonoid release.

4.1.2.3 Plant genes and nodulation

The development of nitrogen fixing nodule involves the synthesis of a group of plant proteins, called nodulins. The nodulins, nodule specific host gene products, are induced following infection of host plant by rhizobia and are known to play role in nodule morphogenesis and function. The nodulins are divided in two groups, early and late nodulins. Early nodulin genes are involved in the synthesis of proteins playing role in the process of nodulation, while the late nodulin genes synthesize protein involved, in N_2 fixation or maintenance of nodules. About 200 nodulin genes have been identified in last 10 years in addition to leghaemoglobin genes in different legumes host. Leghaemoglobin is also a nodulin gene product. A red pigment akin to haemoglobin of blood is found in nodules between bacteroids and the membrane envelop surrounding them.

Leghaemoglobin, the prefix 'leg' indicating its presence in legume nodules, is a haemoprotein having a haeme moiety attached to peptide chain which represents globin part of the molecule. The molecular weight of leghaemoglobin is of the order 16,000 to 17,000 daltons only, while that of blood haemoglobin is of the order of 66,000 daltons, because leghaemoglobin has a one peptide chain linked with one haeme moiety, whereas blood hemoglobin has four peptide chains, each linked with one haeme moiety. The amount of leghaemoglobin in nodule has a direct relationship with the amount of N_2 fixed by legumes.

4.1.3 Nodule structure development

The basic nodule structure, there is an outer, uninfected cortical region containing the peripheral vasculature. This is separated from the inner, infected zone by a nodule endodermis. The central zone contains the infected cells where active N_2 fixation takes place. The shape and size of nodule is determined by host plant. Leguminous nodules are of two types, determinate and indeterminate with peculiar differences in nodule shape. Determinate nodules grow for a fixed period, all parts of the nodule differentiating at the same time, thus senescence also occurs at one time and they have a finite life span. Nodules in mungbean and urdbean are examples of determinate type. In contrast, indeterminate nodules have an apical meristem which continues to be active throughout the life time of the nodule. *Chickpea and pigeonpea nodules are indeterminate. Determinate nodules are always spherical whereas, indeterminate nodules are cylindrical to coralloid.**

There are a number of detailed reviews on infection process and subsequent nodule development. (Nap and Bisseling, 1990, Brewin 1998 and Cohn *et al.*, 1998).

The nodule life span can be categorized into three stages : In early nodule stage both bacteroids and host cells are actively dividing and specific symbiotic proteins, nitrogenase and leghaemoglobin are in the initiation stage. Probably at this stage N is provided to the nodule from other plant parts. At mature nodule stage N_2 fixing activity and leghaemoglobin content are at peak. In determinate nodules, cell division has almost ceased and at nodule senescence stage N_2 fixing activity and leghaemoglobin content are on decline.

Finally peribacteroid membrane (PBM) breaks down leaving bacteroids in direct contact with degenerating host plant cytoplasm. Nodule receives carbohydrates from plant and export in return.

4.1.3.1 Auto regulation of nodulation

Auto regulation means suppression of root nodulation and its exact mechanism of autoregulation is not yet understood very well. It is proposed that once a critical number of the subepidermal cell divisions (SCPs) in the root cortex are initiated, a precursor molecule from the root is transported to the shoot where it is converted into shoot derived inhibitor (SDI). Thus SDI transported back into the nod to suppress the subepidermal cell division formed from the developing emergent nodule (Shenga Harper 1997). They suggested that leaf is the main site for production of regulatory signals to control nodule number. The signals both inhibitory as well as promotory are involved in regulating the nodule number. Auto regulation of nodulation is an important mechanism to control the number of nodules required by the plant to fix sufficient nitrogen and avoid unwanted drain on plants resources.

4.2 Yield response to *Rhizobium* inoculation in legumes

Rhizobium inoculation with effective, competitive and infective strains helps in the establishment of right kind of rhizobia in plant rhizosphere to harness maximum benefit of BNF (Table 4). Field experiments carried out under AICRP on pulses at research stations as well as at farmer's fields showed response to *Rhizobium* inoculation in pulses (Pareek & Chandra, 2005). At research station, increase in yield due to *Rhizobium* inoculation over uninoculated control ranged between 0-49.7 percent (pigeonpea), 1.1 to 47.8 percent (mungbean) and 0-60 percent (urdbean). *Rhizobium* inoculation trials conducted at farmer's fields increase in yield was 6.9, 17.9, 11.1, 18.3 and 16.3 percent in chickpea, pigeonpea, lentil, mungbean and urdbean, respectively over uninoculated control. Native rhizobial population, host and environmental factors greatly influenced the efficiency of introduced *Rhizobium*. The extent of response is greatly influenced by site, season, crop management and crop sequence. The use of quality and viable inoculant is very important to harness the benefit of inoculation under field conditions. High populations of native rhizobia, mineral nitrogen and moisture stress at the time of sowing are other important factors for the poor response of rhizobial inoculation in pulses. *Rhizobium* inoculation has increased yields in several yield experiments of groundnut in India (Sundara Rao, 1971; Bajpai, *et al.*, 1974; Iswaran & Sen, 1974; Nagaraja Rao, 1974). In field, where groundnut has previously been grown *Rhizobium* inoculation has not generally been recommended (Subba Rao, 1976). However, nodules formed by the native strains may not fix nitrogen and their fixation rates may often be inadequate (Nambiar *et al.*, 1982). Gaur *et al.*, (1974) suggested that lack of response to inoculation and low yield in groundnut are probably due to competition from ineffective strains in the soil. Experiments conducted by ICRISAT and other centres of All India Coordinated Research Project on Oilseeds (AICORPO) showed that inoculation with an effective *Rhizobium* strain NC 92 increased the yield of groundnut cultivar 'Robut 33-1' in such fields where ineffective native strains were present (Nambiar & Dart, 1980; AICORPO, 1983). Inoculation with NC 92 increased yield of the cultivar 'JL 24' (Joshi & Kulkarni, 1983, 1983a). Two strains of *Bradyrhizobium* IGR 6 and IGR 40 increased the groundnut yield by 11% whereas, TNAU 14 (a strain of *Rhizobium* at Tamil Nadu) and Tt 9 (*Bradyrhizobium*) have been commonly used for commercial manufacture of inoculants.

Table 4. Major N-fixing micro-organisms and beneficiary plants (Sharma 2007)

Microorganisms	*Association*	*Beneficiary crops*	*N-fixation (Kg N/ha/ Crop)*	*Increase over control*
Rhizobium	Symbiotic with root nodules	Legumes	Up to 360	0-20 per cent
Azotobacter	Free-Living in soil	Cereals, Vegetables, cotton	20-30	8-120 per cent
Azospirillum	Assocciative symbiosis	Maize, Millets, Sugarcane	20-30	10.6-16.7 per cent
Blue Green Algae	Free-living in water	Rice	10-80	
Anabaena azollae	Symbiotic with azolla	Rice	20-100	

4.3 Nitrogen Mineralization

Nitrogen containing complexes are decomposed and converted into inorganic compounds for use by plants. During mineralization proteins, nucleic acids and their components are degraded by micro-organisms with the eventual liberation of NH_3 and this process is called ammonification. All bacteria, fungi and actinomycetes are involved in ammonification. NH_3 produced by ammonification is utilized in several ways. viz. micro-organisms and plants used for their growth, mycorrhizae are capable of utilizing ammonical N and make it available for plant growth.

4.3.1 Nitrification

Ammonia is converted into nitrate and this process is called nitrification. Certain bacteria have capacity to oxidize NH_3^+ as source of energy. Nitrification is a two step process.

1st Step : Ammonia (NH_3^+) is converted into nitrite (NO_2^-) by bacterium *Nitrosomonas*

$NH_3^+ \rightarrow NO_2^-$

2nd Step: Nitrite (NO_2^-) is converted into nitrate (NO_3^-) by bacterium Nitrobacter

$NO_2^- \rightarrow NO_3^-$
(nitrite) (nitrate)

$NO_3^-$is utilized by plants for their growth.

4.3.2 Denitrification

This is the reverse process of nitrification i.e. nitrate is reduced to nitrites and then to nitrogen gas and ammonia. This process occurs under anaerobic condition. Certain bacteria under anaerobic condition use NO_3^-as terminal e-acceptor. NO_3^-respiration leads to reduce $NO_3^-$to N_2 gas or NO_3^-respiration tends to reduce $NO_3^-$to N_2 gas or NO_2^- or NO. This process leads to the loss of N from the soil.

NO_3^- (nitrate) → NO_2^- (nitrite) → NO (nitrous oxide) → N_2O (Nitric oxide) → N_2 gas

Denitrification is not a desirable reaction as it depletes nutrients of soils which are essential for plant growth. Denitrification mostly occurs under water logged anaerobic soils with a high organic matter content. Bacteria which are important in denitrification are—*Thiobacillus denitrificans, Micrococcus denitrificans, Bacillus* and *Pseudomonas* species. Although, denitrification is an undesirable reaction from plant nutrition point of view but it is ecological important reaction because without supply of N_2 on the earth it would have depleted and NO_3^-would have accumulated in the soil because high concentration of $NO_3^-$are toxic. Denitrification is a mechanism by which some N goes back to the atmosphere.

4.4 Mechanism of nitrogen fixation

Biological systems have the enzyme complex known as nitrogenase which catalyzes the reduction of nitrogen to ammonia in combination with necessary reactions. Nitrogenase has been purified from all known types of nitrogen fixing organisms except *Archaebacteria* (Sprent & Sprent, 1990).

4.4.1 Nitrogenase enzyme complex

Nitrogenase accepts electrons from reduced ferredoxin or other effective agents as it catalyzers N_2 fixation—Ferredoxin, Flavodoxin, Fe protein, Mo-Fe protein. Each electron transferred requires two ATP. Unfortunately nitrogenase is not perfectly efficient. It also reduces other compounds like H^+, N_2O, $N_2^-$and CN^-. Hydrogen evaluation is a general by product of N_2 fixation particularly at low electron flux.

$$2H^+ + 2e^- + 4ATP \rightarrow H_2 + 4ADP + 4Pi$$

Nitrogenase is a multimeric protein complex made up of two proteins of different sizes. The larger protein in nitrogenase is called MoFe protein (component I). It is a tetramer consisting of two pairs of identical subunits with total molecular mass of approximately 220KD. The smaller protein is a dimer consisting of two identical subunits of four iron ions bound to four sulphur groups Fe_4S_4 (component II) as reported by Moat & Foster (1995). The smaller subunit ranges from 24 to 36 KD depending upon the bacterial species (Postgate 1971). Each MoFe protein contains two molybdenum ions in the form of an iron-molybdenum sulphur co-factors (Kennedy *et al.*, 1976; Hopkins, 1999). The MoFe proteins also contain Fe_4S_4 cluster although the exact number is uncertain (Kim & Rees, 1992).

The reduction of dinitrogen occurs, when a primary electron donor reduces Fe protein usually Ferredoxin a small (14 to 24 KD) protein containing iron and sulphur group. Electrons are carried by iron moiety, which can exist in either reduced ferrous (Fe^{2+}) or oxidized Ferric (Fe^{3+}) states (Albrecht *et al.*, 1979). The reduced Fe protein passes electrons to Mo, Fe proteins, which catalyzes the reduction of both dinitrogen gas and hydrogen (Moat & Foster, 1995).

The unique property of nitrogenase is that it becomes active in the absence of oxygen. Inactivation of nitrogenase in the presence of O_2 may be reversible or irreversible (Sprent & Sprent, 1990). Reversible inactivation of nitrogenase may be a way of regulating its activity. The half life/time to reduce activity by one half of isolated Fe-protein in air is 30 to 45 seconds and half life of MoFe protein is 10 minutes (Dixon & Wheeler, 1986). This extreme sensitivity of dinitrogenase to oxygen, raises a problem for nitrogen fixing organisms. The different organisms differ in their relation of nitrogenase sensitivity to oxygen (Soupene *et al.*, 1995).

Bacteriods require O_2 for respiration and for synthesis of ATP, but high level of O_2 not only inactivates the nitrogenase complex in the bacteriod but also inhibits the expression of rhizobial genes encoding for this complex (Mylona *et al.*, 1995). The photosynthetic bacteria do not have problem of internally generated O_2 but they may have to guard against the low levels of O_2 from outside. The non-photosynthetic nitrogen fixing organisms have nothing to do with nitrogenase oxygen relation. They do not possess enzymes viz. catalase and superoxide dismutase,which enable aerobic organisms to cope with toxic peroxide and other oxygen radicals (Morris, 1976).

The oxygen supply is regulated to a large extent by an oxygen binding protein called leghaemoglobin in leguminous root nodules (Bauer, 1960). Evans & Barber (1977) have put forward a hypothesis that leghaemoglobin is somewhat linked with the facilitated transport of oxygen and at the same time maintaining a low concentration of O_2 within the dense nodular tissue. Bergersen & Goodchild (1973) demonstrated the doubling of oxygen uptake by adding freshly prepared leghaemoglobin to respiring bacteroidal suspension with a 20-fold increase in nitrogenase activity thus providing a direct evidence for involvement of leghaemoglobin in oxygen transport. Shechy *et al.*, (1985) in a comprehensive report proposed that leghaemoglobin was essential for maintaining a fairly uniform level of nitrogenase activity within the infected cell by keeping a slow but steady supply of oxygen Carrol *et al.*, (1987) have proposed that abnormal O_2 concentration is an important factor limiting nitrogenase activity and leghaemoglobin probably assured the desired oxygen concentration. Suganuma *et al.*, (1988) studied the effects of leghaemoglobin on respiration of bacteroids and reported that oxyhaemoglobin prepared from soybean nodules, promoted the uptake of oxygen by bacteroids. Leguminous nodules contain a myoglobin like heme protein (Lb) which reversibly binds molecular O_2. The role of Lb in nodules is to facilitate the diffusion of O_2 at low concentration but rapid fluxes, to the microaerophilic bacteroid (Appleby, 1992). In the root nodules the O_2 concentrations is maintained at low levels by a diffusion barrier in the inner cortex and by bacteroidal respiration (Hunt & Layzell, 1993). Bacteroids require O_2 for respiration for the ATP synthesis, but high levels of O_2 not only inactivates the nitrogenase complex in the bacteroids, but also inhibit the expression of the rhizobial genes encoding for this complex (Delgado *et al.*, 1998). In leguminous nodules a family of genes encodes Lbs, with the number of detectable Lbs varying among the legume species. Eight isoLbs have been detected in soybean nodules of which four are post translationally modified (Appleby, 1974). Cowpea nodules contain three isoLbs, I, II and III (Dakora *et al.*, 1991) coded by atleast two genes, Lb I and Lb II have been cloned and sequenced in Lupin and Cowpea respectively by Sikorski *et al.*, (1995) and Arredondo Peter *et al.*, (1997). However, the physiological significance of multiple isoLbs in leguminous nodules is still unknown.

Besides leghaemoglobin, the host genes are known to code for a number of other proteins tentatively termed as 'nodulins' which are only produced in effective nodules (Auger & Verma, 1981). Their role is as yet obscure, although Thummler & Verma (1987) have identified nodulin-100 from soybean nodules, having properties of sucrose synthase and suggested that this enzyme played a key role in maintaining carbon economy of the nodules and the free heme may be involved in the flow of carbon to the bacteroids. With the degradation of leghaemoglobin during senescence, they also observed a concomitant decrease in sucrose synthase activity. Using experimental perturbations it has been demonstrated in soybean, that nodule sucrose synthase gene had quickly down regulated in response to stresses that simultaneously reduced nitrogen fixation (Gonzalez *et al.*, 1995; Gordon & James, 1997). Therefore, reduction in nodule sucrose synthase activity reduced the supply of carbohydrates for nitrogen fixation and assimilation (Gordon *et al.*, 1999).

4.4.2 Nodule Physiology : The host plant supplies carbon (C) and regulates O_2. Nitrogen fixation takes a lot of energy and is often limited by the C supply. Theoretically, upto 22 moles of glucose are required for 1 mole of N_2 fixed, so a substantial amount of the plants photosynthetic output 7 to 12% goes for sustaining its nodules (Paul & Kucey, 1981). When grain filling starts nodules die for want of C.

Waterlogging is detrimental to N_2 fixation in legumes. Although nitrogenase requires anaerobic or O_2 limited conditions, plant and rhizobia requires O_2 for respiration. So plant juggles for both of these requirements. Leghaemoglobin accumulates between bacteriod and peribacteroid membrane and gives cut nodules their red color. Leghemoglobin has a high affinity for O_2 and lowers the O_2 concentration in nodules sufficiently to allow respiration and N_2 fixation to occur simultaneously. Actually leghemoglobin is too efficient, because N_2 fixation in environments containing more than 20 percent O_2 is greater than at atmospheric O_2 levels. Thus implying that the rhizobia are O_2 limited. Rhizobia get C and shelter and supply N to plants in turn. The bacteroids excrete ammonia (NH_3) that is quickly fixed by glutamate dehydrogenase (GDH) or GS/GOGAT. The products leave the nodule via xylem. The NH_3 that rhizobia produce is transformed to asparagine and glutamine in temperate legumes and allantonin and allantoic acid in tropical legumes. Overall the cost for legumes to form nodules and fix N_2 is about 3 to 6g of C for every g of N fixed. Consequently when NO_3 and organic N or NH_3 are present, nodule formation and nitrogenase activity are inhibited.

4.5 Assimilation and Translocation

Assimilation or immobilization, the synonymous terms are used in the incorporation of inorganic N into organic N by the soil microorganisms. The N is immobilized with respect to plants, being turned into new microbial cells, it cannot be taken up by plants until those microbial cells decompose or mineralize. The consequences of inadequate N in soil can be devastating for plants. There are two major pathways in which NH_4^+ ions are assimilated by microganism. The first NH_4^+ assimilation pathway is directly into glutamate by the enzyme glutamate dehydrogenase (GDH). This pathway operates at relatively high NH_4^+ concentration (> 1 mM in solution or >0.5 mg NH_4^+/kg soil). At lower NH_4^+ concentrations (< 1 mM NH_4^+or 0.1 mg NH_4^+/kg soil) another pathway GS/GOGAT (GS= glutamine synthetase; GOGAT = glutamine 2-oxoglutarate amino transferase [Glutamate Synthase]). In this two step process the primary assimilation of NH_4^+ occurs which is catalysed by GS. The second step of the process forms glutamate from the transamination of α ketoglutrate by glutamine. Glutamine synthetase catalyzed reaction requires ATP, but the affinity of GS for NH_4^+ is greater than that of GDH. The Km is lower which means that GS allows microorganisms to utilize lower concentrations of environmental NH_4^+, when both pathways are present. GS is represented by high NH_4^+ concentrations. Once N_2 is incorporated as glutamate, it is transferred via transamination reaction to keto acids to form amino acids.

The NH_4^+ is translocated out of the bacteriods before it can be further metabolized and used by the host plant. In the cytosol of bacteriod-containing cells (external to the peribacteriod membrane), NH_4^+ is converted into glutamine, glutamic acid, asparagines and in many species nitrogen rich compounds called ureides. The two principal ureides in legumes are allantoin ($C_4H_4H_6O_3$) and allantoic acid ($C_4N_4C_8O_4$) like asparagines ($C_4N_2H_7O_4$) they have relatively high C:N ratio. Each of these three compounds represents a major form of nitrogen translocated from nodules to the other plant parts. Asparagine predominates in legumes of temperate origin, including peas, alfalfa and Lupins ; ureides predominates in legumes of tropical origin including soybeans, cowpeas and various beans. The varajah *et al.*, (2005) reported that chickpea can be classified as both an amide and urcide exporter on the basis of concentration of both types of N product found in shoot.

Asparagine and ureides move from bacteroid containing cells into pericycle cells adjacent to the vascular bundles that surround the nodule proper. These pericycle cells are modified into transfer cells and they seem to actively secrete nitrogen components into the conducting xylem cells. From here the compounds move into the xylem of the root and shoot to which the vascular bundles of the nodules are connected. Roots of nodulated plants that have not themselves become nodulated seems to receive appreciable nitrogen only after it has moved into the leaves and then back through the phloem along with sucrose. Thus a cyclic process, from nodules upward to shoots in the xylem and then return of excess nitrogen downward to roots via the phloem.

Factors that enhance photosynthesis like adequate moisture, warm temperature, bright sunlight and high CO_2 levels are found to enhance fixation consistent with this, the rate of fixation is unusually maximal in the early afternoon when the translocation of sugars from the leaves to nodules occurs rapidly. Early afternoon is also a time when transpiration is especially rapid, and the transpiration stream aids removal of nitrogen compounds from the roots and root nodules. Several genetic factors control the rates of nitrogen fixation and yield of legumes. One of the factors concerns how effectively nodulation occurs and this depends on the genetically controlled recognition process between the *Rhizobium* and the legume species or variety and the stage of growth also influences N_2 fixation. Efforts should be made to increase the efficiency of nodulation by altering genes in various rhizobia and by selecting for more compatible host varieties. Another genetic factor relates to the ability of nitrogenase from all organisms to reduce H^+ in competition with N_2. According to the estimations one third of the electrons available for reduction of N_2 in legume nodules are used instead to reduce H^+ to H_2 and the H_2 simply escapes into soil atmosphere carrying away wasted energy. Nevertheless certain strains of *Rhizobium* species contain a hydrogenase enzyme which oxidized much of the H_2 to H_2O before it escapes. During this oxidation, ATP is produced from ADP and Pi. There is evidence that soybeans and a few other legumes that contain *Rhizobium* strains with an active hydrogenase (Hup^+) yield slightly more than legumes that contain a mutant or Hup-strain, without hydrogenase probably because of less energy waste. Perhaps even more effective rhizobia species or strains with active hydrogenase can be formed or developed to increase legume yields through genetic engineering techniques.

Three important grain legumes soybeans, pigeonpea and peanuts, all show maximum fixation rates after flowering when the demand for nitrogen in the developing seeds and fruits increases. (Carroll & Gresshoff, 1983). These species, as is common to legumes, contain seeds that are especially protein rich. Quantitatively about 90 percent of the N_2 fixation in these species occurs during the period of reproductive development and roughly 10 percent during first two months of vegetative growth. Surprisingly, N_2 fixation provides only about one fourth to one half to three fourths absorbed as NO_3^-or NH_4^+ from the soil mainly during the period of vegetative growth. Nevertheless, yields of grain legumes cannot usually be increased with nitrogen fertilizers because N_2 fixation is decreased in proportion to the added amount of fertilizer. For nitrate fertilizer, this decrease results from inhibition of attachment of rhizobia to root hairs, abortion of infection threads, slowing of nodule growth, inhibition of fixation within established nodules and more rapid senescence of nodules when either NO_3^-or NH_4^+ is added.

4.6 Energetics of Nitrogen reduction

Nitrogen fixation being a reductive process requires a powerful reductant and energy in the form of ATP. Ferredoxin might be the immediate donor of electrons to nitrogenase and the reductant was supplied by bacteroid (Rokosh, *et al.*, 1973). During the nitrogenase reaction, the energy from ATP and electrons from the reductant cause a conformational change in Fe-protein component and convert it to a very powerful reductant with a redox potential measuring 400 mv (Brill, 1977). This fully reduced iron protein transferred electrons to the Mo Fe protein, which in turn reduced nitrogen to ammonia (Schubert & Evans, 1976). It is postulated that nitrogen remains bound to Mo Fe protein throughout the reaction and ammonia is released as end product (Albrecht *et al.*, 1979). No intermediate compound has been isolated as yet (Schubert, 1986). However, during this process, it was estimated that 4 to 5 molecules of ATP are hydrolysed per two electrons transferred and since the reduction of N to 2 moles of NH_3 involves transfer of 6 electrons, therefore 12 to 15 moles of ATP are required (Hadfield & Bulen, 1969). The energy burden on plant as imposed by the process of N_2 fixation has been reported by Vance & Heichel (1991), to cost approximately 6 milligram of C per milligram of N reduced. They further reported that reduction of one molecule of nitrogen into two molecules of NH_3 required 16 molecules of ATP. Keeping in view the high energy demands of nitrogen fixation, upto 40 molecules of ATP may be required in vivo for reduction of one molecule of dinitrogen to ammonia (Hill, 1992).

The overall reaction for the reduction of dinitrogen to ammonia as given by Taiz & Zeiger (1991) and Hopkins (1999) is as follows :

$$N_2 + 8H^+ + 8e^- + 16\ ATP \rightarrow 2NH_3 + H_2 + 16\ ADP + 16Pi$$

Under natural conditions, the reduction of H^+ to H_2 gas can be substantial and can compete with N_2 for electrons from nitrogenase (Evans *et al.*, 1987). In *Rhizobium* 30 to 60% of energy supplied to nitrogenase may be lost to H_2 (Schubert *et al.*, 1977). Therefore, this reaction would limit the efficiency of nitrogen fixation by these microorganisms. However, some *Rhizobium* strains contain an enzyme hydrogenase that can spilt H_2 and recycle electrons for N_2 reduction, thus increasing the efficiency of nitrogen fixation (Marschner, 1986). Most aerobic N_2 fixers rarely evolve H_2 because a membrane-bound, uptake hydrogenase system recycle the H_2 produced by nitrogenase and produce ATP through respiration to help and support ATP needs of the system (Moat & Foster, 1995).

Nitrogenase can reduce a number of substrates, although, under natural condition it reacts only with N_2 and H^+. One of the reactions catalyzed by nitrogenase i.e. the reduction of acetylene is used in estimating nitrogenase activity by gas chromatography (Jones 1977).

$$\underset{\text{Acetylene}}{C_2H_2} + 2H + 2e^- \rightarrow \underset{\text{Ethylene}}{C_2H_4}$$

This is used to estimate N_2 fixation ratio

$$2e^- + C_2H_2 \rightarrow C_2H_4$$

$$6e^- + N_2 \rightarrow 2NH_3$$

Ideal ratio of acetylene reduced to nitrogen fixed is 3:1, which is not maintained because reduction of alternative substrates is not taken into account.

Expenses at the cost of nitrogen fixation are (i) There is synthesis of nitrogenase and hydrogenase (if organism has an uptake hydrogenase) (ii) Cost of NH_3 assimilation, which begins with glutamine synthetase, an ATP requiring process, (iii) the lost ATP to fuel nitrogenase that could otherwise be used for growth. Fig. 1 summarizes the pathway of biological N_2 fixation.

Photosynthate $(CH_2O)_n$	12 Mg ATP	Dinitrogenase	Dinitrogenase	$2NH_3$	H_2	
	Ferredoxin (reduced)	reductase (reduced)	(reduced)			C_2H_4
↓	↑	↑ ↓	↑ ↓	↑↓	↑	↑
Metabolism (CO_2)	Ferredoxin (oxidized) 12 Mg ADP + 12Pi	Dinitrogenase reductase (Oxidized)	Dinitrogenase (Oxidized)	N_2	$2H^+$	C_2H_2

Fig. 1 : The Pathway of electrons through nitrogenase during nitrogen fixation.

Direct measurement of N_2 fixation requires expensive instrumentation, but reduction of acetylene to ethylene can easily be measured. Acetylene reduction involves two electrons, as opposite eight electrons required for reduction of N_2 and $2H^+$. Therefore, four ethylene molecules correspond to reduction of one N_2 molecule (Taiz & Zeiger, 1991). The acetylene reaction has some limitations to enable some microorganisms to produce ethylene gas. The acetylene method is valuable for comparative studies, whereas direct measurement of N_2 reduction by mass spectroscopy is required for precise quantification of nitrogen fixation.

4.7 Regulation of Carbohydrate Metabolism

Sucrose is the major product of photosynthesis which is transported from leaves to nodules of legumes. Sucrose is not utilized directly by isolated bacteroids but is hydrolysed in the nodules by enzyme invertase of host plant origin. Pathway of carbohydrate metabolism in bacteroids is not completely understood, but it is known that bacteroids contain some of the glycolytic enzymes and utilize a variety of organic acids. Kouchi *et al.*, (1988) measured the activities of enzymes involved in carbon and nitrogen metabolism in cytoplasmic fractions of purified protoplasts from soybean root nodules as well as in the bacteroids and found higher specific activities of enzymes of glycolysis particularly invertase in cortical and uninfected cells than in the infected cells. However, activity of sucrose synthase was highest in the cytosol of infected cells, which also exhibited assimilation *i.e.* glutamine synthetase, aspartate and alanine amino transferase. Similar reports have also come while studying the activities of enzymes of carbohydrate metabolism in plant and bacteroidal fractions of soybean nodules, reported that the enzymes required to convert sucrose to pyruvate and to oxaloacetate as well as enzymes of starch synthesis and degradation were present in the plant fraction of the nodules and the bacteroids had only limited capacity for carbohydrate metabolism and appeared to lack the complete glycolytic pathway. Further, the conversion of sucrose to organic acids might take place to a larger extent in the cortical tissue of the nodules. Bowsher (1989) while working on *Pisum sativum* reported the operation of both glycolysis and the oxidative pentose phosphate pathways during the metabolism of glucose-6-phosphate and reduction of inorganic nitrite during carbohydrate metabolism and nitrite reduction respectively, thus correlating the results with regard to regulation of pathways of carbohydrate oxidation and nitrogen assimilation.

The initial hydrolysis of sucrose is a key step in N_2 fixation. Sucrose synthase was encoded by a class of gene termed as nodulins which are expressed in the "nodules". Gordon & Emerich (1990) observed significant reduction in the activities of sucrose synthase, glutamine synthase and alcohol dehydrogenase in ineffective nodules of soybean as compared to effective nodules. The developmental profile of carbon metabolizing enzymes suggested that carbohydrates were metabolized by sucrose synthase and pentose phosphate pathway, which probably reduced the requirement of ATP of carbohydrate catabolism and generate NADPH for the biosynthetic reaction.

Sucrose is synthesized in the leaves and exported in the phloem to sinks such as nodules. It is unloaded in nodule cortex and should diffuse into the infected region of the nodule to be metabolized. The product of sucrose metabolism particularly malic acid, which are then used by bacteroid to fuel fixation of nitrogen. Gordon *et al.*, (1999) studied the role of sucrose synthase (SS) in the fixation of N_2 in rug 4 mutant of pea in which SS activity was severely reduced and reported that low access activity in nodules, resulted in lower amounts of total soluble protein and leghaemoglobin with only a trace of nitrogenase activity. The ratio of ADP/ATP (energy supplying pool) regulates the activity of nitrogenase and electron flow from Fe-protein to Mo Fe protein, which declined with ADP/ATP ratio falling below 2:1. Further the competitive inhibition by structurally related compounds has also been seen to inhibit/regulate nitrogenase activity, while compounds like ammonia, glutamine, asparagine and glutamate also exerted non-specific inhibition, thereby regulating the activity of nitrogenase. It is well known that ammonia produced through nitrogenase system is assimilated by the bacterium to the level of glutamate by the enzymes glutamine synthetase and glutamate synthase. The higher glutamine and lower glutamate content found in the bacteroid of *Rhizobium* etli, glutamine mutant (LM 16) in comparison with bacteroid of wild type strain indicated that glutamine degradation by glutamine played an important role during symbiosis between *R. etli* and *Phaseolus vulgaris*. Glutamine synthase (GS) catalyzed the ATP dependent condensation of NH_3 with glutamate to produce glutamine thus suggesting that GS levels were controlled both at transcription and translational/post translational levels in alfalfa. A strong constitutive positive element essential for the ammonium regulated root and root nodules. Specific expression of soybean gene encoding for cytosolic glutamine synthetase. The localization of expression of two functional cytosolic glutamine synthatase (GS) genes MtGSa and MtGSb in root nodules of *Medicago truncatula* has been studied and it has been reported that transcriptional regulation of mRNA synthesis played an important part in controlling the GS protein levels, indicating their involvement in other physiological processes in addition to the primary assimilation of ammonia related by bacterial nitrogenase.

4.8 Factors Affecting Legume-*Rhizobium* Symbiosis

Legume-*Rhizobium* symbiosis is a complex dynamic and diverse process. Predominant environmental stresses which affect BNF is governed by edaphic factors (temperature, moisture and gases), chemical factors (Soil salinity/alkalinity/acidity), nutrient deficiencies, agrochemicals and biotic factors which include synergistic and antagonistic interaction of soil rhizosphere with N_2 fixing bacteria.

4.8.1 Edaphic Factors (Temperature, Moisture, Drought)

Temperature

High root temperature strongly affect bacterial infection and N_2 fixation in several legume species, including soybean, clover, gaur, peanuts and cowpea (Naidu *et al.*, 2003).

Events of symbiotic process have different optima and limiting temperatures. Normally for temperate legumes optimum temperature for N_2 fixation is 15-25°C and for tropical and subtropical it ranges between 25-35°C. Soybean plant appear some what more tolerant, nitrogen fixation only being severely inhibited by daytime temperatures greater than 41°C and enhanced at 36°C (La Favre & Eaglesham, 1987). Excessive soil temperature kill majority of bacteria in the surface layers of soil, though some rhizobia can survive at 70°C in dry soil. Survival of the rhizobia in soils at high temperature can be improved by soil organic matter. In general, bacteria are less tolerant to high temperatures in moist than in dry soil. Whereas in chickpea, root temperature at 32°C inhibited nodulation and nitrogenase activity. The minimum and maximum temperature ranges for nodulation in tropical and subtropical legumes are 15-18°C and 35-40°C, respectively. At higher temperature respiration rate increased which resulted in less availability of carbon for symbiotic process. Low temperatures lead to delay in development of plants, including delays in formation of nodules and so decreased rates of N_2 fixation. The optimum temperature range for rhizobia in culture medium is 28-31°C and many are unable to grow at 37°C. However, environment of the Sahel Savannan grew well at 40°C (Eaglesham & Ayanaba, 1984). Strain adaptation to high temperature has been reported by Hartel & Alenander and Karanja & Wads high of the strains persisted at 45°C lost their infectiveness due to plasmid curing 90% of cowpea *Rhizobium* strains obtained from hot dry soil.

Drought

Drought stress has drastic effects on N_2 fixation. These effects could be due to (i) reduced photosynthesis (less availability of carbohydrates), (ii) less water for the transport of N-products from the nodules. (iii) Direct effect on nodule permeability and (iv) alteration in metabolic activity of nodule. Grain legumes with deep rooting systems, e.g. cowpea can withstand drought for prolonged period as the roots have managed to penetrate sufficient deep into soil before the drought beings. Legumes with indeterminate root nodules may be able to recover from drought more rapidly than plants with determinate nodules. (Eaglesham & Ayanab, 1984) because they retain an active meristerm. Ureide-exporting legumes are more sensitive to water stress than legumes that export their products of N_2 fixation as amides (Serraj *et al.*, 1999).The number of rhizobia in soil decline drastically as soil dries. *Bradyrhizobium* strains are more tolerant to drought than strains of *Rhizobium* because later retain less water with in the cells (Al-Rashidi *et al.*, 1982).

Soil Moisture

Sufficient soil moisture is essential for seed germination at sowing time. Water logging condition induced soil anaerobiosis which is harmful for nitrogen fixing activities and rhizobial growth. Rhizobia are normally aerobic organisms, some strains of *Bradyrhizobium* and *S. meliloti* possess a dissimilatory nitrate reducatse that can function as an electron acceptor and thus enable the bacteria to survive under anaerobic conditions

(Zablotowicz *et al.*, 1978). Large reductions in the numbers of rhizobia nodulating chickpea which are generally fast growing rhizobia and may not possess a dissimilatory nitrate-reductase have been found after paddy rice (Toomsan *et al.*, 1982). The degree of tolerance to waterlogging may differ widely between cultivars of legumes. In pigeonpea nodules developed thick cortex and enlarged lenticels under waterlogged conditions, which help in gaseous exchange across the nodule surface.

4.8.2 Chemical Factors

It includes soil acidity, alkalinity and salinity which effect the biological nitrogen fixation.

Soil Salinity and Alkalinity

Salt stress is known to depress greatly the growth and symbiotic performance of nodulated legumes. Indeterminate nodules are more tolerant to salt stress than determinate nodules (Sinclair & Serraj, 1995). Salt tolerance of legumes varies with the crop; *Pisum sativum, Phaseolus spp.* and *Cicer arietinum* are highly sensitive, while *Vigna unguiculata* and *Vicia faba* are relatively less sensitive. The process of root hair infection of legumes in particular sensitive to saline stress, which might be due to the inhibition of root hair growth in saline condition (Sprent & Zaharan, 1988). The reduction of N_2 fixing activity by salt stress is usually attributed to a reduction in respiration of nodules and a reduction in cytosolic protein production, particularly leghaemoglobin in nodules (Delgado *et al.*, 1994). Reduction in photosynthetic activity might also affect N_2 fixation by legumes under salt stress (Heckman *et al.*, 1989).

Although the root nodule-colonizing bacteria of genera *Rhizobium* and *Bradyrhizobium* are more salt tolerant than their legume hosts, they show marked variation in salt tolerance. Growth of a number of rhizobia has been inhibited at 10 mM Nacl, while some rhizobia are tolerant to 300 to 700 mM Nacl (Mahommed *et al.*, 1991). Slow growing rhizobia are less tolerant than fast growing rhizobia as reported by Ghittoni & Bueno (1996).

Soil Acidity

Soil acidity affects all aspects of nodulation and nitrogen fixation from survival and multiplication of the rhizobia in the soil, through infection and nodulation to nitrogen fixation. The failure of legumes to nodulate under acid soil condition is common especially in soils with pH less than 5.0. Low soil pH is often associated with aluminum and manganese toxicity and calcium deficiency. Critical pH for nodulation varies with legume crop and rhizobial strains. Chickpea root nodule bacteria survived well under soil pH 4.2 (Scott & Howieson, 2000). Whereas, reduction in nodulation at pH 5.5 was observed in *Phaseolus vulgaris* (Evans *et al.*, 1980). Species of *Rhizobium* are known to differ in their tolerance to soil acidity. Normally, slow growing rhizobia are relatively more tolerant than fast growing rhizobia (Munns, 1986).

Gases

Levels of oxygen (O_2) and carbondioxide (CO_2) play important role in symbiosis. The level of CO_2 in soil varies from 0.1 to 10.0% (Russel, 1950). Elevated CO_2 is likely to affect carbon cycling and crop productivity by stimulating photosynthesis. (Zanethi *et al.*, 1996). Increased atmospheric CO_2 increases the photosynthesis as a result of which more photosynthates are diverted to nodules thus increasing the nitrogenase activity (Hardy & Havelka, 1976). Low level of oxygen in soil either due to water logging or poor soil structure, may limit symbiotic BNF.

4.8.3 Nutritional Factors

Nitrogen

In pulse crops generally application is not required because under favourable conditions, these crops are able to grow well on soil N, stored N (colyledons) and N sequestered through BNF. However, experimentation have revealed that a starter dose of 15 to 20 kg N ha^{-1} is beneficial to stimulate seedling growth during early phase, when nodules have not fully developed. Well developed seedlings bear more nodules with better size leading to higher capability of BNF which in turns increases the productivity. Seed inoculation with *Rhizobium* proves effective, while the starter dose of 15 kg N ha^{-1} is helpful in improving the BNF. Excessive N application viz. 22.5 kg N ha-1 and above could cause considerable inhibition in glutamine synthetase activity, nitrogenase and carbohydrate content in mungbean and even could reduce cold tolerance in crops (Malhotra *et al.*, 1995). The effect of fertilizer N on nitrogen fixation in pulses depends on host species, *Rhizobium* strain, time of application and the time of sowing (Sekhon *et al.*, 1978; Pareek & Chandra, 2005). The method of application has a profound effect on response to applied N_2. Application of N as starter dose was found to be beneficial as compared to N_2 applied at other stages of crop growth in mungbean and urdbean (Mahadkar & Saraf, 1988), however the split application of N_2 does not improve productivity over basal dose in pigeonpea and chickpea (Ali & Mishra, 2000). Nodules of pulses disintegrate and becomes ineffective in the post-flowering period. Sap bleeding studies in pea and other legumes revealed a sharp decline in amino N. Split dose of N *i.e.* 15 kg N ha^{-1} at sowing and the second dose at flower initiation could yield higher in fieldpea than 30 kg N ha^{-1} at the time of sowing. This could be due to source sink relationship. During early pod development, there exists a competition for photosynthate between developing fruit and the nodules. Stronger sinks i.e. developing pods are closer to the source (leaves) therefore they receive assimilates at the cost of nodules. Thus, a supplement dose of 15 kg N ha^{-1} as top dressing decreases the competition for photosynthates (Aulakh & Sharma, 2005). Nitrate inhibited every phase of symbiosis, retarded nodule growth and nodule contained less leghaemoglobin. Most of the studies on N application in pulses are on individual crop basis and the information on their N requirement in cropping systems is lacking.

Phosphorus

Phosphorus (P) the most critical nutrient which is involved in various metabolic activities and its deficiency leads to low yield. In legumes, the process of nitrogen fixation has a very high energy requirement of adenosine triphosphate (ATP) for the reduction of N_2 to NH_3. Approximately 21 mole of ATP are converted to adenosine diphosphate (ADP) while reducing one mole of N_2 (Shanumgam *et al.*, 1978). Thus, there is scarcity of soluble P in the soil, which acts as a critical factor in legumes. The increase in the level of P increases the number of effective nodules per plant and subsequently the efficiency of BNF is improved. Studies conducted at Ludhiana showed that in lentil the combination of 20 kg P_2O_5 ha^{-1} with *Rhizobium* inoculation gave yield equivalent to those produced with 40 kg P_2O_5 ha^{-1} without *Rhizobium* inoculation (Dhingra *et al.*, 1988). The treatment of *Rhizobium* inoculation and 20 kg P_2O_5 ha^{-1} produced nitrogenase activity of 27117 n mole $h^{-1}g^{-1}$ dry weight of nodules while it was 18164, 22285 and 23682 n mole $h^{-1}g^{-1}$ in the case of uninoculated treatment with 20, 40 and 60 kg P_2O_5 ha^{-1}, respectively. Increased efficiency of P with *Rhizobium* inoculation could be due to the growth of mycorrhizal fungi, which are associated with the legumes, is assumed that soil microorganisms, stimulates the mycelial growth of the mycorrhizal fungi via production of certain compounds, which may increase P availability in soils.

Potassium

Soils of arid and semi-arid regions are medium to high in available K because of K bearing mineral (illite) so the crops do not respond to K fertilizers. The importance of root cation exchange capacity in absorption of divalents and monovalents has been well recognized. Legumes have higher root cation exchange capacity than oilseeds and cereals (Tiwari, 1988). High values of cation exchange capacity in legumes were associated with lower monovalent to divalent ratio, thereby increasing absorption by legumes as compared to other crops. Cereals have high K:Ca ratio than pulses. Thus the relative abundance of divalent cations on root surface of pulse crops may be interfering with K absorption, thus making them more sensitive to K stress in soil. Potassium application improves the crop quality as protein and sugar contents in grain are increased. The addition of KCl (source of K), the chloride could be so high to accelerate the disintegration of proteins.

Calcium

Calcium (Ca) has direct role in infection. Though exact function of Ca is not known, it helps in attachment of rhizobia to root hair tip. In Ca deficient environment rhizobia lack rhicadhesin (surface protein), playing role in root adhesion. Root growth, root hair development, infection thread elongation and nodule growth are less sensitive to Ca deficiency than initiation of infection thread as calcium is involved in plant cell wall synthesis (Pareek & Chandra 2005).

Sulphur

Sulphur (S) promotes nodulation and BNF and is an essential constituent of S containing amino acids. The increasing S deficiency in soils has been mainly due to extensive use of S free fertilizers, adoption of high yielding crop cultivars which demand a high fertility level resulting in greater exploitation of soil reserve nutrients and removal of much larger quantities of nutrients in harvested crop, intensive cultivation and low retension of SO_4^{2-} ions in such soils leads to low organic matter. In adequate supply of S, plants accumulate large amounts of amine, nitrate and amino N. A significant reduction in non-protein N fractions, with concomitant increase in protein N, due to applied S demonstrated the acceleration of metabolic pathway of protein synthesis in the plant. Therefore, adequate S fertilisation of different crops improves the quality of produce. Sulphur being an important consituent of nitrogenase enzyme and its severe deficiency may effect the synthesis of nitrogenase (Aulakh & Sharma, 2005).

Zinc

Zinc (Zn) is involved in the initiation of nodules, increasing leghaemoglobin content and N uptake is well known. Important functions of Zn in plant nutrition includes enzyme activation, protection of biomembranes and hormone metabolism (Zaidi, *et al.*, 1997). Deficiency of Zn causes several physiological disorders such as loss of chlorophyll, senescence and nutrient imbalances impairing normal functioning of plant. Zn application increases the content of amino acids particularly lysine, histidine and arginine in addition to crude protein.

Iron

All the energy producing and utilizing processes and many redox reactions involves the iron (Fe) there by forming an integral part of nitrogenase, ferrodoxin, leghaemoglobin and cellular enzyme systems in nodules. When Fe and Zn are in short supply in soils, rhizobia fail to function and fix N_2.

Molybdenum

Molybdenum (Mo) is important for the nutrition of pulse crops as it is essential for the activity of enzyme nitrogenase. Application of molybdenum improved the nodulation, N content and productivity in most of the pulse crops (Ali & Mishra, 2000).

Boron

Boron (B) is required for the synthesis of glutamine, nodules development and pollen tube growth. Without sufficient B in the environment of pollen tube prior to flowering, the pollen tube does not extend and grow from the pollen grains down through the style to enable fertilization. Thus the complete or partial failure of fertilization leads to fruit or seed drop, deformed fruit and partially filled grain pods. Deficiency of this nutrient affects nodulation, nodules remain small, lack vascular strands and bacteroids (Munni 1977).

Copper and Cobalt

Copper (Cu) and cobalt (Co) are required relatively in large amount by nodulated legume than plants growing with combined nitrogen (Raj 1987). These elements are essential for proper functioning of nodules.

Chlorine

Chlorine (Cl) may adversely affect the symbiosis, however their direct role in symbiotic fixation has not been established.

4.8.4 Agrochemicals

The use of pesticides to overcome the economic losses has become an essential part of agriculture. One of the most important and potentially limiting factors to BNF is the use of herbicides, fungicides and other pesticides. Herbicides are used for controlling weeds in the crops. However, some herbicides may adversely affect biological nitrogen fixation in legumes whereas, some other herbicides have been reported to favourably influence nitrogen fixation. Effect of herbicides on BNF varies with the crop variety, soil type, environmental conditions, *Rhizobium* strain and the herbicide, its dose and time of application. Seed bacterization of soybean seeds with *Rhizobium* and application of fluchloralin (1000 ga δ ha^{-1}) as pre-emergence herbicide did not inhibit nodulation (Gopal *et al.,* 2002). On the other hand, application of oxyfluorfen and oxadiazon at recommended rates adversely affected nodulation and N_2 fixation in lentil (Sandhu *et al.,* 1999).

Fungicides normally at their recommended dose did not show undesirable effect on symbiosis. *Rhizobium* strains show differential tolerance to fungicides. Carbandazin fungicide showed no significant effect on nodulation in cowpea (Singh *et al.,* 1986). Yet, the same compound decreased nodulation and yield in soybean (Catroux & Arnand, 1991). Carbandazin and thiram had neither negative or positive effects on chickpea nodulation (Welty *et al.,* 1988).

Variation among legume species with regard to nodulation and N_2 fixation with insecticide treatment may depend on the type and dose of insecticide, species of *Rhizobium* and legume and stage of development of *Rhizobium* legume symbiosis. Insecticide dimethoate at higher concentration reduced nodulation, leghaemoglobin concentration, nitrogenase activity and size of bacterial region in urdbean and fieldpea. At lower concentration, the insecticide phorate enhanced the nitrogenase activity but at higher rates it was inhibited significantly (Nayak *et al.,* 1980). The increase in nitrogenase activity at higher concentration of phorate was due to the increased concentration of gibberellins (Singh *et al.,* 1982).

4.8.5 Biological factors

Survival of rhizobia in soil is adversely influenced by antagonism and competition by other organisms. Soil bacteria and fungi have been found inhibitory as well as stimulatory to rhizobia in culture media. Majority of the rhizobacteria isolates were

Pseudomonas (Singh, 1999). Lysis of rhizobiophages, predation by protozoa and parasitism by *Bdellovibiro* are also important in regulating the population of rhizobia in soil. Such microbial interactions in soil depends on many edaphic and chemical factors. Susceptibility of rhizobial strains to bacteriophages may also result in their poor survival (Barnet, 1980). Specific damage to nodules of pigeonpea by larvae of *Rivellia angulata* has been reported in vertisol soils.

4.9 Nutrient uptake by legume

Legume crops are generally energy rich crops and they contain 20-28% protein in them, so they remove sizeable quantity of nutrients from the soil (Table 5). There is a considerable variations among the crops due to variations in length of growing season and yield.

Table 5. Nutrient uptake by pulse crops (Ganeshamurthy *et al.*, 2003)

Crop	*Yield (kg/ha)*	*Total uptake (kg/ha)*			
		N	*P*	*K*	*S*
Pigeonpea	1200	85.0	8.0	16.0	9.0
Urdbean	890	70.0	5.6	50.1	5.1
Mungbean	1000	106.0	21.0	90.4	6.5
Soybean	2500	125.0	43.0	101.0	22.0
Chickpea	1500	191.0	6.0	49.0	13.0
Lentil	2000	114.0	13.0	36.0	6.0

4.10 Status of Native *Rhizobium* in Soil

Status of *Rhizobium* is quite variable ranging from low to high in pulse growing areas. Under the All India Co-ordinated Pulses Improvement Project (AICPIP), 79% of total area had low-medium status of rhizobia in soil. As regards crop-wise distribution: 86.1, 73.4, 85.7 and 75.2 percent area showed low to medium population of *Rhizobium* in chickpea, lentil, pea and lathyris (Pareek & Chandra, 2005) respectively. This clearly indicates that the population of native rhizobia is not enough to capitalise upon the potential of pulse crops to fix atmospheric N. Poor to moderate nodulation at farmers fields could be due to the excessive use of fertilizers, agrochemicals and burning of straw, etc., which decreases the population of native rhizobia. There is a great need to create awareness amongst the farmers regarding the use of *Rhizobium* as a 'Cheap insurance policy' by which superior, competitive and efficient rhizobia can be introduced in the soil. Repeated inoculation with effective strains of *Rhizobium* can enhance the yield of legumes. The strain should survive and persist in soil in the absence of host, be compatible with agro chemicals and tolerate environmental stress. Sindhu & Dadarwal (2000) suggested that competition among rhizobial strains in the rhizosphere to dominate nodulation of legume root is a complex phenomenon as it involves the interaction of biotic and abiotic components of soil ecosystem.

5.0 BNF Contributions

Estimates of the proportion of plant N-derived from N_2 fixation (P fix) and the total nitrogen fixed by different pulse crops vary to great extent (Table 4) as these are determined by a variety of factors including the crop, environment, the rhizobial strain and soil properties. In winter legumes, the 50% N has been derived from atmosphere. In Kharif and summer pulses, the variability in the proportion of N-derived from fixation varied very widely. This is mainly because of fluctuation in soil temperature and moisture during monsoon (Ali & Mishra, 2000). Apart from the above factors, genotypes of crop may also influence P fix and the amounts of nitrogen fix could be quite high. Long duration varieties are known to fix nitrogen more as compared to extra short duration varieties. In pigeonpea short duration varieties fixed nitrogen upto 7 kg/ha where as in medium and long duration it fixes upto 68 kg N /ha (Kumar Rao & Dart, 1987). Soil moisture, *Rhizobium* inoculation, crop management practices such as use of P fertilizer and nitrogen status influence BNF contribution. Legumes are grown not only as sole crop but also as inter crop and the BNF contribution by the legumes in intercrops depends on species, plant morphology, proportion of intercrop and management practices. Sole pigeonpea fixed 88 kg N/ha, whereas when it intercrop with cereal it fixed upto 75 kg N/ha. (Kumar Rao *et al.*, 1987). Direct estimation of N_2 fixed in field situation increased nodulation and biomass yield of the crop are also measures of BNF and widely used under AICRP programme by microbiologists in India. Nodulation surveyed under AICRP on pulses revealed moderate to poor nodulation in pulses at different locations, suggesting BNF contribution through pulse crops.

5.1 Residual Effect

BNF of legumes normally benefits the associated or succeeding crops in terms of N contribution. Pulses grown as a sole and intercrop with cereal crops showed N economy equivalent to 16-68 kg N/ha for succeeding cereals (Pareek & Chandra, 2005). N requirement of wheat was much lower when preceeding pigeonpea was intercropped with urdbean or cowpea. Short duration legumes like mungbean, urdbean and cowpea showed higher production of succeeding wheat with N economy of 40 kg N/ha (Ali, 1989). Maize after chickpea, lentil, peas and wheat needed 57, 95, 93 and 125 kg N/ha to produce 45 q/ha grains indicating BNF contribution of 30-68 kg N/ha to maize (Ahlawat *et al.*, 1981)

Table 6. Effect of *Rhizobium* inoculation and P application on nodulation and yield of chickpea and their residual effect on maize/sorghum (Sharma *et al.*, 1995)

Treatment	*Nodule number per plant*	*Grain yield (kg/ha)*	*Succeeding crop yield*	
			Maize (kg/ha)	*Sorghum (t/ha)*
Control	10.5	1510	2195	32.4
Inoculation	18.7	2379	2773	43.0
40 kg P_2O_5/ha	16.1	2180	2426	38.0
Inoculation + 40 kg P_2O_5 /ha	20.7	2567	2740	46.5

BNF contribution of pulses as a residual effect can not be taken granted as sometimes negative N balance was observed. The favourable effect of the legumes, however, may not be exclusive, due to BNF, but they increase availability of other plant nutrients, improve soil structure, enhances level of growth promoting substances and reduces disease-insects incidence.

5.2 Phytohormones mediated response to nodulation and nitrogen fixation

It is well documented that some of the Phytohormones are specific in their effects on nodulation and nitrogen fixation but some also exhibit overlapping functions with other hormones. The response of plants to exogenous application may depend on the particular legume species, stage of development of nodule and external environmental conditions. Such responses imply that their normal internal levels may be regulated by some factors. For example, nod factors from *Rhizobium* species act as a signal leading to mitotic activity in nodule meristem and resemble several well known plant hormones (Van Spronsen *et al.*, 1995).

The plant hormones or growth regulators are extremely important agents in integration of developmental activities and are also concerned with response of plants to external environment. Environmental factors often exert inductive effects by evoking changes in the hormone metabolism, distribution within plant and are the principal agents which regulate the expression of intrinsic genetic potential of plants (Moore, 1980). Various endogenous growth substances probably play an important role in the determination of nodulation components and their interaction. In recent years, it has been suggested that PGR's (Plant Growth Regulator) might play an important role in determination of nitrogen fixation in legumes (Hirsch & Fang, 1994).

Five major classes of PGR's have been recognized viz. auxins, gibberellins, cytokinins, abscisic acid and ethylene. Many other growth regulating substances are awaiting their grouping as plant hormones, since they are of plant origin and active at extremely low concentrations. Cutler (1988) has reviewed the literature and stated that all the five hormones are present in microorganisms. Thimann (1935) discovered IAA in *Rhizopus suinis*. It was earlier called as rhizopin. Gibberellins were isolated from *Gibberella fujikuori* (Brian, 1957), whereas ethylene has been emanating from numerous fungi (Ilag, 1970). Cytokinins were isolated from autoclaved yeast DNA (Miller, 1956). Lui & Carns (1961) isolated a crystalline material called Abscision I from the native cotton fruit balls. This substance was later known as abscisic acid (Addicot *et al.*, 1968).

Plant physiologist have invariably been using growth regulators to manoeuvre important aspects of plant metabolism vis-à-vis growth and development. The literature pertaining to the subject has been reviewed briefly under the following sub-heads.

5.2.1 *Auxins*

The effects of tryptophan and an antiauxin 2,3,5 trioidobenzoic acid (TIBA) had been studied by Bond (1948) on pea roots and the division of parenchymatous cells were stimulated resulting in an increase in nodule number. However, Kefford *et al.*, (1960) reported that napthalene acetic acid (NAA) at low concentrations delayed nodule initiation. Somehow, the nodules with this treatment were larger in size as compared to the control. Further different concentrations of tryptophan showed that higher concentration delayed nodule initiation whereas the lower concentrations of NAA hastened nodulation. Biproduction of indole-3-acetic acid was reported in root nodules and roots of *Lupinus luteus* by Dullart (1967). Wang *et al.*, (1982) identified IAA in culture medium of nodulating and non-nodulating strains, though the level of IAA was stimulated through exogenous application of tryptophan. This was also supported by the work of Nandwal & Bharti (1982), where the effect of IAA was studied on nodulation and nitrogen fixation in pea. Low concentration increased the nodule weight and nitrogen fixing efficiency of nodules. But the higher concentration of IAA led to the reduction in these parameters. It has been well demonstrated that all growth regulating substances particularly indole-3-acetic acid (IAA), gibberellins (GA) and Cytokinins (Ck) intensified the rate of photosynthesis in soybean leaves possibly by increasing sources-sink translocation of ^{14}C assimilates. IAA application enhanced the cell wall plasticity as well as the elongation rate of hypocotyl segments in mungbean (Sharma & Neag, 1982, Prat 1987). Higher endogenous levels of IAA like substance were observed in mature nodules of *Phaseolus aureus* by Dangar & Basu (1987), further exhibiting higher activities of nitrogenase and low levels of phenolics. Exogenous application of 0.05 mM napthalene acetic acid, significantly promoted the symbiotic nitrogen fixation in berseem 2,3,5-triiodobenzoic acid or NPA, which blocks the auxin transport, when applied elicited the formation of nodule like structure i.e. pseudonodules in alf-alfa (Hirsch *et al.*, 1989). Afghanistan pea roots (Scheres *et al.*, 1992) or sweet cloversym mutants (Wu *et al.*, 1996). Maximum quantities of endogenous IAA has been reported from root nodules of cowpea (*Vigna ungiculata* (L.) Walp) by Nisha *et al.*, (1993) at 50 days after sowing followed by a decline at later stages of growth. Similar observation have been reported in the root nodules of leguminous medicinal herb *Tephrosea purpurea* pers by De & Basu

(1995). IAA amount increased in root nodules when tryptophan (precursor of IAA) was supplemented. Foliar application of IAA (300 M) significantly increased the nodulation in cowpea (Nisha *et al.*, 1996) and cajanus cajan L. (Mill sp.) (Raghuwanshi & Dudeja, 1996, Datta & Basu, 1997). It has been reported by Yi *et al.*, (1999) that indole 3-acetic acid, increased ethylene production by inducing expression of three-1-amino cyclopropane 1-carboxylate (ACC) synthase in mungbean (*Vigna radita* L). Bhattacharya & Pati (2000) have observed high amount of IAA in culture from tryptophan supplemented yeast extract mannitol medium in root nodules of *Alysicarpus vaginalis* D.C. and tried to establish a relationship between rhizobial IAA production and legume-*Rhizobium* symbiosis. Foliar application of NAA (40 ppm) after 30 DAS and there after a week resulted in an increase in nodulation of mungbean. Further the foliar spray of NAA @ 20 ppm enhanced nitrate redutase activity in mungbean cultivar ML 613. (Annonymous 2000).

Sohail *et al.*, (2004) isolated and characterized rhizobial strains from root nodules of cultivated legumes viz. chickpea, mungbean and pea. The plant infectivity test and acetylene reduction assay showed effective root nodule formation by all the isolates on their respective host except chickpea, ^{18}Ca that failed to infect its original host. The strain ^{18}Ca was characterized by its phosphate solubilization and IAA production. PGR's and Mg-salts provided to the plant through foliar sprays had promotary effect on nodulation and nitrogen fixation.

5.2.2 *Gibberellins*

The presence of GA like substances in the nitrogen fixing root nodules has received scanty attention. While using dwarf pea bioassay. Radley (1961) demonstrated that nodules of *Pisum sativum* and *Phaseolus vulgaris* contained easily detectable levels of GA-like substances. Similarly, Dullart (1967) found that nodules of *Lupinus luteus* had higher levels of GA-like activity as compared to roots. Keeping in view, the stimulating effect of GA_3 on auxin production from tryptophan by cell free nodule extracts, it has been speculated that the presence of significant amounts of GAs in the nodules of *Lupinus luteus* might be responsible for synthesising higher levels of auxins in the nodules. Similar reports have been found in *Alnus glutinosa* by Hensen & Wheeler (1977). No effect on nodulation in *Trifolium repens* and *Medicago sativa* by sodium gibberellate has been reported, whereas same concentration reduced nodule number by 50%in *Trifolium hybridum, T. carnatum* and *T. subterranean* though the quality of nitrogen fixing capacity of remaining nodules was unimpaired. Effect of GA treatment was dependent upon the age of plant at which the treatment was given e.g. detrimental effects were seen when the sprays were given during the early phases of plant growth, whereas promotory effects could be seen with the late spray. Parkash (1966) showed inhibition of nodulation in *Trifolium alexandrianum* by GA (25-250 ppm). Gibberallic acid 20 ppm enhanced root nodulation and nitrogen fixation in *Cicer arietinum*, while 100 ppm of GA showed negative effects in ultimate analysis (Swaraj & Garg, 1970). No significant effect on GA

(5ppm) was found on nodulation by Kulkarni *et al.*, (1978). Foliar application of GA_3, GA_4 and GA_7 delayed nodulation and reduced the number, mass and specific activity of soybean nodules by 43, 31 and 47 percent respectively. This negative effect was overcome by reducing the endogenous level of gibberellin content with the help of cycocel spray, which resulted in an increase of nodule number by 56 percent (Williams & De Mallorca, 1984). Gibberellin @ 500, 100 and 10 mg/l, led to the accumulation of large quantity of soluble ^{14}C photosynthates in the petioles of soybean and pea leaves as reported by Karpov & Belozerova (1985) and Pereto & Beltran (1987) respectively. When the endogenous level of gibberellin like substances were compared in the dwarf and normal cultivars of pea, it was observed that these growth substances were higher in cotyledons of normal cultivars than the dwarf ones (Pesme & Israelstəm, 1986). GA_3 has been reported to promote hypocotyls growth in *Vigna radiata* (Sharma 1986). However, synergistic effects of gibberellins and cytokinins have been reported in delaying leaf senescence and in maintaining leaf functions by Nooden (1986). Application of gibberellic acid mainly activated the synthesis of soluble chloroplast protein in 12-day old pea plants (Maksyntova *et al.*, 1987). Effect of different concentration of GA_3 and kinetin on nitrogen fixation and yield in chickpea (*Cicer arietinum* L.) has been studied by Garg, *et al.*, (1992) and significant enhancement in nitrogen fixing efficacy with both GA_3 and Kinetin has been reported. Higher amount of gibberellin like substance in the root nodules of leguminous medicinal herb *Tephrosea purpurea* pers by De & Basu (1995). Foliar application of 30 M gibberellic acid increased nodulation in cowpea significantly as reported by Nisha *et al.*, (1996). Number of genes encoding for biosynthetic pathway enzymes have been cloned from pea which included those genes coding for C-20 oxidases (Thomas *et al.*, 1999), Kaurene synthase A (Ait-Ali *et al.*, 1997) and 3-hydroxy lase (Lester *et al.*, 1999; Martin *et al.*, 1996). Dey & Srivastava (2003) have reported the presence of IAA, GA_3 and kinetin along with $MgSO_4$ $7H_2O$ in the free living culture of *Rhizobium* sp. enhanced its growth, nitrate reductase, nitrite reductase, nitrogenase and soluble proteins. Further their foliar application promoted nodulation and nitrogen fixation.

5.2.3 *Cytokinins*

Kinetin treatment (10^{-8} M) prevented half the plants from nodulating, but those plants which nodulated produced the first nodule at the same time as the control. Nodule number per plant was unaffected by kinetin at a concentration of 10^{-6} M to 10^{-9} M. (Kefford *et al.*, 1960). However, favourable effects of kinetin on root nodulation has been reported by Lona (1961). Liquid endosperm of coconut had a stimulating effect on nodulation in soybean as reported by Iswaran *et al.*, (1977). Nodules as a source of cytokinin in *Alnus glutinosa* has been studied by Wheeler & Hensen (1978) and it was seen that endogenous levels of cytokinin activity were found to vary due to growth differences rather than due to nodulation. Thus suggesting that although a part of cytokinin (Ck)from the nodules was exported to the aerial parts of plant, it was not enough for major contribution to shoot cytokinin while assessing the role of cytokinin in relation to legume *Rhizobium* symbiosis. Kinetin resulted in an early nodule initiation and a considerable increase in fresh and dry matter of nodules in *Pisum sativum*. Further studies by Nandwal & Bharti (1982) demonstrated maximum efficiency of nodules to fix nitrogen with 220 M kinetin in *Pisum sativum*. Endogenous levels of cytokinin were estimated from young and old nodules of *Phaeolus aureus* and it was observed that high levels of Ck like substances in mature nodules possessed higher nitrogenase activity while low levels in the old nodules. Significant enhancement of nitrogen fixing efficacy in chickpea was reported with both GA_3 and kinetin, however kinetin proved to be more effective than GA. Further different concentrations of GA_3 and kinetin were tested and their effects were visualized. The results were similar to the earlier findings of Garg *et al.*, (1992). Nod factors of *Rhizobium* species induced perturbations in auxin-cytokinin balance and enhanced "pseudo" nodulation on *Medicago atropurpureum* (Relic *et al.*, 1993). Similarly Spaink (1994) observed the interaction between soil bacteria belonging to genera, *Rhizobium, Bradyrhizobium* and *Azorhizobium* and leguminous plants and reported that bacterial signal lead to mitotic activity in nodule meristem which resemble several well known plant hormones like auxin and cytokinins. De & Basu (1995) have also reported higher amounts of cytokinins and other hormones like substances in the root nodules of leguminous medicinal herbs. Nodulation increased in cowpea with the foliar application of indole acetic acid (IAA), gibberellic acid (GA) and kinetin (kin) (Nisha *et al.*, 1996) Hirsch *et al.*, (1997) observed higher concentration of auxins, cytokinins and ethylene during nodule development in alfalfa roots. Further, they proposed a model for nodule showing involvement of cytokinin and auxin in stimulating cell division in the inner cortex, which ultimately led to nodule formation. Cytokinin induced expression of ENOD 40, an early gene following inoculation with *Rhizobium meliloti* in roots of transgenic alfalfa (Fang & Hirsh, 1998). Rajani *et al.*, (2000) reported 20-fold increase in endogenous cytokinin in xylem sap and leaves of cow-pea (*Vigna ungiculata*), treated with IC-3342 strain which is known to cause hormonal imbalance or over production of cytokinin. Treated plants showed significant increase in chlorophyll content and number of branches, both are the promotory effects of cytokinins. Benzyladenine (BA) improved reproductive efficiency both by improving the biomass production and its destruction into economic yield, which is reflected in the improved harvest index. The requirement of cytokinin as a limiting factors which is provided by exogenous application is shown. Exogenous application of BA improved reproductive efficiency by either counteracting the abscission promoters or augmenting endogenous cytokinin levels or both. BA may be acting by improving the ability of the treated plant to mobilize nutrients and their partitioning during the period of pod and seed set in *Cicer arietinum* (L). A significant increase in the nitrogen in the nitrogen fixing efficiency was observed with GA_3 and Kinetin in case of *Cicer arietinum* (Garg *et al.*, 1992). Presence of (IAA, GA_3 and kinetin) along with magenisum sulphate in the free-living culture of *Rhizobium* species enhanced its growth, NR activity, NiR activity, N_2ase activity and soluble proteins as observed by Dey & Srivastava (2003). It was thus suggested that if

PGRs along with Mg salt are provided to plant through foliar spray, these have promotory effect on nodulation and nitrogen fixation.

5.2.4 *Abscisic acid*

References related to the role of abscisic acid (ABA)in relation to nodulation and nitrogen fixation are scantly ABA resulted in a marked suppression of nodule formation by reducing the number of polyploidy cells in the root tissue of pea has been reported by Phillips (1971a). He further proposed (1971b) that ABA was translocated from the cotyledons of *Pisum sativum* to the root system where it inhibited nodulation. Application of ABA (10^{-5}M) in Pea, inhibited the shoot growth with a concomitant increase in organic solutes and proline accumulation. A correlation between low nitrogenase activity of old nodules and high endogenous levels of ABA has been worked out in *Phaseolus aureus* by Dangar & Basu (1987). A significant increase in the concentration of ABA in the leaves of pea plant flooded with water, which led to slower transpiration and increased leaf water potential is shown. Higher amount of ABA like substance has been reported by De & Basu (1995) in the root nodule of leguminous medicinal herb *Tephrosea purpurea* pers. More work is required on the effect of ABA on nodulation and N fixation in legumes.

5.2.5 *Ethylene*

Ethylene, a gaseous hormone has been shown to inhibit nodulation in the excised roots of *Phaseolus vulgaris* (Grobbelaar *et al.*, 1971) and *Trifolium subterraneum* grown in vitro conditions (Day, 1972). Foliar sprays of ethrel reduced nodulation (Drennan & Norton 1972). In confirmation with this, application of ethylene in alfalfa plants inhibited root nodulation, and a large quantity of ethylene was released from roots on application of fixed nitrogen to legume culture, there by indicating a positive correlation between nitrate concentration and the quantity of ethylene released according to Ligero *et al.*, (1987).

Ethylene regulates many aspects of growth and development and in general, studies have shown that ethylene production has a negative effect on nodule formation and its activity (Abeles *et al.*, 1992). Many workers have reported that ethylene production significantly increased in roots infected by *Rhizobium* or *Bradyrhizobium* and the exogenous application of ethylene decreased the number of nodules formed on infected legume plants. Treatment with amino ethyoxy vinyl glycine (AVG) or Ag^+ inhibitors of ethylene formation and perception stimulated the nodule production in *Medicago truncatula* roots (Ligero *et al.*, 1987) and Caba *et al.*, (1998). Further many workers have reported that these inhibitors partially restored nodulation in unique *vicia sativa* sub. sp. *nigra* symbiosis. Nod factors produced by *Rhizobium leguminosarum* biovar. viciae induced ethylene related changes in the root cortical cells of *Vicia sativa* sub. sp. *nigra* thus indicating that at some stages of nodule formation were oversensitive to ethylene inhibition as observed by Gurnel & LaRue, (1992). Soybean when treated with *B. japonicum* caused an increase in ethylene production, while exogenous treatment with ethylene and AVG could not inhibit or suppress nodulations reported by Suganuma *et al.*, (1995). Hoffman *et al.*, (1999) have isolated number of soybean lines with decreased responsiveness to ethylene. These lines could provide a tool for analysis of many plant functions that were influenced by ethylene, like studies on root growth and symbiotic root nodules formation. Effect of auxin on ethylene production in mungbean (*Vigna radiata* L.) was studied by Yi *et al.,* (1999). It was postulated that auxin differentially regulated the expression of three members of 1-aminocyclopropane-1-carboxylate (ACC) synthase. Thus the multi-regulatory pathway for expression of ACC synthase multigene family by auxin was deplicated.

It may be concluded that Nod-factor is a primary morphogenic signal for nodulation as it triggers on a compatible host, the earliest stages of nodule development including root hair deformation, curling and cortical cell division (Truchet *et al.,* 1991). Nod factors also elicit the expression of several early nodulin (ENOD) genes, such as ENOD 12 and ENOD 40 (Fang & Hirsch, 1998). Nodulins are plant-encoded proteins which are expressed during nodule development. ENOD 40, one of the earlier nodulin to be expressed upon *Rhizobium* inoculation has been cloned from a number of legumes.

The effect of nod factors on legume plants can be copied by modulating the balance of endogenous plant hormones for example by the application of 2,3,5 triodobenzoic acids or NPA, both of which presumably block auxin polar transport in the root, elicits the formation of nodule like structures (pseudonodules) on alalfa, Afganistan Pea roots and or Sweet clover sym mutants as reported by Wu *et al.,* 1996). Trancripts for early nodulin genes viz. ENOD 8 and ENOD 12 were detected in these pseudonodules. Similarly, when alfalfa roots were inoculated with R. *meliloti*, nodulation mutants carrying the *Agrobacterium tumefaciens* tzs gene (encoding DNA transfease, an enzyme associated with transZeatin biosynthesis), small-uninfected nodules developed (Beaty *et al.,* 1986). Likewise, cytokinin application triggers ENOD 2 expression in *Sesbania robusta* roots (Bauer *et al.,* 1996). These results suggest that a change in endogenous hormone level, may be a consequence of nod factor perception in a responsive host plant. It is likely that the induction of expression of nod genes (ENOD 2, ENOD 5, ENOD 12 etc.) by cytokinin and other endogenous molecules upon *Rhizobium* inoculation could serve as an amplification mechanism, thereby triggering a localized hormone imbalance, which initiates cell division in the root cortex. How this change in hormone (presumably auxin/cytokinin) balance is brought about, still remains undeciphered. However, the fact remains that some early nodulin genes are hormone resparsive, which might lead to the identification of specific transcription factors or cis-acting elements that are involved in gene regulation upon nod factor application. There is a dire need for more research in order to achieve a better understanding of the signal transduction pathways in the nitrogen fixing symbiosis as well as plant hormone regulation of plant genes.

Constraints

It can be concluded that legumes particularly are endowed with unique capability of fixing atmospheric nitrogen which varies considerably with change in management practices, genotypes and inoculation techniques. There is a need for new and improved strains of both N_2 fixing bacteria which can be competitive with native strains. Shorter shelf-life, contamination in the carrier and non-availability of carrier material are the main reasons responsible for poor quality of *Rhizobium*. These improvements in strain quality can come through natural selection or by genetic engineering methods. Much current research is targeted at selecting or genetically manipulating *Rhizobium* strains to produce tailored strains that will out compete indigenous rhizobia. The lack of suitable methodology is one of the key factors limiting the proper assessment and success of individual *Rhizobium* strain in nodule occupancy with different cultivar, soil and agroclimatic conditions. The development and use of marker gene technology will enable soil ecologists to rapidly screen the competitive ability of a number of potential inoculum strain. In addition, application technique is being modified in order to deliver high viable inoculum at the site of seed germination.

An alternative approach could also be employed which involves the alteration of host cultivars through breeding programme that can restrict the nodulation by the indigenous competitive strains and allow nodulation by the inoculated strains. Through these strategies unique combinations of efficient nitrogen fixing rhizobial inoculants with agronomically desirable legume cultivars can be developed for high BNF.

Development of common inoculant of beneficial rhizobacteria, PSB and VAM in consortia with *Rhizobium* is essential for their easy delivery in field. Selection of efficient strains of such micro-organism having synergistic interaction and compatibility in broth and carrier is essential to develop a inoculant having consortia of microbes useful to BNF. Systematic research on the role of different microorganism with *Rhizobium* in different legume crops under varied soil environment and its exact mechanism is required to exploit such symbiosis under field conditions by developing their commercial inoculants. Considering the tremendous scope of biofertilisers in sustainable nutrient management system due importance should be given to eco-farming.

Systematic and indepth studies are required regarding the role of phytohoromones and growth regulating substances on nodulation and nitrogen fixation in legumes. Further the mechanisms underlying the phytohormonal regulation needs to be investigated. To achieve the potential economic benefits by rhizobial inoculation and growth regulating substances there is a need to analyze abiotic soil factors such as soil texture, pH, temperature, moisture content and nutrient availability which largely determine the survival and activity of the introduce microorganisms.

REFERENCES

Abeles GB Morgan PW Saltiveit ME 1992 Ethylene in plant biology Academic Press San Diego C A

Addicott FT Carns HR Cornfoth JW Lyon JL Milborrow BV Onkuma K Kayback G Smith OE Thiessen WE Wareing PF 1968 Abscisic acid: A proposal for designation of Abscison II (dormin) In : Biochemistry and Physiology of Plant Regulators The Range Press Ltd Ottawa Canola pp-1527-1529

Ahlawat IPS Singh A Saraf CS 1981 Effect of winter legumes on the nitrogen economy and productivity of succeeding cereals *Experimental Agriculture* **17** 57-62

AICORPO 1983 Annual Progress Report Rajendranagar Hyderabad India All India Co-ordinated Research Project on Oilseeds **2** 241-248

Ait-Alit T Swain SM Reid JB Sun TP Kamiya Y 1997 The LS locus of pea encodes the gibberellin bionsynthesis enzme ent-kaurene synthase-A *Planta* **11** 443-454

Albrecht SL Maier RJ Hanus FJ Russel SA Emerich DW Evans HJ 1979 Hydrogenase in *Rhizobium japonicum* increases nitrogen fixation by nodulated soybean *Science* **203** 1255-1257

Ali M 1989 Consolidated Report on kharif Pulses Directorate of Pulses Research Kanpur

Ali Masood and Mishra J P 2000 Nutrient management in Pulses and pulse-based cropping systems *Fertilizer News* **45** : 59 61-67 69

Al Rashidi RK Loyna Chan TE Fredrick LR 1982 Dessication tolerance of four strains of *Rhizobium japonicum* Soil Biol and Biochem **14** 489-493

Anonymous 2000 Plant Physiology Annual Progress Report for Kharif Pulses Punjab Agricultural University Ludhiana

Appleby CA 1974 Leghemoglobin In A Quispel ed The Biology of Nitrogen fixation American Elsevier Publishing Co New York pp 521-554

Appleby CA 1992 The origin and functions of haemoglobin in plants *Sci Progress* **76** 365-398

Arredondo-Peter R Moran JF Sarath G Luan P Klucas RV 1997 Molecular Cloning of cowpea leghaemoglobin II gene and expression of its cDNA in *Escherichia coli Plant Physiol* **114** 493-500

Auger S Verma DPS 1981 Induction and expression of nodule-specific host genes in effective and ineffective root nodules *Biochem* **20** 1300-1306

Aulakh M S Sharma Pushp 2005 Nutrients Dynamics and their efficient use in pulse crops pp 241-278 In: Pulses (Eds Guriqbal Singh H S Sekhon J S Kolar) Agrotech Publishing Academy Udaipur

Bajpai PD Lehri LK Pattak AN 1974 Effect of seed inoculation with *Rhizobium* strains on the yield of leguminous crops *Proc Ind Nat Sci Acad* **40B** 571-573

Barnet YM 1980 The effect of rhizophages on population of *Rhizobium trifolii* in the root zone of clover plants *Cand J of Microbiol* **26** 572-576

Bauer NA 1960 A probable free radical mechanism for nitrogen fixation *Nature* (London) **188** 471

Bauer P Patel P Crespi MD Schultze M Kondorosi A 1996 Nod factors and cytokinis induce similar cortical cell division amploplast deposition and M_5ENOD 12A expression in alfa-alfa roots Plant J **10** 91-105

Beaty JS Powell GK Licia L Rigies DA MacDonald EMS Hommes NG Morris RO 1986 T_2S a nopaline Ti plasmid gene from *Agrobacterium* tumefaciens associated with transzeatin biosynthesis Mol Gene Genet **203** 274-280

Beijerinck MW 1888 Die Bacterien der Papillionaceen Knollche *Botanische Zeitung* **46** 797-804

Bergersen FJ Goodchild DJ 1973 Cellular location and concentration of leghemoglobin in soybean root nodules *Aust J Biol Sci* **26** 741-756

Bhattacharya RN Pati BR 2000 Growth behaviour and indole acetic acid (IAA) production by *Rhizobium* isolated from root nodules of *Alysicarpus vuginalis* DC *Acta Microbiol Immunal Hung* **47(1)** 41-51

Bond L 1948 Origin and development of morphology of root nodules of *Pisum sativum* L *Bot Gaz* **103** 411-435

Bowsher CG Hucklesby DP Emes MJ 1989 Nitrate reduction and carbohydrate metabolism in plastids purified from the roots of *Pisum sativum L Planta* **177** 359-366

Brain BW 1957 Role of gibberellin like hormones in regulation of plant growth and flowering *Nature* **181** 1122-1123

Brewin NJ 1988 Tissue and cell invasion by *Rhizobium* : the structure and development of infection threads and symbiosomes. In the Rhizobiaceae (eds HP Spanik A Kondorasi PJ Hooykaas pp 417-429Kluwer, Dordrecht, The Netherlands

Brill WJ 1977 Biological nitrogen fixation *Sci Amer* **236** 68-81

Burns RC Hardy RWF 1975 Nitrogen Fixation In Bacteria and Higher Plants Berlin : Springer-Verlag Fred EB

Baldwin IL McCo E 1932 Root Nodule Bacteria and Leguminous Plants Madison Wisconsin : University of Wiscons

Caba JM Recalde L Ligero F 1998 Nitrate induced ethylene biosynthesis and the control of nodulation in alf-alfa *Plant cell Environ* **21** 87-93

Carrol BJ Hansen AP McNeil DL Gresshoff PM 1987 Effect of oxygen supply on nitrogenase activity of nitrate and dark-stressed soybean (*Glycine max* (L) Merr) plants *Aust J Plant Physiol* **14** 679-689

Carroll BJ Gresshoff PM 1983 Nitrate inhibition of nodulation and nitrogen fixation in white clover *zeitschrift für Pflanzen physiologic* **110** 77-88

Catroux G Arnand F 1991 Compatibility of a soybean peat inoculant with some seed applied fungicides and micro granular insecticides *Toxicol Environ Chew* **30** 229-239

Cohn J Day RB Stacey G 1998 Legume nodule organogensis Trends in Plant Science **3** 105-110

Cutler HG 1988 Usual plant-growth regulators from micro-organisms *Crit Rev Plant Sci* **69** 323-343

Dakora FD Appleby CA Atkins CA 1991 Effect of O_2 on the formation and status of leghemoglobin nodules of cowpea and soybean *Plant Physiol* **95** 723-730

Dangar TK Basu PS 1987 Studies on plant growth substance IAA metabolism and nitrogenase activity in root nodules of *Phaseolus aureus* Roxb var mungo *Biol Plantarum* (Praha) **29(5)** 350-354

Datta C Basu PS 1997 Contents of indole acetic acid and its metabolism in root nodules of cajanus cajan (Linn) Millsp *Plant Physiol Biochem* **24(2)** 71-74

Day JM 1972 Studies on the role of light in legume symbiosis Ph D Thesis Univ of London

De PS Basu PS 1995 Content of hormones and indole acetic acid and metabolism in root nodules of *Tephrosea purpurea* Pers *Ind J Plant Physiol* **3** 228-232

Delgado MJ Bedmar EJ Downie JA 1998 Genes involved in the formation and assembly of rhizobial cytochromes and their role in symbiotic nitrogen fixation *Adv Microb Physiol* **40** 191-231

Delgado MJ Ligero F Lluch C 1994 Effect of salt stress on growth and nitrogen fixation by pea fababean common bean and soybean plants *Soil Biol Biochem* **26** 371-376

Dey B Srivastava RC 2003 Cumulative effects of magnesium sulphate and Plant Growth hormones on growth and nitrogen metabolizing enzymes and soluble protein in *Rhizobium* sp. Ind J Plant Physid (special issue) 588-591

Dixon ROD Wheeler CT 1986 Nitrogen fixation in plants Chapman and Hall New York

Drennam BSH Norton C 1972 The effect of ethrel on nodulation in *Pisum sativum* L *Plant Soil* **36** 53-57

Dullart J 1967 Quantitative estimation of IAA and indole carboxylic acid in root nodule and roots of *Lupinus luteus* L *Acta Bot Neerl* **19** 877-883

Eaglesham ARJ Ayanaba A 1984 Tropical stress ecology of rhizobia root nodulation and legume fixation In Current Development in Biological Nitrogen Fixation (Eds N Subba Rao) pp 1-35 Edward Arnold Publishers London United Kingdom

Evans HJ Barber LE 1977 Biological nitrogen fixation for food and fibre production What are some immediate feasible possibilities ? *Science* **197** 332-339

Evans HJ Harber AR Papen H Russel AA Hanus EJ Zuber M 1987 Physiology biochemistry and genetics of uptake hydrogenase in rhizobia *Ann Rev Microbiol* **41** 335-361

Evans LS Lewin KF Vella FA 1980 Effect on nutrient medium pH on symbiotic nitrogen fixation by *Rhizobium leguminosarum* and *Pisum sativum Plant* and *Soil* **56** 71-80

Fang Y Hirsch AM 1998 Studying early nodulin gene ENOD 40 expression and induction by nodulation factor and cytokinin in transgenic alf-alfa *Plant Physiol* **116** 553-68

Frank B 1879 Uber die parasiten in den Wuzelanshwellunigen der Papilion ceen *Botanische Zeitung* **37** 377-388 393-400

Frank B 1889 uber die Plizsymbiose der Leguminosen Berichte der Deutschen Butanischen Gasell schaft **7** 332-346

Fred EB Baldwin IL Mccoy E 1932 Root Nodule Bacteria and Leguminous Plants University of Wisconsin Press Madison

Ganeshamurthy AN Subba Rao A Rupa TR 2003 Integrated nutrient management in pulse based cropping systems in Pulses in New Perspective (eds M Ali BB Singh Shiv Kumar Vishwa Dhar) pp 287-300 IIPR Kanpur

Garg N Garg OP Dua IS 1990 Promotion of symbiotic nitrogen fixation and economic yield by exogenous application of naphthalene acetic acid (NAA) in *Trifolium alexandrianum* (L) Berseen *Indian J Plant Physiol* **33(1)** 1-6

Garg N Garg OP Dua IS 1992 Effect of Gibberellia acid and kinetin on nodulation and nitrogen fixation in chickpea (*cicer arietinum L*) Ind J Plant Physiol **35(4)** 401-40

Gaur YD Sen AN Subba Rao NS 1974 Promiscuity in ground nut *Rhizobium* association Proc *Ind Nat Sci Acad* **40** 562-570

GhittoniNE Bueno MA 1996 Changes in the cellular content of trehalose in four peanut rhizobia strains cultured under hyper salinity *Symbiosis* **20** 117-127

Gonzalez EM Gordon AJ James CL Arreselgor C 1995 The role of sucrose synthase in response of soybean nodules to drought *J Exp Bot* **49** 1515-1523

Gopal M Gupta A May SP 2002 Effect of fluochloralin on N_2 fioxation and growth of rhizobia and flourescent pseudomonas in soybean Tropical Agric **79** 173-79

Gordon AJ Emerich DW 1990 Developmental regulation of enzymes of sucrose and hexose metabolism in effective and ineffective soybean nodules *Plant Physiol* (Bethesda) **92(2)** 346-351

Gordon AJ James CL 1997 Enzymes of carbohydrate and amino acid metabolism in developing and mature nodules of white clover *J Exp Bot* **48** 895-903

Gordon AJ Minchin FR James CL Komina O 1999 Sucrose synthase in legume nodules is essential for nitrogen fixation *Plant Physiol* **120(3)** 867-877

Gouffi K N Pica Pichereau Blanco C 1999 Diassacharides as new class of non accumulated osmoprotectants for sinorhizobium *meliloti Appl Environ Microbiol* **65** **1491-1500**

Grobbelar N Clarke B Hough M 1971 The nodulation and nitrogen fixation of isolated roots of *Phaseolus vulgaris* L III The effect of Carbon dioxide and ethylene *Plant Soil* (Sp vol) 215-221

Guinel FC La Rue TA 1992 Ethylene inhibitors partly restore nodulation to pea mutant E 107 (brz) *Plant Physiol* **99** 515-518

Hadfield KL Bulen WA 1969 Adenosine triphosphate requirement of nitrogenase from *Azotobacter vinelandii Biochem* **8** 1503-1508

Hardy RWF Havelka UD 1976 Photosynthate as major limiting nitrogen fixation by field grown legumes with emphasis on soybean In Symbiotic N_2 fixation (Ed P S Nutman) pp 421-436 Cambridge Univ Press Cambridge U K

Heckman MO Drevon JJ Sanglio P Salsac L 1989 Effect of oxygen and malate on No-3 inhibition of nitrogenase in soybean nodules *Plant Physiol* **90** 224-229

Hellriegel H Wilfarth H 1888 unter such ungenuber die stick snff nah rung der Gramineen and Legminosen Beliageneft Zuder vereins der Ruebenzucker-Industrie Deutschen Reiche pp 234

Hensen IE Wheeler CT 1977 Hormones in plants bearing nitrogen fixing root nodules. Gibberellin like substances in *Alnus glutinosal* Gaerth *New Phytol* **78** 378-381

Hill S 1992 Physiology of nitrogen fixation in free living heterotrophs Chapman Hall New York pp: 87-134

Hirsch AM Fang Y 1994 Plant hormones and nodulation What's the connection? *Plant Mol Biol* **26** 5-9

Hirsch AM Fang Y Asad S Kapulnik Y 1997 The role of phytohormones in plant-microbe symbiosis. *Plant Soil* **19(4)** 171-184

Hirsch AM Bhuvaneswari TV Terry JG Bisselong T 1989 Early nodulin genes are induced in alflfa out growth elicited by auxin transport inhibitors *Proc Natl Acad Sci* (USA) **86** 1244-1248

Hoffman T Schmidt JS Zheng X Bent AF 1999 Isolation of ethylene insensitive soybean mutants that are altered in pathogen susceptibility and gene for gene disease resistance *Plant Physiol* **119** 935-949

Hopkins WG 1999 Introduction to plant physiology John Wiley Sons pp 99-118

Huntt S Layzell DB 1993 Gas exchange of legume nodules and the regulation of nitrogenase activity *Ann Rev Plant Physiol Plant Mol Biol* **44** 483-511

Ilag VL 1970 Ethylene production by fungi Ph D Thesis Purgue university Latayette Ind pp-93

Iswaran V Sen A 1974 Some studies on the groundnut *Rhizobium* symbiosis Zentl Bakt Parasit In fekt Hyp **129** 477-480

Iswaran V Jayasanker NP Najhai KJ Suba Rao NS 1977 Nodulation of Soybean as influenced by liquid endosperm of coconut *Proc Linneam Soc* N S Wales **72** 265-291

Joshi PK Kulkarni JH 1983a Research Project on oilseeds **2** 241-248

Joshi PK Kulkarni JH 1983 Response of groundnut to single and mixed *Rhizobium* strains Ind J *Agric Sci* **53** 376-378

Jones K 1977 Acetylene reduction in blue-green algae in sub-tropical grassland *New Phytol* **78** 421-426

Karpov EA Belozerova OL 1985 Effects of growth regulation on photosynthesis and out flow of labelled 14 C-photosynthesis from leaves of soybean plants during the reproductive period pp 429-433 Translated from *Fiziolosiya Restenii* **32(2)** 539-544

Kefford NP Brockwell J Zwar JA 1960 The symbiotic synthesis of auxin by legumes and nodule bacteria and its role in nodule development *Aust J Biol Sci* **13** 456-467

Kennedy C Eady RR Kondorosi E Rekosh DK 1976 The molybdenum iron protein of *Klebsiella pneumoniae* nitrogenase *Biochem J* **155** 383-389

Kim J Rees DC 1992 Structural models for metal centres in the nitrogenase molybdenum iron protein *Science* **257** 1677-1682

Kouchi H Fukai K Katogiri H Minamisawa K Tajima S 1988 Isolation and enzymological characterization of infected and unifected cell protoplasts from root nodules of *Glycine max Physiol Plant* **73** 327-34

Kulkarni AP Dev DV Deshpande DS 1978 Studies of some condensed thiazole systems on nodulation *Curr Sci* **47(19)** 728-730

Kumar Rao JVDK Thompson JA Sastry PVss Giller KE Day JM 1987 Measurement of N_2 fixation in field grown pigeonpea (*Cajanus cajan* (L Millsp) *J of Agron and Crop Sci* **175** 15-28

Kumar Rao JVDK Dart PJ 1987 Nodulation nitrogen fixation and nitrogen uptake in pigeonpea (Cajanus Cajan (L) Millsp) of different maturity groups *Plant und soil* **99** 255-266

Kumar Rao JVDK Thompson JA Sastry PVSS Giller KE Day JM 1987 Measurement of N2 fixation in field grown pigeonpea (Cajanus cajan (L) Millsp) using ISN labeled fertilizer *Plant and soil* **101** 107-113

La Favre AK Eaglesham ARJ 1987 Effects of high soil temperatures and starter nitrogen on the growth and nodulation of soybean *Crop Sci* **27** 742-745

Ladha Reddy DM 1995 Extension of nitrogen fixation to rice-necessity and possibilities *Geo Journal* **35** 363-372

Lester DR Ross JJ Davies PJ Reid JB 1999 Mendels stem length gene (Le) encodes a gibberellin 3-hydroxylase *Plant Cell* **9** 14350-1443

Ligero F Lluch C Olivares J 1987 Evalution of ethylene from roots and nodulation rate of alfalfa (*Medicago sativa* L) plants inoculated with *Rhizobium meliloti* as effected by presence of nitrate *J Plant Physiol* **129** 461-467

Lona F 1961 Inhibition of root nodulation in meteorpea by nictophasic treatment with far red radiation First general assembly and first scientific meeting of italyain society of plant physiol *Pavia Nouvog Bot Ital* **68(1/2)** 228-232

Lui WC Carns HR 1961 Isolation of Abscisin-an abscission accerlating substance *Science* **134** 385

Mahadkar UV Saraf CS 1988 Nitrate reductase in relation to nodulation and nitrogen nutrition of green gram J of Maharashtra Agri Uni **13** 291-293

Maksyutova NN Martynova TB Tarchevaskii IA 1987 Effect of Gibberellic acid on the synthesis of soluble pea chloroplast proteins *Fizial Biokhim Kulf Rast* **19(4)** 348-352

Malhotra RS Singh KB Saxena MC 1995 Effect of nitrogen fertilizer application on cold tolerance in chickpea international chickpea and pigeonpea Newsletter **2** 24-25

Marscher H 1986 Mineral nutrition of higher plants Academic Press London

Martin DN Proebstung WM Park TD Dougherty WG Lange K T Lewis MJ Gaskin P Hedden P 1996 Feedback regulating of gibberellin biosynthesis and gene expression in *Pisum sativum L Planta* **200** 159-166

Miller CO 1956 Evidence for natural occurrence of zeatin and derivatives compounds from maize which promote cell division *Proc Natl Acad Sci* (USA) **54** 1052-1058

Moat AG Foster JW 1995 Microbial physiology (3rd Ed) John Wiley and Sons pp 436-461

Mohammed RM Akhavan Kharazian M Campbell WF Rumbaugh MD 1991 Identification of salt and drought tolerant *Rhizobium meliloti* L strains *Plant and Soil* **134** 271-276

Moore TC 1980 Biochemical and physiology of Plant hormones Springer-vertage New York

Morris M 1976 Oxygen and obligate anaerobes J *App Bact* **40** 229-244

Munns DN 1977 Mineral nutrition and the legume symbiosis In a Treatise on Dinitrogen Fixation section 4 (eds RWF Hardy AH Gibson) pp 353-391 New York Wiley

Munns DN 1986 Acid soils tolerance in legumes and rhizobia *Adv Plant Nutr* **2** : 63-91

Mylona P Pawlovski P Bisseling T 1995 Symbiotic nitrogen fixation *Plant Cell* **7** 865-869

Nagaraja Rao H S 1974 Response of groundnut variety 'TMVZ' to rhizobial inoculation *Proc Ind Nat Sci Acad* **40B** 650-651

Nambiar PTC Dart PJ 1980 Studies on nitrogen fixation on groundnuts at ICRISAT In Proceedings of the International Workshop on Groundnuts Ed R W Gibbons PP 110-124 ICRISAT Patancheru Andhra Pradesh

Nambiar PTC Dar PJ Rao BS Ravi Sankar HN 1982 Response of groundnut (*Arachis* hypogea) to inoculation In Biological Nitrogen Fixation Technology for Tropical Agriculture Eds PH grahanand SC Harris PP 241-248 Centro International the Agriculture Tropical Cali Colambia

Nandwal AS Bharti S 1982 Effects of kinetin and indole-acetic acid on growth yield and nitrogen fixing efficiency of nodules in pea (*Pisum sativum L) Ind J Plant Physiol* **25(4)** 358-363

Naidu VSG R Panwar JDS Saikia SP 2003 Symbiotic nitrogen fixation response to abiotic factors in plant physiol Mol. Biol. New Millennium (ed A Hemantaranjan) pp 165-175 Scientific Publishers (India) Jodhpur

Nap JP Bisseling T 1990 Development biology of plant-prokaryote symbioses : the legume root nodule Sci **250** 948-954

Nayak DN Pasalu IC Rao VR 1980 Influence of natural and synthetic insecticides on nitrogen fixation in rice-rhizosphere *Curr Sci* **49** 118-119

Nisha R Kumar P Rhaghava RP 1993 Quantitative estimation of IAA content in root nodules of cowpea *Plant Physiol Biochem* **20(2)** 73-76

Nisha R Kumar P Rhaghava RP 1996 Leaf growth nodulation and yield performance in cowpea as influenced by growth regulators *Plant Physiol Biochem* 23(1 85-90

Nooden LD 1986 Synergism between gibberellins and cytokinns in delaying leaf senescence in soybean (Glycine max) explants *Plant Cell Physiol* **27(3)** 577-580

Pareek RP Chandra R 2005 Biological Nitrogen Fixation In: Pulses (eds G Singh H S Sekhon and J S Kular) pp 279-312 Agrotech Publishing Academy Udaipur

Parkash V 1966 Effect of 2-chloroethyl trimethyl ammonium chloride (CCC) and gibberellic acid on nodulation in *Triflolum alexandriamum Ind J Expt Bot* **4** 251-253

Parsons R 1997 Contrasting C supply N assimilation and N transport across a range of symbiotic plants In Biological Fixation of Nitrogen for Ecology and Sustainable Agriculture (Eds A Legocki H Bothe and A Puhler) pp 211-214 Berlin : Springer-Verlog

Paul EA Kucey RMN 1981 Carbon flow in plant microbial associations *Science* **213** 473-474

Peoples MB Ladha JK Herridge DF 1995 Enhancing legume N_2 fixation through plant and soil management *Plant and Soil* **174** 83-101

Pereto JG Beltran JP 1987 Hormone directed sucrose transport in developing dwarf and normal s cultivars of pea (*Pisum sativum*) *Biol Plant* (Praha) **28** 338-344

Pesme WS Israelstan GF 1986 levels of Gibberellin like substances and their possible transport in developing dwarf and normal cultivars of pea (*Pisum sativum*) *Biol Plant* (Praha) **28** 338-344

Phillips DA 1971a) Abscisic acid inhibition of root nodulation in *Pisum sativum (L) Planta* **100** 181-190

Phillips DA 1971b) A cotyledonary inhibitor of root nodulation in *Pisum sativum Physiol Plant* **25** 482-487

Postgate JR 1971 Chemistry and Biochemistry of Nitrogen fixation Plenum Press London

Prat R 1987 Inter-relation between IAA or proton induced cell elongation and cell wall extension in mungbean hypocotyl segment *Plant Physiol Biochem* (Paris) **25(4)** 401-408

Quispel A 1988 Hellriegel and Wilfarths discovery (symbiotic) nitrogen fixation hundred years ago In Nitrogen Fixation : Hundred years After (Eds H4 Bothe F J de Brujin WE Newton) pp 3-12 Suttgart : Gustav Tischer

Radely M 1961 Gibberellin like substance in plant *Nature* **191** 684-685

Raghuwanshi A Dudeja SS 1996 Effect of flavonoids on indole-acid production by fast and slow growing pigeonpea rhizobia *Indian J Plant Physiol* **1(4)** 229-223

Raj AS 1987 Cobalt nutrition of pigeonpea and groundnut in relation to growth and yield J of Plant Nutrition **10** 2137-2145

Rajani LM Upadhya NM Prasad TG Sadharshana L Sashidhar VR 2000 Influence of nodulation by cytokinin producing *Rhizobium* strain IC 3342 on growth and endogenous cytokinin contents in compatible host cowpea (*Vigna unguiculata*) *J Plant Biol* **27(1)** 61-66

Rangaswami G Bagyaraj DJ 2002 Microorganisms in soil In Agricultural Microbiology (eds G Rangaswami DJ Bagyaraj) pp-200-264 Published by Prentince-Hall of India Pvt Ltd New Delhi

Relic B Talmont F Kopcinska J Golinowski W Proma JC Broughton WJ 1993 Biological activity of *Rhizobium* sp NGR 234 nod factor on *Macroptilium atropurpureum Mol Plant Microbe Inter act* **6(6)** 769-774

Rokosh DA Kurz WGW Wetter LR 1973 Reluctant for symbiotic nitrogen fixation *Plant Physiol* (Suppl) **51** 185

Russell EJ 1950 Soil conditions and Plant Growth 8th edition Essex : Longman

Sandhu PS Dhingra KK Bhansari SC Gupta RP 1999 Effect of hand hoeing and application of herbicides on nodulation nodule activity and grain yield of *Lens culinaris* Medic Plant and Soil **135** 293-296

Scheres B Mckhann HI Zalensky A Lobler M Bisseling T Hirsch AM 1992 The Ps ENOD 12 gene is expressed at two different sites in Afghanistan pea pseudonodules induced by auxin transport inhibitors *Plant Physiol* **100** 1649-1655

Schilling G 1988 Hellriegel Wilfarth and their discovery of nitrogen fixation at Bernburg In Nitrogen Fixation Hundred year After (Eds H Bothe F J de Brujin and WE Newton) pp 13-19 Stuttgart: Gustav Fischer

Schubert KR Evans HJ 1976 Hydrogen evolution : A major factor affecting the efficiency of nitrogen fixation in nodulated symbionts Proc Nat Acad Sci (USA) **73** 1207-1211

Schubert KR Engelke JA Russel SA Evans HJ 1977 Hydrogen reactions of nodulated leguminous plants I effects of Rhizobial strain and plant age Plant Physiol **60** 651-654

Schubert KR 1986 Product of biological nitrogen fixation in higher plants Synthesis transport and metabolism *Ann Rev Plant Physiol* **37** 539-574

Scott J Howieson J 2000 Pulse research and industry development in Western Australia : Chickpea nodulation survey In : Crop Updates pp 32

Sekhon HS Kaul JN Dahiya BS 1978 Response of lentil to *Rhizobium* inoculation and N fertilizer J of Agri Sci Cambridge **90** 325-327

Sekhon HS Singh Guriqbal Sharma Poonam 2002 Biological Nitrogen Fixation In : Recent Advances in Agronomy (eds Singh G Kolar J S Sekhon HS) 141-166 Indian Society of Agronomy IARI New Delhi India

Serraj R Sinclair TR Purcell IC 1999 Symbiotic N_2 fixation response to drought J of Expl Bot **50** 143-155

Shanumugam KTO Gara F Anderson K Valentine RC 1978 Biological nitrogen fixation Ann Rev of Plant Physiol **29** 263-276

Sharma CM Neag A 1982 Synergistic effects of Indole-3 acetic acid and Benzyladenine on elongation of bean hypocotyl segments Ind *J Plant Physiol* **25(1)** 102-105

Sharma P 2007 Soil microorganisms Role in organic farming In Low cost biological approaches for sustainable and organic agriculture (eds R Kumar K Chandra) pp-79-84 Department of Agriculture and Cooperation, Regional Centre of organic farming Hissar

Sharma Poonam Khurana A S Sekhon HS 1996 Effect of different environments and nutrition of BNF growth and yield of different pulse crops In plant Microbe Interaction in Sustainable Agriculture (Eds R K Behl A L Khurana Dogra) pp 101-114 Hisar Bioscience Publishes

Sharma V 1986 Effect of Rustin GA3 on seedling growth and hydrolase activities in *Vigna radiata* cv P 57 *Plant Physiol Biochem* **13(1)** 13-19

Shechy JE Minchin FR Witty JF 1985 Control of nitrogen fixation in legume nodule: an analysis of role of oxygen diffusion in relation to nodule structure *Annals Bol* **55** 549-562

Sheng C Harper JE 1997 Shoot versus root signal involvement in nodulation and vegetatius growth in wild-type and hypernodulating soybean genotype Plant Physiol **113** 825-831

Sikorski MM Topunov AF Strozycki PM Vorgias CE Wilson KE Legocki AB 1995 Cloning and expression of Plant leghemoglobinc DNA of *Lupinus luteus* and *Escherichiacoli* and purification of recombinant protein *Plant Sci* **108** 1109-117

Sinclair T R Serraj R 1995 Dinitrogen fixation sensitivity to drought among legume species *Nature* **378** 344-349

Sindhu SS Dadarwal KR 2000 Competition for nodulation among rhizobia in legume *Rhizobium* symbiosis *Indian J Microbiol* **40** : 211-246

Singh AP 1999 Role of rhizobacteria in increasing the efficiency of inoculated *Rhizobium* in lentil MSc Thesis Development of soil Science GBPUA T Pantnagar

Singh G Sekhon HS 2005 Pulse-based cropping systems in Pulses (eds G Singh HS Sekhon and JS Kolar) pp-33-58 Agrotech Publishing Academy Udaipur

Singh GR Sharma SR Anil Kumar TB 1986 Effects of biloxazol carbondazim and triadimefon on nodulation and nitrogen fixation in cowpea *Zentralbi Mikrobil* **141** 35-37

Singh VK Mathur M Mathur SN 1982 Phytotoicity of insecticide phorate on germination of *Vigna mungo Agril Biol Chem* **46** 1681-1682

Sohail H Sumera Y Malik KA Yusuf Z Hafees FY 2004 *Rhizobium* Bradyrhizobium and Agrobacterium strains isolated from cultivated legumes Biol and Fertility of Soil **39(3)** 179-185

Soupene E Faussard M Boistard P Truchet G Batit J 1995 Oxygen as key developmental regulator of *Rhizobium meliloti* nitrogen fixation gene regulation within alf-alfa nodule *Proc Natl Acad Sc* (USA) **92(9)** 3759-3763

Spaink HP 1994 The molecular basis of host specificity of *Rhizobium* bacteria *Antoine Van Leeuwenhoek* **65(2)** 81-98

Sprent JI Sprent P 1990 Nitrogen fixing organisms : Pure and Applied Aspects Chapman and Hall London

Sprent JL Zaharan 1988 Infection development and functioning of nodules under drought and salinity In Nitrogen Fixation by Legumes in Mediterranean Agriculture (Eds DP Beck and LA Materon) pp 145-151 Martinus Nijhott Bordrecht ISBN 90-247 2405-9

Subba Rao NS 1976 Field response of legumes in India to inoculation and fertilizer application In Symbiotic Nitrogen fixation in Punjab Ed P S Nutman PP 255-268 Cambridge U K

Subba Rao N 2000 *Rhizobium* and Legume root nodulation In Soil Microbiology (ed NS Subba Rao) pp-166-232 Oxford and IBH Publishing Co Pvt Ltd New Delhi

Suganuma N Yomauchi H Yomamoto K 1995 Enhanced production of ethylene by soybean roots after inoculation with *Bradyrhizobium japonicum Plant Sci* **111** 163-168

Suganuma N Makota K Yukio Y 1988 Carbon metabolism in relation to cellular organization of soybean root nodules and respiration of mitochandria aided by leghemoglobin *Plant Cell Physiol* **28(1)** 113-122

Sundara Rao WVB 1971 Field experiments on nitrogen fixation by nodulated legumes *Plant and soil* (Special volume) 287-291

Swaraj K Garg OP 1970 Effect of gibberellic acid when applied to the rooting medium on nodulation and nitrogen fixation in grains (*Cicer arietinum L*) *Physiol Plant* **32** 747-754

Taiz L Zeiger E 1991 *Plant Physiol*ogy The Benjamin/ cumming Inc: California pp 292-315

Thavarajah D Ball R A Schnoenau JF 2005 Nitrogen fixation amino acid and ureide associations in chickpea *Crop Sci* **45** 2497-2502

Thiessen WE Wareing PF 1968 Absisic acid: A propiosal for designation of Abicisn II (dormin) In: Biochemistry and Physiologyof Plant Regulators The Range Press Ltd Ottawa Canola 1500-1529

Thimann KV 1935 On the plant growth hormone produced by *Rhizopus suinus J Biol Chem* **109** 279-291

Thomas SG Phillips AL Hedden P 1999 Molecular cloning and functional expression of gibberellin 2-oxidases multifunctional enzymes involved gibberllin deactivation *Proc Natl Acad Sci* (USA) **13** 96(8) 4698-4703

Thummler F Verma DPS 1987 J *Biol Chem* **26(30)** 1473-14736

Tiwari KN 1988 Potassium in soils and crop response to Potassium Application in U.P. Annual Report PR II Sponsored Project Kanpur Department of soils and Agricultural Chemistry CS Azad University of Agriculture and Technology

Toomsan B Rupela OP Dart PJ 1982 Studies on soil and rhizosphere populations of *Rhizobium* sp nodulating cicer arietinum In : Proceeding of the National Symposium on Biological Nitrogen Fixation pp 517-531 IARI Delhi

Truchet G Roche P Lerange P Vasse J Camut Sd Billy F Pron C Denaire J 1991 Sulphated lipo-oligosaccharide signals of *Rhizobium meliloti* elicit root nodule organogensis in alfalfa Nature London **351** 670-673

Van Spronsen PC Van Brussel AA Kijne JW 1995 Nod factor produced by *Rhizobium leguminosarum* biovar vicial induce thylene related changes in root cortical cells of *vicia sativa* spp nigra Evr *J Cell Biol* **68(4)** 463-469

Vance CP Heichel GH 1991 Carbon in nitrogen fixation: limitation or exquistic adaptation *Ann Rev Plant Physiol Mol Biol* **42** 373-392

Wang TL Wood EA Brewin NF 1982 Growth regulators *Rhizobium* and nodulation and pea indole 3-acetic acid from culture medium of nodulating and non-nodulating strains of R Leguminosarum *Planta* **155** 345-349

Wani SP Rupela OP Lee KK 1995 Sustainable agriculture in semi arid tropics through biological nitrogen fixation in grain legumes *Plant and soil* **174** 29-49

Welty LE Presbyte LS Hall JA Mathre DE Dilterline RL 1988 Effect of fungicide treatment with *Rhizobium* inoculation on chickpea production *App Agric Res* **3** 17-20

Wheeler CT Hensen IE 1978 Hormones in plants bearing nitrogen fixing root nodules The nodules as a source of cytokinin in *Alnus glutinosa* L Gaertn *New Physiol* **80** 557-565

Williams PM De Mallorca MS 1984 Effect of gibberellins and the growth retardant CCC on the nodulation of Soya *Plant Soil* **77** 53-60

Wu C Dickstein R Cary AJ Norris JI I 1996 The auxin transport N-(1-napthyl) phthalmic acid elicits pseudo nodules on non-nodulating mutants of white sweet clover *Plant Physiol* **110** 501-510

Yi YC Joo S Nam KH Lee JS Kang BG Kim WT 1999 Auxin and brasinosteroid differentially regulate the expression of three members of the 1-aminocyclo-1-carboxylate synthase gene family in mungbean (*Vigna radiata L*) *Plant Mol Biol* **41(4)** 443-454

Young JPN Haukka KE 1996 Diversity and Phylogeny of rhizobia *New Physiologists* **133** 87-94

Zablotowicz RM DL Eskew Focht DO 1978 Denitrification in *Rhizobium Canadian Journal of Microbiology* **103** 806-808

Zaidi PH Rafiq S Singh BB 1997 Physiology of Zinc nutrition in field crops fertiliser News **42(11)** 63-66

Zanethi S Hartwig VA Lusher A Hebeisen T Freshner M Fisher BU Handrey GR Blum H Nosberger J 1996 Stimulation of symbiotic N_2 fixation in *Trifolium repens* under elevated atmospheric CO_2 in grassland ecosystem *Plant Physiol* **112** 575-583

INDEX

Q

R

S